D1653057

Nanotechnology

Volume 5: Nanomedicine
Edited by Viola Vogel

Related Titles

Nanotechnologies for the Life Sciences

Challa S. S. R. Kumar (ed.)

Volume 1: Biofunctionalization of Nanomaterials

2005

978-3-527-31381-5

Volume 2: Biological and Pharmaceutical Nanomaterials

2005

978-3-427-31382

Volume 3: Nanosystem Characterization Tools in the Life Sciences

2005

978-3-527-31383-9

Volume 4: Nanodevices for the Life Sciences

2006

978-3-527-31384-6

Volume 5: Nanomaterials - Toxicity, Health and Environmental Issues

2006

978-3-527-31385-3

Volume 6: Nanomaterials for Cancer Therapy

2006

978-3-527-31386-0

Volume 7: Nanomaterials for Cancer Diagnosis

2006

978-3-527-31387-7

Volume 8: Nanomaterials for Biosensors

2006

978-3-527-31388-4

Volume 9: Tissue, Cell and Organ Engineering

2006

978-3-527-31389-1

Volume 10: Nanomaterials for Medical Diagnosis and Therapy

2007

978-3-527-31390-7

Nanotechnology

Günter Schmid (ed.)

Volume 1: Principles and Fundamentals

2008

978-3-527-31732-5

Harald Krug (ed.)

Volume 2: Environmental Aspects

2008

978-3-527-31735-6

Rainer Waser (ed.)

Volume 3: Information Technology I

2008

978-3-527-31738-7

Rainer Waser (ed.)

Volume 3: Information Technology II

2008

978-3-527-31737-0

Viola Vogel (ed.)

Volume 5: Nanomedicine and Nanobiotechnology

2009

978-3-527-31736-3

Harald Fuchs (ed.)

Volume 6: Nanoprobes

2009

978-3-527-31733-2

Michael Grätzel, Kuppuswamy Kalyanasundaram (eds.)

Volume 7: Light and Energy

2009

978-3-527-31734-9

Lifeng Chi (ed.)

Volume 8: Nanostructured Surfaces

2009

978-3-527-31739-4

www.wiley.com/go/nanotechnology

G. Schmid, H. Krug, R. Waser, V. Vogel, H. Fuchs,
M. Grätzel, K. Kalyanasundaram, L. Chi (Eds.)

Nanotechnology

Volume 5: Nanomedicine

Edited by Viola Vogel

WILEY-VCH Verlag GmbH & Co. KGaA

The Editor

Prof. Viola Vogel
ETH Zürich
Laboratory for Biologically Oriented Materials
Department of Materials
HCI F443
Wolfgang-Pauli-Str. 10
8093 Zürich
Switzerland

Cover: Nanocar reproduced with kind permission of Y. Shirai/Rice University

Library of Congress Card No.: applied for

British Library Cataloguing-in-Publication Data
A catalogue record for this book is available from the British Library.

Bibliographic information published by the Deutsche Nationalbibliothek
The Deutsche Nationalbibliothek lists this publication in the Deutsche Nationalbibliografie; detailed bibliographic data are available in the Internet at http://dnb.d-nb.de.

Typesetting Thomson Digital, Noida, India
Printing Strauss GmbH, Mörlenbach
Binding Litges & Dopf Buchbinderei GmbH, Heppenheim

Printed in the Federal Republic of Germany
Printed on acid-free paper

ISBN: 978-3-527-31736-3

Contents

Nanotechnology, Volume 5: Nanomedicine. Edited by Viola Vogel

ISBN: 978-3-527-31736-3

List of Contributors

Kathryn T. Applegate
The Scripps Research Institute
Laboratory for Computational Cell Biology
La Jolla, CA 92037
USA

Matthew L. Baker
National Center for Macromolecular Imaging
Baylor College of Medicine
One Baylor Plaza
Houston, TX 77030
USA

Gang Bao
Georgia Institute of Technology and Emory University
Department of Biomedical Engineering
Atlanta, GA 30332
USA

Ramille M. Capito
Northwestern University
Institute for BioNanotechnology in Medicine
303 E. Superior St.
Chicago, IL 60611
USA

and

Northwestern University
Department of Materials Science and Engineering
Evanston, IL 60208
USA

Shelton D. Caruthers
Washington University School of Medicine
Consortium for Translational Research in Advanced Imaging and Nanotechnology
660 S. Euclid Avenue
CB 8215 Saint Louis, MO 63110
USA

Wah Chiu
National Center for Macromolecular Imaging
Baylor College of Medicine
One Baylor Plaza
Houston, TX 77030
USA

Gaudenz Danuser
The Scripps Research Institute
Laboratory for Computational Cell Biology
10550 N. Torrey Pines Road
La Jolla, CA 92037
USA

Nanotechnology, Volume 5: Nanomedicine. Edited by Viola Vogel

ISBN: 978-3-527-31736-3

Paolo Decuzzi
University of Texas Health Science
Center at Houston
School of Health Information Sciences
Houston, TX 77030
USA

and

BioNEM–Center of Bio-Nanotechnology
and Engineering for Medicine
University of Magna Graecia
88100 Catanzaro
Italy

Dennis E. Discher
University of Pennsylvania
Biophysical Engineering Laboratory
Philadelphia, PA 19104
USA

Iain E. Dunlop
Max Planck Institute for Metals
Research
Department New Materials and
Biosystems
Heisenbergstr. 3
70569 Stuttgart
Germany

and

University of Heidelberg
Department of Biophysical Chemistry
Im Neuenheimer Feld 253
69120 Heidelberg
Germany

Michael L. Dustin
The Helen L. and Martin S. Kimmel
Center for Biology and Medicine at the
Skirball Institute for Biomolecular
Medicine
540 First Avenue
NYC 10016
USA

Adam J. Engler
University of Pennsylvania
Biophysical Engineering Laboratory
Philadelphia, PA 19140
USA

and

Present Address:
University of California San Diego
Department of Bioengineering
La Jolla, CA
USA

Mauro Ferrari
University of Texas Health Science
Center at Houston
Division of Nanomedicine
Department of Biomedical Engineering
1825 Pressler, Suite 537D
Houston, TX 77030
USA

and

The University of Texas MD Anderson
Cancer Center
Department of Experimental
Therapeutics
Houston, TX 77030
USA

and

Rice University
Department of Bioengineering
Houston, TX 77030
USA

Peter Fratzl
Max Planck Institute of Colloids and
Interfaces
Department of Biomaterials
Research Campus Golm
14424 Potsdam
Germany

Biana Godin
University of Texas Health Science Center at Houston
Division of Nanomedicine
Department of Biomedical Engineering
Houston, TX 77030
USA

Anita Goel
Nanobiosym Labs
200 Boston Avenue, Suite 4700
Medford, MA 02155
USA

and

Harvard University
Department of Physics
Cambridge, MA 02138
USA

Himadri S. Gupta
Max Planck Institute of Colloids and Interfaces
Department of Biomaterials
Research Campus Golm
14424 Potsdam
Germany

Michael S. Hughes
Washington University School of Medicine
Consortium for Translational Research in Advanced Imaging and Nanotechnology
660 S. Euclid Avenue
CB 8215 Saint Louis
MO 63110
USA

Satoshi Jinno
Harvard-MIT Division of Health Sciences and Technology,
Massachusetts Institute of Technology
Cambridge, MA 02139
USA

and

Center for Biomedical Engineering
Department of Medicine
Brigham and Women's Hospital
Harvard Medical School
Cambridge, MA 02139
USA

Ali Khademhosseini
Massachusetts Institute of Technology
Harvard-MIT Division of Health Sciences and Technology
PRB 252
65 Landsdowne Street
Cambridge, MA 02139
USA

and

Center for Biomedical Engineering
Department of Medicine
Brigham and Women's Hospital
Harvard Medical School
Cambridge, MA 02139
USA

Klaus Klaushofer
Ludwig Boltzmann Institute of Osteology at Hanusch Hospital of WGKK and AUVA Trauma Centre Meidling
4th Medical Department
Hanusch Hospital
Heinrich Collin Street 30
1140 Vienna
Austria

Philipp Kukura
ETH Zurich
Laboratory of Physical Chemistry
and Zurich Center for Imaging Science
and Technology (CIMST)
Nano-Optics Group
8093 Zürich
Switzerland

Robert Langer
Harvard-MIT Division of Health
Sciences and Technology
Massachusetts Institute of Technology
Cambridge, MA 02139
USA

and

Department of Chemical Engineering
and Department of Biological
Engineering
Massachusetts Institute of Technology
Cambridge, MA 02139
USA

Gregory M. Lanza
Washington University School of
Medicine
Consortium for Translational Research
in Advanced Imaging and
Nanotechnology
660 S. Euclid Avenue
CB 8215 Saint Louis
MO 63110
USA

Alvaro Mata
Northwestern University
Institute for BioNanotechnology in
Medicine
Chicago, IL 60611
USA

and

Nanotechnology Platform
Parc Cientific de Barcelona
Baldiri Reixac 10–12
08028 Barcelona
Spain

Michael P. Marsh
National Center for Macromolecular
Imaging
Baylor College of Medicine
One Baylor Plaza
Houston, TX 77030
USA

Jon N. Marsh
Washington University School of
Medicine
Consortium for Translational Research
in Advanced Imaging and
Nanotechnology
660 S. Euclid Avenue
CB 8215 Saint Louis
MO 63110
USA

Nitin Nitin
Georgia Institute of Technology and
Emory University
Department of Biomedical Engineering
Atlanta, GA 30332
USA

Bimal Rajalingam
Harvard-MIT Division of Health
Sciences and Technology
Massachusetts Institute of Technology
Cambridge, MA 02139
USA

and

Center for Biomedical Engineering
Department of Medicine
Brigham and Women's Hospital
Harvard Medical School
Cambridge, MA 02139
USA

Florian Rehfeldt
Georg-August Universitiat Göttingen
III. Physikalisches Institut
Friedrich-Hund-Platz 1
37077 Göttingen
Germany

Alois Renn
ETH Zurich
Laboratory of Physical Chemistry
and Zurich Center for Imaging Science
and Technology (CIMST)
Nano-Optics Group
8093 Zürich
Switzerland

Won Jong Rhee
Georgia Institute of Technology and
Emory University
Department of Biomedical Engineering
Atlanta, GA 30332
USA

Paul Roschger
Ludwig Boltzmann Institute of
Osteology at Hanusch Hospital of
WGKK and AUVA Trauma Centre
Meidling
4th Medical Department
Hanusch Hospital
Heinrich Collin Street 30
1140 Vienna
Austria

Jason Sakamoto
University of Texas Health Science
Center at Houston
Division of Nanomedicine
Department of Biomedical Engineering
Houston, TX 77030
USA

Vahid Sandoghdar
ETH Zurich
Laboratory of Physical Chemistry
and Zurich Center for Imaging Science
and Technology (CIMST)
Nano-Optics Group
8093 Zürich
Switzerland

Phillip Santangelo
Georgia Institute of Technology and
Emory University
Department of Biomedical Engineering
Atlanta, GA 30332
USA

Rita E. Serda
University of Texas Health Science
Center at Houston
Division of Nanomedicine
Department of Biomedical Engineering
Houston, TX 77030
USA

Michael P. Sheetz
Columbia University
Sherman Fairchild Center
Department of Biological Sciences
1212 Amsterdam Avenue, Room 713
New York 10027
USA

Joachim P. Spatz
Max Planck Institute for Metals Research
Department New Materials and Biosystems
Heisenbergstr. 3
70569 Stuttgart
Germany

and

University of Heidelberg
Department of Biophysical Chemistry
Im Neuenheimer Feld 253
69120 Heidelberg
Germany

Samuel I. Stupp
Northwestern University
Institute for BioNanotechnology in Medicine
303 E. Superior
Chicago, IL 60611
USA

and

Northwestern University
Department of Materials Science and Engineering
Evanston, IL 60208
USA

and

Northwestern University
Department of Chemistry
Evanston, IL 60208
USA

and

Northwestern Universtiy
Department of Medicine
303 E. Superior, Suite 11-129
Chicago, IL 60611
USA

Viola Vogel
ETH Zurich
Laboratory for Biologically Oriented Materials
Department of Materials
8049 Zürich
Switzerland

Kirk D. Wallace
Washington University School of Medicine
Consortium for Translational Research in Advanced Imaging and Nanotechnology
660 S. Euclid Avenue
CB 8215 Saint Louis, MO 63110
USA

Samuel A. Wickline
Washington University School of Medicine
Consortium for Translational Research in Advanced Imaging and Nanotechnology
660 S. Euclid Avenue
CB 8215 Saint Louis, MO 63110
USA

Ge Yang
The Scripps Research Institute
Laboratory for Computational Cell Biology
La Jolla, CA 92037
USA

Part One:
Nanomedicine: The Next Waves of Medical Innovations

Nanotechnology, Volume 5: Nanomedicine. Edited by Viola Vogel

ISBN: 978-3-527-31736-3

1
Introduction

Viola Vogel

1.1
Great Hopes and Expectations are Colliding with Wild Hype and Some Fantasies

What is nanomedicine? Will nanomedicine indeed help to cure major diseases and live up to the great hopes and expectations? What innovations are on the horizon and how can sound predictions be distinguished from wild hype and plain fantasy? What are realistic timescales in which the public might benefit from their ongoing investments?

When first exploring whether nanotechnology might reshape the future ways of diagnosing and treating diseases, the National Institutes of Health stated in the report of their very first nanoscience and nanotechnology workshop in 2000 (http://www.becon.nih.gov/nanotechsympreport.pdf. Bioengineering Consortium):

> *Every once in a while, a new field of science and technology emerges that enables the development of a new generation of scientific and technological approaches. Nanotechnology holds such promise.*

Our macroscopic bodies and tissues are highly structured at smaller and smaller length scales, with each length scale having its own secrets as to how life-supporting tasks are mastered. While we can still touch and feel our organs, they are all composed of cells which are a little less than one million times smaller and only visible under the light microscope (microscopic). Zooming further into the cell, about one thousand times, we find the nanoscale molecular machineries that drive and control the cellular biochemistry, and thereby distinguish living systems from dead matter. Faced with a rest of new technologies that has enabled researchers to visualize and manipulate atoms and molecules, as well as to engineer new materials and devices at this tiny length scale [1], major think tanks have begun since the late 1990's to evaluate the future potential of nanotechnology [2], and later at the interface to medicine [3–11]. These efforts were paralleled by a rapid worldwide increase in funding and research activities since 2000. The offset of a gold rush into the 'nano', by which the world of the very small is currently discovered, will surely also lead to splendid new entrepreneurial opportunities. Progress impacting on human health came much faster than expected.

Nanotechnology, Volume 5: Nanomedicine. Edited by Viola Vogel

ISBN: 978-3-527-31736-3

1.2
The First Medical Applications are Coming to the Patients' Bedside

The public most commonly associates nanomedicine with engineered nanoparticles in the context of drug delivery devices or advanced medical imagine applications. Novel is that the molecules which are coassembled into nanoparticles can nowadays carry many different chemical functionalities. It has thereby become feasible to integrate multiple tasks into drug delivery device, from targeting specific tissues to releasing drugs, from enhancing contrast to probing their environment. How is this all done? First of all, the nanoparticles are loaded with drugs. The particles might then carry molecules on their surfaces that bind with great specificity to complementary molecules that are unique to cancers or to other diseased tissues, as reviewed later in this volume [12, 13]. For example, by using antibody– antigen recognition such engineered nanoparticles can be accumulated in the targeted tissues rather than being distributed over the entire body. Local accumulation greatly enhances the efficiency of drugs and reduces any unintended adverse side effects that might otherwise harm other organs. The selectivity by which disease can be treated by using engineered nanoparticles is thus in stark contrast to how conventional drugs operate; as conventional drugs lack the capacity to target specific tissues, they are distributed much more uniformly over the entire body, and must therefore be administered at much higher doses. Beyond tissue targeting, the nanoparticles might further be engineered to absorb therapeutic radiation, which might heat them up when they have reached the diseased tissues to damage the local tissue, either by heat or by the release of drugs [11, 12, 14]. Alternatively, the nanoparticles might hold on to drugs by bonds that can be locally cleaved by enzymes or be broken by light or radiation, thereby releasing the drug under the control of a physician (as reviewed elsewhere in this volume [12, 13]). The goal here is to design new strategies to inflict damage only to the aberrant cells, while leaving the surrounding tissues unharmed. This multifunctional integration of many different diagnostic and therapeutic tasks in single particles thereby enables applications that go far beyond those of conventional drugs.

Engineered nanoparticles can also change the future of medical imaging, as they enable us to combine structural imaging with spatially resolved diagnostics and interventions. Only eight years after wondering whether nanotechnology will revolutionize medicine, the US National Cancer Institute (NCI) stated that [15]:

> *Nanodevices are used in detecting cancer at its earliest stages, pinpointing its location within the body, delivering anticancer drugs specifically to malignant cells, and determining if these drugs are killing malignant cells.*

Increasingly sophisticated medical imaging technologies continue to revolutionize medicine. X-ray imaging, which was later complemented by ultrasound, positron emission tomography (PET) and magnetic resonance imaging (MRI), opened the possibility to visualize *nonintrusively* first bones, and then the inner organs, of our bodies. The objectives now are to obtain images of the structure of organs, at much higher resolution, together with spatially resolved biochemical information which is reflective of how well cells and organs function. This includes probing noninvasively whether certain organs produce the hormones and enzymes at normal rates, and whether other metabolic activities might deviate from the norm. Major advances are about to come from the usage of nanoparticles that are engineered to serve both, as drug delivery systems and to enhance the contrast in ultrasound, PET and MRI images (for a review, see Ref. [16] and elsewhere in this volume [12, 13]). Enhancing the contrast and spatial resolution of images will enable the detection of cancers and other structural abnormalities of organs at much earlier stages, which in turn will enhance the chances of an effective therapy. Such multitasking approaches might also one day substitute for a variety of surgical interventions. Today, many books and articles have been published discussing the various medical applications of such engineered nanoparticles, while the pharmaceutical industry continues to invest heavily in their development (for reviews, see Refs [6, 15–22]).

1.3
Major Advances in Medicine Have Always been Driven by New Technologies

During the past few decades, the deciphering of the molecular origins of many diseases has had a most profound impact on improving human health. One historical step was the deciphering of the first protein structure in 1958 [23]. This opened a new era in medicinal chemistry, as drugs could since then be designed in a rational manner – that is, drugs that fit tightly into essential binding pockets thereby regulating protein and DNA functions. The invention of how to harness DNA polymerase in order to amplify genetic material in the test tube – which we now know as the polymerase chain reaction (PCR) [24] – then opened the field of molecular biology during the 1980s. PCR also enabled targeted genetic alterations of cells to identify the functional roles of many proteins, and this in turn led to the discovery that cell signaling pathways of many interacting proteins existed, and could be altered by diseases. The explosion of knowledge into how cell behavior is controlled by biochemical factors opened the door to target drugs to very specific players in cell signaling pathways. This also led to a host of new biotechnology start-up companies, the first of which became profitable only around 2000.

The next major breakthrough came with the solving of the human genome in 2001 [25–29]. Access to a complete genetic inventory of more than 30 000 proteins in our body, combined with high-resolution structures for many of them, enables a far more systematic search for correlations between genetic abnormalities and diseases. Finally, various diseases could for the first time be traced to inherited point mutations of proteins. In achieving this, much insight was gained into the regulatory roles of these proteins in cell signaling and disease development [30]. This includes recognizing genetic predispositions to various cancers [31], as well as to inherited syndromes where larger sets of seemingly uncorrelated symptoms could finally be explained by the malfunctioning of particular proteins or cell signaling pathways [32–38], including ion channel diseases [39–42].

1.4
Nanotechnologies Foster an Explosion of New Quantitative Information How Biological Nanosystems Work

Far less noticed by the general public are the next approaching waves of medical innovations, made possible by an explosion of new quantitative information how biological systems work.

The ultimate goal is to achieve an understanding of all the structural and dynamic processes by which the molecular players work with each other in living cells and coregulate cellular functions. Driven by the many technologies that have emerged from the physical, chemical, biological and engineering sciences, to visualize (see elsewhere in this volume [43, 44]) and manipulate the nanoworld, numerous discoveries are currently being made (as highlighted in later chapters [43–51]). These findings result from the new capabilities to create, analyze and manipulate nanostructures, as well as to probe their nanochemistry, nanomechanics and other properties within living and manmade systems. New technologies will continuously be developed that can interrogate biological samples with unprecedented temporal and spatial resolution [52]. Novel computational technologies have, furthermore, been developed to simulate cellular machineries in action with atomic precision [53]. New engineering design principles and technologies will be derived from deciphering how natural systems are engineered and how they master all the complex tasks that enable life. Take the natural machineries apart, and then learn how to reassemble their components (as exemplified here in Chapter 8 for molecular motors [48]).

How effectively will these novel insights into the biological nanoworld are recognized in their clinical significance, and translated into addressing grand medical challenges? This defines the time that it takes for the emergence of a next generation of diagnostic and therapeutic tools. As these insights change the way we think about the inner workings of cells and cell-made materials, totally new ways of treating diseases will emerge. As described in detail elsewhere in this volume, new developments are already under way of how to probe and control cellular activities [45, 47, 49–51, 54, 55]. This implicates the emergence of new methodologies of how to correct tissue and organ malfunctions. Clearly, we need to know exactly how each disease is associated with defects in the cellular machinery before medication can be rationally designed to effectively cure them.

Since every one of the new (nano)analytical techniques has the potency of revealing something never seen before, a plethora of opportunities can be envisioned. Their realization, however, hinges on the scientists' ability to recognize the physiological and medical relevance of their discoveries. This can best be accomplished in the framework of interdisciplinary efforts aimed at learning from each other what the new technologies can provide, and how this knowledge can be effectively translated to address major clinical challenges. Exploring exactly how these novel insights into the nanoworld will impact medicine has been the goal of many recent workshops [3–11, 56–58]). This stimulated the creation of the NIH

Roadmap Initiative in Nanomedicine [57, 58], and is the major focus of this volume.

1.5 Insights Gained from Quantifying how the Cellular Machinery Works will lead to Totally New Ways of Diagnosing and Treating Disease

Which are some of the central medical fields that will be impacted? Despite these stunning scientific advances, and the successful suppression or even eradication of a variety of infectious diseases during the past 100 years, the goal has not yet been reached of having medication at hand to truly cure many of the diseases that currently kill the largest fraction of humans per year, including cancers, cardiovascular diseases and AIDS. Much of the current medication against these diseases fights symptoms or inhibits their progression, often inflicting considerable side effects. Unfortunately, however, much of the medication can slow down but it cannot *reverse* disease progression in any major way – all of which contributes to healthcare becoming unaffordable, even in the richest nations of the world. For instance, intense cancer research over the past decades has revealed that the malignancy of cancer cells progresses with the gradual accumulation of genetic alterations [12, 50, 59–65]. Yet, little remains known as to how cancer stem cells form, in the first place, and about the basic mechanisms that trigger the initiation of their differentiation into more malignant cancer cells after having remained dormant, sometimes for decades [66–69]. While much has been learned in the past about the molecular players and their interactions, the above-mentioned shortcomings in translating certain advances in molecular and cell biology into more effective medication reflect substantial gaps in our knowledge of how all these components within the cells work in the framework of an integrated system. How can so many molecular players be tightly coordinated in a crowded cellular environment to generate predictable cell and tissue responses? Whilst lipid membranes create barriers that enclose the inner volumes of cells and control which molecules enter and leave (among other tasks), it is the proteins that are the 'workhorses' that enable most cell functions. In fact, some proteins function as motors that ultimately allow cells to move, as discussed in different contexts in the following chapters [46, 48, 50]. Other proteins transcribe and translate genetic information, and efforts to visualize these in cells have been summarized in Chapter 6 [45]. Yet other sets of proteins are responsible for the cell signaling through which all metabolic functions are enabled, orchestrated and regulated. But what are the underlying rules by which they play and interact together to regulate diverse cell functions? How do cells sense their environments, integrate that information, and translate it to ultimately adjust their metabolic functions if needed? Can this knowledge help to develop interventions which could possibly reverse pathogenic cells such that they performed their normal tasks again? Deciphering how all of these processes are regulated by the physical and biochemical microenvironment of cells is key to addressing various biomedical challenges with new perceptions, and is described as one of the major foci in this volume. But, how can this be accomplished?

1.6
Engineering Cell Functions with Nanoscale Precision

The engineering of nanoenvironments, nanoprobes and nanomanipulators, together with novel modalities to visualize phenomena at this tiny scale, have already led to the discovery of many unexpected mechanisms of how cellular nanoparts function, and how they cooperate synergistically when integrated into larger complex systems [43, 46–51]. Today, nanotechnology tools are particularly well suited to explore and quantify the physical aspects of biology, thereby complementing the tool chests of biochemists, molecular biologists and geneticists. Such nanotechnology approaches could already reveal that not only biochemical factors but also mechanical aspects as well as the micro- and nanoscale features of a cell's microenvironment, play pivotal roles in regulating their fate. The insights and implications thereof are described in chapters 9 to 14 [47, 49–51].

These discoveries are particularly relevant since most of our biomedical knowledge of how cells function has been derived in the past from the study of cells cultured on flat polymer surfaces (Petri dishes or on multi-welled plates). Cells in a more tissue-like environment, however, often show a vastly different behavior [70–75] (and chapters 9 to 14). With a still poorly understood cell signaling response system, cells in tissues thus 'see' and 'feel' an environment that is poorly mimicked by the common cell culturing conditions or scaffolds used in tissue engineering. Nanotechnologies will thus be pivotal to deciphering how cells sense and integrate a broad set of cues that regulate cell fate, from the moment that a sperm fertilizes an egg, to sustained, normal tissue functions. Moreover, these dependencies must be known in order to develop far more efficient drugs and treatment methods. However, ultimately it is the combination of many different technologies – some of which may originate from the physical sciences and others from biology – that must be combined to understand and quantify biology. Unfortunately, today an insufficient number of research workers are trained to perform these tasks [76].

1.7
Advancing Regenerative Medicine Therapies

Virtually any disease that results from malfunctioning, damaged or failing tissues may be potentially cured through regenerative medicine therapies, as was recently stated in the first NIH report on Regenerative Medicine [4]. But, how will nanotechnology make a difference? The repair – or ultimately replacement – of diseased organs, from larger bone segments to the spinal cord, or from the kidneys to the heart, still poses major challenges as discussed in chapters 9 to 14 [47, 49, 50, 54, 55, 77]. The current need for organ transplants surpasses the availability of suitable donor organs by at least an order of magnitude, and the patients who finally receive an organ transplant must receive immune suppressant drugs for the rest of their life. Thus, one goal will be to apply the mounting insights into how cells work, and how their functions are controlled by matrix interactions, to design alternate therapies that stimulate regenerative healing processes of previously irreparable organs. In a most promising approach, some molecules have been designed that can self-assemble in the body into provisional matrices [55]. If these are injected shortly after injury, they help to repair spinal cord injuries and heart tissues damaged by an infarction. And if such strategies do not work, then another possibility might be to seed the patient's cells or stem cells into engineered biohybrid matrices to grow simple tissues *ex vivo* – that is, in the laboratory – and later implant them to support or regenerate failing organ functions [51, 54]. This could provide new ways of treating diabetes, liver and kidney failures, cardiovascular and many other diseases, or of replacing or repairing organs damaged in accidents or removed during surgery. Learning how to control the differentiation of stem cells in engineered matrices is therefore central to advancing our technical abilities in tissue engineering and regenerative medicine, and the challenges ahead as discussed in chapters 9 to 14 [50, 51, 54]. Nanofibers thereby mimic much better the fibrous nature of extracellular matrices [54], and the nanoscale patterning of ligands can control cell activation [49], including the activation of cells that play central roles in the immune response system (for a review, see [49]). In summary, the insights derived with the help of nanotechnology will enable the engineering of tissue-mimetic scaffolds that better control and regulate tissue function and repair. Improving human health will thus critically hinge upon translating nanotechnology-derived insights about cellular and tissue functions into novel diagnostic and therapeutic technologies.

1.8 Many More Relevant Medical Fields Will be Innovated by Nanotechnologies

Whilst the major focus of this volume is to outline the biomedical implications derived from revealing the underpinning mechanisms of how human cells function, it should also be mentioned that fascinating developments that are prone to alter medicine are being made in equally relevant other biomedical sectors. Future ways to treat infection will change when the underpinning mechanisms of how microbial systems function are deciphered, and how they interact with our cells and tissues. Many beautiful discoveries have already been made that will help us for example to interfere more effectively with the sophisticated machinery that bacteria have evolved to target, adhere and infect cells and tissues. Nanotechnology tools have revealed much about the function of the nano-engines that bacteria and other microbes have evolved for their movement [78–80], how bacteria adhere to surfaces [81–83], and how microbes infect other organisms [81–85]. Equally important when combating infection is an ability to exploit micro- and nanofabricated tools in order to understand the language by which microorganisms communicate with each other [86, 87] and how their inner machineries function [88–90]. A satisfying understanding of how a machine works can only be reached when we are capable of reassembling it from its components. It is thus crucial to learn how these machineries can be reassembled *ex vivo*, potentially even in nonbiological environments, as this should open the door to many technical and medical applications [84, 91–93]. Today, we have only just started along the route to combining the natural and synthetic worlds, with the community seeking how bacteria might be used as 'delivery men' for nanocargoes [94], or in manmade devices to move fluids and objects [95, 96].

Finally, microfabricated devices with integrated nanosensors, nanomonitors and nanoreporters – all of which are intrinsic to a technology sector enabled by (micro/nano)biotechnology – will surely also lead to changes in medical practice. In the case of chemotherapies and many other drugs, it is well known that they may function well in some patients, but fail in others. It is feasable that this 'one-size-fits-all' approach might soon be replaced by a more patient-specific system. *Personalized medicine* refers to the use of genetic and other screening methods to determine the unique predisposition of a patient to a disease, and the likelihood of them responding to particular drugs and treatments [30, 97–100]. Cheap diagnostic systems that can automatically conduct measurements on small gas or fluid volumes, such as human breath or blood, will furthermore enable patients to be tested rapidly, without the need to send samples to costly medical laboratories. Needless to say, portable integrated technologies that will allow the testing and treatment of patients on the spot (point-of-care) will save many lives, and are urgently needed to improve human health in the Third World [101–104].

Faced with major challenges in human healthcare, an understanding what each of the many nanotechnologies can do – and how they each can best contribute to address the major challenges ahead – is crucial to drive innovation forwards. An improved awareness of how new technologies will help to unravel underpinning mechanisms of disease is crucial to setting realistic expectations and timescales, as well as to prepare for the innovations to come.

Since ultimately, thriving towards providing access to efficient and affordable healthcare, by improving upon technology, is not just an intellectual luxury, but our responsibility.

Acknowledgments

Many of the thoughts about the future of nanomedicine were seeded during exciting discussions with Drs Eleni Kousvelari (NIH) and Jeff Schloss (NIH), as well as with Dr Mike Roco (NSF), when preparing for the interagency workshop on Nanobiotechnology [9]. The many inspirations and contributions from colleagues and friends worldwide are gratefully acknowledged, several of whom have contributed chapters here, as well as from my students and collaborators and the many authors whose articles and conference talks have left long-lasting impressions.

References

1 Nalwa, H.S. (2004) *Encyclopedia of Nanoscience and Nanotechnology*, American Scientific Publishers.

2 Roco, M.C. (2003) National Nanotechnology Initiative: From Vision to Implementation. http://www.nano.gov/nni11600/sld001.htm.

3 National Institutes of Health (2000): Nanoscience and Nanotechnology: Shaping Biomedical Research. Organized

by the Bioengineering Consortium. Available at http://www.becon.nih.gov/nanotechsympreport.pdf.

4 National Institutes of Health (2003) 2020: A New Vision – A Future for Regenerative Medicine, U.S. Department of Health and Human Services. Available at http://www.hhs.gov/reference/newfuture.shtml

5 National Institutes of Health (2003) Nanotechnology in Heart, Lung, Blood, and Sleep Medicine. http://www.nhlbi.nih.gov/meetings/nano_sum.htm.

6 Duncan, R. (2005) Nanomedicine. Forward Looking Report. http://www.esf.org/publication/214/Nanomedicine.pdf, European Science Foundation (ESF)-European Medical Research Council (EMRC). Report Number 2-912049-52-0.

7 NIBIB/DOE (2005) Workshop on Biomedical Applications of Nanotechnology. Organized by the National Institute of Biomedical Imaging and BioEngineering, Department of Energy, USA. Available at http://www.capconcorp.com/nibibdoenanotech/

8 Vogel, V. and Baird, B. (2005) 'BioNanotechnology'. National Nanotechnology Initiative Grand Challenge Workshop Report. Available at http://nano.gov/nni_nanobiotechnology_rpt.pdf.

9 European-Technology-Platform (2006) Nanomedicine: Nanotechnology for Health. Available at ftp://ftp.cordis.europa.eu/pub/nanotechnology/docs/nanomedicine_bat_en.pdf

10 Nantional Institute of Health (2008) Roadmap Initiative in Nanomedicine. Available at http://nihroadmap.nih.gov/nanomedicine/index.

11 Schmid, G., Brune, H., Ernst, H., Grünwald, W., Grunwald, A., Hofmann, H., Janich, P., Krug, H., Mayor, M., Rathgeber, W., Simon, U., Vogel, V. and Wyrwa, D. (2006) *Nanotechnology. Assessment and Perspectives*, Springer Verlag.

12 Godin, B., Serda, R.E., Sakamoto, J., Decuzzi, P. and Ferrari, M. (2009) Nanoparticles for cancer detection and therapy, in *Nanomedicine* (ed. V. Vogel), Ch. 3, Wiley-VCH, Weinheim.

13 Wallace, K.D., Hughes, M.S., Marsh, J.N., Caruthers, S.D., Lanza, G.M. and Wickline, S.A. (2009) From *in vivo* ultrasound and MRI imaging to therapy: Contrast agents based on target-specific nanoparticles, in *Nanomedicine* (ed. V. Vogel), Ch. 2, Wiley-VCH, Weinheim.

14 Johannsen, M., Gneveckow, U., Thiesen, B., Taymoorian, K., Cho, C.H., Waldofner, N., Scholz, R., Jordan, A., Loening, S.A. and Wust, P. (2007) Thermotherapy of prostate cancer using magnetic nanoparticles: feasibility, imaging, and three-dimensional temperature distribution. *European Urology*, **52** (6), 1653–1661.

15 National Cancer Institute (2008). Alliance for Nanotechnology in Cancer. http://nano.cancer.gov/resource_center/tech_backgrounder.asp.

16 Phelps, M.E. (2004) *PET: Molecular Imaging and Its Biological Applications*, Springer.

17 National Cancer Institute (2004) Cancer Nanotechnology. Going Small for Big Advances, NIH.

18 Jain, K.K. (2007) Applications of nanobiotechnology in clinical diagnostics. *Clinical Chemistry*, **53** (11), 2002–2009.

19 Nie, S., Xing, Y., Kim, G.J. and Simons, J.W. (2007) Nanotechnology applications in cancer. *Annual Review of Biomedical Engineering*, **9**, 257–288.

20 Kumar, C.S.S.R. (ed.) (2008) *Nanomaterials for Cancer Diagnosis*, Nanotechnologies for the Life Sciences, Wiley-VCH, Weinheim.

21 Kumar, C.S.S.R. (ed.) (2008) *Nanomaterials for Cancer Therapy*. Nanotechnologies for the Life Sciences, Wiley-VCH, Weinheim.

22 Kumar, C.S.S.R. (ed.) (2008) *Nanomaterials for Medical Diagnosis and*

Therapy Nanotechnologies for the Life Sciences, Wiley-VCH, Weinheim.

23 Perutz, M.F. (1962) X-ray analysis of haemoglobin. Nobel Lecture.

24 Mullis, K.B. (1993) The Polymerase Chain Reaction. Nobel Lecture.

25 Cheung, V.G., Nowak, N., Jang, W., Kirsch, I.R., Zhao, S., Chen, X.N., Furey, T.S., Kim, U.J., Kuo, W.L., Olivier, M., Conroy, J., Kasprzyk, A., Massa, H., Yonescu, R., Sait, S., Thoreen, C., Snijders, A., Lemyre, E., Bailey, J.A., Bruzel, A., Burrill, W.D., Clegg, S.M., Collins, S., Dhami, P., Friedman, C., Han, C.S., Herrick, S., Lee, J., Ligon, A.H., Lowry, S., Morley, M., Narasimhan, S., Osoegawa, K., Peng, Z., Plajzer-Frick, I., Quade, B.J., Scott, D., Sirotkin, K., Thorpe, A.A., Gray, J.W., Hudson, J., Pinkel, D., Ried, T., Rowen, L., Shen-Ong, G.L., Strausberg, R.L., Birney, E., Callen, D.F., Cheng, J.F., Cox, D.R., Doggett, N.A., Carter, N.P., Eichler, E.E., Haussler, D., Korenberg, J.R., Morton, C.C., Albertson, D., Schuler, G., de Jong, P.J. and Trask, B.J. (2001) Integration of cytogenetic landmarks into the draft sequence of the human genome. *Nature* **409** (6822), 953–958.

26 Helmuth, L. (2001) Genome research: map of the human genome 3.0. *Science*, **293** (5530), 583–585.

27 McPherson, J.D., Marra, M., Hillier, L., Waterston, R.H., Chinwalla, A., Wallis, J., Sekhon, M., Wylie, K., Mardis, E.R., Wilson, R.K., Fulton, R., Kucaba, T.A., Wagner-McPherson, C., Barbazuk, W.B., Gregory, S.G., Humphray, S.J., French, L., Evans, R.S., Bethel, G., Whittaker, A., Holden, J.L., McCann, O.T., Dunham, A., Soderlund, C., Scott, C.E., Bentley, D.R., Schuler, G., Chen, H.C., Jang, W., Green, E.D., Idol, J.R., Maduro, V.V., Montgomery, K.T., Lee, E., Miller, A., Emerling, S., Kucherlapati, Gibbs, R., Scherer, S., Gorrell, J.H., Sodergren, E., Clerc-Blankenburg, K., Tabor, P., Naylor, S., Garcia, D., de Jong, P.J., Catanese, J.J., Nowak, N., Osoegawa, K., Qin, S., Rowen, L., Madan, A., Dors, M., Hood, L., Trask, B., Friedman, C., Massa, H., Cheung, V.G., Kirsch, I.R., Reid, T., Yonescu, R., Weissenbach, J., Bruls, T., Heilig, R., Branscomb, E., Olsen, A., Doggett, N., Cheng, J.F., Hawkins, T., Myers, R.M., Shang, J., Ramirez, L., Schmutz, J., Velasquez, O., Dixon, K., Stone, N.E., Cox, D.R., Haussler, D., Kent, W.J., Furey, T., Rogic, S., Kennedy, S., Jones, S., Rosenthal, A., Wen, G., Schilhabel, M., Gloeckner, G., Nyakatura, G., Siebert, R., Schlegelberger, B., Korenberg, J., Chen, X.N., Fujiyama, A., Hattori, M., Toyoda, A., Yada, T., Park, H.S., Sakaki, Y., Shimizu, N., Asakawa, S., Kawasaki, K., Sasaki, T., Shintani, A., Shimizu, A., Shibuya, K., Kudoh, J., Minoshima S., Ramser J., Seranski P., Hoff C., Poustka A., Reinhardt R. and Lehrach H. (2001) A physical map of the human genome. *Nature* **409** (6822), 934–941.

28 Olivier, M., Aggarwal, A., Allen, J., Almendras, A.A., Bajorek, E.S., Beasley, E.M., Brady, S.D., Bushard, J.M., Bustos, V.I., Chu, A., Chung, T.R., De Witte, A., Denys, M.E., Dominguez, R., Fang, N.Y., Foster, B.D., Freudenberg, R.W., Hadley, D., Hamilton, L.R., Jeffrey, T.J., Kelly, L., Lazzeroni, L., Levy, M.R., Lewis, S.C., Liu, X., Lopez, F.J., Louie, B., Marquis, J.P., Martinez, R.A., Matsuura, M.K., Misherghi, N.S., Norton, J.A., Olshen, A., Perkins, S.M., Perou, A.J., Piercy, C., Piercy, M., Qin, F., Reif, T., Sheppard, K., Shokoohi, V., Smick, G.A., Sun, W.L., Stewart, E.A., Fernando, J., Tejeda, J., Tran, N.M., Trejo, T., Vo, N.T., Yan, S.C., Zierten, D.L., Zhao, S., Sachidanandam, R., Trask, B.J., Myers, R.M. and Cox, D.R. (2001) A high-resolution radiation hybrid map of the human genome draft sequence. *Science*, **291** (5507), 1298–1302.

29 Reich, D.E., Cargill, M., Bolk, S., Ireland, J., Sabeti, P.C., Richter, D.J.,

Lavery, T., Kouyoumjian, R., Farhadian, S.F., Ward, R. and Lander, E.S. (2001) Linkage disequilibrium in the human genome. *Nature*, **411** (6834), 199–204.

30 Ideker, T. and Sharan, R. (2008) Protein networks in disease. *Genome Research*, **18** (4), 644–652.

31 King, M.C., Marks, J.H. and Mandell, J.B. (2003) Breast and ovarian cancer risks due to inherited mutations in BRCA1 and BRCA2. *Science*, **302** (5645), 643–646.

32 Lin, M.T. and Beal, M.F. (2006) Mitochondrial dysfunction and oxidative stress in neurodegenerative diseases. *Nature*, **443** (7113), 787–795.

33 Kato, T. (2007) Molecular genetics of bipolar disorder and depression. *Psychiatry and Clinical Neurosciences*, **61** (1), 3–19.

34 Madsen, E. and Gitlin, J.D. (2007) Copper and iron disorders of the brain. *Annual Review of Neuroscience*, **30**, 317–337.

35 Oti, M. and Brunner, H.G. (2007) The modular nature of genetic diseases. *Clinical Genetics*, **71** (1), 1–11.

36 Abrahams, B.S. and Geschwind, D.H. (2008) Advances in autism genetics: on the threshold of a new neurobiology. *Nature Reviews. Genetics*, **9** (5), 341–355.

37 Ferrell, R.E. and Finegold, D.N. (2008) Research perspectives in inherited lymphatic disease: an update. *Annals of the New York Academy of Sciences*, **1131**, 134–139.

38 Judge, D.P. and Dietz, H.C. (2008) Therapy of Marfan syndrome. *Annual Review of Medicine*, **59**, 43–59.

39 Weinreich, F. and Jentsch, T.J. (2000) Neurological diseases caused by ion-channel mutations. *Current Opinion in Neurobiology*, **10** (3), 409–415.

40 Wilde, A.A. and van den Berg, M.P. (2005) Ten years of genes in inherited arrhythmia syndromes: an example of what we have learned from patients, electrocardiograms, and computers. *Journal of Electrocardiology*, **38** (4 Suppl), 145–149.

41 Fiske, J.L., Fomin, V.P., Brown, M.L., Duncan, R.L. and Sikes, R.A. (2006) Voltage-sensitive ion channels and cancer. *Cancer Metastasis Reviews*, **25** (3), 493–500.

42 Terrenoire, C., Simhaee, D. and Kass, R.S. (2007) Role of sodium channels in propagation in heart muscle: how subtle genetic alterations result in major arrhythmic disorders. *Journal of Cardiovascular Electrophysiology*, **18** (8), 900–905.

43 Baker, M.L., Marsh, M.P. and Chiu, W. (2009) Electron cryo-microscopy of molecular nanomachines and cells, in *Nanomedicine* (ed. V. Vogel), Ch. 4, Wiley-VCH, Weinheim.

44 Kukura, P., Renn, A. and Sandoghdar, V. (2009) Pushing optical microscopy to the limit: from single-molecule fluorescence microscopy to label-free detection and tracking of biological nano-objects, in *Nanomedicine* (ed. V. Vogel), Ch. 5, Wiley-VCH, Weinheim.

45 Bao, G., Santangelo, P., Nitin, N. and Rhee, W.-J. (2009) Nanostructured probes for *in vivo* gene detection, in *Nanomedicine* (ed. V. Vogel), Ch. 6, Wiley-VCH, Weinheim.

46 Applegate, K.T., Yang, G. and Danuser, G. (2009) High-content analysis of cytoskeleton functions by fluorescent speckle microscopy, in *Nanomedicine* (ed. V. Vogel), Ch. 7, Wiley-VCH, Weinheim.

47 Fratzl, P., Gupta, H.S., Roschger, P. and Klaushofer, K. (2009) Bone nanostructure and its relevance for mechanical performance, disease and treatment, in *Nanomedicine* (ed. V. Vogel), Ch. 12, Wiley-VCH, Weinheim.

48 Goel, A. and Vogel, V. (2009) Transport, assembly and proof-reading: harnessing the engineering principles of biological nanomotors, in *Nanomedicine* (ed. V. Vogel), Ch. 8, Wiley-VCH, Weinheim.

49 Dunlop, I.E., Dustin, M.L. and Spatz, J.P. (2009) The Micro- and Nanoscale

Architecture of the Immunological Synapse, in *Nanomedicine* (ed. V. Vogel), Ch. 11, Wiley-VCH, Weinheim.

50 Vogel, V. and Sheetz, M.P. (2009) Mechanical Forces Matter in Health and Disease: From Cancer to Tissue Engineering, in *Nanomedicine* (ed. V. Vogel), Ch. 9, Wiley-VCH, Weinheim.

51 Rehfeldt, F., Engler, A.J. and Discher, D.E. (2009) Stem Cells and Nanomedicine: Nanomechanics of the Microenvironment, in *Nanomedicine* (ed. V. Vogel), Ch. 10, Wiley-VCH, Weinheim.

52 Shorokhov, D. and Zewail, A.H. (2008) 4D electron imaging: principles and perspectives. *Physical Chemistry Chemical Physics*, **10** (20), 2879–2893.

53 Sotomayor, M. and Schulten, K. (2007) Single-molecule experiments in vitro and *in silico*. *Science*, **316** (5828), 1144–1148.

54 Khademhosseini, A., Rajalingam, B., Jinno, S. and Langer, R. (2009) Nanoengineered Systems for Tissue Engineering and Regeneration, in *Nanomedicine* (ed. V. Vogel), Ch. 13, Wiley-VCH, Weinheim.

55 Capito, R.M., Mata, A. and Stupp, S.I. (2009) Self-Assembling Peptide-Based Nanostructures for Regenerative Medicine, in *Nanomedicine* (ed. V. Vogel), Ch. 14, Wiley-VCH, Weinheim.

56 Alper, M.D. and Stupp, S.I. (2003) Biomolecular Materials. Basic Energy Sciences Advisory Committee to the Office of Science (DOE) Report http://www.sc.doe.gov/bes/besac/BiomolecularMaterialsReport.pdf.

57 NIH (2008) Roadmap Initiative in Nanomedicine. http://nihroadmap.nih.gov/nanomedicine/index.

58 NIH (2008) US National Cancer Institute (NCI), Alliance for Nanotechnology in Cancer.

59 Sekido, Y., Fong, K.M. and Minna, J.D. (2003) Molecular genetics of lung cancer. *Annual Review of Medicine*, **54**, 73–87.

60 Ishikawa, T., Zhang, S.S., Qin, X., Takahashi, Y., Oda, H., Nakatsuru, Y. and Ide, F. (2004) DNA repair and cancer: lessons from mutant mouse models. *Cancer Science*, **95** (2), 112–117.

61 Sogn, J.A., Anton-Culver, H. and Singer, D.S. (2005) Meeting report: NCI think tanks in cancer biology. *Cancer Research*, **65** (20), 9117–9120.

62 Makrantonaki, E. and Zouboulis, C.C. (2007) Molecular mechanisms of skin aging: state of the art. *Annals of the New York Academy of Sciences*, **1119**, 40–50.

63 Wren, B.G. (2007) The origin of breast cancer. *Menopause (New York, NY)*, **14** (6), 1060–1068.

64 Frey, A.B. and Monu, N. (2008) Signaling defects in anti-tumor T cells. *Immunological Reviews*, **222**, 192–205.

65 Savage, S.A. and Alter, B.P. (2008) The role of telomere biology in bone marrow failure and other disorders. *Mechanisms of Ageing and Development*, **129** (1–2), 35–47.

66 Indraccolo, S., Favaro, E. and Amadori, A. (2006) Dormant tumors awaken by a short-term angiogenic burst: the spike hypothesis. *Cell Cycle (Georgetown, Tex)*, **5** (16), 1751–1755.

67 Townson, J.L. and Chambers, A.F. (2006) Dormancy of solitary metastatic cells. *Cell Cycle (Georgetown, Tex)*, **5** (16), 1744–1750.

68 Vessella, R.L., Pantel, K. and Mohla, S. (2007) Tumor cell dormancy: an NCI workshop report. *Cancer Biology & Therapy*, **6** (9), 1496–1504.

69 Riethdorf, S. and Pantel, K. (2008) Disseminated tumor cells in bone marrow and circulating tumor cells in blood of breast cancer patients: current state of detection and characterization. *Pathobiology*, **75** (2), 140–148.

70 Goodman, S.L., Sims, P.A., Albrecht, R.M. (1996) Three-dimensional extracellular matrix textured biomaterials. *Biomaterials*, **17** (21), 2087–2095.

71 Friedl, P. and Brocker, E.B. (2000) The biology of cell locomotion within three-dimensional extracellular matrix. *Cellular and Molecular Life Sciences*, **57** (1), 41–64.

72 Cukierman, E., Pankov, R., Stevens, D.R. and Yamada, K.M. (2001) Taking cell-matrix adhesions to the third dimension. *Science*, **294** (5547), 1708–1712.

73 Grinnell, F. (2003) Fibroblast biology in three-dimensional collagen matrices. *Trends in Cell Biology*, **13** (5), 264–269.

74 Li, S., Moon, J.J., Miao, H., Jin, G., Chen, B.P., Yuan, S., Hu, Y., Usami, S. and Chien, S. (2003) Signal transduction in matrix contraction and the migration of vascular smooth muscle cells in three-dimensional matrix. *Journal of Vascular Research*, **40** (4), 378–388.

75 Larsen, M., Artym, V.V., Green, J.A. and Yamada, K.M. (2006) The matrix reorganized: extracellular matrix remodeling and integrin signalling. *Current Opinion in Cell Biology*, **18** (5), 463–471.

76 Stryer, L. (2003) Bio2010: Transforming Undergraduate Education for Future Research Biologists. National Research Council Report website.

77 Ratner, B.D. and Bryant, S.J. (2004) Biomaterials: where we have been and where we are going. *Annual Review of Biomedical Engineering*, **6**, 41–75.

78 Berg, H.C. (2003) The rotary motor of bacterial flagella. *Annual Review of Biochemistry*, **72**, 19–54.

79 Weibel, D.B., Diluzio, W.R. and Whitesides, G.M. (2007) Microfabrication meets microbiology. *Nature Reviews Microbiology*, **5** (3), 209–218.

80 Jarrell, K.F. and McBride, M.J. (2008) The surprisingly diverse ways that prokaryotes move. *Nature Reviews Microbiology*, **6** (6), 466–476.

81 Sokurenko, E., Vogel, V. and Thomas, W.E. (2008) Catch bond mechanism of force-enhanced adhesion: counter-intuitive, elusive but . . . widespread? *Cell Host & Microbe, October.*

82 Thomas, W.E., Vogel, V. and Sokurenko, E. (2008) Biophysics of catch bonds. *Annual Review of Biophysics*, **37**, 399–416.

83 Biais, N., Ladoux, B., Higashi, D., So, M. and Sheetz, M. (2008) Cooperative retraction of bundled type IV pili enables nanonewton force generation. *PLoS Biology*, **6** (4), e87.

84 Fletcher, D.A. and Theriot, J.A. (2004) An introduction to cell motility for the physical scientist. *Physical Biology*, **1** (1–2), T1–T10.

85 Brandenburg, B. and Zhuang, X. (2007) Virus trafficking – learning from single-virus tracking. *Nature Reviews Microbiology*, **5** (3), 197–208.

86 Balagadde, F.K., Song, H., Ozaki, J., Collins, C.H., Barnet, M., Arnold, F.H., Quake, S.R. and You, L. (2008) A synthetic *Escherichia coli* predator-prey ecosystem. *Molecular Systems Biology*, **4**, 187.

87 Welch, M., Mikkelsen, H., Swatton, J.E., Smith, D., Thomas, G.L., Glansdorp, F.G. and Spring, D.R. (2005) Cell-cell communication in Gram-negative bacteria. *Molecular BioSystems*, **1** (3), 196–202.

88 Xie, X.S., Choi, P.J., Li, G.W., Lee, N.K. and Lia, G. (2008) Single-molecule approach to molecular biology in living bacterial cells. *Annual Review of Biophysics*, **37**, 417–444.

89 Johansson, M., Lovmar, M. and Ehrenberg, M. (2008) Rate and accuracy of bacterial protein synthesis revisited. *Current Opinion in Microbiology*, **11** (2), 141–147.

90 Zorrilla, S., Lillo, M.P., Chaix, D., Margeat, E., Royer, C.A. and Declerck, N. (2008) Investigating transcriptional regulation by fluorescence spectroscopy, from traditional methods to state-of-the-art single-molecule approaches. *Annals of the New York Academy of Sciences*, **1130**, 44–51.

91 Forero, M., Thomas, W., Bland, C., Nilsson, L., Sokurenko, E. and Vogel, V. (2004) A catch-bond based nano-adhesive sensitive to shear stress. *Nano Letters*, **4**, 1593–1597.

92 Sleytr, U.B., Huber, C., Ilk, N., Pum, D., Schuster, B. and Egelseer, E.M. (2007) S-layers as a tool kit for nanobiotechnological applications. *FEMS Microbiology Letters*, **267** (2), 131–144.

93 Chevance, F.F. and Hughes, K.T. (2008) Coordinating assembly of a bacterial macromolecular machine. *Nature Reviews Microbiology*, **6** (6), 455–465.

94 Diao, J.J., Hua, D., Lin, J., Teng, H.H. and Chen, D. (2005) Nanoparticle delivery by controlled bacteria. *Journal of Nanoscience and Nanotechnology*, **5** (10), 1749–1751.

95 Darnton, N., Turner, L., Breuer, K. and Berg, H.C. (2004) Moving fluid with bacterial carpets. *Biophysical Journal*, **86** (3), 1863–1870.

96 Hiratsuka, Y., Miyata, M., Tada, T. and Uyeda, T.Q. (2006) A microrotary motor powered by bacteria. *Proceedings of the National Academy of Sciences of the United States of America*, **103** (37), 13618–13623.

97 Turner, S.T., Schwartz, G.L. and Boerwinkle, E. (2007) Personalized medicine for high blood pressure. *Hypertension*, **50** (1), 1–5.

98 Katsanis, S.H., Javitt, G. and Hudson, K. (2008) Public health. A case study of personalized medicine. *Science*, **320** (5872), 53–54.

99 Zhong, J.F., Chen, Y., Marcus, J.S., Scherer, A., Quake, S.R., Taylor, C.R. and Weiner, L.P. (2008) A microfluidic processor for gene expression profiling of single human embryonic stem cells. *Lab on a Chip*, **8** (1), 68–74.

100 van't Veer, L.J. and Bernards, R. (2008) Enabling personalized cancer medicine through analysis of gene-expression patterns. *Nature*, **452** (7187), 564–570.

101 Toner, M. and Irimia, D. (2005) Blood-on-a-chip. *Annual Review of Biomedical Engineering*, **7**, 77–103.

102 Yager, P., Edwards, T., Fu, E., Helton, K., Nelson, K., Tam, M.R. and Weigl, B.H. (2006) Microfluidic diagnostic technologies for global public health. *Nature*, **442** (7101), 412–418.

103 Park, J.Y. and Kricka, L.J. (2007) Prospects for nano- and microtechnologies in clinical point-of-care testing. *Lab on a Chip*, **7** (5), 547–549.

104 Phillips, K.A., Liang, S.Y. and Van Bebber, S. (2008) Challenges to the translation of genomic information into clinical practice and health policy: Utilization, preferences and economic value. *Current Opinion in Molecular Therapeutics*, **10** (3), 260–266.

Part Two:
Imaging, Diagnostics and Disease Treatment by Using Engineered Nanoparticles

Nanotechnology, Volume 5: Nanomedicine. Edited by Viola Vogel

ISBN: 978-3-527-31736-3

2
From *In Vivo* Ultrasound and MRI Imaging to Therapy: Contrast Agents Based on Target-Specific Nanoparticles

Kirk D. Wallace, Michael S. Hughes, Jon N. Marsh, Shelton D. Caruthers, Gregory M. Lanza, and Samuel A. Wickline

2.1
Introduction

Advances and recent developments in the scientific areas of genomics and molecular biology have created an unprecedented opportunity to identify clinical pathology in pre-disease states. Building on these advances, the field of molecular imaging has emerged, leveraging the sensitivity and specificity of molecular markers together with advanced noninvasive imaging modalities to enable and expand the role of noninvasive diagnostic imaging. However, the detection of small aggregates of precancerous cells and their biochemical signatures remains an elusive target that is often beyond the resolution and sensitivity of conventional magnetic resonance and acoustic imaging techniques. The identification of these molecular markers requires target-specific probes, a robust signal amplification strategy, and sensitive high-resolution imaging modalities.

Currently, several nanoparticle or microparticle systems are under development for targeted diagnostic imaging and drug delivery [1]. Perfluorocarbon (PFC) nanoparticles represent a unique platform technology, which may be applied to multiple clinically relevant imaging modalities. They exploit many of the key principles employed by other imaging agents. Ligand-directed, lipid-encapsulated PFC nanoparticles (with a nominal diameter of 250 nm) have inherent physicochemical properties which provide acoustic contrast when the agent is bound to a surface layer. The high surface area of the nanoparticle accommodates 100 to 500 targeting molecules (or ligands), which impart high avidity and provides the agent with a robust 'stick and stay' quality (Figure 2.1). The incorporation of large payloads of lipid-anchored gadolinium chelate conjugates further extends the utility of the agent to detect sparse concentrations of cell-surface biochemical markers with magnetic resonance imaging (MRI) [2]. Moreover, for MRI the high fluorine signal from the nanoparticle core allows the noninvasive quantification of ligand-bound particles, which enables clinicians to confirm tissue concentrations of drugs when

Nanotechnology, Volume 5: Nanomedicine. Edited by Viola Vogel

ISBN: 978-3-527-31736-3

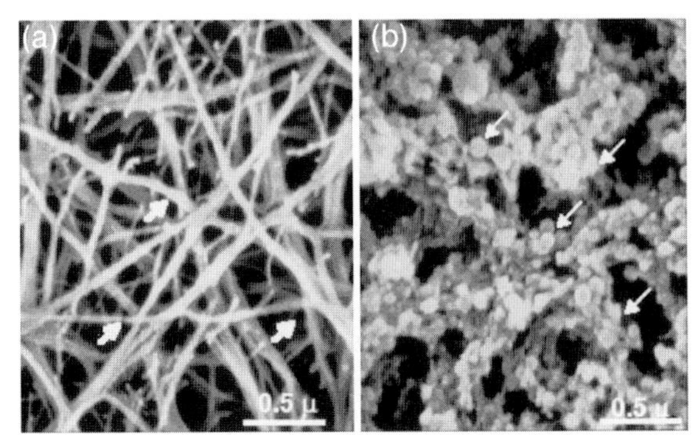

Figure 2.1 Scanning electron microscopy images (original magnification, ×30 000) of (a) a control fibrin clot and (b) fibrin-targeted paramagnetic nanoparticles bound to the clot surface. Arrows in (a) indicate a fibrin fibril; arrows in (b) indicate fibrin-specific nanoparticle-bound fibrin epitopes.

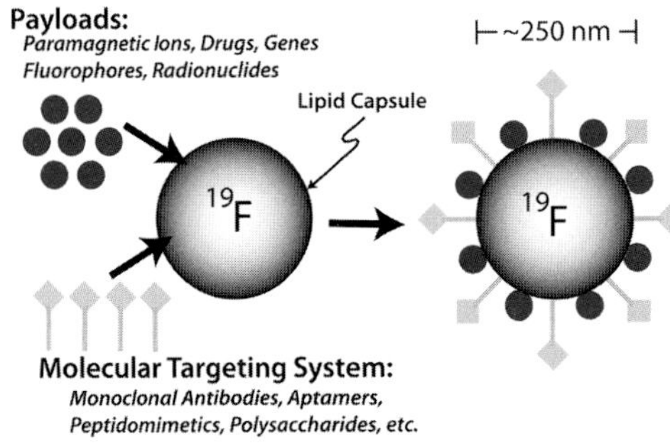

Figure 2.2 Paradigm for targeted liquid perfluorocarbon-based nanoparticle contrast agent. This example has a payload of Gd^{3+} chelates and monoclonal antibodies. The platform is extremely versatile, applicable to almost any imaging modality, and capable of carrying other payloads such as drugs or genes.

the functionality of the nanoparticles is extended to include targeted therapy (Figure 2.2). The detection of sparse concentrations of cell-surface biochemical markers is also possible with ultrasound [3]; however, novel signal processing is required for this application [4–6].

2.2
Active versus Passive Approaches to Contrast Agent Targeting

The passive targeting of a contrast agent is achieved by exploiting the body's inherent defense mechanisms for the clearance of foreign particles. Macrophages of the macrophage phagocytic system are responsible for the removal of most of these contrast agents from the circulation; these are produced, in size-dependent fashion, from the lung, spleen, liver and bone marrow. Phagocytosis and accumulation within specific sites can be enhanced by biologic tagging (i.e. *opsonization*) with blood proteins such as immunoglobulins, complement proteins or nonimmune serum factors. In general, sequestration in the liver appears to be complement-mediated, while the spleen removes foreign particulate matter via antibody F_c receptors [8]. This natural process of nondirected and nonspecific uptake of particles is generally referred to as '*passive targeting*' (e.g. Feridex in the liver, or iron oxide in the sentinel lymph nodes [9]).

Distinguished from passive contrast agents, targeted (i.e. 'ligand-directed') contrast agents are designed to enhance specific pathological tissue that otherwise might be difficult to distinguish from the surrounding normal tissues. Here, an extensive array of ligands can be utilized, including monoclonal antibodies and fragments, peptides, polysaccharides, aptamers and drugs. These ligands may be attached either covalently (i.e. by direct conjugation) or noncovalently (i.e. by indirect conjugation) to the contrast agent. Engineered surface modifications, such as the incorporation of polyethylene glycol (PEG), are used to delay or avoid the rapid systemic removal of the agents, such that ligand-to-target binding is allowed to occur.

The effectiveness of this concept of contrast agent targeting is demonstrated with the application of paramagnetic MRI contrast agents. Paramagnetic agents influence only those protons in their immediate vicinity, and removal of these contrast agents by the macrophage phagocytic system during passive targeting may decrease their effectiveness via two mechanisms: (i) an accumulation of contrast agent in specific organs that are distal to region of interest; and (ii) endocytosis, which further decreases their exposure to free water protons. By targeting the contrast agent, the paramagnetic ions can be brought in close proximity to the region of interest with sufficient accumulation to overcome the partial dilution effect that plagues some MRI contrast agents. Its efficacy is further enhanced with some targeting platforms by delivering multiple contrast ions per particle [2].

2.3
Principles of Magnetic Resonance Contrast Agents

The fundamental physics underpinning MRI is grounded in the quantum mechanical magnetic properties of the atomic nucleus. All atomic nuclei have a fundamental property known as the *nuclear magnetic momentum* or *spin quantum number*. Individual protons and neutrons are fermions that possess an intrinsic angular momentum, or 'spin', quantized with a value of 1/2 [10, 11].

The overall spin of a nucleus (a composite fermion) is determined by the numbers of neutrons and protons. In nuclei with even numbers of protons and an even numbers of neutrons, these nucleons pair up to result in a net spin of zero. Nuclei with an odd number of protons or neutrons will have a nonzero net spin which, when placed in a strong magnetic field (with magnitude B_0), will have an associated net magnetic moment, $\vec{\mu}$, that will orient either with ('parallel') or against ('anti-parallel') the direction of B_0. For a nucleus with a net spin of 1/2 (e.g. ^1H), this results in two possible spin states with an energy of separation $\Delta E = h\gamma B_0/2\pi$ (where $h = 6.626 \times 10^{-34}$ J s is Planck's constant and γ is the *gyromagnetic ratio*, which for hydrogen is equal to 42.58 MHz T^{-1}).

For a given population of nuclei in a static magnetic field, an equilibrium exists, described by Maxwell–Boltzmann statistics, in which only a slight majority of nuclei are oriented in the 'parallel' position (i.e. a lower energy state); however, this small difference in spin distribution results in a net magnetization that is perceptible. Absorption of the appropriately tuned radiofrequency (RF) radiation by the nuclei can alter the equilibrium distribution of 'anti-parallel' states. On a macroscopic level, this is equivalent to tilting the net magnetization away from the direction of the main magnetic field (B_0). Once the RF energy is removed, decay to the previous lower energy state takes place and occurs in two distinct and independent processes known as longitudinal relaxation (T_1) and transverse relaxation (T_2). The relaxation times, T_1 and T_2, as well as the proton density of the nuclei of interest, determine the signal intensity for various types of tissue in MRI.

Magnetic resonance contrast agents function by accelerating the longitudinal and/or transverse relaxation rates. The most commonly used nontargeted MR contrast agents are paramagnetic ions (e.g. gadolinium chelates), and these predominantly shorten T_1 relaxation to result in a bright signal on T_1-weighted images. The mechanism by which paramagnetic ions affect T_1 relaxation depends upon close nuclear interaction with protons (^1H) in water molecules (H_2O). Therefore, T_1 agents only influence protons proximate to themselves and are highly dependent on local water flux [12].

Superparamagnetic and ferromagnetic compounds have a high magnetic susceptibility, and when placed in a magnetic field (B_0) they concentrate the field; this results in a large local net positive magnetization [13]. This large magnetic susceptibility heterogeneity induces spin dephasing in tissue and results in a loss of the T_2-weighted signal. In contrast to T_1 contrast agents, superparamagnetic agents disturb the magnetic field and have a net effect far beyond their immediate vicinity.

2.3.1
Mathematics of Signal Contrast

To elucidate the source of image contrast, let us assume that two adjacent tissue types (A and B) manifest identical longitudinal (T_1) and transverse (T_2) relaxation times prior to nanoparticle binding, but only one tissue (say, type B) expresses the molecular epitope of interest that binds the targeted paramagnetic nanoparticles. The bound paramagnetic nanoparticles affect the relaxation times in the targeted tissue according to the following equations [14]:

$$\frac{1}{T_{1B}} = \frac{1}{T_{1A}} + r_{1P}\langle NP\rangle \quad (2.1)$$

$$\frac{1}{T_{2B}} = \frac{1}{T_{2A}} + r_{2P}\langle NP\rangle \quad (2.2)$$

where T_{1B} and T_{2B} are the observed relaxation times after the nanoparticle binding, T_{1A} and T_{2A} are the original relaxation times, r_{1P} and r_{2P} are the particle-based relaxivities, and $\langle NP \rangle$ represents the average nanoparticle concentration within the imaging voxel. For the purpose of this example, the assumption is made that targeted binding does not affect particle relaxivity (i.e. r_{1p} and r_{2p} are constant).

The contrast-to-noise ratio (CNR) between the two tissues for a given sequence is calculated as the absolute difference between their signal intensities. If I_A and I_B represent the signal intensities of tissue A and B respectively, and N is the expected level of noise in the resulting image, the CNR ratio is given by:

$$\mathrm{CNR} = \frac{I_A - I_B}{N} \tag{2.3}$$

For a spin echo pulse sequence, the signal intensity of each tissue is related to the chosen scan parameters (echo time, TE, and repetition time, TR) as well as its magnetic properties (T_1 and T_2), which change due to binding of the contrast agent, and is described with the following relationships for tissues A and B [15]:

$$I_A = k_A(1 - 2e^{-(TR-TE/2)/T_{1A}} + e^{-TR/T_{1A}})e^{-TE/T_{2A}} \tag{2.4}$$

$$I_B = k_B(1 - 2e^{-(TR-TE/2)/T_{1B}} + e^{-TR/T_{1B}})e^{-TE/T_{2B}} \tag{2.5}$$

The constants k_A and k_B incorporate factors such as proton density, RF excitation and coil sensitivity. As these tissues are assumed identical, except for binding of the contrast agent, k_A and k_B are identical except for relative coil sensitivity to the positional differences between the two tissues for this simulation. Substituting Equations (2.4) and (2.5) into Equation (2.3) and optimizing the resulting equation for TR provides a relationship between the T_1 values for the two tissues and the repetition time that will create the highest CNR [15].

$$TR_{\mathrm{opt}} = \frac{T_{1A}T_{1B}}{T_{1B} - T_{1A}} \log\left(\frac{k_A T_{1B}}{k_B T_{1A}}\right) \tag{2.6}$$

With use of the field-dependent input parameters specified, model predictions for the minimum concentration of contrast agent required to generate visually apparent contrast between the two tissues may easily be determined [16]. As a point of reference, visually apparent contrast is typically defined as a $\mathrm{CNR} \geq 5:1$ [17].

2.3.2
Perfluorocarbon Nanoparticles for Enhancing Magnetic Resonance Contrast

For use as a T_1-weighted paramagnetic contrast agent, perfluorocarbon nanoparticles can be functionalized by surface incorporation of homing ligands and more than 50 000 gadolinium chelates (Gd^{3+}) per particle [7]. In addition, all of the paramagnetic ions are present in the outer aqueous phase to achieve maximum relaxivity of T_1 [18]. The result is a perfluorocarbon nanoparticle that is capable of overcoming the diluting partial volume effects that plague most magnetic resonance contrast agents [19]. The efficiency of an magnetic resonance contrast agent can be described by its relaxivity ($\mathrm{m}M^{-1}\,\mathrm{s}^{-1}$), which is simply calculated as the change in relaxation rate ($1/T_1$ or $1/T_2$) divided by the concentration of the contrast agent. The relaxivity of Gd^{3+} in saline ($4.5\,mM^{-1}\,s^{-1}$) [20] is lower when compared to Gd^{3+} bound to the surface of a PFC nanoparticle ($33.7\,mM^{-1}\,s^{-1}$) [18] at a field strength of 1.5 Tesla. Considering that each nanoparticle carries approximately 50 000 to 100 000 Gd^{3+}, the 'particle' relaxivity has been measured at over 2 000 000 $mM^{-1}\,s^{-1}$ [18]. The high level of relaxivity achieved using this paramagnetic liquid PFC nanoparticle allows for the detection and quantification of nanoparticle concentrations as low as 100 picomolar, with a CNR of 5 : 1 [16].

2.3.3
Perfluorocarbon Nanoparticles for Fluorine (^{19}F) Imaging and Spectroscopy

The intensity of a magnetic resonance signal is directly proportional to the gyromagnetic ratio (γ) and the number of nuclei in the volume of interest [21]. Although there are seven medically relevant nuclei, the ^{1}H proton is the most commonly imaged nuclei in clinical practice because of its high γ and natural abundance. The isotopes, their γ-values, natural abundance and relative sensitivity compared to ^{1}H with a constant field are listed in Table 2.1. With a gyromagnetic ratio second only to ^{1}H and a natural abundance of 100%, ^{19}F is an attractive nucleus for MRI [22]. Its sensitivity is 83% (when compared to ^{1}H) at a constant field strength and with an equivalent number of nuclei. In biological tissue, low ^{19}F concentrations (in the range of micromoles) makes MRI impractical at clinically relevant field strengths without ^{19}F-specific contrast agents [23]. Perfluorocarbon nanoparticles are 98% perfluorocarbon by volume, which for perfluoro-octylbromide ($1.98\,g\,ml^{-1}$, $498.97\,g\,mol^{-1}$) equates to an approximately 100 M concentration of fluorine within a nanoparticle. The paucity of endogenous fluorine in biological tissue allows the use of exogenous PFC nanoparticles as an effective ^{19}F MR contrast agent, without any interference from significant background signal. When combined with local drug delivery, detection of the ^{19}F signal serves as a highly specific marker for the presence of nanoparticles that would permit the quantitative assessment of drug dosing.

^{19}F has seven outer-shell electrons rather than a single electron (as is the case for hydrogen); as a result, the range and sensitivity to the details of the local environment of chemical shifts are much higher for fluorine than hydrogen. Consequently,

Table 2.1 Medically relevant MRI nuclei.

Isotope	γ (MHz T^{-1})	Natural abundance (%)	Relative sensitivity
^{1}H	42.58	99.98	1.00
^{19}F	40.05	100	0.83
^{23}Na	11.26	100	0.093
^{31}P	17.24	100	0.066
^{13}C	10.71	1.11	0.015
^{2}H	6.54	0.015	0.0097
^{15}N	3.08	4.31	0.0010

distinct spectra from different PFC species can be obtained and utilized for simultaneous targeting of multiple biochemical markers.

For use as a clinically applicable contrast agent, the biocompatibility of PFC nanoparticles must be considered. Liquid PFCs were first developed for use as a blood substitute [24], and no toxicity, carcinogenicity, mutagenicity or teratogenic effects have been reported for pure fluorocarbons within the 460 to 520 molecular-weight range. Perfluorocarbons, which inert biologically, are removed via the macrophage phagocytic system and excreted primarily through the lungs and in small amounts through the skin, as a consequence of their high vapor pressure relative to their mass [25]. The tissue half-lives of PFCs range from 4 days for perfluoro-octylbromide up to 65 days for perfluorotripropylamine. The prolonged systemic half-life of PFC nanoparticles, in conjunction with the local concentrating effect produced by ligand-directed binding, permits ^{19}F spectroscopy and imaging studies to be conducted at clinically relevant magnetic field strengths.

2.3.4
Fibrin-Imaging for the Detection of Unstable Plaque and Thrombus

Of the over 720 000 cardiac-related deaths that occur each year in the United States, approximately 63% are classified as sudden cardiac death [26]. Unfortunately, for the majority of patients, this is the first and only symptom of their atherosclerotic heart disease [27]. Atherosclerosis manifests initially as a fatty streak but, without proper treatment, it can progress to a vulnerable plaque that is characterized by a large lipid core, a thin fibrous cap and macrophage infiltrates [28]. These vulnerable plaques are prone to rupture, which can lead to thrombosis, vascular occlusion and subsequent myocardial infarction [29] or stroke. Routine angiography is the most common method of diagnosing atherosclerotic heart disease, with the identification of high-grade lesions (>70% stenosis) being referred for immediate therapeutic intervention. Ironically, most ruptured plaques originate from coronary lesions classified as nonstenotic [28]. Even nuclear and ultrasound-based stress tests are only designed to detect flow-limiting lesions. Because the most common source of thromboembolism comes from atherosclerotic plaques with 50–60% stenosis [30], diagnosis by traditional techniques remains elusive. In addition, there appears to be a 'window of opportunity' that exists between the detection of a vulnerable or ruptured plaque and acute myocardial infarction (measured in a few days to months) [31], when intervention could prove to be beneficial.

The acoustic enhancement of thrombi using fibrin-targeted nanoparticles was first demonstrated *in vitro* as well as *in vivo* in a canine model at frequencies typically used in clinical transcutaneous scanning [32]. The detection of thrombi was later expanded to MRI in a study by Flacke *et al.* [7] Fibrin clots were targeted *in vitro* with paramagnetic nanoparticles and imaged using typical low-resolution T_1-weighted proton imaging protocols with a field strength of 1.5 Tesla. Low-resolution images show the effect of increasing the amount of Gd^{3+} incorporated in the nanoparticles: a higher gadolinium loading results in brighter T_1 signals from the fibrin-bound PFC nanoparticles (Figure 2.3). In the same study, *in vivo* MR images were obtained of

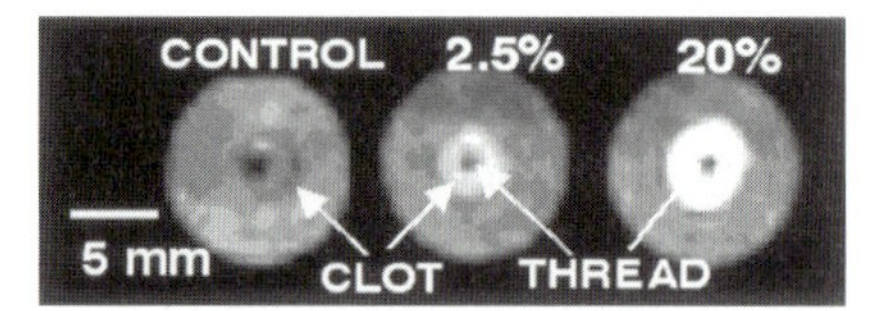

Figure 2.3 Low-resolution images (three-dimensional, T_1-weighted) of control and fibrin-targeted clot with paramagnetic nanoparticles presenting a homogeneous, T_1-weighted enhancement.

fibrin clots in the external jugular vein of dogs. Enhancement with fibrin-targeted PFC nanoparticles produced a high signal intensity in treated clots (1780 ± 327), whereas the control clot exhibited a signal intensity (815 ± 41) similar to that of the adjacent muscle (768 ± 47).

This method was extended to the detection of ruptured plaque in human carotid artery endarterectomy specimens resected from a symptomatic patient (Figure 2.4). Fibrin depositions ('hot spots') were localized to microfissures in the shoulders of the ruptured plaque in the targeted vessel (where fibrin was deposited), but this was not appreciated in the control. Further investigation towards the molecular imaging of small quantities of fibrin in ruptured plaque may someday detect this silent pathology sooner in order to pre-empt stroke or myocardial infarction.

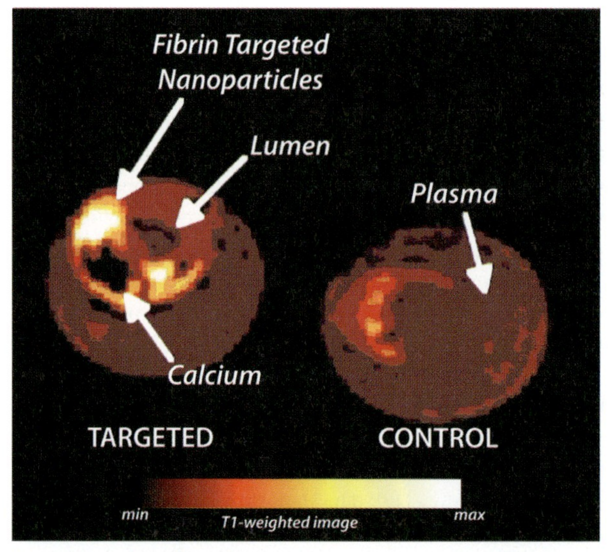

Figure 2.4 Color-enhanced magnetic resonance imaging of fibrin-targeted and control carotid endarterectomy specimens, revealing contrast enhancement (white) of a small fibrin deposit on a symptomatic ruptured plaque. The black area shows a calcium deposit. Three-dimensional, fat-suppressed, T_1-weighted fast gradient echo.

The high fluorine content of fibrin-targeted PFC nanoparticles, as well as the lack of background signal, can also be exploited for ^{19}F MRI and spectroscopy. In a recent study conducted by Morawski *et al.*, several methods were described for quantifying the number of nanoparticles bound to a fibrin clot using the ^{19}F signal [16]. First, fibrin-targeted paramagnetic perfluoro-crown-ether nanoparticles and trichlorofluoromethane ^{19}F spectra were obtained (Figure 2.5a). The relative crown ether signal intensity (with respect to the trichlorofluoromethane peak) from known emulsion volumes provided a calibration curve for nanoparticle quantification (Figure 2.5b). The perfluorocarbon (crown ether) nanoparticles then were mixed in titrated ratios with fibrin-targeted nanoparticles containing safflower oil and bound to plasma clots *in vitro*. As the competing amount of nonsignaling safflower-oil agent was increased, there was a linear decrease in the ^{19}F and Gd^{3+} signal. The number of bound nanoparticles was calculated from the ^{19}F signal and the calibration curve described above, and compared with mass of Gd^{3+} as determined by neutron activation analysis. As expected, there was excellent agreement between measured Gd^{3+} mass and number of bound nanoparticles (calculated from the ^{19}F signal) (Figure 2.5c).

In addition, clots were treated with fibrin-targeted nanoparticles containing either of two distinct PFC cores: crown ether and perfluoro-octyl bromide (PFOB) [33]. These exhibited two distinct ^{19}F spectra at a field strength of 4.7 Tesla, and the signal from the sample was highly related to the ratio of PFOB and crown ether emulsion applied. These findings demonstrated the possibility of simultaneous imaging and quantification of two separate populations of nanoparticles, and hence two distinct biomarkers.

These quantification techniques were applied to the analysis of human carotid endarterectomy samples (see Figure 2.6). An optical image of the carotid reveals extensive plaques, wall thickening and luminal irregularities. Multislice ^{19}F images showed high levels of signal enhancement along the luminal surface due to the binding of targeted paramagnetic nanoparticles to fibrin deposits (not shown in Figure 2.6). The ^{19}F projection images of the artery, taken over approximately 5 min, showed an asymmetric distribution of fibrin-targeted nanoparticles around the vessel wall, corroborating the signal enhancement observed with ^{1}H MRI. Concomitant visualization of ^{1}H and ^{19}F images would permit the visualization of anatomical and pathological information in a single image. In theory, the atherosclerotic plaque burden could be visualized with paramagnetic PFC contrast-enhanced ^{1}H images, while ^{19}F could be used localize identify plaques with high levels of fibrin and thus prone to rupture.

2.3.5 Detection of Angiogenesis and Vascular Injury

As described previously, ligand-directed PFC nanoparticles are well suited to the detection of very sparse biomarkers, such as integrins involved in the process of *angiogenesis* [12, 33, 34]. Although angiogenesis is a critical physiological process in wound healing, inflammation and organ development, it also contributes to the

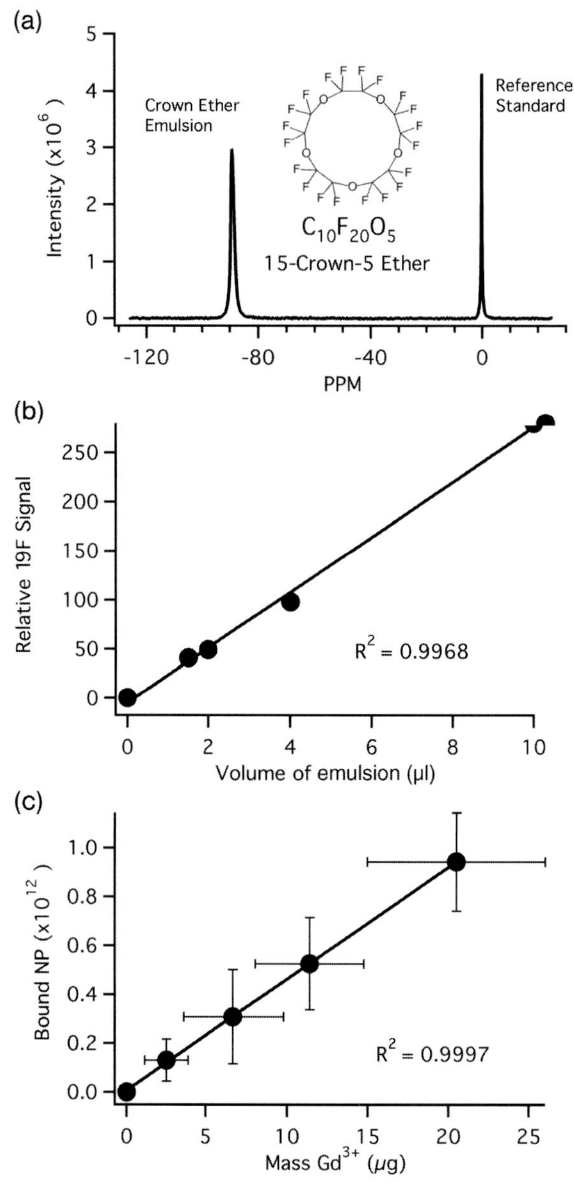

Figure 2.5 (a) Representative spectrum, taken at a field strength of 4.7 T, showing crown ether emulsion (~90 ppm) and trichlorofluoromethane (0 ppm) references; (b) The calibration curve for the crown ether emulsion has a slope of 28.06 with an R^2 of 0.9968; (c) Number of bound nanoparticles (mean ± SE) as calculated from ^{19}F spectroscopy versus the mass of total gadolinium (Gd^{3+}) in the sample as determined by neutron activation analysis, showing excellent agreement as independent measures of fibrin-targeted nanoparticles binding to clots. The linear regression line has an R^2 of 0.9997.

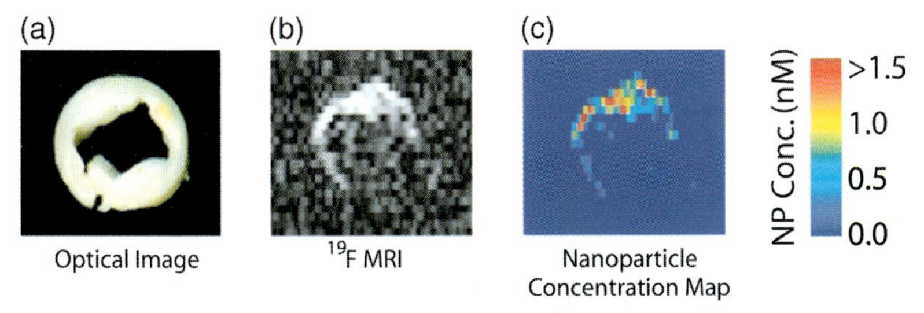

Figure 2.6 (a) Optical image of a 5 mm cross-section of a human carotid endarterectomy sample. The section showed moderate lumenal narrowing, and several atherosclerotic lesions; (b) A ^{19}F projection image acquired at 4.7 T through the entire carotid artery sample, showing a high signal along the lumen due to nanoparticles bound to fibrin; (c) Concentration map of bound nanoparticles in the carotid sample.

pathology of many disease processes such as diabetic retinopathy, rheumatoid arthritis, cancer and atherosclerosis. The process of angiogenesis depends on the adhesive interactions of vascular cells, and the integrin $\alpha_v\beta_3$ has been identified as playing a vital role in angiogenic vascular tissue. The functions of integrin $\alpha_v\beta_3$ includes vascular cell apoptosis (i.e. cell death), smooth muscle cell (SMC) migration and proliferation, and vascular remodeling [35]. The integrin is expressed on the luminal surface of activated endothelial cells, but not on mature quiescent cells. These findings support the fact that the role of $\alpha_v\beta_3$ in pathological conditions characterized by neovascularization may be an important diagnostic and therapeutic target. In fact, the use of a monoclonal antibody against $\alpha_v\beta_3$ has demonstrated an inhibition of angiogenesis, without affecting mature vessels [36]. Although $\alpha_v\beta_3$ integrins are expressed on other mature cells, such as SMCs, tissue macrophages, and on neovascular tissues in gut and developing bone, the PFC-based nanoparticles are too large to escape the normal vasculature and bind in sufficient quantities to create a detectable signal.

Perfluorocarbon nanoparticles have been developed to detect the sparse expression of the $\alpha_v\beta_3$ integrin on the neovasculature, and to deliver anti-angiogenic therapy (Figure 2.7) [37]. This approach has been used to visualize tumor-related angiogenesis in New Zealand White rabbits bearing Vx-2 tumors (<1.0 cm) using a 1.5 Tesla field [38]. MRI signals monitored at 2 h post-injection of $\alpha_v\beta_3$-targeted nanoparticles showed an enhancement of 126%, predominantly in an asymmetrical distribution along the tumor border. These results were consistent with the immunohistochemical staining results. Moreover, *in vivo* competitive blocking with $\alpha_v\beta_3$-targeted nonparamagnetic nanoparticles resulted in a decreased signal enhancement to a level attributable to local extravasation.

In a similar study, athymic nude mice bearing human melanoma tumors (C32, ATCC; 33 mm^3) were injected with $\alpha_v\beta_3$-targeted PFC nanoparticles and imaged at 2 h [39]. MR enhancement was apparent within 30 min and had increased by 173% at 2 h (Figure 2.8). Again, MRI results were correlated with

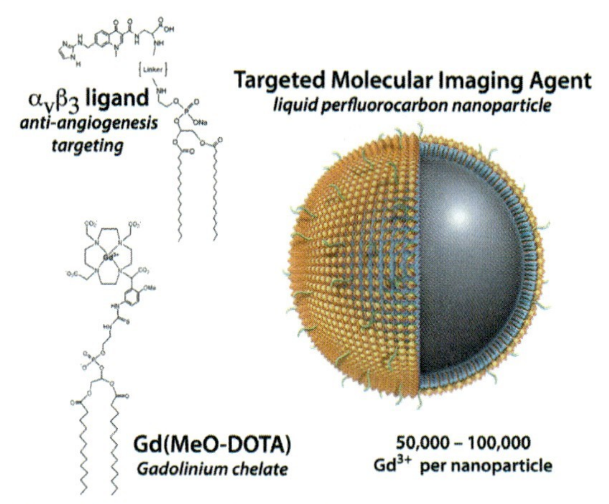

Figure 2.7 Schematic depicting the $\alpha_v\beta_3$ targeting aspect and the Gd^{3+} component that are incorporated into the lipid shell of the liquid perfluorocarbon nanoparticle agent.

histological results. In both studies, *in vivo* competitive blocking with $\alpha_v\beta_3$-targeted nonparamagnetic nanoparticles showed a 50% decrease of signal enhancement. These findings demonstrated the high specificity achievable with $\alpha_v\beta_3$-targeted nanoparticles.

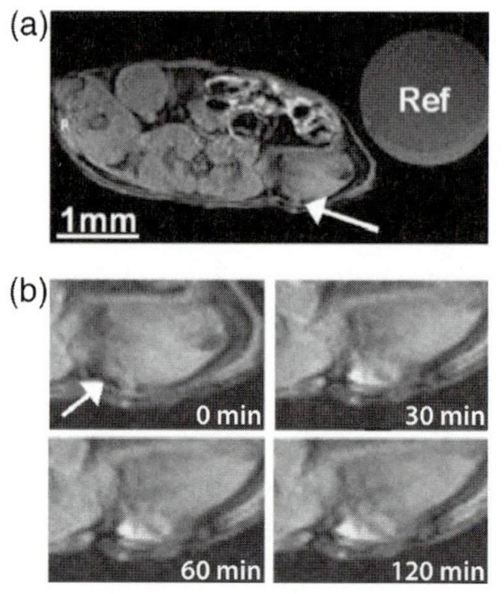

Figure 2.8 (a) T_1-weighted MR image (axial view) of an athymic nude mouse before injection of paramagnetic $\alpha_v\beta_3$-targeted nanoparticles. The arrow indicates a C-32 tumor that is difficult to detect. The reference (Ref) is Gd^{3+} in a 10 ml syringe; (b) Enlarged section of an MR image showing T_1-weighted signal enhancement of angiogenic vasculature of early tumors over 2 h, as detected by $\alpha_v\beta_3$-targeted nanoparticles.

2.4
Perfluorocarbon Nanoparticles as an Ultrasound Contrast Agent

Using the bubbles produced by agitating saline, Gramiak and Shah introduced the concept of an ultrasonic contrast agent in 1968 [40]. Today, commercially available ultrasound contrast agents are based on gas-filled encapsulated microbubbles (average diameter 2–5 μm) that transiently enhance the blood pool signal, which is otherwise weakly echogenic. When insonified by an ultrasound wave, microbubbles improve the gray scale images and Doppler signal via three distinct mechanisms [41–43]. First, at lower acoustic power, microbubbles are highly efficient scatterers due to their large differences in acoustic impedance ($Z = \rho c$, where ρ is the mass density and c is the speed of sound) compared to the surrounding tissue or blood [44]. With increasing acoustic energies, microbubbles begin nonlinear oscillations and emit harmonics of the fundamental (incident) frequency, thus behaving as a source of sound, rather than as a passive reflector [44, 46]. As biological tissue does not display this degree of harmonics, the contrast signal can be exploited to preferentially image microbubbles and improve signal-to-noise ratios (SNRs).

At even higher acoustic power, the destruction of microbubbles occurs allowing the release of free gas bubbles. Although not desirable for most forms of imaging, this results in a strong but transient scattering effect and provides the most sensitive detection of microbubbles. To emphasize these strong echogenic properties, it has been shown that even one microbubble can be detected with medical ultrasound systems [47]. Interestingly, the destruction and cavitation of microbubbles by ultrasound waves have been shown to facilitate drug delivery by 'sonoporating' membranes and allowing drugs and gene therapy to enter the cell [48, 49]. When this process occurs in capillary beds, the permeability increases allowing a subset of particles access to surrounding tissue for further drug deposition [50].

The wide use of microbubbles in everyday clinical applications highlight its effectiveness as a blood pool agent [45]. For example, microbubbles enhance the blood–tissue boundary of the left ventricular cavity, allowing for better diagnostic yield in resting as well as stress echocardiograms [51]. Improved Doppler signals are beneficial in the diagnosis of valvular stenosis and regurgitation [52]. Additionally, microbubbles are removed from the circulation via the macrophage phagocytic system and accumulate in the liver and spleen – that is, passive targeting. This mechanism can be employed for the detection of focal liver lesions and malignancies [48, 53]. When used as targeted contrast agents, microbubbles have been conjugated with ligands for a variety of vascular biomarkers including integrins expressed during angiogenesis, the glycoprotein IIb/IIIa receptor on activated platelets in clots, and L-selectin for the selective enhancement of peripheral lymph nodes, *in vivo* [54–56]. One disadvantage associated with the targeting of microbubbles is the 'tethering' of these particles to a surface. This interaction with a solid structure limits the ability of insonified microbubbles to oscillate, and dampens its echogenicity.

Unlike microbubble formulations that are naturally echogenic, liquid PFC nanoparticles have a weaker inherent acoustic reflectivity, and suspensions of them have been shown to exhibit backscattering levels 30 dB below that of whole blood [57]. However, when collective deposition occur on the surfaces of tissues or a cell in a layering effect, these particles create a local acoustic impedance mismatch that produces a strong ultrasound signal, without any concomitant increase in the background level [58]. The echogenicity of nanoparticles does not depend upon the generation of harmonics, and therefore is not affected by binding with molecular epitopes. Due to their small size and inherent *in vivo* stability, PFC nanoparticle emulsions have a long circulatory half-life compared to microbubble contrast agents. This is accomplished without modification of their outer lipid surfaces with PEG or the incorporation of polymerized lipids, which may detract from the targeting efficacy. Acquired data have suggested that the PFC nanoparticles remain bound to the tissues for up to 24 h. In additionally, nongaseous PFOB-filled nanoparticles neither easily deform nor cavitate with ultrasound imaging.

The successful detection of cancer *in vivo* depends on a variety of factors when using molecularly targeted contrast agents. The number of epitopes to which the ligand can bind must be sufficient to allow enough of the contrast agent to accumulate for detection, while the ligand specificity must be maintained to ensure that nonspecific binding remains negligible. As stated above, the background signal from unbound, circulating contrast agent is low enough (or even absent) so as to not interfere with the assessment of bound, targeted agent. Previous studies have already demonstrated the use of high-frequency ultrasound in epitope-rich pathologies, such as fibrin in thrombus, where targeted PFC nanoparticles can act as a suitable molecular imaging agent by modifying the acoustic impedance on the surface to which they bind in a configuration that is well-approximated by a reflective layer [59]. However, at lower frequencies and for sparse molecular epitopes, in the typically tortuous vascular bed associated with the advancing front of a growing tumor, the clear delineation between nontargeted normal tissue and angiogenic vessels remains a challenge. The imaging technology itself must be highly sensitive and capable of detecting and/or quantifying the level of contrast agent bound to the pathological tissue. In clinical ultrasonic imaging, the sensitivity of detection depends on a physical difference in the way sound interacts with a surface covered by targeted contrast agent versus one that is not. The data presented below show that, in many cases, the sensitivity of this determination can be improved by applying novel and specific signal-processing techniques based on thermodynamic or information-theoretic analogues.

Site-targeted nanoparticle contrast agents, when bound to the appropriate receptor, must be detected in the presence of bright echoes returned from the surrounding tissue. One approach to the challenge of detecting the acoustic signature of site-specific contrast is through the use of novel signal receivers (i.e. mathematical operations that reduce an entire RF waveform, or a portion of it, to a single number) based on information-theoretic quantities, such as Shannon entropy (H), or its counterpart for continuous signal (H_f). These receivers have been shown to be sensitive to diffuse, low-amplitude features of the signal that often are obscured by noise, or else are lost in large specular echoes and, hence, not usually perceivable by a human observer [60–64].

Although entropy-based techniques have a long history in image processing for image enhancement and the post-processing of reconstructed images, the approach we take is different in that entropy is used directly as the quantity defining the pixel values in the image. Specifically, images are reconstructed by computing the entropy (or a limiting form of it: H_f) of segments of the individual RF A-lines that comprise a typical medical image by applying a 'moving window' or 'box-car' analysis. The computation of an entropy value for each location within an image is therefore possible, and the results can be superimposed over the conventional grayscale image as a parametric map.

2.4.1
Entropy-Based Approach

Radiofrequency data are obtained by sampling a continuous function $y = f(t)$. For an 8-bit digitizer, the sampled waveform is quantized into 256 (2^8) different levels. If we compute the probability, p_k of the kth digitizer value appearing in the digital waveform, then we may compute the Shannon entropy of the resultant probability distribution

$$H - \sum_{k=0}^{255} p_k \log(p_k). \tag{2.7}$$

While this quantity has demonstrated utility for signal characterization [60], it also has the undesirable feature that it depends critically on the attributes of the digitizer used to acquire the data. This dependence may be removed by taking the limit where the sampling rate and dynamic range are taken to infinity [61, 62]. In that case, the probabilities, p_k, are replaced by density function, $wf(y)$, of the signal $f(t)$. While the Shannon entropy H_S becomes infinite in this limit, we may extract a finite portion of it, called H_f, that is also useful for signal characterization.

This well-behaved quantity can be expressed as

$$H_f = \int_{f\,min}^{f_{max}} w_f(y) \log w_f(y) dy. \tag{2.8}$$

This quantity has been shown to be very sensitive to local changes in backscattered ultrasound that arise from the accumulation of targeted nanoparticles in the acoustic field of view [4–6, 57, 65]. In contrast to most methods used to construct medical images, the waveform $f(t)$ does not directly enter the expression used to compute pixel values. Instead, the density function of the waveform is used.

2.4.2
The Density Function $w_f(y)$

The density function $w_f(y)$ corresponds to the density functions that are the primary mathematical objects in statistical signal processing and the description from which other mathematical quantities are subsequently derived (e.g. mean values, variances, covariances) [66–68]. In that setting, the density function constitutes the most fundamental unit of information that the experimentalist has about a measured variable. It is important to note that the density function $w_f(y)$ may be used to compute not only the entropy H_f but also the signal energy E_f, and hence all conventional energy-based signal analysis may be placed within this same mathematical framework.

Without loss of generality, we may adopt the convention that the domain of $f(t)$ is over the unit interval [0,1], then $w_f(y)$, the density function of $f(t)$, can be defined by the basic integral relationship

$$\int_0^1 \phi(f(t))dt = \int_{f\min}^{f\max} \phi(y)\, w_f(y)dy \tag{2.9}$$

for any continuous function $\phi(y)$. This should be compared with the expression for the expectation value of a function ϕ of a random variable X with density $p_X(x)$, which is given by

$$\int \phi(x)_{p_X}(x)dx,$$

which explains why $w_f(y)$ is referred to as the density function for $f(t)$ [69]. If we chose $\phi(x) = x^2$, then

$$\int_0^1 f(t)^2 dt = \int_{f\min}^{f\max} y^2 w_f(y)dy, \tag{2.10}$$

an expression which represents the signal energy.

Many applications of either probability or information theory to signal processing proceed, usually very early in the discussion, by an *a priori* assumption of a specific underlying density function [70–72]. In contrast, the analysis steps detailed for H_f begin with the measured time-domain waveform data and proceed to calculate the density functions without imposing any additional assumptions.

2.4.3
Ultrasound in a Precancerous Animal Model

The capabilities of entropy-based signal processing for the acoustic detection of nanoparticles targeted to neovasculature has been demonstrated in several animal models [5, 6, 57]. One relevant example was obtained using the transgenic K14-HPV16 mouse [6]. This animal model contains human papilloma virus (HPV)-16 oncoproteins driven by a keratin promoter, so that lesions develop in the skin. Typically, the ears exhibit squamous metaplasia, a precancerous condition, associated with abundant neovasculature that expresses the $\alpha_v\beta_3$ integrin. Eight of these transgenic mice [73, 74] were treated intravenously with 1.0 mg kg^{-1} of either $\alpha_v\beta_3$ integrin-targeted nanoparticles ($n = 4$) or untargeted nanoparticles ($n = 4$), and subsequently imaged subsequently using a research ultrasound system (Vevo 660; Visualsonics, Toronto, Canada). Imaging was accomplished with the mouse ears positioned in the focal zone of a 40 MHz single-element 'wobbler' sector-scan transducer, with an F-number of 2 (diameter 3 mm, focal length 6 mm). Radio-frequency ultrasonic backscatter waveforms corresponding to a region 80 mm wide × 30 mm deep, were digitized at time points 0, 15, 30 and 60 min after administration of the nanoparticle contrast agent. All of these RF data were processed off-line to reconstruct images using information theoretic (entropy-based) and conventional (energy-based) receivers. Image segmentation was performed automatically using the threshold, which excluded 93% of the area under the composite histogram for all data sets. The mean value of segmented pixels was computed at each time point post injection.

A diagram depicting the placement of transducer, gel standoff and mouse ear is shown in the left side of Figure 2.9, together with a representative B-mode grayscale image (i.e. logarithm of the analytic signal magnitude). The labels indicate the location of the skin (top of image insert), the structural cartilage in the middle of the ear, and a short distance below this, the echo from the skin at the bottom of the ear. To the right of this ultrasound image, is an histological view of a HPV mouse ear that has been magnified 20-fold to permit a better assessment of the thickness and architecture of the sites where the $\alpha_v\beta_3$ integrin-targeted nanoparticle might attach (red by β_3 staining). Both, the skin and tumor are both visible in the image. On either side of the cartilage (center band in image), extending to the dermal–epidermal junction, is the stroma, which is filled with neoangiogenic microvessels. These microvessels are also decorated with targeted $\alpha_v\beta_3$ nanoparticles, as indicated by the fluorescent image (labeled, upper right of Figure 2.9) of a bisected ear from an $\alpha_v\beta_3$-injected K14-HPV16 transgenic mouse. It is in this region that the $\alpha_v\beta_3$-targeted nanoparticles are expected to accumulate, as indicated by the presence of red β_3 stain in the magnified image of an immunohistological specimen also shown in the image.

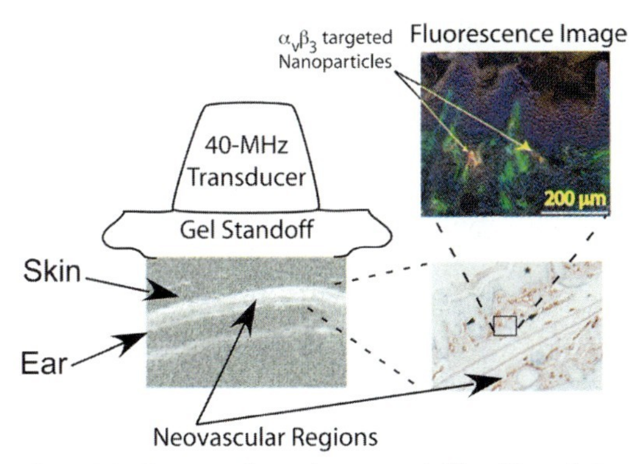

Figure 2.9 Close-up of transducer, standoff and B-mode image of ear, with an enlarged histological view showing the location of the binding sites, and a fluorescent image from the same anatomical region. This shows that the $\alpha_v\beta_3$-targeted nanoparticles accumulate in this portion of the mouse ear.

2.4.3.1 **Image Analysis**

For this study, in which the same portion of the anatomy was imaged at successive intervals, the objective was to quantify changes in image features as a function of time. The first step in this process was the creation of a composite image from the images obtained at 0 through 60 min. Next, an estimate of the probability density function (PDF) of this composite image was computed by normalizing the pixel value histogram to have unit area. It must be emphasized that this function is not related to the density functions $w_f(y)$ as were defined in Equation (2.8). Rather, it is a calculational device used to objectively segment H_f and $\log[E_f]$ images into 'enhanced' and 'unenhanced' regions. A typical histogram is shown in Figure 2.10. The first, larger maxima, corresponds to the relatively homogeneous gray background visible in most H_f images, while the smaller peak corresponds to tissue interfaces, which appear also as bright features in grayscale B-mode images, such as that shown in the inset of Figure 2.9 [76].

Several different methods of image segmentation based on the PDF were investigated. In all of these a specific value, or threshold, in the histogram was chosen and the images divided into two regions: (i) those having pixel values above the threshold (considered to be unenhanced); and (ii) those having pixel values below (referred to subsequently as enhanced pixels). The PDF of all composite images exhibited a two- peak structure with a large and small peak. Thresholds were set at the second minimum, and at the half-way point between the large and small peaks. The full width at half maximum (FWHM) was also computed, and thresholds set at: 4.5, 3.5, 3.25, 3, 2.75 and 2.5 FWHM below the large peak. Thresholds were also set at points such that 97, 95, 93, 90, 87 and 80% of the pixel values were above the threshold. After selection of a threshold value, regions of interest (ROI) were

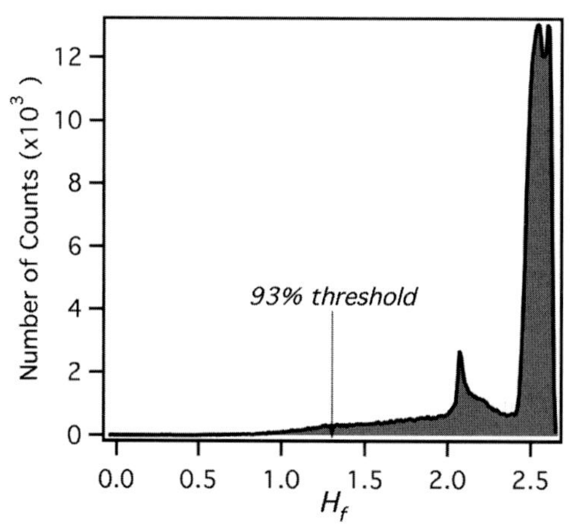

Figure 2.10 A histogram from the composite H_f images acquired at 0, 15, 30, 45 and 60 min post injection. The histogram has two peaks; these have been characteristically observed in several studies using different equipment and animal models. The 93% threshold level used is indicated by the arrow.

selected using NIH ImageJ (http://rsb.info.nih.gov/ij/), and the mean value of the pixels lying below the threshold were computed for each of the images acquired at 0, 15, 30, 45 and 60 min post injection. The mean value at zero minutes was subtracted from the values obtained for all subsequent times, to obtain a sequence of changes in receiver output as a function of time post injection. This was done for all four animals injected with targeted nanoparticles, and also for the four control animals. The sequences of relative changes were then averaged over the targeted and control groups to obtain a sequence of time points for change in receiver output for both groups of animals. The threshold of 93% was finally chosen as it produced the smallest p-value (0.00 043) for a t-test comparing the mean values of the ROI at 15 min as compared to 60 min. The corresponding p-value for the control group was 0.27.

The average change, with time after injection, of the mean value of the enhanced regions of H_f images obtained from all eight of the animals reported in the study are compared in Figure 2.11a [6]. As these data show, the mean value, or enhancements, obtained in the targeted group increased steadily with time. After 30 min the mean

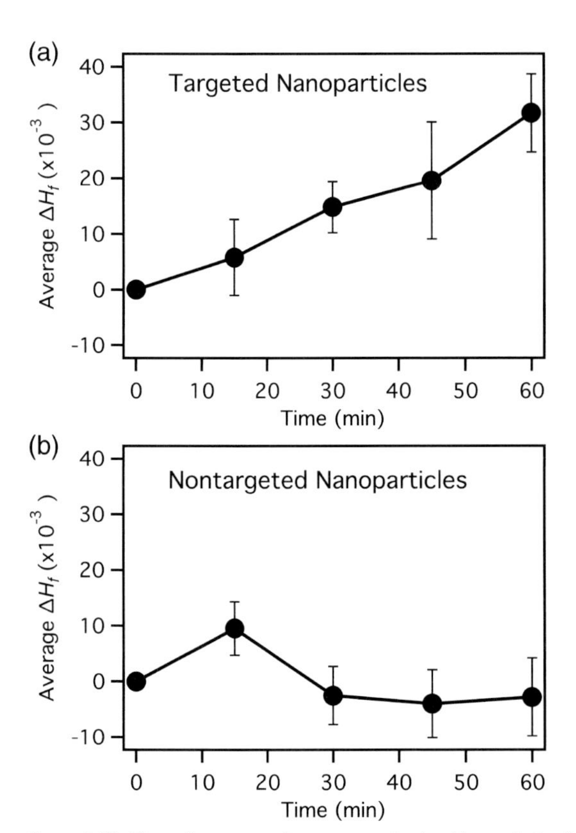

Figure 2.11 Plots of average enhancement obtained by analysis of H_f images from (a) four HPV mice injected with $\alpha_v\beta_3$-targeted nanoparticles, and (b) four HPV mice injected with nontargeted nanoparticles.

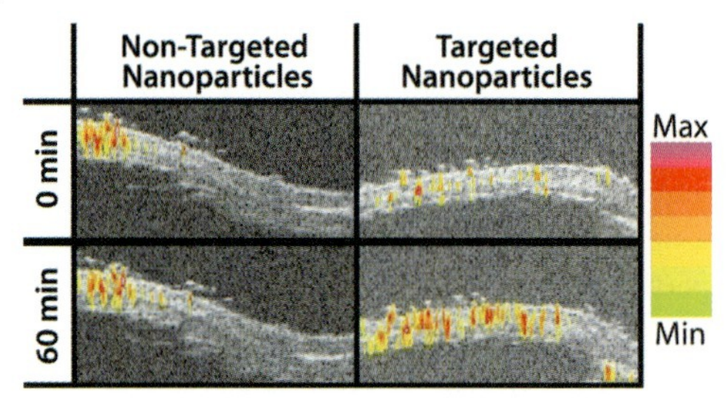

Figure 2.12 Cropped images of transverse cross-sections of HPV mouse ear showing of H_f-enhanced conventional images comparing the effects of targeting (right column) versus nontargeted (left column) both at 0 min (top row) and 60 min (bottom row) after injection.

value of enhancement was measurably different from baseline values ($p < 0.005$). Moreover, the values at 15 and 60 min were also statistically different ($p < 0.005$). The corresponding results obtained from control animals that were injected with nontargeted nanoparticles are shown in Figure 2.11b. There was no discernible trend in the group, and the last three time points were not statistically different from zero. A comparison of the enhancement measured at 15 and 60 min yielded a p-value ≈ 0.27. The enhancement in H_f observed after 60 min for representative instances of targeted and nontargeted nanoparticles is shown in Figure 2.12. These images were generated by overlaying a 93% thresholded version of the H_f using a look-up table (LUT) on top of the conventional grayscale B-mode image.

For comparison, and to illustrate the potential value of the entropy-based analysis, the corresponding results obtained using the $\log[E_f]$ analysis are shown in Figure 2.13a for same data used to generate Figure 2.11. These data were obtained by computing the mean value of pixels lying below the 93% threshold at each time point (0, 15, 30, 45 and 60 min) for each animal (four injected with targeted nanoparticles, and four with nontargeted nanoparticles), as discussed above. Unlike the entropy case, the values at 15 and 60 min were not statistically different ($p = 0.10$). Figure 2.13b shows the corresponding result obtained from the control group of animals that were injected with nontargeted nanoparticles. There was no discernible trend in the group, and the last three time points were not statistically different from zero.

2.4.4 Targeting of MDA-435 Tumors

In a separate investigation, nascent MDA 435 tumors implanted in athymic nude mice reproducibly stimulated neovascular growth and the expresses ion of $\alpha_v\beta_3$ integrin [75]. Human MDA 435 cancer cells were implanted, by injection, in the left hindquarters of athymic nude mice between 10 and 22 days prior to acquisition of the data. Figure 2.14 shows an immunohistologically stained (β_3 staining) section of an excised MDA 435 tumor. β_3 expression, a marker for $\alpha_v\beta_3$-integrin, was found in

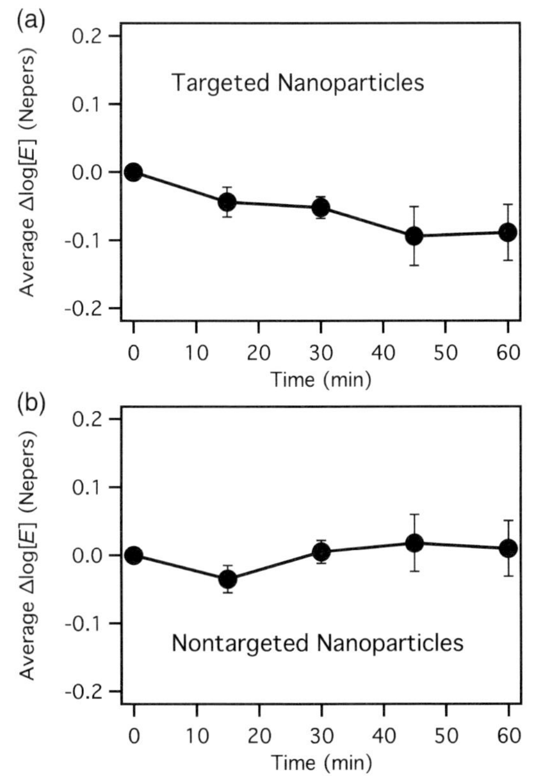

Figure 2.13 Plots of average enhancement obtained by analysis of log [E_f] images from (a) four HPV mice injected with $\alpha_v\beta_3$-targeted nanoparticles and (b) four control HPV mice given nontargeted nanoparticles.

abundance (red regions), although not exclusively, between the skin and the tumor capsule. The close proximity of these binding sites to the skin–transducer interface is one of the primary obstacles that must be overcome by any quantitative detection scheme intended to determine the extent of this region. Accordingly, the acoustic portion of the experiment was designed to maximize system sensitivity near this interface. This was carried out in order to maximize the opportunity to detect nanoparticles targeted towards angiogenic neovasculature. It also provided a stringent test of the H_f entropy-based metric's ability to separate signals near the confounding skin–tissue interface, which was one of the primary goals of this study.

Nine animals were injected with targeted nanoparticles, and seven with nontargeted nanoparticles to serve as controls. Each mouse was preanesthesitized with ketamine, after which an intravenous catheter was inserted into the right jugular vein to permit the injection of nanoparticles (either $\alpha_v\beta_3$-targeted or untargeted). The mouse was then placed on a heated platform maintained at 37 °C, and anesthesia administered continually with isoflurane gas through a nose cone.

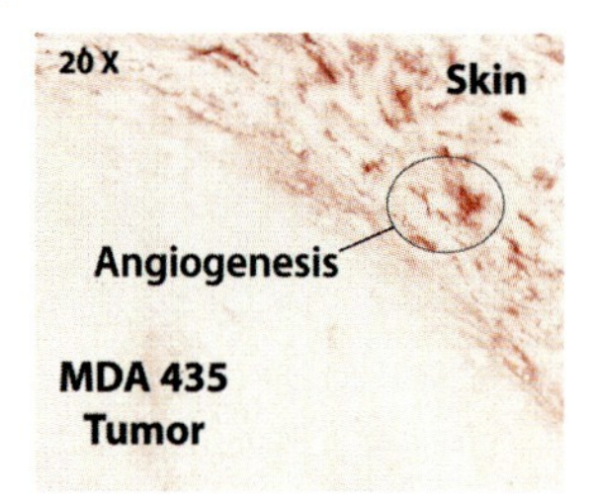

Figure 2.14 A histological specimen extracted from a MDA 435 mouse model, magnified 20-fold, to permit a better assessment of the thickness and architecture of the sites (red by β_3 staining) where $\alpha_v\beta_3$-targeted nanoparticle might attach. The skin and tumor are both visible in the image. The close proximity of neovasculature to the skin–transducer interface is one of the primary obstacles that must be overcome by any quantitative detection scheme intended to determine the extent of this region.

Subsequently, the mouse was injected with 0.030 ml of nanoparticle emulsion (equivalent to a whole-body dose of $1\,\mathrm{mg\,kg^{-1}}$ body mass). Ultrasound data were then acquired at 0, 15, 30, 60 and 120 min intervals.

No evidence of change was observed between the zero-minute and 120-min conventional grayscale B-mode images, while there was a slight (but nonsignificant) change with the color LUT (Figure 2.15). The lower part of Figure 2.15 shows images reconstructed from the same raw RF data, again, with a grayscale LUT in the top row and color LUT in the bottom row. Significant changes in the size of the brightest (red) region, located between the skin–tumor capsule boundary, were observed as expected. Unlike the conventional B-mode processing case, here the H_f processed data showed the region to comprise pixel values far brighter than the mean pixel value that occurred in the rest of the image.

A loss of spatial resolution between skin and tumor capsule was also observable in the image, as expected; this resulted from the smoothing effect of the moving window analysis. In view of this smoothing, which tends to reduce the variations in magnitude of a function, it was somewhat surprising that the image showed a greater separation between the background and enhanced regions. From the images, it could also be seen that the magnitude of values in the region between the skin and the tumor boundary increased with post-injection time. Moreover, the shape and location of the regions were consistent with a brightening effect due to an accumulation of nanoparticles in the angiogenic neovasculature [77, 78].

The data in Figure 2.16 compare the change in mean value of the 3% thresholded region (i.e. enhancement) for the B-mode images (logarithm of signal envelope) as a function of time post injection for controls versus targeted. The plots show that the conventional B-mode images cannot be used to distinguish between 0 and 120 min post injection.

Corresponding results, obtained using entropy imaging, are shown in Figure 2.17, using the same vertical and horizontal scales as in Figure 2.16, for ease of comparison. The plots show that only the entropy-based receiver was able to distinguish between 0 and 120 min post injection (paired t-test produces $p < 0.05$). Moreover,

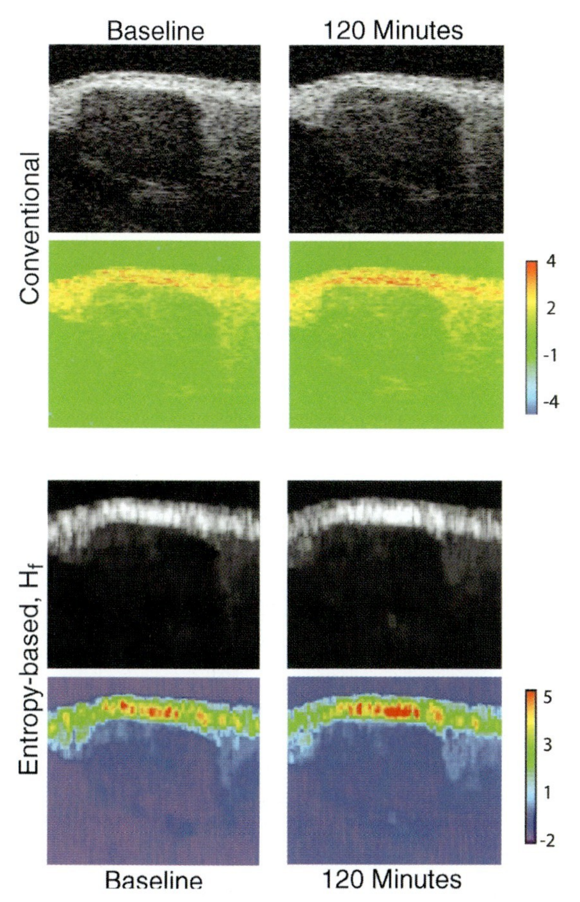

Figure 2.15 A comparison of 0-min and 120-min (post-injection) images obtained from conventional energy-based signal processing (upper part of figure) and the entropy-based H_f metrics (lower part of figure). Grayscale images of before and after data are presented for both types of signal processing. No change is evident in the B-mode images, and at most there is a slight change in the images. However, the application of a color look-up table (LUT) to the images revealed more detail. The same color LUT mapping was applied to both B-mode images to facilitate comparisons. The calibration bar of the mapping is shown to the right of the images in both cases. The images show a greater change with time and a greater separation of the neovascular region from the rest of the image than the B-mode images. Replication of this experiment in nine animals showed the change to be statistically significant only for the H_f-processed images.

the mean values increased in an approximately linear fashion versus time. The plot of control experiments showed there was no significant change in enhancement with time in these animals. However, a careful visual inspection of the image sequence revealed measurable changes in tumor shape and position that most likely were induced by respiration and relaxation of the animal over the 2-h experiment.

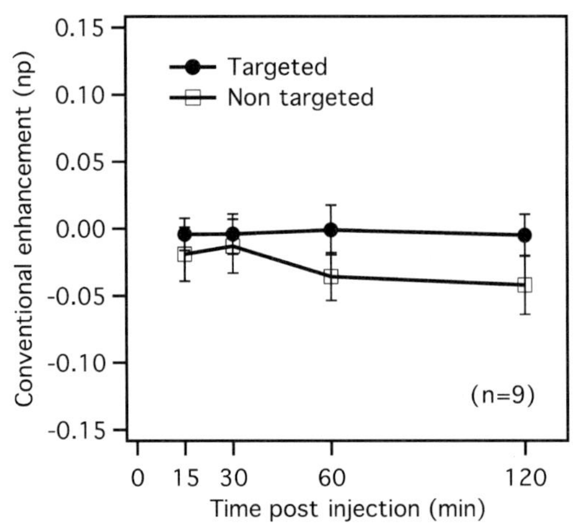

Figure 2.16 Quantitative comparison of enhancement and B-mode images obtained using targeted nanoparticles. No significant changes were observed with either untargeted or targeted nanoparticles. All plots have vertical units of nepers, as the image analysis was performed on noise-scaled or normalized images (which are unitless). As explained in the text, this does not alter the quantitative conclusion presented in these plots.

2.4.5
In Vivo Tumor Imaging at Clinical Frequencies

A separate ultrasound study utilizing targeted nanoparticles was designed to assess the feasibility of image-based angiogenic neovasculature using backscattered

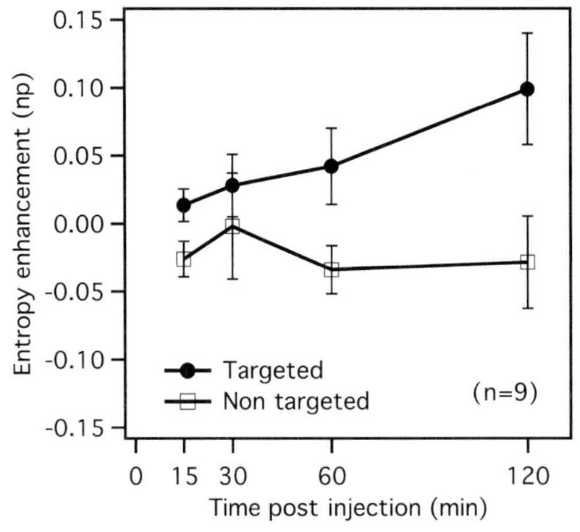

Figure 2.17 Quantitative comparison of enhancement for a thermodynamic receiver (H_f) obtained using targeted and untargeted nanoparticles. Only the targeted nanoparticles produced a significant change in enhancement ($p < 0.05$).

ultrasound in the frequency range between 7 and 15 MHz [79]. These investigators employed a liquid–PFC nanoparticle conjugation of an $\alpha_v\beta_3$ peptidomimetic to target the expression of $\alpha_v\beta_3$ in Vx-2 tumors implanted in the hindquarters of New Zealand White rabbits ($n = 9$). Anesthesia was administered continually with isoflurane gas, the model was injected with a whole-body dose of 0.66 ml kg^{-1} of nanoparticle emulsion, after which the ultrasound data were acquired at 0, 15, 30, 60 and 120 min. Six control rabbits were also imaged using the same methodology, but were not injected with nanoparticles. Beam-formed RF data were acquired using a modified research version of a clinical ultrasound system (Philips HDI-5000). Data were analyzed for all rabbits at all times post injection, using three different techniques: (i) conventional grayscale; (ii) H_f (an entropy-based quantity); and (iii) log[E_f] (i.e. the logarithm of the signal energy, E_f). Representative image data are shown in Figure 2.18, depicting the tumor in cross-section. A paired t-test comparing H_f image enhancement obtained at 0 and 120 min for the rabbits injected with targeted nanoparticles indicated a significant difference ($p < 0.005$). For control rabbits there was no significant difference between 0 and 120 min ($p = 0.54$). Conventional grayscale imaging at the fundamental frequency and log[E_f] imaging failed to detect a coherent signal, and did not show any systematic pattern of signal change.

2.5 Contact-Facilitated Drug Delivery and Radiation Forces

2.5.1 Primary and Secondary Radiation Forces

Acoustic radiation force is a phenomenon associated with the propagation of acoustic waves through a dissipative medium. It is caused by a transfer of momentum from the wave to the medium, arising either from absorption or reflection of the wave [80, 81]. For particles suspended in a liquid medium, these forces manifest themselves in two ways. The first way, which is referred to as the *primary radiation force*, tends to accelerate the suspended particles away from the source. The second way, referred to as the *secondary force*, is an interparticle force that can be completely attractive, if the particles lie in contours perpendicular to the incident field and can be completely repulsive if the particles are oriented parallel to the incident field. One very useful form for these primary and interparticle (secondary) forces is given by [82]

$$F_{\text{primary}} = \frac{V_0 P_A^2}{4\rho_0 c_0^2} k \sin(2k) f\left(\frac{\rho}{\rho_0}\right) \tag{2.11}$$

where

$$f\left(\frac{\rho}{\rho_0}\right) = \frac{\rho_0 c_0^2}{\rho c^2} - \frac{5\rho - 2\rho_0}{3\rho + \rho_0}, \tag{2.12}$$

Here, V_0 is the sphere volume, P_A is the acoustic pressure amplitude, ρ and ρ_0 are the sphere and fluid densities, respectively, and c and c_0 are the sound velocity in the

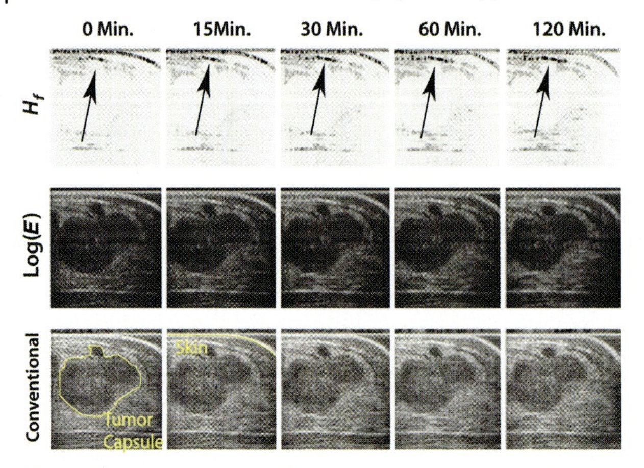

Figure 2.18 Images produced from beam-formed RF acquired from a rabbit injected with $\alpha_v\beta_3$-targeted nanoparticles. The three rows show composite images formed by the application of three different signal processing techniques: H_f, $\log[E_f]$ signal receiver (both applied with a moving window), and conventional image processing. Each composite image is comprised of five sub-images reconstructed from beam-formed RF acquired at 0, 15, 30, 60 and 120 min post injection, as indicated by the labels. Only the H_f composite image showed any evidence of change after injection (black arrows).

sphere and fluid, respectively. The interparticle force given by

$$F_{i.p.} = \xi r^{-4} f(\theta) \tag{2.13}$$

where

$$\xi = \frac{2\pi}{3} \frac{(\rho - \rho_0)^2}{\rho_0} a^3 b^3 u_0^2, \tag{2.14}$$

and a, b are the sphere radii, r is their separation distance, and u_0 is the velocity amplitude of the suspending medium.

The action of these forces *in vivo* is to concentrate the suspended nanoparticles and push them away from the acoustic source (i.e. away from the center of arterial flow and onto the capillary wall), as shown in Figure 2.19. This effect increases the potential to increase their therapeutic efficacy.

2.5.2 *In Vitro* Results

Besides detecting sparse epitopes for noninvasive imaging, PFC nanoparticles are capable of specifically and locally delivering drugs and other therapeutic agents through a novel process known as *contact-facilitated drug delivery* [12]. The direct transfer of lipids and drugs from the nanoparticles' surfactant layer to the cell membrane of the targeted cell is usually a slow and inefficient process. However, through ligand-directed targeting this process can be accelerated by minimizing the separation of the lipids and surfaces, and increasing the frequency and duration

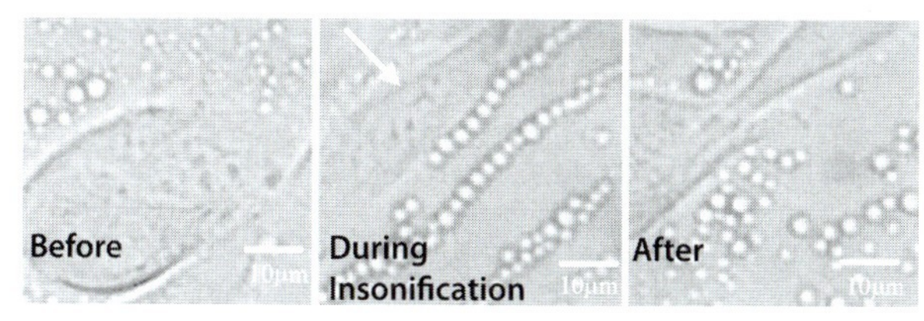

Figure 2.19 Still images of C32 melanoma cells with nanoparticles before, during and after insonification. These show primary and secondary radiation forces acting on the perfluorocarbon nanoparticles. The direction of acoustic insonification, as indicated by the arrow in the center panel, is the same in all three cases. Left panel: pre-insonification; the particles are arranged randomly. Center panel: during insonification, the particles line up on an axis perpendicular to the direction of insonification and move away from the source. Right panel: after insonification, the particle configuration is re-randomized by Brownian motion. During ultrasound exposure, the alignment of nanoparticles relative to the acoustic field (arrow) demonstrates conclusively that acoustic radiation forces (primary and secondary) influence the nanoparticles. This mechanism was observed to be a reversible and safe process; after ultrasound treatment the nanoparticles were no longer aligned, but had been neither destroyed nor altered.

of the lipid–surface interactions (see Figures 2.20 and 2.21). Spatial localization (via high-resolution ^{19}F-enhanced MRI) and quantification of the nanoparticles (via ^{19}F spectroscopy) permits the local therapeutic concentrations to be estimated. Thus, PFC nanoparticles can be used for detection, therapy and treatment monitoring.

As an example, *in vitro* vascular smooth muscle cells were treated with tissue factor-targeted PFC nanoparticles containing 0, 0.2 or 2.0 mol% doxorubicin or paclitaxel, or an equivalent amount of drug in buffer solution alone [83]. After targeting for only for 30 min, proliferation was inhibited for three days, while *in vitro* dissolution studies revealed that the nanoparticles' drug release persisted for more than one week. High-resolution MRI with a 4.7 Tesla field strength showed that the image intensity of the targeted vascular smooth muscle cells was twofold higher compared to nontargeted cells. In addition, the fluorine signal amplitude at 4.7 Tesla was unaffected by the presence of surface gadolinium, and was linearly

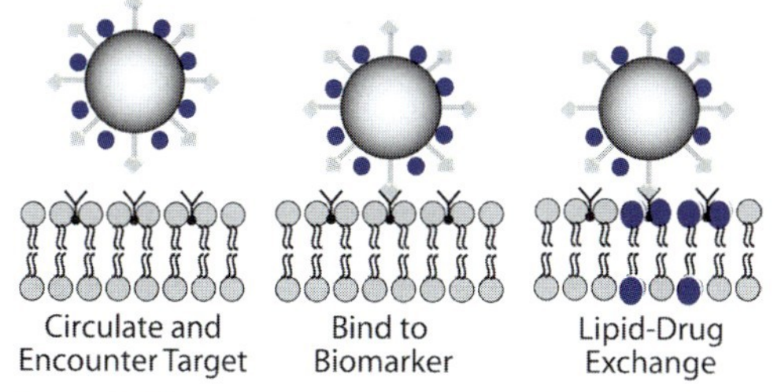

Figure 2.20 Schematic representation illustrating contact-facilitated drug delivery. The phospholipids and drugs within the nanoparticles surface-exchange with the lipids of the target membrane through a convection process, rather than diffusion, as is common among other targeted systems.

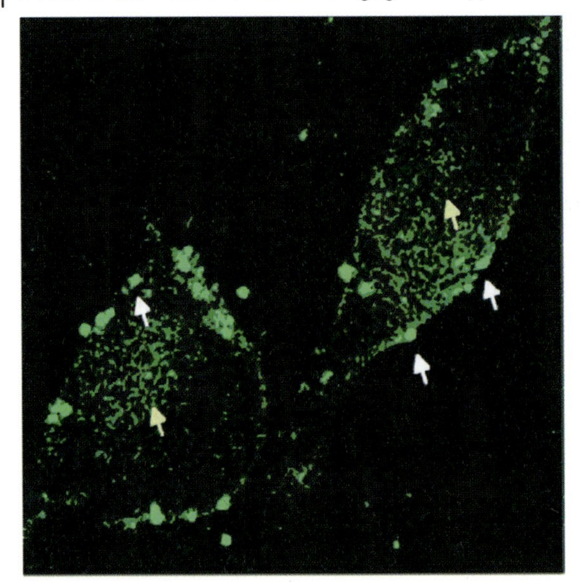

Figure 2.21 *In vitro* targeting of fluoroscein isothiocyanate (FITC)-labeled nanoparticles (white arrows) targeted to $\alpha_v\beta_3$ integrin expressed by C-32 melanoma cells. This illustrates the delivery of FITC-labeled surfactant lipids into target cell membranes (yellow arrows).

correlated to the PFC concentrations which, by direct inference could be related to the nanoparticles' number.

2.6 Conclusions

Targeted liquid PFC nanoparticles represent an extremely versatile platform which has been successfully employed in conjunction with ultrasound, single positron-emission tomography and MRI. These nanoparticles are capable of aiding in the detection of sparse biomarkers, such as integrins in angiogenesis, as well as high-density epitopes such as fibrin. They are unique in that they can be used to diagnose, treat and monitory therapy in a one-step process. Hence, their ongoing and future impact in the fields of cardiology and oncology are predicted to be substantial.

The field of engineered contrast agents continues to grow steadily and advance in line with the rapid developments in nanotechnology. At research centers worldwide, multidisciplinary teams have been assembled which combine expertise in the areas of physics, chemistry, biology, engineering and medicine, to focus on the challenges of creating this next generation of agents. Today, this field is progressing along a path that embraces the prediction summarized within the oft-quoted title of Richard Feynman's presentation, on the world of the nano-scale, to the American Physical Society on December 29th 1959: "There's plenty of room at the bottom."

The potential for significant contributions to paradigms of patient care have been reinforced in recent years via specific funding mechanisms from granting agencies, such as the US National Institutes of Health (NIH) initiatives creating the Programs of Excellence in Nanotechnology (NHLBI-PEN) and the Centers of Cancer Nanotechnology Excellence (NCI-CCNE). Currently, a number of agents are moving towards clinical trials under this aegis. In order to gain acceptance, the approval process by the US Food and Drug Administration is typically on the order of four to eight years for imaging agents (slightly longer for therapeutics). At the time of this writing, some agents have already advanced to Phase 1 and 2 clinical trials, and several new biotechnology start-up companies have also been launched that are devoted to the same goal.

References

1 Cyrus, T. (2006) *Nanomaterials for Cancer Therapy and Diagnosis* (eds S. Challa and S. Kumar), Wiley-VCH, Weinheim, p. 121.

2 Lanza, G.M., Lorenz, C.H., Fischer, S.E., Scott, M.J., Cacheris, W.P., Kaufmann, R.J., Gaffney, P.J. and Wickline, S.A. (1998) *Academic Radiology*, **5**, S173.

3 Marsh, J.N., Partlow, K.C., Abendschein, D.R., Scott, M.J., Lanza, G.M. and Wickline, S.A. (2007) *Ultrasound in Medicine and Biology*, **33**, 950.

4 Hughes, M.S., Marsh, J.N., Hall, C.S., Savery, D., Scott, M.J., Allen, J.S., Lacy, E.K., Carradine, C., Lanza, G.M. and Wickline, S.A. (2004) Proceedings IEEE Ultrasonics Symposium, 04CH37553, p. 1106.

5 Hughes, M.S., Marsh, J.N., Zhang, H., Woodson, A.K., Allen, J.S., Lacy, E.K., Carradine, C., Lanza, G.M. and Wickline, S.A. (2006) *IEEE Transactions on Ultrasonics, Ferroelectrics, and Frequency, Control*, **53**, 1609.

6 Hughes, M.S., McCarthy, J., Marsh, J.N., Arbeit, J., Neumann, R., Fuhrhop, R., Wallace, K., Znidersic, D., Maurizi, B., Baldwin, S., Lanza, G.M. and Wickline, S.A. (2007) *Journal of the Acoustical Society of America*, **121**, 3542.

7 Flacke, S., Fischer, S., Scott, M.J., Fuhrhop, R.J., Allen, J.S., McLean, M., Winter, P., Sicard, G.A., Gaffney, P.J., Wickline, S.A. and Lanza, G.M. (2001) *Circulation*, **104**, 1280.

8 Moghimi, S. and Patel, H. (1989) *Biochimica et Biophysica Acta*, **984**, 384.

9 Harisinghani, M., Saini, S., Weisleder, R., Hahn, P., Yantiss, R., Tempany, C., Wood, B. and Mueller, P. (1999) *American Journal of Roentgenology*, **172**, 1347.

10 Keeler J. (2005) *Understanding NMR Spectroscopy*, J.W. Wiley & Sons, Chichester.

11 Slichter, Charles P. (1996) *Principles of Magnetic Resonance: Springer Series in Solid State Sciences*, Springer-Verlag, New York.

12 Lanza, G., Winter, P., Caruthers, S., Morawski, A., Schmieder, A., Crowder, K. and Wickline, S. (2004) *Journal of Nuclear Cardiology*, **11**, 733.

13 Nelson, K. and Runge, V. (1995) *Topics in Magnetic Resonance Imaging*, **7**, 124.

14 Kirsch, J.E. (1991) *Topics in Magnetic Resonance Imaging*, **3**, 1.

15 Ahrens, E., Rothbacher, U., Jacobs, R. and Fraser, S. (1998) *Proceedings of the National Academy of Sciences of the United States of America*, **5**, 8443.

16 Morawski, A.M., Winter, P.M., Crowder, K.C., Caruthers, S.D., Fuhrhop, R.W., Scott, M.J., Robertson, J.D., Abendschein, D.R., Lanza, G.M. and Wickline, S.A. (2004) *Magnetic Resonance in Medicine*, **51**, 480.

17 Rose, A. (1948) *Journal of the Optical Society of America*, **38**, 196.
18 Winter, P.M., Caruthers, S.D., Yu, X., Song, S.K., Chen, J.J., Miller, B., Bulte, J.W.M., Robertson, J.D., Gaffney, P.J., Wickline, S.A. and Lanza, G.M. (2003) *Magnetic Resonance in Medicine*, **50**, 411.
19 Gupta, H. and Weissleder, R. (1996) *Magnetic Resonance Imaging Clinics of North America*, **4**, 171.
20 Stanisz, G. and Henkelman, R. (2000) *Magnetic Resonance in Medicine*, **44**, 665.
21 Bushong, S. (2003) *Magnetic Resonance Imaging: Physical and Biological Principles*, 3rd edn, St. Louis, Mosby.
22 Longmaid, H., Adams, D., Neirinckx, R., Harrison, G., Brunner, C.P., Seltzer, S., Davis, M., Neuringer, L. and Geyer, R. (1985) *Investigative Radiology*, **20**, 141.
23 McFarland, E., Koutcher, J., Rosen, B., Teicher, B. and Brady, T. (1985) *Journal of Computer-Assisted Tomography*, **9**, 8.
24 Sloviter, H. and Mukherji, B. (1983) *Progress in Clinical and Biological Research*, **122**, 181.
25 Joseph, P., Yuasa, Y., Kundel, H., Mukherji, B. and Sloviter, H. (1985) *Investigative Radiology*, **20**, 504.
26 Zheng, Z., Croft, J., Giles, W. and Mensah, G. (2001) *Circulation*, **104**, 2158.
27 Kuller, L. and Lilienfeld, A. and Fisher, R. (1966) *Circulation*, **34**, 1056.
28 Naghavi, M., Libby, P., Erling, F., Casscells, S., Litovsky, S., Rumberger, J., Badimon, J., Stefanadis, C., Moreno, P. and Pasterkamp, G. (2003) *Circulation*, **108**, 1664.
29 Davies, M. and Thomas, A. (1985) *British Heart Journal*, **53**, 363.
30 Ambrose, J., Tannenbaum, M., Alexopoulos, D., Hjemdahl-Monsen, C., Leavy, J., Weiss, M., Borrico, S., Gorlin, R. and Fuster, V. 1988 *Journal of the American College of Cardiology* **12**, 56.
31 Ojio, S., Takatsu, H., Tanaka, T., Ueno, K., Yokoya, K., Matsubara, T., Suzuki, T., Watanabe, S., Morita, N., Kawasaki, M., Nagano, T., Nishio, I., Sakai, K., Nishigaki, K., Takemura, G., Noda, T., Minatoguchi, S. and Fujiwara, H. (2000) *Circulation*, **102**, 2063.
32 Lanza, G.M., Wallace, K.D., Scott, M.J., Cacheris, W.P., Abendschein, D.R., Christy, D.H., Sharkey, A.M., Miller, J.G., Gaffney, P.J. and Wickline, S.A. (1996) *Circulation*, **94**, 3334.
33 Morawski, A.M., Winter, P.M., Crowder, K.C., Caruthers, S.D., Fuhrhop, R.W., Scott, M.J., Robertson, J.D., Abendschein, D.R., Lanza, G.M. and Wickline, S.A. (2004) *Magnetic Resonance in Medicine*, **52**, 1255.
34 Lanza, G. and Wickline, S. (2001) *Progress in Cardiovascular Diseases*, **44**, 13.
35 Sajid, M. and Stouffer, G. (2002) *Thrombosis and Haemostasis*, **87**, 187.
36 Brooks, P., Clark, R. and Cheresh, D. (1994) *Science*, **264**, 569.
37 Anderson, S., Randall, K., Westlin, W., Null, C., Jackson, D., Lanza, G., Wickline, S. and Kotyk, J. (2000) *Magnetic Resonance in Medicine*, **44**, 433.
38 Winter, P., Caruthers, S., Kassner, A., Harris, T., Chinen, L., Allen, J., Zhang, H., Robertson, J., Wickline, S. and Lanza, G. (2003) *Cancer Research*, **63**, 5838.
39 Schmieder, A.H., Winter, P.M., Caruthers, S.D., Harris, T.D., Williams, T.A., Allen, J.S., Lacy, E.K., Zhang, H., Scott, M.J., Hu, G., Robertson, J.D., Wickline, S.A. and Lanza, G.M. (2005) *Magnetic Resonance in Medicine*, **53**, 621.
40 Gramiak, R. and Shah, P. (1968) *Investigative Radiology*, **3**, 356.
41 Dalla Palma, L. and Bertolotto, M. (1999) *European Radiology*, **9**, S338.
42 Correas, J., Bridal, L., Lesavre, A., Mejean, A., Claudon, M. and Helenon, O. (2001) *European Radiology*, **11**, 1316.
43 McCulloch, M., Gresser, C., Moos, S., Odabashian, J., Jasper, S., Bednarz, J., Burgess, P., Carney, D., Moore, V., Sisk, E., Waggoner, A., Witt, S. and Adams, D. (2000) *Journal of the American Society of Echocardiography*, **13**, 959.

44 Szabo, T.L. (2004) *Diagnostic Ultrasound Imaging: Inside Out*, Elsevier Academic Press, Burlington, MA.

45 Goldberg, B.B. Raichlen, J.S. and Forsberg, F. (eds) (2001) *Ultrasound Contrast Agents: Basic Principles and Clinical Applications*, Martin Dunitz, London.

46 Leighton, T.G. (1997) *The Acoustic Bubble*, Academic Press, San Diego.

47 Klibanov, A.L., Rasche, P.T., Hughes, M.S., Wojdyla, J.K., Galen, K.P., Wible, J.H. and Brandenburger, G.H. (2004) *Investigative Radiology*, **39**, 187.

48 Blomley, M., Cooke, J., Unger, E., Monaghan, M. and Cosgrove, D. (2001) *British Medical Journal*, **322**, 1222.

49 Shohet, R., Chen, S., Zhou, Y., Wang, Z., Meidell, R., Unger, R. and Grayburn, P. (2000) *Circulation*, **101**, 2554.

50 Price, R., Skyba, D.M.P., Kaul, S.M. and Skalak, T.C.P. (1998) *Circulation*, **98**, 1264.

51 Cheng, T.D., S.C. and Feinstein, S. (1998) Contrast echocardiography: review and future directions. *American Journal of Cardiology*, **81**, 41.

52 Terasawa, A., Miyatake, K., Nakatani, S., Yamagishi, M., Matsuda, H. and Beppu, S. (1993) *Journal of the American College of Cardiology*, **21**, 737.

53 Harvey, C., Blomley, M., Eckersley, R., Cosgrove, D., Patel, N., Heckemann, R. and Butler-Barnes, J. (2000) *Radiology*, **216**, 903.

54 Leong-Poi, H., Christiansen, J., Klibanov, A., Kaul, S. and Lindner, J. (2003) *Circulation*, **107**, 455.

55 Schumann, P., Christiansen, J., Quigley, R., McCreery, T., Sweitzer, R., Unger, E., Lindner, J. and Matsunaga, T. (2002) *Investigative Radiology*, **37**, 587.

56 Hauff, P., Reinhardt, M., Briel, A., Debus, N. and Schirner, M. (2004) *Radiology*, **231**, 667.

57 Hughes, M.S., Marsh, J.N., Arbeit, J., Neumann, R., Fuhrhop, R.W., Lanza, G.M. and Wickline, S.A. (2005) Proceedings IEEE Ultrasonics Symposium, 05CH37716, p. 617.

58 Lanza, G.M., Trousil, R.L., Wallace, K.D., Rose, J.H., Hall, C.S., Scott, M.J., Miller, J.G., Eisenberg, P.R., Gaffney, P.J. and Wickline, S.A. (1998) *Journal of the Acoustical Society of America*, **104**, 3665.

59 Hall, C.S., Marsh, J.N., Scott, M.J., Gaffney, P.J., Wickline, S.A. and Lanza, G.M. (2000) *Journal of the Acoustical Society of America*, **108**, 3049.

60 Hughes, M.S. (1992) *Journal of the Acoustical Society of America*, **91**, 2272.

61 Hughes, M.S. (1992) Proceedings IEEE Ultrasonics Symposium, 92CH31187, p. 1205.

62 Hughes, M.S. (1993) *Journal of the Acoustical Society of America*, **93**, 892.

63 Hughes, M.S. (1993) Proceedings IEEE Ultrasonics Symposium, 93CH33019, p. 697.

64 Hughes, M.S. (1994) *Journal of the Acoustical Society of America*, **95**, 2582.

65 Hughes, M.S., Marsh, J.N., Hall, C.S., Savy, D., Scott, M.J., Allen, J.S., Lacy, E.K., Carradine, C., Lanza, G.M. and Wickline, S.A. (2005) *IEEE Transactions on Ultrasonics, Ferroelectrics, and Frequency, Control*, **52**, 1555.

66 Bucy, R.S. and Joseph, P.D. (1987) *Filtering for Stochastic Processes with Applications to Guidance*, Chelsea Publishing Company, New York.

67 Weiner, N. (1949) *Extrapolation, Interpolation, and Smoothing of Stationary Time Series: with Engineering Applications*, M.I.T. Press, Cambridge, MA.

68 Grenander, U. and Rosenblatt, M. (1984) *Statistical Analysis of Stationary Time Series*, Chelsea Publishing Company, New York.

69 Wheeden, R.L. and Zygmund, A. (1977) *Measure and Integral: An Introduction to Real Analysis*, Marcel-Dekker, New York.

70 Cover, T.M. and Thomas, J.A. (1991) *Elements of Information Theory*, Wiley-Interscience, New York.

71 Kullback, S. (1997) *Information Theory and Statistics*, Dover, New York.

72 Reza, F.M. (1994) *An Introduction to Information Theory*, Dover, New York.

73 Arbeit, J.M., Riley, R.R., Huey, B., Porter, C., Kelloff, G., Lubet, R., Ward, J.M. and Pinkel, D. (1999) *Cancer Research*, **59**, 3610.

74 Arbeit, J.M., Manger, K., Howley, P.M. and Hanahan, D. (1994) *Journal of Virology*, **68**, 4358.

75 Hughes, M.S., Marsh, J.N., Woodson, A.K., Lacey, E.K., Carradine, C., Lanza, G.M. and Wickline, S.A. (2005) Proceedings IEEE Ultrasonics Symposium, 05CH37716 p. 373.

76 Hughes, M., Marsh, J., Hall, C., Fuhrhop, R.W., Lacy, E.K., Lanza, G.M. and Wickline, S.A. (2005) *Journal of the Acoustical Society of America*, **117**, 964.

77 Winter, P.M., Morawski, A.M., Caruthers, S., Harris, T., Allen, J.S., Zhang, H.Y., Fuhrhop, R.W., Lacy, E.K., Williams, T.A., Lanza, G.M. and Wickline, S.A. (2003) *Circulation*, **108**, 168.

78 Winter, P.M., Morawski, A.M., Caruthers, S.D., Fuhrhop, R.W., Zhang, H., Williams, T.A., Allen, J.S., Lacy, E.K., Robertson, J.D., Lanza, G.M. and Wickline, S.A. (2003) *Circulation*, **108**, 2270.

79 Hughes, M.S., Marsh, J.N., Allen, J., Brown, P.A., Lacy, E.K., Scott, M.J., Lanza, G.M., Wickline, S.A. and Hall, C.S. (2004) Proceedings of the IEEE Ultrasonics Symposium, 04CH37553, p. 1106.

80 Torr, G.R. (1984) The acoustic radiation force. *American Journal of Physics*, **52**, 402.

81 Nyborg, W.L. (1965) Acoustic streaming, in *Physical Acoustics* Vol. **IIB** (ed. W. Mason), Academic Press, p. 265.

82 Ter Haar, G. and Wyard, S. (1978) *Ultrasound in Medicine and Biology*, **4**, 111.

83 Lanza, G., Yu, X., Winter, P., Abendschein, D., Karukstis, K., Scott, M., Chinen, L., Fuhrhop, R., Scherrer, D. and Wickline, S. (2002) *Circulation*, **106**, 2842.

3
Nanoparticles for Cancer Detection and Therapy

Biana Godin, Rita E. Serda, Jason Sakamoto, Paolo Decuzzi, and Mauro Ferrari

3.1
Introduction

3.1.1
Cancer Physiology and Associated Biological Barriers

Cancer is a major public health problem in developed countries, accounting for nearly one-fourth of deaths in the United States, exceeded only by heart diseases. According to a 2008 report by the American Cancer Society, estimated numbers for US cancer cases are 745 and 692 thousands for men and women, respectively [1, 2], with the lifetime probability of developing cancer higher in men (45%) than in women (38%). *Cancer* is a general term used to define any disease characterized by the uncontrolled proliferation of abnormal cells. Due to a widely used generalization of the condition, it is easy to overlook the fact that cancer is not a single disease, but rather a conglomerate of many diseases. During the past five decades, the complexity of cancer has been rendered more tangible by a large body of knowledge accumulated on the common principles of pathogenesis. It is now clear that cancer is a complex ailment caused by accumulation of multiple molecular alterations in the genetic material.

The disease can be divided into two broad categories of hematological malignancies (which affect circulating cells) and solid tumors. Solid tumors can be considered as an organ, and are divisible into three main subcompartments: vascular; interstitial; and cellular [3, 4]. Each of these subcompartments accounts for several biological barriers (or 'biobarriers') that a therapeutic agent should bypass to treat the disease effectively [5, 6]. Later in this section, we will describe the tumor compartments as well as related and other intrinsic 'biobarriers', which severely impede the localization of chemicals, biomolecules and particulate systems at their intended site of action. Biobarriers are sequential in nature, and therefore the probability of an active agent of reaching its therapeutic goal is the interrelated result of the individual probabilities of overcoming each one of the challenges it faces [7, 8].

Nanotechnology, Volume 5: Nanomedicine. Edited by Viola Vogel

ISBN: 978-3-527-31736-3

The *tumor vasculature* is extremely heterogeneous, with necrotic and hemorrhagic areas neighboring regions with a dense vascular network formed as a result of angiogenesis triggered to sustain a sufficient supply of oxygen and nutrients necessary for tumor growth and progression [5, 9]. Tumor blood vessels are architecturally and structurally different from their normal counterparts. The vascular networks that are formed in response to tumor growth are not organized into definitive venules, arterioles and capillaries – as for the normal circulation – but rather share chaotic features of all of them. Furthermore, the blood flow in tumor vessels is irregular, sluggish, and sometimes oscillating. Angiogenic vessels possess several abnormal features such as a comparatively high percentage of proliferating endothelial cells, an insufficient number of pericytes, an enhanced tortuosity, and the formation of an atypical basal membrane. As a result, tumor vasculature is more permeable, with the pore cut-off size ranging from 380 to 780 nm in different tumor models [10, 11]. The hemoglobin in the erythrocytes is 'oxygen-starved', which makes the microenvironment profoundly hypoxic. The tumor environment is also nutrient-deficient (e.g. glucose), acidic (owing to lactate production from anaerobic glycolysis), and under oxidative stress [3, 9]. Although the molecular controls of the above abnormalities are not fully elucidated, these may be attributed to the imbalanced expression and function of angiogenic factors. Various mediators can affect angiogenesis as well as vascular permeability. Among these are vascular endothelial growth factor (VEGF), nitric oxide, prostaglandins and bradykinin. Macromolecules can traverse through neoplastic vessels using one of the following pathways: vasculature fenestrations; interendothelial junctions; transendothelial channels (open gaps); and vesicular vacuolar organelles [9]. The tumor vasculature, in being formed *de novo* during the angiogenic process, possesses a number of characteristic markers which are not seen on the surface of normal blood vessels, and can serve as therapeutic targets (these will be discussed later).

The *interstitial compartment* of solid tumors is mainly composed of a collagen and elastic fiber, crosslinked structure. Interstitial fluid and high-molecular-weight gelling constituents, such as hyaluronate and proteoglycans, are interdispersed within the above network. The characteristic feature of the interstitium, which distinguishes it from the majority of normal tissues, is the intrinsic high pressure resulting from the absence of an anatomically well-defined and operating lymphatic network, as well as an apparent convective interstitial fluid flow. These parameters present additional biobarriers towards the penetration of a therapeutic agent into the cancer cells, as the transport of an anticancer molecule or nanovector in this tumor subcompartment will be governed by physiological (pressure) and physico-chemical (charge, lipophilicity, composition, structure) properties of the interstitium and the agent itself [4, 5].

The *cellular subcompartment* accounts for the actual cancerous cell mass. The barriers directly related to the cellular compartment are generally categorized in terms of alterations in the biochemical mechanisms within the malignant cells making them resistant to anticancer medications. Among these biochemical shifts are the P-glycoprotein efflux system, which is responsible for multidrug resistance and the impaired structure of specific enzymes (i.e. topoisomerase). Moreover, in order to efficiently treat the disease, a cytotoxic agent should be able to cross the cytoplasmic and nuclear membranes – a far from trivial deed for basic drugs that are ionizable within an acidic tumor environment [12, 13].

As mentioned above, following their administration, therapeutic agents encounter a multiplicity of biological barriers that adversely impact their ability to reach the intended target at the desired concentrations [5–8, 14]. This problem is considerably decoupled from the ability of agents to recognize and selectively bind to the target, that is, by the use of antibodies, aptamers or ligands. In other words, despite their high specificity these agents invariably present with concentrations at target sites that are vastly inferior to what is expected on the basis of molecular recognition alone. The biodistribution profiles for conventional chemotherapeutic agents are evenly adverse, if not worse, leading to a plethora of unwanted toxicities and collateral effects at the expense of the therapeutic action (i.e. a decreased 'therapeutic index'). The reticuloendothelial system (RES), which comprises immune cells and organs such as the liver and spleen, presents an important physiological biobarrier, causing an efficient clearance of the agent from the bloodstream. Other barriers of epithelial and endothelial nature, for example the blood–brain barrier, are based on tight-junctions, which significantly limit the paracellular transport of agents that owe their molecular discrimination to several mechanisms and proteins (occludin, claudin, desmosomes, zonula occludens).

To summarize, some of the most challenging biobarriers as the main cause for tumor resistance to therapeutic intervention, include physiological noncellular and cellular barriers, such as the RES, epithelial/endothelial membranes and drug extrusion mechanisms, and biophysical barriers, which include interstitial pressure gradients, transport across the extracellular matrix (ECM), and the expression and density of specific tumor receptors.

3.1.2
Currently Used Anticancer Agents

Since the pathology of cancer involves the dysregulation of endogenous and frequently essential cellular processes, the treatment of malignancies is extremely challenging. The vast majority of presently used therapeutics utilize the fact that cancer cells replicate faster than most healthy cells. Thus, most of these agents do not differentiate greatly between normal and tumor cells, thereby causing systemic toxicity and adverse side effects. More selective agents – which include monoclonal antibodies and anti-angiogenic agents – are now available, and the efficiency of these medications is still under evaluation in various types of tumor. Since cancer arising from certain tissues – including the mammary and prostate glands – may be inhibited or stimulated by appropriate changes in hormone balance, several malignancies may also respond to hormonal therapy. Various groups of anticancer therapeutics are exemplified below.

3.1.2.1 Chemotherapy

Chemotherapy, or the use of chemical agents to destroy cancer cells, is a mainstay in the treatment of malignancies. The modern era of cancer chemotherapy was launched during the 1940s, with the discovery by Louis S. Goodman and Alfred Gilman of nitrogen mustard, a chemical warfare agent, as an effective treatment for blood malignancies [15, 16].

Through a variety of mechanisms, chemotherapy affects cell division, DNA synthesis, or induces apoptosis. Consequently, more aggressive tumors with high growth fractions are more sensitive to chemotherapy, as a larger proportion of the targeted cells are undergoing cell division at any one time. A chemotherapy agent may function in only one phase (G_1, S, G_2 and M) of the cell cycle (when it is called cell cycle-specific), or be active in all phases (cell cycle-nonspecific). The majority of chemotherapeutic drugs can be categorized as alkylating agents (e.g. cisplatin, carboplatin, mechlorethamine, cyclophosphamide, chlorambucil), antimetabolites (e.g. azathioprine, mercaptopurine), anthracyclines (daunorubicin, doxorubicin, epirubicin, idarubicin, valrubicin), plant alkaloids (vinca alkaloids and taxanes) and topoisomerase inhibitors (irinotecan, topotecan, amsacrine, etoposide) [17–19].

The lack of any great selectivity by chemotherapeutic agents between cancer and normal cells is apparent when considering the adverse effect profiles of most chemotherapy drugs [18, 19]. Hair follicles, skin and the cells that line the gastrointestinal tract are some of the fastest growing cells in the human body, and therefore are most sensitive to the effects of chemotherapy. It is for this reason that patients may experience hair loss, rashes and diarrhea, respectively. As these agents do not possess favorable pharmacokinetic profiles to localize specifically into the tumor tissue, they become evenly distributed throughout the body, with resultant adverse side effects and other toxic reactions that greatly limit their dosage.

3.1.2.2 Anti-Angiogenic Therapeutics

The publication of Judah Folkman's imaginative hypothesis in 1971 launched the current research area of anti-angiogenic therapy for cancer [20], although more than three decades elapsed before the Food and Drug Administration (FDA) approved the first anti-angiogenic drug, bevacizumab (a humanized monoclonal antibody directed against VEGF) [21, 22]. The first clinical trials with this agent, when used in combination with standard chemotherapy, resulted in an enhanced survival of metastatic colorectal cancer and advanced non-small-cell lung cancers [23, 24]. Another group of anti-angiogenic therapeutics, also approved by the FDA, is based on small-molecule receptor tyrosine kinase inhibitors (RTKIs) which target VEGF receptors, platelet-derived growth factor (PDGF) and other tyrosine kinase-dependent receptors [25]. Examples of agents in this group are sorafenib and sunitinib; these orally administered medications have been shown to be effective in the treatment of metastatic renal cell cancer and hepatocellular carcinoma, when used as monotherapy [26–28]. When used as monotherapy, the survival benefits of these treatments are relatively modest (usually measured in months). Additionally, the treatments are also costly [29] and have toxic side effects [30–34].

3.1.2.3 Immunotherapy

While tumor cells are ultimately derived from normal progenitor cells, transformation to a malignant phenotype is often accompanied by changes in antigenicity. Antibodies are amazingly selective, possessing the natural ability to produce a cytotoxic effect on target cells. The immune system was first appreciated over 50 years ago for its ability to recognize malignant cells and defend against cancer, when Pressman and Korngold [35] showed that antibodies could distinguish efficiently between normal and tumor tissues. These results were confirmed by Burnet [36] during the 1960s, who also showed that neoplasms are actually formed only when lymphocytes lose the capability of differentiating between normal and malignant cells. These studies grounded the foundation for modern monoclonal antibody (mAb) -based cancer therapy. The expression of tumor-associated antigens can arise due to a variety of mechanisms, including alterations in glycosylation patterns [37], the expression of virally encoded genes [38], chromosomal translocations [39], or an overexpression of cellular oncogenes [40, 41]. The first challenge in the development of efficient mAb-based therapeutics is the detection of an appropriate and specific tumor-associated antigen. Some examples of mAbs used for cancer therapy are given below.

Hematologic malignancies, which possess fewer barriers capable of preventing mAbs from accessing their target antigens, are well suited for mAb therapy. Following intravenous injection and distribution throughout the vascular space, therapeutic antibodies may easily access their targets on the surface of blood malignant cells. Many of these B- and T-cell surface antigens, such as CD20, CD22, CD25, CD33 or CD52, are expressed only on a particular family of hematopoietic cells [42, 43]. These antigens are also expressed at high levels on the surface of various populations of malignant cells, but not on normal tissues or hematopoietic progenitor cells. The chimeric antibody which binds to CD20 B lymphocyte surface antigens, rituxan (rituximab; Genentech) was among the first of the mAbs to receive FDA approval for the treatment of nonHodgkin's lymphoma [44]. Alemtuzumab, which recognizes CD52 antigens present on normal B and T lymphocytes (Campath-1; Ilex Oncology) has also received FDA approval for the treatment of patients suffering from chronic lymphocytic leukemia.

The successful treatment of solid tumors with mAb therapeutics has proved to be more elusive compared to hematological malignancies, although some significant therapeutic benefits have been achieved. The failure of mAbs in the treatment of these malignancies is primarily attributable to an insufficient level of injected mAb that actually reaches its target within a tumor mass. The results of several studies using radiolabeled mAbs have suggested that only a very small percentage of the original injected antibody dose (0.01–0.1% g^{-1} tumor tissue) is able to ever reach target antigens within a solid tumor [45–47]. These low *in vivo* concentrations are due to the series of biobarriers (see above) that an intravenously administered mAb encounters *en route* to its specific antigens on the surface of cancer cells. Herceptin (trastuzumab; Genentech) is a humanized antibody marketed for the treatment of metastatic breast cancer. This mAb recognizes an extracellular epitope of the HER-2 protein, which is highly overexpressed in approximately 25–30% of invasive breast tumors [40, 41]. It is noteworthy that HER-2 expression on breast cancer cells can be as high as 100-fold in comparison to normal breast epithelial cells. Clinical trials with herceptin have shown it to be well tolerated, both as a single agent for second- or third-line therapy or in combination with chemotherapeutic agents as a first line of therapy. A combination therapy resulted in a 25% improvement in overall survival among patients with HER-2-overexpressing tumors that are refractory to other forms of treatment [48, 49].

The levels of prostate-specific membrane antigen (PSMA), a transmembrane protein expressed primarily on the plasma membrane of prostatic epithelial cells [50], are elevated in virtually all cases of prostatic adenocarcinoma, with maximum expression levels observed in metastatic disease and androgen-independent tumors [50–53]. Due to this behavior, PSMA has become an important biomarker for prostate cancer, and antibodies to PSMA are currently being developed for the diagnosis and imaging of recurrent and metastatic prostate cancer, as well as for the therapeutic management of malignant disease [53–56].

Another mAb used for the treatment of colorectal cancer is elecolomab (panorex; GlaxoSmith-Kline), the anti-epithelial cellular adhesion molecule. Today, many other immunotherapeutics are being used in the clinic or are undergoing various stages of clinical trials. Beyond their pronounced therapeutic potential, these agents can be efficiently combined with nanovectors to enhance targeting of the latter to cancer tissues.

3.1.2.4 Issues and Challenges

As mentioned above, currently used conventional cancer therapies have several drawbacks that result in a pronounced toxicity and poor treatment efficacy. On the other hand, current diagnostic techniques do not allow for the competent detection of various malignancies, and do not reflect the vast clinical heterogeneity of the condition. Targeted approaches will ultimately increase the treatment efficiency, while decreasing toxicity to normal cells and tissues; thus, specific drug delivery in cancer treatment is of prime importance. As opposed to cancers of the blood, solid malignancies possess several unique characteristics, such as extensive blood vessel growth (angiogenesis), damaged vascular architecture and enhanced permeability, and impaired lymphatic flow and drainage. All of the above can serve as effective therapeutic targeting mechanisms, as well as for the passive homing of agents into the tumor tissue by means of various delivery systems.

To summarize, current issues and unmet needs in translational oncology include:

- Improved strategies for early cancer diagnostics and imaging.
- Advanced technologies to overcome the toxicity and adverse effects of chemotherapeutic agents.
- An accumulation of new knowledge on cancer biology, allowing for the design of more efficient therapeutics for more aggressive and lethal cancer phenotypes.

Progress in the above listed fields will sculpt the major cornerstones for a yet-to-come 'personalized' tumor therapy and early and predictive diagnosis of the disease.

Later in this chapter we will describe the currently available and under-development carriers and vectors from a 'nano-toolbox', and critically discuss the benefits and weaknesses of these systems for the design of specific, personalized and targeted medications. The benefits of rational design of the nanovectors to efficiently negotiate biobarriers and various aspects of a preclinical characterization for the nanoscale systems will be argued.

3.2 Nanotechnology for Cancer Applications: Basic Definitions and Rationale for Use

Nanoscience involves investigations to learn new behaviors and properties of materials on a submicron scale. Various important functions of living organisms and biological processes take place at the nanoscale. As an example, a typical protein such as hemoglobin, which carries oxygen through the bloodstream, is 5 nm (i.e. five-billionths of a meter) in diameter, while gamma-globulin accounts for a diameter of about 10 nm. For a comparison, the diameter of an erythrocyte – the smallest cell in the human body – ranges from 5 to 7 μm.

Research on the nanoscale has been a missing dimensional link, among an atomic scale which provides the basics for chemistry and physics, and micro-scale technologies, such as electronics. This issue was addressed by a Nobel Laureate Richard Feynman in his legendary lecture, "There is a plenty room in the bottom," in 1959 [57]. Almost four decades later, Richard Smalley, who received his Nobel Prize in 1996 for the discovery of the foundational in nanoscience and nanotechnology carbon-60 molecules, said "...human health has always been determined on the nanometer scale; this is where the structure and properties of the machines of life work in every one of the cells in every living thing." Nowadays, nanotechnology is a rapidly growing multidisciplinary field involving support from scientists in academia, industry and regulatory as well as federal sectors. As an example, the National Nanotechnology Initiative (NNI) program was established in 2001 to coordinate Federal Nanotechnology Research and Development [58]. The 2009 budget request provides US$ 1.5 billion for the NNI, with major investment in nanotechnology research and development over the past decade, reflecting a broad support of the US Congress for this program. Nanotechnology can offer impressive resolutions, when applied to medical challenges such as cancer, diabetes, Parkinson's or Alzheimer's disease, cardiovascular problems and inflammatory or infectious diseases.

Nanotechnology is more than simply throwing together a batch of nanoscale materials – it requires the ability to manipulate and control them in a useful way. The definition of nanotechnology pertains to synthetic and engineerable objects which are nanoscale in dimensions, or have critical functioning components of such a size, and that therefore possess special emergent properties [59]. This is a general and operational definition involving the following interrelated constituents: nanoscale dimensions of the whole system or its vital components; man-made nature; and the unique characteristics of new material that arise due to its nanoscopic size, with each element in this three-part description being equally essential for an object to be defined as 'nanotechnological'. Another vital component in this definition is that the unique features and emerging properties of the nanomaterial must be backed up by the correct mechanism of action (e.g. mathematical modeling). Other definitions of nanotechnology can be found in the literature and, according to some agencies, the word 'nanoscale' should be interpreted to encompass the range of 1 to 100 nm. For example, the National Cancer Institute defines nanotechnology as

> *"The field of research that deals with the engineering and creation of things from materials that are less than 100 nanometers (one-billionth of a meter) in size, especially single atoms or molecules."*

Nanotechnology has already occupied its niche for quite a few years in medicine, being known as 'nanomedicine', particularly in oncology [60–62]. The most studied and commercially available drug-delivery nanoparticle is the liposome, with liposomal doxorubicin having been granted FDA approval since 1996 for use against Kaposi's sarcoma. Later, it was also approved for use in metastatic breast cancer and recurrent ovarian cancer. Cancer-related issues of nanomedicine are supported by major funding programs; for example in 2005, the National Cancer Institute launched a US$ 144 million Alliance for Nanotechnology in Cancer.

The use of nanoparticles as carriers for therapeutic and imaging contrast agents is based on the simultaneous, anticipated advantages of drug localization at cancer lesions, and the ability to circumvent the biological barriers encountered between the point of administration and the projected target. Although physical localization at the tumor site is frequently defined as 'targeting' among the drug-delivery community, this term has a different scientific connotation, referring to the preferential activity of the agent on tumor-associated biological pathways. Due to this discrepancy, we will here use the term 'targeting' only when referring to the specific recognition between particles and the lesion (e.g., due to the presence of mAb on the particle's surface), while referring to passive concentration governed by physical laws as 'localization' or 'direction'.

A '*nanovector*' is a nanoscale particle or system having nanoscale components for the delivery of therapeutic or contrast agent. Currently used and investigated nanovectors can be generally organized into three main categories or 'generations' as shown schematically in Figure 3.1 [6, 8, 63]:

- The first generation (Figure 3.1a) comprises a delivery system that homes into the action site governed by passive mechanisms. In the case of liposomes as a nanovector, the mode of tumor localization is based on the enhanced permeation and retention (EPR) effect, which drives the system to localize into tumor through fenestrations in the adjacent neovasculature. Some of these carriers are surface-modified with a stealth layer [e.g. polyethylene glycol (PEG)] which prevents their uptake by the RES, and thus substantially prolongs the particles' circulation time [63].
- The second generation in this classification is thus defined as having specific additional functionalities on each individual particle, allowing for molecular recognition of target tissue (Figure 3.1b), or for the active or triggered release of the payload at the disease site. The best examples of the first subclass of nanovectors

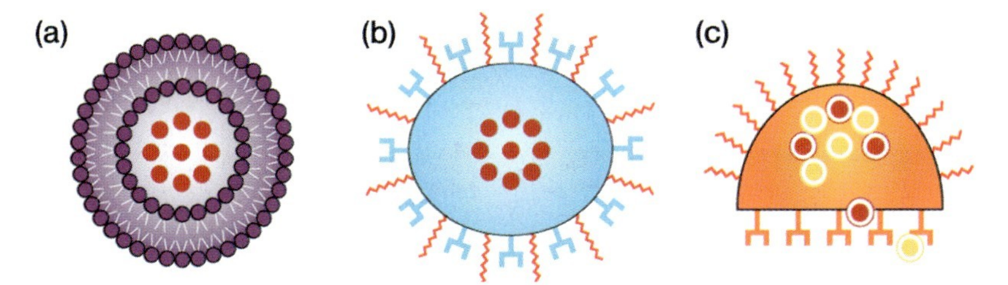

Figure 3.1 (a) First-generation nanovectors (e.g. clinical liposomes) comprise a container and an active principle, and localize in the tumor by enhanced permeation and retention (EPR), or the enhanced permeability of the tumor neovasculature; (b) Second-generation nanovectors further possess the ability to targeting their therapeutic action via antibodies and other biomolecules, remote activation, or responsiveness to environment; (c) Third-generation nanovectors (e.g. multistage agents) are capable of more complex functions, such as time-controlled deployment of multiple waves of active nanoparticles, deployed across different biological barriers and with different subcellular targets.

in this category are antibody-targeted nanoparticles, such as mAb-conjugated liposomes [64–69].

- Third-generation nanovectors, such as multistage agents, are capable of more complex functions which enable sequential overcoming of multiple biobarriers. An example is the time-controlled release of multiple payloads of active nanoparticles, negotiating different biological barriers and with different subcellular targets [7].

Later in this chapter we will focus on each of the three generations of nanovectors, discussing the 'pros' and 'cons', and presenting various examples of these technologies.

3.3 First-Generation Nanovectors and their History of Clinical Use

Today, the first-generation nanovectors that passively localize into tumor sites represents the only generation of nanomedicines broadly represented in the clinical situation. These systems are generally designed to achieve long circulation times for therapeutics and an enhanced accumulation of the drug into the target tissue. This is achieved through a pronounced extravasation of the carrier-associated therapeutic agent into the interstitial fluid at the tumor site, exploiting the locally increased vascular permeability, the EPR effect (Figure 3.2). An additional physiological factor which contributes to the EPR effect is that of impaired lymphatic function impeding clearance of the nanocarriers from their site of action [69–71]. The localization in this case is driven only by the particles' nanodimensions, and is not related to any specific recognition of the tumor or neovascular targets.

In order to prolong their circulation time, these systems are generally decorated on their surface by a 'stealth' layer (e.g. PEG) which prevents their uptake by phagocytic blood cells and organs of the RES system [63, 72, 73]. The most pronounced

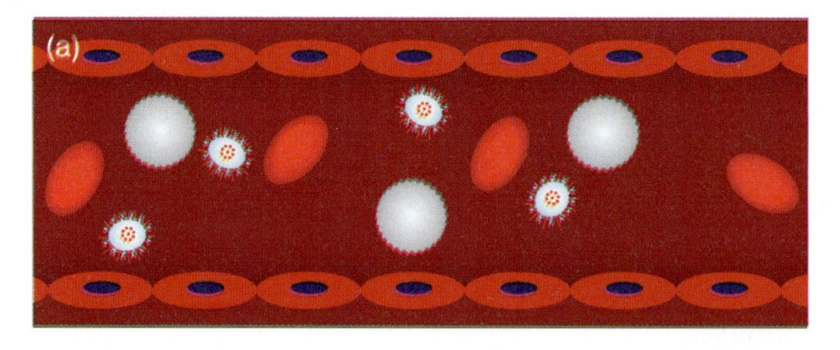

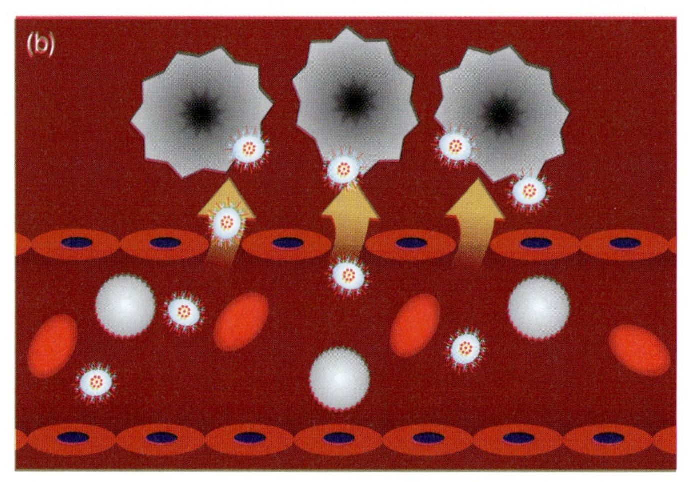

Figure 3.2 Mechanism of passive tumor targeting by enhanced permeation and retention (EPR).

representatives of this generation in clinical use are *liposomes*, which are the leaders among nanocarriers used in clinics. These self-assembling structures, which were first discovered by Bangham in 1965 [74], are composed of one or several lipid bilayers surrounding an aqueous core. This structure imparts an ability to encapsulate molecules that possess different degrees of lipophilicity; lipophilic and amphiphilic drugs will be localized in the bilayers while water-soluble molecules will concentrate into the hydrophilic core. The first drug to benefit from being encapsulated within this delivery system was doxorubicin. As of today, various companies market doxorubicin liposomal formulations, but Myocet (non-PEGylated liposomes) and Doxil (PEGylated liposomes) were among the first systems in clinical use [71, 75]. The pronounced advantages of liposomally encapsulated doxorubicin can be illustrated in its pharmacokinetic performance: an elimination half-life for the free drug is only 0.2 h, but this increases to 2.5 and 55 h, respectively, when non-PEGylated and PEGylated liposomal formulations are administered. Moreover, the area under the time–plasma concentration profile (the AUC), which indicates the bioavailability of an agent following its administration, is increased 11- and 200-fold for Myocet and Doxil, respectively, compared to the free drug [76]. Encapsulation into the liposomal carrier also causes a significant reduction in the most significant adverse side effect of doxorubicin, namely cardiotoxicity, as demonstrated in clinical trials [71, 75–77].

Liposomal doxorubicin is currently approved for the treatment of various malignancies, including Kaposi's sarcoma, metastatic breast cancer, advanced ovarian cancer and multiple myeloma. Other liposomal drugs which are either currently in use or are being evaluated in clinical trials include non-PEGylated liposomal daunorubicin (DaunoXome) and vincristine (Onco-TCS), PEGylated liposomal cisplatin (SPI-77) and lurtotecan (OSI-211) [78, 79].

Other systems in this category include metal nanoparticles for use in diagnostics, albumin paclitaxel nanoparticles approved for use in metastatic breast cancer, and drug–polymer constructs.

Nanoscale particles can act as contrast agents for all radiological imaging approaches. Iron-oxide particles provide a T_2-mode negative contrast for magnetic resonance imaging (MRI), while gold nanoparticles can be used to enhance the contrast in X-ray and computed tomography (CT) imaging, in a manner which is essentially proportional to their atomic number. Mechanical impendence disparity is the origin for the contrast in ultrasound imaging provided by the materials that are either more rigid (metals, ceramics) or much softer (microbubbles) than the surrounding tissue. The very existence of better contrast agents can drive the development of new imaging modalities. The emergence of nanocrystalline quantum dots has generated great interest in novel optical imaging technologies. The architecture and composition of quantum dots provide tunable emission properties that resist photobleaching. By concentrating preferentially at tumor sites following an EPR mechanism, nanoparticles which comprise a contrast material can provide an enhanced definition of anatomical contours and location, as well as the extent of disease. In addition, if coupled with a biological recognition moiety they can further offer molecular distribution information for the diagnostician [78].

Albumin-bound paclitaxel (Abraxane) was granted FDA approval in 2005. Paclitaxel is a highly lipophilic molecule that was previously formulated for injection with Cremophor, a toxic surfactant, under the trade name Taxol®. In a multicenter Phase II clinical trial involving 4400 women with metastatic breast cancer, Abraxane (30-min infusion, $260\,\mathrm{mg\,m^{-2}}$) was proven to be more beneficial in terms of treatment efficiency and reduction in side effects than the free drug (3-h infusion, $175\,\mathrm{mg\,m^{-2}}$) [80]. Albumin-bound methotrexate is currently being evaluated in the clinical situation.

Although the next group to be discussed does not have a particulate nature, these agents – drug–polymeric cleavable constructs – have also been considered as nanoengineered objects. In 1975, Ringdorf proposed a new concept of drug–polymer constructs that could be conjugated by using a linker with a certain degree of selectivity, and which would be stable in blood but cleaved in an acidic or enzymatic environment of a tumor site, or within an acidic intracellular compartment (e.g. endosomes) [81]. Some 20 years later, in 1994, doxorubicin conjugated to poly(*N*-(2-(hydroxypropyl)methacrylamide) (PHPMA), through an enzymatically cleavable tetrapeptide spacer (GFLG), was the first polymeric construct to enter clinical trials [70, 78]. This system significantly improved the therapeutic index of the drug, as indicated by a 45-fold higher maximum tolerated dose of the drug–polymer conjugate when compared to doxorubicin alone [82].

Other examples in this subcategory include PEG-L-asparaginase for lymphoblastic leukemia, PSMA-bound neocarcinostatin (which has been approved in Japan for the treatment of liver cancer), and PLGA-conjugated paclitaxel, which is currently undergoing Phase III evaluation for ovarian and non-small-cell lung cancer. In addition to such conventional polymer–drug, polymer–protein and protein–drug conjugates, several novel types of polymeric nanomedicines have also recently entered clinical trials, including cationic polyplexes for DNA and small interfering RNA (siRNA) delivery, dendrimers and polymeric micelles [71, 78, 79, 83].

3.4 Second-Generation Nanovectors: Achieving Multiple Functionality at the Single Particle Level

As defined above, the second generation of nanoparticles has specific additional functionalities on each individual particle, thus allowing for the molecular recognition of target tissues or for the active or triggered release of a payload at the disease site. The best examples of the targeting moieties utilized for homing the first subclass of nanovectors in this category (e.g. liposomes and other nanoparticles) are antibodies [64, 83–88]. Another example is the targeting through folate-receptors overexpressed on the membrane of some cancer cells.

Currently, a variety of targeting moieties besides antibodies are under investigation worldwide. These include ligands, aptamers and small peptides binding to specific target cell-surface markers or surface markers expressed in the disease microenvironment [89, 90].

By using active targeting, ligands can be attached to drugs to act as homing devices for binding to receptor structures expressed at the target site. Antibody–drug conjugates targeted to, for example, CD20, CD25 and CD33, which are (over) expressed in non-Hodgkin's lymphoma, T-cell lymphoma and acute myeloid leukemia, respectively, have been successfully used to deliver radionuclides (Zevalin), immunotoxins (Ontak) and antitumor antibiotics (Mylotarg) more selectively to tumor cells. Three platforms – immunoconstructs, immunoliposomes and immunopolymers – that utilize immune functional groups for targeting are presented schematically in Figure 3.3 [83].

A still unresolved question when targeting the solid tumor is the choice between high or low binding affinity of the ligand for its antigen or receptor. When the binding affinity is high, there is some evidence that the penetration of targeted therapeutics into a solid tumor is decreased due to the 'binding-site barrier'. In this case the targeted therapeutics binds strongly to the first targets encountered, but fails to diffuse further into the tumor. On the other hand, for targets in which most of the cells are readily accessible to the delivery system – for example, tumor vasculature and certain hematological malignancies – a high binding affinity might be desirable.

Another recently reported new system achieves targeting and detection based on PEGylated gold nanoparticles and surface-enhanced Raman scattering (SERS) (Figure 3.4a) [91]. These pegylated SERS nanoparticles have a significantly higher

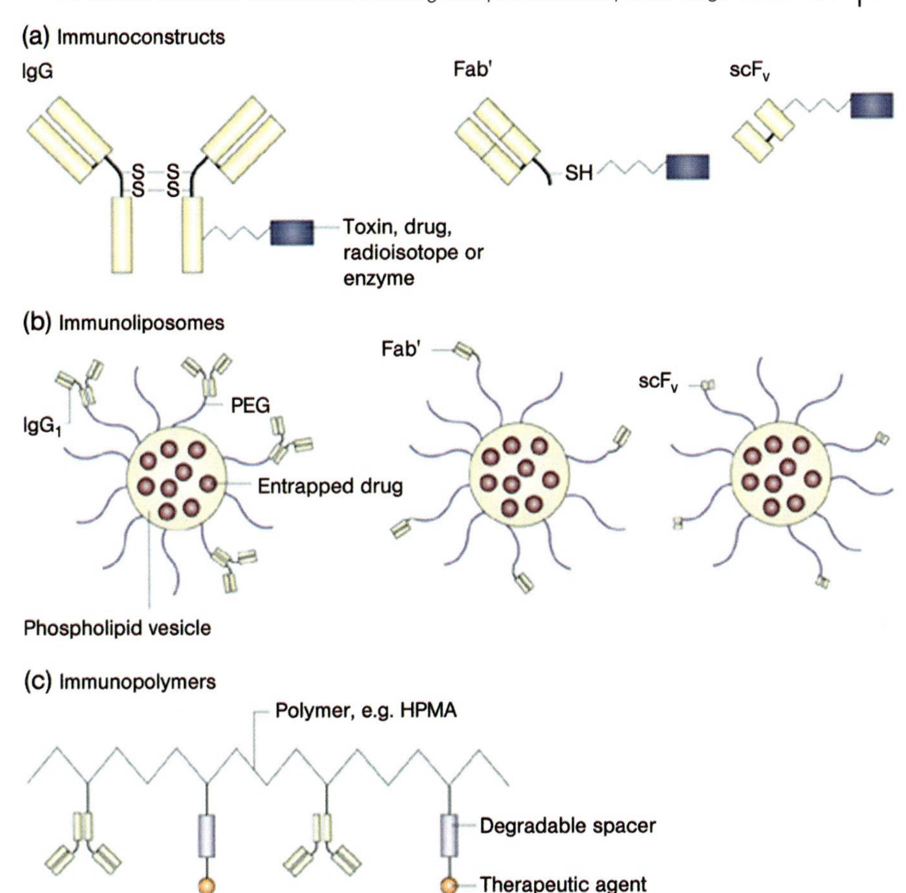

Figure 3.3 (a) Examples of targeted therapeutics constructs. Immunoconstructs are formed by the linking of antibodies, antibody fragments or nonantibody ligands to therapeutic molecules, such as toxins (immunotoxins), radioisotopes (radioimmunotherapy), drugs (immunoconjugates) or enzymes (ADEPT). Drug release, if required (immunotoxins and immunoconjugates), occurs through intracellular degradation of the peptide linker; (b) Immunoliposomes are formed by the attachment of multivalent arrays of antibodies, antibody fragments or nonantibody ligands to the liposome surface or, as in the example, to the terminus of hydrophilic polymers, such as polyethylene glycol (PEG), which are grafted at the liposome surface. The liposomes contain up to several million molecules of the therapeutic, the release of which occurs gradually by diffusion down its concentration gradient; (c) Immunopolymers are formed by linking both therapeutic agents and targeting ligands to separate sites on water-soluble, biodegradable polymers, such as hydroxypropylmethacrylamide (HPMA), with the use of appropriate degradable spacers to allow for drug release. ADEPT = antibody-directed enzyme–prodrug therapy; LTT = ligand-targeted therapeutic [83].

fluorescent intensity than quantum dots, with light emission in the near-infrared window, which is very appropriate for *in vivo* imaging. When conjugated to tumor-targeting ligands such as single-chain variable fragment (ScFv) antibodies, the conjugated nanoparticles were able to target tumor biomarkers such as epidermal

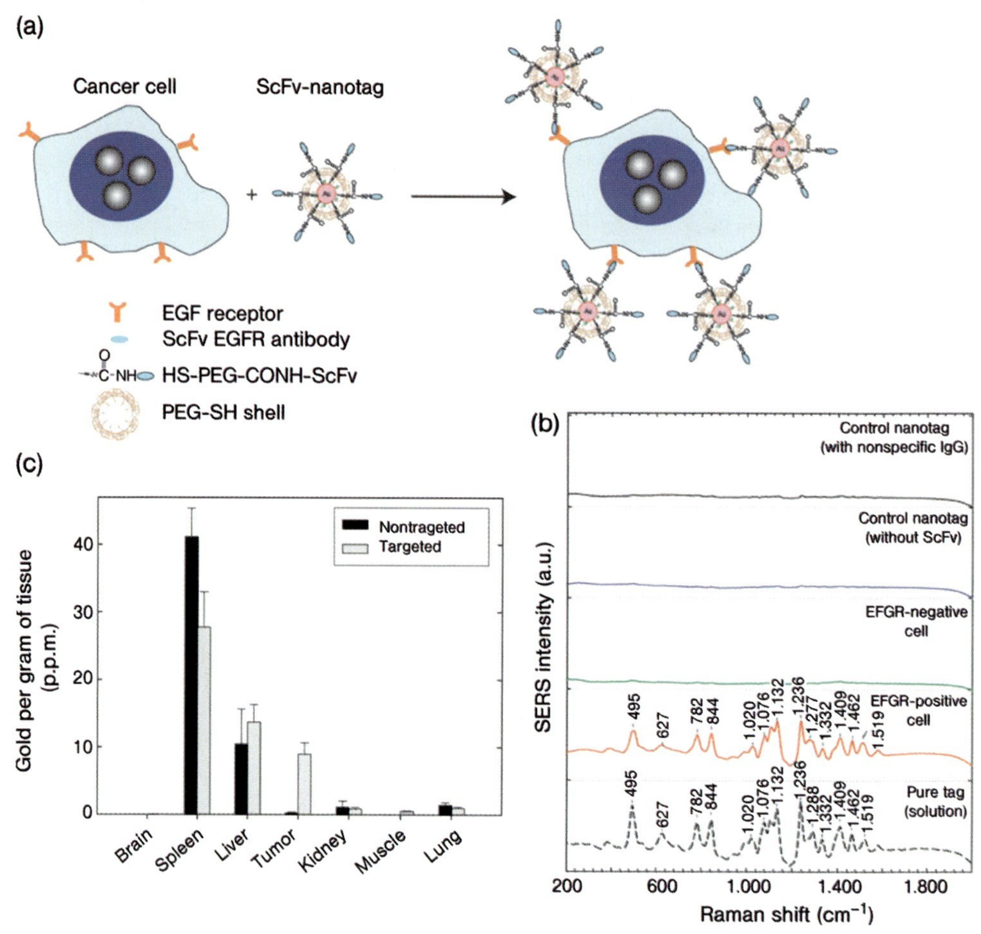

Figure 3.4 (a) Preparation of targeted SERS nanoparticles by using a mixture of SH-PEG and a heterofunctional PEG (SH-PEG-COOH). Covalent conjugation of an endothelial growth factor receptor (EGFR)-antibody fragment occurs at the exposed terminal of the heterofunctional PEG; (b) SERS spectra obtained from EGFR-positive cancer cells (Tu686) and EGFR-negative cancer cells (human non-small-cell lung carcinoma NCI-H520), together with control data and the standard tag spectrum. All spectra were taken in cell suspension with 785-nm laser excitation, and were corrected by subtracting the spectra of nanotag-stained cells by the spectra of unprocessed cells. The Raman reporter molecule is diethylthiatricarbocyanine (DTTC), and its distinct spectral signatures are indicated by wave numbers (cm^{-1}); (c) Biodistribution data of targeted and nontargeted gold nanoparticles in major organs at 5 h after injection, measured using inductively coupled plasma-mass spectrometry (ICP-MS). Reproduced with permission from Ref. [91].

growth factor (EGF) receptor on human cancer cells and in xenograft tumor models, with a 10-fold higher accumulation for targeted particles (see Figure 3.4b).

The nanovectors in the second subclass of this generation include responsive systems, such as pH-sensitive polymers or those activated by the disease site-specific enzymes, as well as a diverse group of remotely activated vectors. Among the most interesting examples here are gold nanoshells that are activated by near-infrared light, or iron oxide nanoparticles triggered by oscillating magnetic fields [92, 93]. Other techniques used to remotely activate the second-generation particulates include ultrasound and radiofrequency (RF) [94–96]. Linking nanoshells to antibodies that recognize cancer cells enables these novel systems to seek out their cancerous targets prior to applying near-infrared light to heat them up. For example, in a mouse model of prostate cancer, nanoparticles activated with 2′-fluoropyrimidine RNA aptamers that recognized the extracellular domain of the PSMA, and loaded with docetaxel as a cytostatic drug, were used for targeting and destroying cancer cells [97, 98]. Another new approach is based on the coupling of nanoparticles to siRNA, used to silence specific genes responsible for malignancies. By using targeted nanoparticles, it was shown that the delivered siRNA can slow the growth of tumors in mice, without eliciting the side effects often associated with cancer therapies. Although the representatives of the second generation have not yet been approved by FDA, there are today numerous ongoing clinical trials involving targeted nanovectors, especially in cancer applications.

3.5 Third-Generation Nanoparticles: Achieving Collaborative Interactions Among Different Nanoparticle Families

The fundamental basis for the administration of drugs is to achieve a favorable therapeutic outcome in the treatment of a medical condition or disease, with minimal detrimental side effects. So far, we have described in detail the first- and second-generation nanoparticle therapeutic strategies. Although each generation has demonstrated incremental improvements relative to conventional intravenously administered chemotherapies, 'blockbuster' drug status has yet to be achieved by any nanobased construct. The second-generation nanoparticles offered new degrees of sophistication compared to their predecessors by employing additional complexities such as targeting moieties, remote activation, and environmentally sensitive components. However, these improvements predominantly represent simply a progressive evolution of the first-generation vectors; these subtle, yet augmenting, particle improvements do not fully address the primary challenge – or set of challenges – presented in the form of sequential biological barriers that continue to impair the efficacy of first- and second-generation nanoparticulates. This fundamental problem has given rise to a nanoparticle paradigm shift with the emergence of a third generation of particle that is specifically engineered to avoid biological barriers and to codeliver multiple nanovector payloads with tumor specificity.

The human body presents a robust bodily defense system that is extremely effective in preventing injected chemicals, biomolecules, nanoparticles and any other foreign agent(s) of therapeutic action from reaching their intended destinations. In addition to these natural biologic defenses, tumor-associated obstacles also exist. As a demonstration of the efficacy of these combined biological barriers, it has been calculated that only one out of every 100 000 molecules of drug successfully reaches the intended site, permitting the overwhelming majority of the highly toxic, nondiscriminating, systemically disbursed poison to manifest in a number of undesirable side effects associated with cancer chemotherapy. This familiar scenario was quantitated in a study of Kaposi's sarcoma study that showed the percentage concentration of doxorubicin in Kaposi's sarcoma lesions to be ~0.001% [99]. This therapeutic phenomenon does not appear to be a tumor-specific challenge, however, and is therefore applicable to the lion's share of malignancies and tumor types [5, 100–102].

Some of the above-mentioned and most notable challenges include physiological barriers (i.e. the RES, epithelial/endothelial membranes, cellular drug extrusion mechanisms) and biophysical barriers (i.e. interstitial pressure gradients, transport across the ECM, expression and density of specific tumor receptors, and ionic molecular pumps). Biobarriers are sequential in nature, and therefore the probability of reaching the therapeutic objective is the product of individual probabilities of overcoming each and every one of them [8]. The requirement for a therapeutic agent to be provided with a sufficient collection of weaponry to conquer all barriers, yet still be small enough for safe vascular injection, is the major challenge faced by nanotechnology [14]. Once injected, nanoscale drug delivery systems (or 'nanovectors') are ideal candidates for the time-honored problem of optimizing the therapeutic index for treatment – that is, to maximize efficacy, while reducing adverse side effects.

The ideal injected chemotherapeutic strategy is envisioned to be capable of navigating through the vasculature after intravenous administration, to reach the desired tumor site at full concentration, and to selectively kill cancer cells with a cocktail of agents with minimal harmful side effects. Third-generation nanoparticle strategies represent the first wave of next-generation nanotherapeutics that are specifically equipped to address biological barriers to improve payload delivery at the tumor site. By definition, third-generation nanoparticles have the ability to perform a time sequence of functions which involve the cooperative coordination of multiple nanoparticles and/or nanocomponents. This novel generation of nanotherapeutics is exemplified through the employment of multiple nanobased products that synergistically provide distinct functionalities. In this chapter, the nanocomponents will include any engineered or artificially synthesized nanoproducts, including peptides, oligonucleotides (e.g. thioaptamers, siRNA) and phage with targeting peptides. Naturally existing biological molecules, such as antibodies, will be excluded from this designation, despite their ability to be synthesized.

Third-generation approaches have been developed to address the numerous challenges responsible for reducing the chemotherapeutic efficacy of earlier strategies. For example, surface modification of the exterior of nanoparticles with PEG has proven to be effective in increasing the circulation time within the bloodstream; however, this preservation tactic proves detrimental to the biological recognition and targeting ability of the nanovector [103]. In order to avoid such paradoxical approaches of employing debilitating improvements to therapeutic delivery systems, many research groups are combining multiple nanotechnologies to exploit the additive contributions of the constituent components. One example of third-generation nanoparticles is the biologically active molecular networks known as 'nanoshuttles'. These self-assemblies of gold nanoparticles within a bacteriophage matrix combine the hyperthermic response to near-infrared radiation of the gold with the biological targeting capabilities of the 4C-RGD sequence presented by the phage [104]. The nanoshuttles also collectively accommodate enhanced fluorescence, dark-field microscopy and surface-enhanced Raman scattering detection.

The next example of third-generation nanoparticles is the disease-inspired approach of the 'nanocell'. Newly emerging chemotherapeutic models utilize combinational therapies that are intended to inhibit tumor neovasculature growth and kill cancer cells. The coadministration of anti-angiogenic agents with conventional cytotoxic agents is a novel concept, but this practice has faced two critical problems. First, it has been shown that anti-angiogenic agents are capable of depleting blood flow to the tumor by interrupting new vessel growth. Unfortunately, however, this shutdown of tumor blood vessels has resulted in the prevention of chemotherapeutic agents from reaching the tumor site at sufficient concentrations. Furthermore, the decreased blood flow elicits intra-tumoral hypoxia which in turn increases the expression of hypoxia-inducible factor-1α (HIF1-α). HIF1-α overexpression is correlated to increased tumor invasiveness and chemotherapy resistance [105]. By using the same combinatorial chemotherapy approach, researchers have developed a nested nanoparticle construct that comprises a lipid-based nanoparticle enveloping a polymeric nanoparticle core called a 'nanocell'. Here, a conventional chemotherapeutic drug (doxorubicin) is conjugated to a polymer core and an anti-angiogenic agent (combretastatin) is then trapped within the lipid envelope. When the nanocells are accumulated within the tumor through the EPR effect, the sequential time release of the anti-angiogenic agent, followed by the cytotoxic drug, causes an initial disruption of tumor vascular growth and effectively traps the drug-conjugated nanoparticle core within the tumor to allow an eventual delivery of the cancer cell-killing agent.

The final example of third-generation nanoparticle technology utilizes a multistage approach that addresses many biological barriers experienced by an injec therapy. Currently, research groups are developing nanoporous silicon microparticles that utilize their unique particle size, shape and other physical characteristics in concert with active tumor biological targeting moieties to efficiently deliver payloads of nanoparticles to the tumor site. The ability to deliver a therapeutic agent to a tumor is analogous to the lunar voyage embarked upon by the Apollo 11 crew. This epic feat was not achieved simply by piloting a single vehicle to the moon and back; instead, it required a sequential execution of numerous steps, which included three stages of the Saturn V rocket to escape the Earth's atmosphere, a lunar module to descend and ascend to and from the lunar landscape, and finally a command module for re-entry and splashdown back to Earth. Similar mission-critical issues must also be addressed in a sequential manner when developing drug delivery systems to fight cancer.

The multistage drug delivery system is predicated upon a Stage 1 nanoporous silicon microparticle that is specifically designed (through mathematical modeling) to exhibit superior margination and adhesion properties during its negotiation through the systemic blood flow en route to the tumor site. Particle characteristics such as size, shape, porosity and charge can be exquisitely controlled with precise reproducibility through semiconductor fabrication techniques. In addition to its

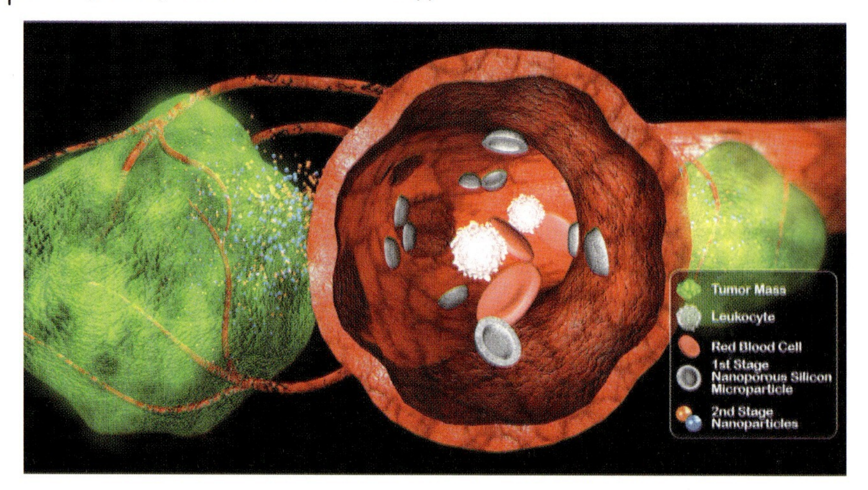

Figure 3.5 Multistage nanovectors. Stage 1 nanoporous silicon microparticles are engineered to exhibit an enhanced ability to marginate within blood vessels and adhere to tumor-associated endothelium. Once positioned at the tumor site, the Stage 1 particle can release its nanoparticle payload to achieve the desired therapeutic effect, prior to complete biodegradation of the carrier particle. The therapeutic outcome can be determined by selection of the nanoparticles loaded within the nanoporous structure of the Stage 1 particle.

favorable physical characteristics, the Stage 1 particle can be surface-treated with such modifications as PEG for RES avoidance and also equipped with biologically active targeting moieties (e.g. aptamers, peptides, phage, antibodies) to enhance the tumor-targeting specificity. This approach decouples the challenges of: (i) transporting therapeutic agents to the tumor-associated vasculature; and (ii) delivering therapeutic agents to cancer cells. The Stage 1 particles shoulder the burden of efficiently transporting a nanoparticle payload to the tumor site within the nanoporous structures of its interior (Figure 3.5). The nanoparticles called Stage 2 particles, generically represent any nanovector construct within the approximate diameter range of 5 to 100 nm. The Stage 1 particles have demonstrated the ability to rapidly load (within seconds) and gradually release (within hours) multiple nanoparticles (i.e. single-walled carbon nanotubes and quantum dots) during *in vitro* experiments, with complete biodegradation within 24–48 h, depending on the pore density [106]. Furthermore, unpublished preliminary data have demonstrated the ability to deliver liposomes and other nanovectors, as well as indications of the successful *in vivo* delivery of Stage 2 nanoparticles to tumor masses in xenograft murine models.

The multistage drug delivery system is emblematic of third-generation nanoparticle technology, since the strategy combines numerous nanocomponents to deliver multiple nanovectors to a tumor lesion. The Stage 1 particle is rationally designed to have a hemispherical shape to enhance particle margination within blood vessels, and to increase particle/endothelium interaction to maximize the probability of active tumor targeting and adhesion [107]. In addition to improved hemodynamic physical properties and active biological targeting by utilizing nanocomponents such as aptamers and phage, the Stage 1 particle can also present with specific surface modifications in order to avoid RES uptake and exhibit degradation rates predetermined by nanopore density. Upon tumor recognition and vascular adhesion, a series of nanoparticle payloads may be released in a sequential order predicated upon Stage 1 particle degradation rates and payload conjugation strategies (e.g. environmentally sensitive crosslinking techniques, pH, temperature, enzymatic triggers). The versatility of this platform nanovector multistage delivery particle allows for a multiplicity of applications. Depending upon the nanoparticle 'cocktail' loaded within the Stage 1 particle, this third-generation nanoparticle system can provide for the delivery not only of cytotoxic drugs but also of remotely activated hyperthermic nanoparticles, contrast agents and future nanoparticle technologies.

3.6 Nanovector Mathematics and Engineering

Third-generation particles are transported by the blood flow and interact with the blood vessel walls, both specifically – through the formation of stable ligand–receptor bonds – and nonspecifically, by means of short-ranged van der Waals, electrostatic and steric interactions. If suitable conditions are met in terms of a sufficiently high expression of vascular receptors and sufficiently low hydrodynamic shear stresses at the wall, particles may adhere firmly to the blood vessel walls and control cell uptake, either by avoiding or favoring, based on their final objective. Such an intravascular 'journey' can be broken down into three fundamental events which form the cornerstone of the rational design, namely: the margination dynamics; the firm adhesion; and the control of internalization. The rational design of particles has the aim of identifying the dominating governing parameters in each of the above-cited events in order to propose the optimal design strategy as a function of the biological target (diseased cell or environment).

In physiology, the term 'margination' is conventionally used to describe the lateral drift of leukocytes and platelets from the core of the blood vessels towards the endothelial walls. This event is of fundamental importance as it allows an intimate contact between the circulating cells and the vessel walls, and in the case of leukocytes it is required for diapedesis. Similarly, the rational particle design should aim at generating a marginating particle, that can spontaneously move preferentially in close proximity to the blood vessel walls. Accumulating the particles in close proximity to the blood vessel walls is highly desirable both in vascular targeting and when the delivery strategy relies on the EPR approach. This occurs for two main reasons:

- The particles can 'sense' the vessel walls for biological and biophysical diversities, as for instance the overexpression of specific vascular markers (vascular targeting) or the presence of sufficiently large fenestrations through which they extravasate (EPR-based strategy).
- The particles can more easily leave the larger blood vessels in favor of the smaller ones, thus accumulating in larger numbers within the microcirculation (Figure 3.6) [108].

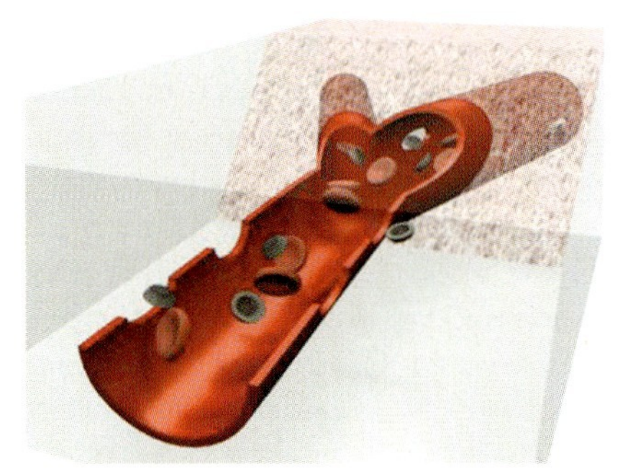

Figure 3.6 Marginating particles can more likely 'sense' the vessel walls for biological and biophysical diversity and more easily leave the vascular compartments through openings along the endothelium.

While leukocyte and platelet margination is an active process requiring an interaction with red blood cells (RBCs) and the dilatation of inflamed vessels with blood flow reduction [109], particle margination can only be achieved by proper rational design.

It should be noted that the RBCs – the most abundant blood-borne cell population – have a behavior opposite to margination, with an accumulation that occurs preferentially within the core of the vessels. This has long been described by Fahraeus and Lindqvist [110], and is referred to as the *plasma skimming* effect. An immediate consequence of this phenomenon is the formation of a 'cell-free layer' in the proximity of the wall, which varies in thickness with the size of the channel and mean blood velocity. For example, it may be as large as few tens of microns in arterioles ($\geq 100\,\mu m$ in diameter) and a few microns in capillaries ($\geq 10\,\mu m$ in diameter [111]). Particles designed to marginate should accumulate and move in a cell-free layer, which is also characterized by an almost linear laminar flow.

The motion of spherical particles in a linear laminar flow has been described by Goldmann *et al.* [112], who showed that the exerted hydrodynamic forces grow with the particle radius, and that no lateral drift would be observed unless an external forces such as gravitational or magnetic, or short-ranged van der Waals and electrostatic interactions were applied [113]. In other words, a neutrally buoyant spherical particle moving in close proximity to a wall can drift laterally only if an external force is applied. Here, it is important to recall that the gravitational force has been shown to be relevant even for submicrometer polystyrene beads (relative density to water of $0.05\,g\,cm^{-3}$), and that margination dynamics can be effectively controlled in horizontal channels by changing the size of the nonbuoyant nanoparticles [114]. On the other hand, nonspherical particles exhibit more complex motions with tumbling and rolling that can be exploited to control their margination dynamics, without any need for lateral external forces. The longitudinal (drag) and lateral (lift) forces, as well as the torque exerted by the flowing blood, depend on the size, shape and orientation of the particle to the stream direction, and change over time as the particle is transported. Considering an ellipsoidal particle with an aspect ratio of 2 (Figure 3.7a) in a linear laminar flow, the particle trajectory and its separation distance from the wall are shown in Figure 3.7b. Clearly, the particle motion is very complex,

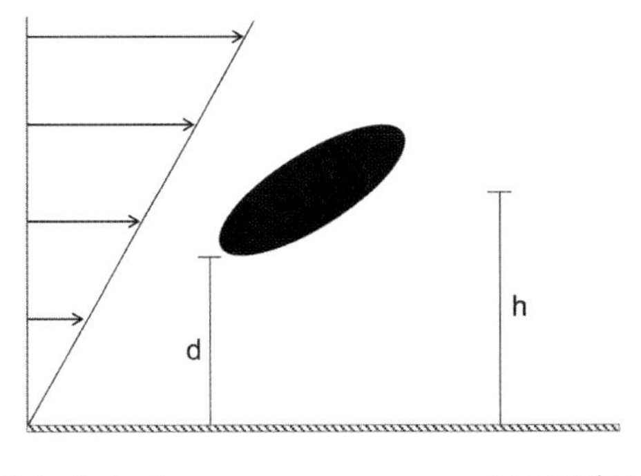

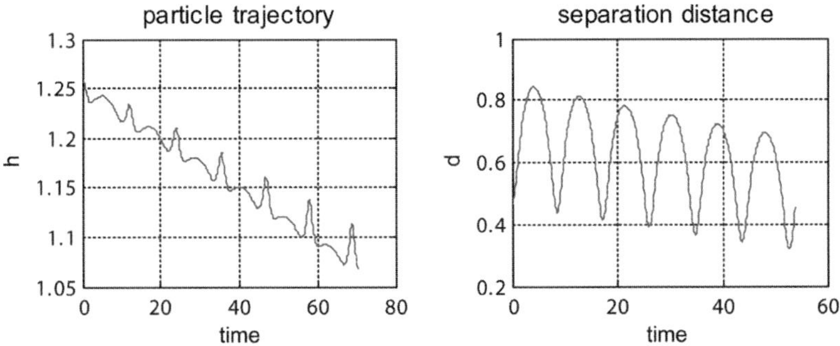

Figure 3.7 (a) An ellipsoidal particle transported within a linear laminar flow at a distance *d* from a rigid wall, as it would be in close proximity to the vessel walls; (b) The trajectory of the ellipsoidal particle with its characteristic oscillatory motion, and its separation distance from the wall reducing with time.

with periodic oscillations towards and away from the wall. Overall, however, the particle would approach the wall and interact with its surface.

For nonspherical particles, it has been shown that the lateral drifting velocity is directly related to their aspect ratio [115, 116], with a maximum between the two extremes: a sphere, with aspect ratio unity, and a disk, with aspect ratio infinity.

More recently, *in vitro* experiments have been conducted using spherical, discoidal and quasi-hemispherical particles with the same weight injected into a parallel plate flow chamber under controlled hydrodynamic conditions [117]. The experiments have shown that discoidal particles tend to marginate more than quasi-hemispherical and more than spherical particles in a gravitational field. Notably, these observations neglect the interaction of the particles with blood cells, in particular RBCs. However, this is a reasonable assumption as long as the particles are sufficiently smaller than RBCs and tend to accumulate within the cell-free layer.

Therefore, with regards to what concerns the design of marginating particles their size and shape, their geometric properties are of fundamental importance.

The marginating particle moving in close proximity to the blood vessels can interact both specifically and nonspecifically with the endothelial cells, and eventually adhere firmly to it. Firm and stable adhesion is ensured as long as the dislodging forces (hydrodynamic forces and any other force acting to release the particle from

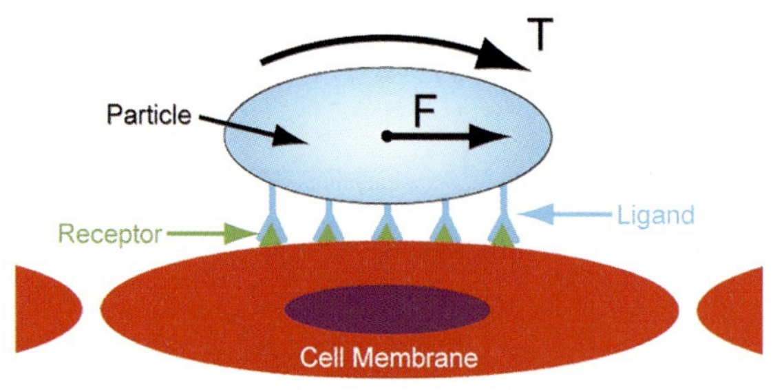

Figure 3.8 The longitudinal (drag) force (F) and the torque (T) exerted over a particle adhering to a cell layer under flow.

the target cell) are balanced by specific ligand–receptor interactions and nonspecific adhesion forces arising at the cell–particle interface (Figure 3.8).

The strength of adhesion must be expressed in terms of an adhesion probability factor, P_a, defined as the probability of having at least one ligand–receptor bond formed under the action of the dislodging forces. The probability of adhesion is decreased as the shear stress at the blood vessel wall μS and as the characteristic size of the particle increase; and grows as the surface density of ligand molecules m_l distributed over the particle surface and of receptor molecules m_r expressed at the cell membrane increases. However, for a fixed volume particle – that is to say, for a fixed payload – oblate particles with an aspect ratio γ larger than unity would have a larger strength of adhesion having fixed all other parameters [118]. Interestingly, for each particle shape, a characteristic size can be identified for which the probability of adhesion has a maximum, as shown in Figure 3.9. For small particles, the hydrodynamic forces are small but the area of interaction at the particle–cell interface is also smaller, leading consequently to a small number of ligand–receptor bonds involved which cannot withstand even a small dislodging force. For large particles, the number

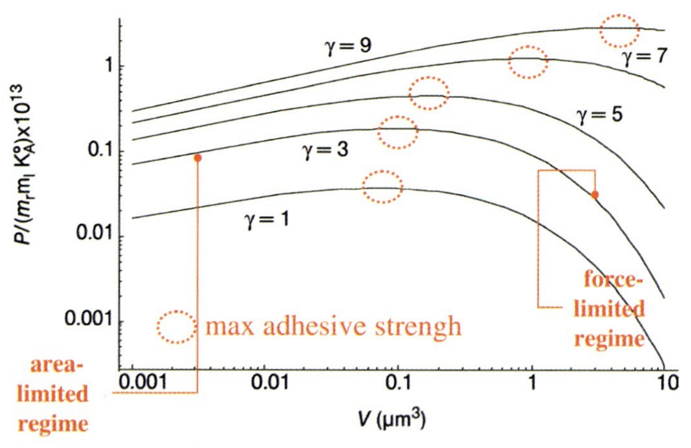

Figure 3.9 Variation of the normalized adhesive probability factor (P) with the volume V of the particle for different values of the aspect ratio γ [118].

of ligand–receptor bonds that can be formed grows, but the hydrodynamic forces grow even more. The optimal size for adhesion – that is, the size for which P_a has a maximum – falls between these two limiting conditions.

As an example, when considering a capillary with a shear stress at the wall of $\mu S = 1$ Pa and a surface density of receptors $m_r = 100\,\mu m^2$, the optimal radius for a spherical particle would be about 500 nm with a total volume of 0.05 μm^3, whereas the optimal volume for an oblate spheroidal particle with an aspect ratio $\gamma = 2$ would be more than 50 times larger (3.5 μm^2) [118].

In particle adhesion, rational design should focus on the shape of the particle and the type and surface density of ligand molecules decorating the particle surface.

Once the particle has adhered to the target cell, it should be internalized if the aim is to release drugs or therapeutic agents within the cytosol or at the nuclear level (gene delivery). Alternatively, it should resist internalization if the target cell is used just as a docking site (vascular targeting) from which are released second-stage particles. The internalization rate is affected by the geometry of the particle and the ligand–receptor bonds involved.

Freund and colleagues [119] developed a mathematical model for receptor-mediated endocytosis based on an energetic analysis. This showed that a threshold particle radius R_{th} exists, below which endocytosis could never occur and, that an optimal particle radius R_{opt} exists, slightly larger than R_{th}, for which internalization is favored with the maximum internalization rate, thus confirming (in theory) the above-cited experimental observations. This analysis was then generalized to account for the contribution of the surface physico-chemical properties that may dramatically affect the internalization process, changing significantly both R_{opt} and R_{th} (Figure 3.10), as shown by Decuzzi and Ferrari [114].

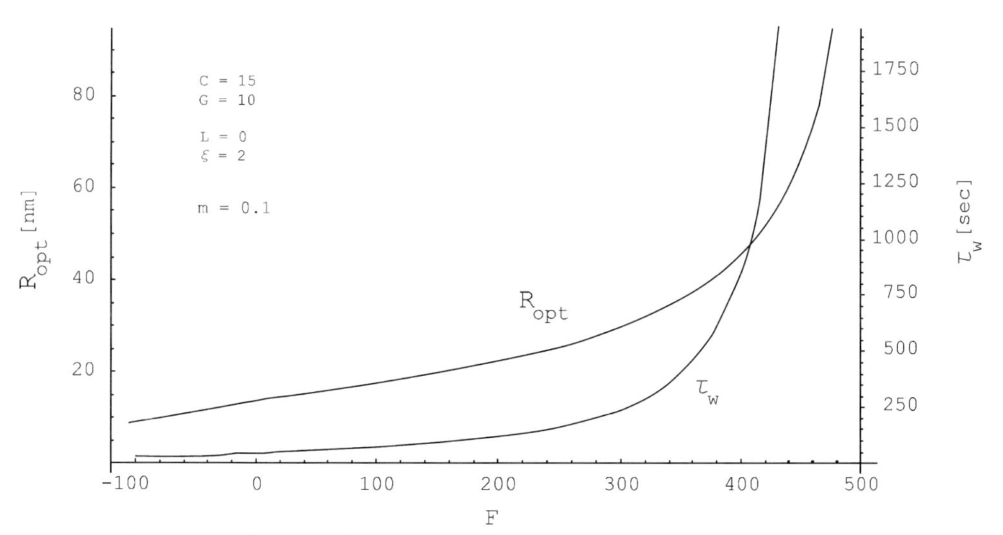

Figure 3.10 The optimal radius R_{opt} and the wrapping time τ_w as a function of the nonspecific parameter F, growing with the repulsive nonspecific interaction at the cell–particle interface.

A more recent theoretical model has been developed by Decuzzi and Ferrari for the receptor-mediated endocytosis of nonspherical particles [120]. This shows how elongated particles laying parallel to the cell membrane are less prone to internalization compared to spherical particles or particles laying normal to the cell membrane. The results show clearly how particle size and shape can be used to control the internalization process effectively, and that particles that deviate slightly from the spherical shape are more easily internalized compared to elongated particles that deviate severely from the classical spherical shape.

Even in the case of particle internalization, a judicious combination of surface physico-chemical properties and particle geometry would lead to a particle with optimized internalization rates, depending on the final biological applications.

Finally, a mathematical model has been recently developed [121] that allows one to predict the adhesive and endocytotic performances of particulate systems based on three different categories of governing parameters: (i) geometric (radius of the particle); (ii) biophysical (ligand-to-receptor surface density ratio; nonspecific interaction parameter; hydrodynamic force parameter); and (iii) biological (ligand–receptor binding affinity). This finding has led to the definition of *Design Maps* through which the three different states of the particulate system can be predicted: (i) no adhesion at the blood vessel walls; (ii) firm adhesion with no internalization by the endothelial cells; or (iii) firm adhesion and internalization (Figure 3.11) [121].

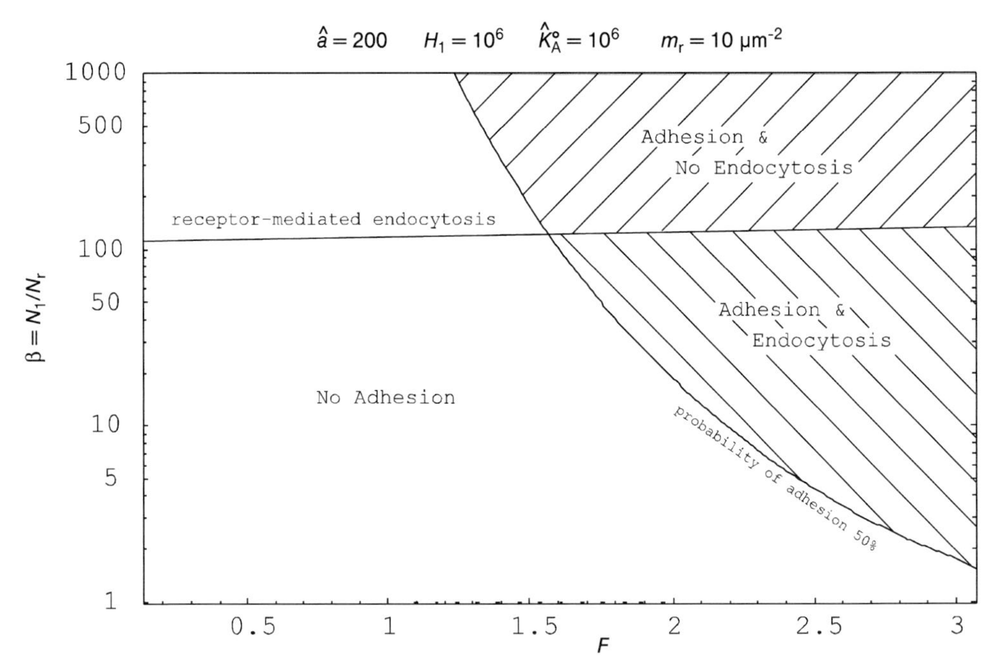

Figure 3.11 A typical design map showing areas for no adhesion; adhesion and no endocytosis; and adhesion and endocytosis [121].

3.7 The Biology, Chemistry and Physics of Nanovector Characterization

The small size, unique physico-chemical properties, and biological activity of nanoparticles create the need for extensive characterization prior to their use in biomedical applications. The National Cancer Institute has established the Nanotechnology Characterization Laboratory (NCL) to standardize and perform preclinical characterizations of nanomaterials designed for cancer therapeutics and diagnostics [122]. The objectives of the NCL are to speed the development of nanotechnology-based products for cancer patients, while reducing the risk to inventors, as well as encouraging private-sector investment. A further aim is to establish an analytical cascade of protocols for nanomaterial characterization. Challenges to creating standardized characterization techniques include the wide variety of materials used to construct nanomedicines. Thus, the characterization strategy is broad and includes physico-chemical characterization, sterility and pyrogenicity assessment, biodistribution (absorption, distribution, metabolism, excretion) and toxicity, both *in vitro* and *in vivo* in animal models [123]. An examination of the biological and functional characteristics of multicomponent/combinatorial platforms is also addressed.

NCL's standardized analytical cascade includes tests for preclinical toxicology, pharmacology and efficacy. The protocols include assays for physical attributes; *in vitro* testing for toxicity or biocompatibility; and *in vivo* testing for safety, efficacy and toxicokinetic properties in animal models (Figure 3.12).

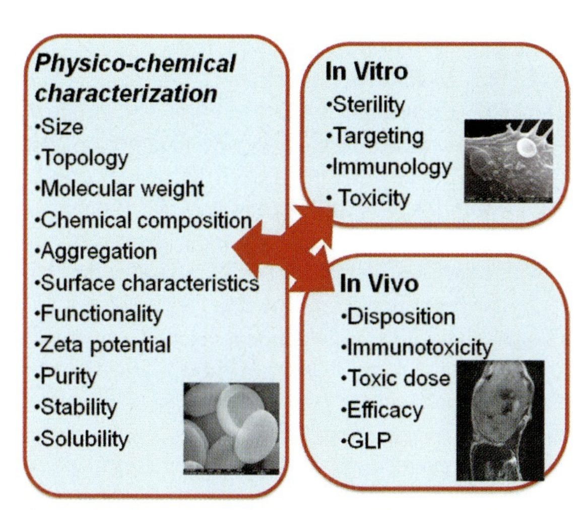

Figure 3.12 Preclinical characterization of nanoparticles involves physico-chemical, *in vitro* and *in vivo* characterization. The list includes assays outlined by the National Cancer Institute Nanotechnology Characterization Laboratory.

3.7.1 Physical Characterization

Physical characterization includes assays for particle size, size distribution, molecular weight, density, surface area, porosity, solubility, surface charge density, purity, sterility, surface chemistry and stability. The mean particle size – that is, the hydrodynamic diameter – is determined by batch-mode dynamic light scattering (DLS) in aqueous suspensions. Care must be taken with these measurements, because they can be affected by other parameters. For example, precautions include the cleaning of cuvettes with filtered, demineralized water; media filtering with 0.1 μm pore size membranes and pre-rinsing cuvettes multiple times; scattering contributions by media in the absence of analyte; optimized sample concentration; and filtering samples in conjunction with loading into the cuvette. The sample concentration should be optimized to avoid signal-to-noise ratio (SNR) deterioration at low concentration and particle interactions and scattering effects at high concentration. Another precaution is to add only small amounts of monovalent electrolyte in order to avoid salt effects on the electrical double-layer surrounding the particles in the media. Again, concentration optimization is necessary for optimal measurements. In order to evaluate instrument performance, latex size standards are commercially available. When analyzing these data, the absolute viscosity and refractive index for the suspending media is required to calculate the hydrodynamic diameter.

3.7.2 *In Vitro* Testing

In vitro testing includes the assessment of sterility, targeting, *in vitro* immunology and toxicity testing. Sterility testing for contaminates includes monitoring for the presence of endotoxins, bacteria, yeast, molds and mycoplasm. As an example, the LAL (limulus amoebocyte lysate) assay is commonly used to test for the presence of bacterial endotoxin. Although standard immunological *in vitro* assays exist, the preclinical immunotoxicity testing of nanoparticles has been hampered due to interference by the nanoparticles within the assay. Whilst a variety of mechanisms of interference exists, the most common occurrences are light absorbance by nanoparticles (which interfere with colorimetric methods) and the catalytic properties of nanoparticles creating false-positive effects in enzyme assays [122].

In vitro targeting assays measure cell binding and the internalization of particles (Figure 3.13). This is particularly relevant for drug delivery systems, as the route of internalization dictates the subcellular localization of nanoparticles. As an example, caveolar-mediated uptake leads to nanoparticles being localized into organelles with a nonacidic pH [124], whereas clathrin-mediated uptake favors their lysosomal entrapment [125], the latter leading to drug degradation. The uptake of larger particles (typically >500 nm) generally occurs by phagocytosis [126, 127]. Phagosomes typically fuse with endosomes, leading to lysosomal accumulation [126]. Common targeting assays include confocal microscopy, transmission and scanning electron microscopy, flow cytometric analysis and quantitative assessment assays (e.g. the BCA protein

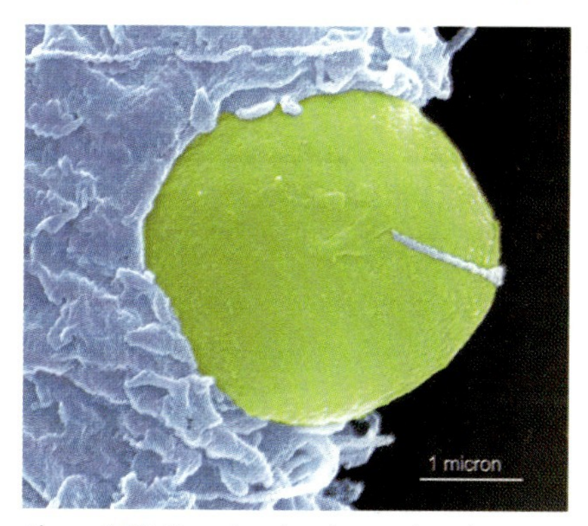

Figure 3.13 Pseudocolored scanning electron microscopy image of an endothelial cell internalizing a nanoporous silicon particle.

assay to assess PLGA uptake [128]). Both, fluorescence microscopy and flow cytometry rely on the attachment of fluorescent probes to the nanoparticle; alternatively, the latter technique may rely on changes in light scattering caused by the presence of internalized nanoparticles [127]. Controls are always essential to ensure that any intracellular fluorescence is not due to the uptake of dye that might have been released from the particles [128]. Factors affecting nanoparticle uptake include nanoparticle concentration, incubation time, nanoparticle size and shape and culture media.

For multicomponent systems, targeting may be difficult to access *in vitro*, especially for systems composed of nested particles, where each particle is targeting a specific and discrete population. Additional problems arise due to the modification of particles with imaging agents. For example, the conjugation of fluorescent probes to the surface of particles alters the surface charge density of the particle, and may also mask the binding of ligands on the particle surface. This in turn alters the ability of particles to bind to the cell-surface receptors that are responsible for their uptake. *In vitro* targeting assays also need to emphasize the impact of serum opsonization on particle uptake [127]. Serum components are signals for immune cells, and may either activate cells or serve as bridges attaching particles to cells. For example, antibodies found in serum may bind to particles and mediate their uptake via Fc receptors found on specific cell populations. The end result is dramatic, however, and may even completely alter which cell populations are able to internalize the particles.

To date, research investigations have shown that nanoparticles can stimulate and/or suppress the immune response [129]. Compatibility with the immune system is affected to a large degree by the surface chemistry. The cellular interaction of nanoparticles is dependent on either their direct binding to surface receptors or binding through the absorption of serum components to particles and their subsequent interaction with cell receptors [127]. Blood-contact properties of the nanomaterial and cell-based assays are used to determine the immunological compatibility of the device.

The blood-contact properties of nanoparticles are characterized by plasma protein binding, hemolysis, platelet aggregation, coagulation and complement activation.

- Plasma protein binding is achieved using two-dimensional gel electrophoresis, with individual proteins being evaluated by mass spectrometry. A drawback here is the need for 1 mg of nanoparticles to complete the assay.
- Hemolysis is assayed by a quantitative colorimetric determination of hemoglobin in whole blood, with the end result expressed as percentage hemolysis.
- Platelet aggregation is expressed as the percentage of active platelets per sample compared to a control baseline sample determined using a Z2 Coulter counter for the analysis of platelet-rich plasma.
- Multiple tests are used to assess the effect of nanoparticles on plasma coagulation, including prothrombin time.
- Complement activation is measured initially by the qualitative determination of total complement activation by Western blot. Anti-C3 specific antibodies recognize both native C3 and its cleaved product, which is a common product of all three complement activation pathways. Positive results elicit additional assays aimed at determining the specific complement activation pathway.

In vitro immunology assays also include cell-based assays including colony-forming units-granular macrophages (CFU-GM), leukocyte proliferation, macrophage/neutrophil function and cytotoxic activity of natural killer (NK) cells. The effect of nanoparticles on the proliferation and differentiation of murine bone marrow hematopoietic stem cells (HSC) is monitored by measuring the number of colony forming units (CFU) in the presence and absence of nanoparticles. The effect of nanoparticles on lymphocyte proliferation is determined in similar manner. The ability of nanoparticles to either induce or suppress proliferation is measured and compared to control induction by phytohemaglutinin. Macrophage/neutrophil function is measured by the analysis of phagocytosis, cytokine induction, chemotaxis and oxidative burst. Similar to earlier targeting studies, nanoparticle internalization is measured, but with respect to classical phagocytic cells rather than to the target populations. A current phagocytosis assay utilizes luminol-dependent chemiluminescence, although alternative detection dyes must be used for nanoparticles that interfere with measurements. Cytokine production induced by nanoparticles is measured using white blood cells isolated from human blood. Following particle incubation, the cell culture supernatants are collected and analyzed for the presence of cytokines, using cytometry beads. The chemoattractant capacity of nanoparticles is measured used a cell migration assay; here, cell migration through a 3 μm filter towards test nanoparticles is quantitated using a fluorescent dye. The final measure of macrophage activation is a measure of nitric oxide production using the Greiss reagent. NK-mediated cytotoxicity can be measured by radioactive release assays, in which labeled target cells release radioactivity upon cytolysis by NK cells. A new label-free assay known as 'xCELLigence' (available from Roche) is used to measure the electrical impedance of cells attached to the bottom of a microtiter plate containing cell sensor arrays. In this system, any changes in cell morphology, number, size or attachment are detected in real time.

3.7.2.1 *In Vitro* Toxicity Testing

Standard assays for toxicity assess oxidative stress, necrosis, apoptosis and metabolic stability. Oxidative stress is quantified as a measure of glutathione (GSH) reduction, lipid peroxidation and reactive oxygen species (ROS) in cells treated with nanoparticles. These are measured using colorimetric and fluorescence assays. Cytotoxicity can be measured by using two assays: reduction of 3-(4,5-dimethyl-2-thiazolyl)-2,5-diphenyl-2*H*-tetrazolium bromide (MTT); and lactate dehydrogenase (LDH) release. The degree of caspase-3 activation is also used as a measure of cytotoxicity as it is an indicator of apoptosis. Assays for metabolic stability include cytochrome P450 (CYP450) and glucuronidation.

3.7.3 *In Vivo* Animal Testing

The final category of assays relies on *in vivo* animal testing. Under this umbrella are included disposition studies, immunotoxicity, dose-range-dependent toxicity and efficacy. The initial disposition of nanoparticles is dependent upon tissue distribution, clearance, half-life and systemic exposure. In the NCL regime, immunotoxicity is measured as a 28-day screen and by immunogenicity testing (repeat-dosing). Dose-dependent toxicity can be evaluated by monitoring blood chemistry, hematology, histopathology and gross pathology. Depending on the nature of the delivery system, the efficacy is measured either by imaging or by therapeutic impact.

One possible route of nanoparticle exposure within the work environment is that of *inhalation*, which in turn creates a need for additional studies that include animal inhalation and intratracheal instillation assays [130, 131]. These additional studies also illicit the need for even more characterization studies, such as determining the dispersion properties of nanoparticles. Hence, new methods to determine not only hazard and risk assessments but also therapeutic efficacy continue to be developed as new areas of concern arise. The careful characterization and optimized bioengineering of both nanoparticles and microparticles represent key contributors to the generation of nanomedical devices with optimal delivery and cellular interaction features.

3.8 A Compendium of Unresolved Issues

Unresolved issues and opportunities live in symbiosis. Programmatically, we welcome even the most daunting challenges, as their mere identification as such – not a simple task in most cases, and invariably one that requires the right timing and knowledge maturation – frequently happens when solutions are conceivable, or well within the reach of the scientific community. With this essentially positive outlook we will list in this section some questions that appear daunting at this time, but are starting to present themselves with finer detail and resolution, indicating in our mind that readers in a few years, if any, will find them to be essentially resolved, and the value of this section, if any, to be basically that of a message in a bottle across the seas of time – and reading patience.

1. The key issue for all systemically administered drugs (nanotechnological, biological, and chemotherapeutical alike) is the management of biological barriers. Biological targeting is always helpful, under the assumption that a sufficient amount of the bioactive agent successfully navigates the sequential presentation of biological barriers. This is a very stringent and daunting assumption – essentially the success stories of the pharmaceutical world correspond to the largely serendipitous negotiation of a subset of biological barriers, for a given indication, and in a sufficiently large subset of the population. The third-generation nanosystems described above are but a first step toward the development of a general, modular system that can systematically address the biological barriers in their sequential totality. We certainly expect that novel generations, and refinements of nanovector generations 1–3 will be developed, to provide a general solution to the chief problem of biobarriers and biodistribution-by-design.

2. There is no expectation that any single, present or future biobarrier management vectoring system will be applicable to all, or even to most. Personalization of treatment is the focus of great emphasis worldwide – with overwhelming bias toward personalization by way of molecular biology. The vectoring problem, on the other hand, is a combination of biology, physics, engineering, mathematics, and chemistry – with substantial prevalence of the non-biological components of vector design. The evolution of nanotechnologies makes it conceivable, that personalization of treatment will develop as combination of biological methods, and vector design based on non-biological sciences. Foundational elements of mathematics-based methods of rational design of vectors were disclosed in the preceding chapters. The missing link to personalized therapy at this time is the refinement of imaging technologies that can be used to identify the characteristics of the target pathologies – lesion-by-lesion, at any given time, and with the expectation that a time evolution will occur – providing the basis for the synthesis of personalized vectors, which may then carry bioactive agents that may be further personalized for added therapeutic optimization. The word 'personalization' does not begin to capture the substance of this proposition; perhaps 'individualization' is a better term, with the understanding that treatment would individualize at the lesion level (or deeper) in a time-dynamic fashion, rather than at the much coarser level of a individual patient at a given time.

3. Hippocrates left no doubt that safety is first and foremost. The conjoined twin of personalized treatment by biodistribution design is the adverse collateral event by drug concentration at unintended location. Safety – and the regulatory approval pathways that are intended to ensure it – in a more advanced sense that the current observation of macroscopic damage requires an accurate determination of the biodistribution of the administered agents. This would be an ideal objective for all drugs, to be sure, but arguably an impossible one in general. Here is where a challenge turns into an opportunity for the nanomedicine: The ability of nanovectors

to be or carry active agents of therapy, while at the same time be or carry contrast agents that allow the tracking and monitoring of their biodistribution in real time provides the nanopharmaceutical world with a unique advantage. Alas reality at this time smiles much less than this vision would entail: In general it proves very difficult at this time to comprise or conjugate nanovectors with contrast agents or nuclear tracers in a manner that is stable in-vivo. Forming a construct that will not separate in their components once systemically administered is a difficult general conjugation chemistry proposition. Less than total success at it means that what is being tracked may be the label rather than the vector or the drug. The problem becomes combinatorially more complex with increasing numbers of nanovector components. Another facet of the same problem is that frequently the conjugation of labels, contrast agents or tracers dramatically alters biodistribution with respect to the construct that is intended for therapeutic applications. One strategic recommendation that naturally emerges for the immediate future is to prioritize nanovectors that are themselves easily traceable with current radiological imaging modalities.

4. Again in common with all drugs, the development and clinical deployment of nanomedicines would greatly benefit from the development of methods for the determination of toxicity and efficacy indicators from non-invasive or minimally invasive procedures such as blood draws. Serum or plasma proteomics and peptidomics are a promising direction toward this elusive goal. The challenge-turned-into-competitive advantage for nanomedicine is in the ability of nanovectors to carry reporters of location and interaction, which can be released into the blood stream and collected therefrom, to provide indications of toxicity and therapeutic effect.

5. Individualization by rational design of carriers together with biological optimization of drug – both informed by imaging and biological profiling – are a dimension of progress toward optimal therapeutic index for all. Another dimension is in the time dynamics of release: the right drug at the right place at the right time is the final objective. With their exquisite control of size, shape, surface chemistry and overall deign parameters, nanovectors are outstanding candidates for controlled release by implanted (nano)devices or (nano)materials; yet another case of challenge turned opportunity, and of synergistic application of multiple nanotechnologies to form a higher generation nanosystem.

6. The last and perhaps most important challenge ahead – and a wide, extraordinarily exciting prairie of opportunities for rides of discovery – is the generation of novel biological hypotheses. With the higher-order nanotechnologies in development, it is possible to reach subcellular target A with nanoparticle species X at time T, and in the same cell and from the same platform then reach subcellular target B with nanoparticle species Y at subsequent time T′. What therapeutic advantages that may bring, for the possible combinations of A, B, X, Y, T and T′ (and extensions-by-induction of the concept) is absolutely impossible to fathom at this time. There is basically little is any science on it – and of course that is the case, since the technology that permits the validating experiment is in its infancy at this time. The

growth of the infant must be accompanied by the co-development of the biological sciences that frame the missing hypotheses and turn their investigation into science. We respectfully suggest that this would change the course of medicine.

References

1 American Cancer Society website, Statistics 08 (2008) http://www.cancer.org/docroot/STT/stt_0.asp (accessed 21 August 2008).

2 Jemal, A., Siegel, R., Ward, E., Murray, T., Xu, J. and Thun, M.J. (2007) Cancer statistics, 2007. *Cancer Journal for Clinicians*, **57**, 43.

3 Brigger, I. Dubernet, C. and Couvreur, P. (2002) Nanoparticles in cancer therapy and diagnosis. *Advanced Drug Delivery Reviews*, **54**, 631.

4 Jain, R.K. (1987) Transport of molecules in the tumor interstitium: a review. *Cancer Research*, **47**, 3039–3051.

5 Jain, R.K. (1999) Transport of molecules, particles, and cells in solid tumors. *Annual Review of Biomedical Engineering*, **1**, 241.

6 Sanhai, W.R., Sakamoto, J.H., Canady, R. and Ferrari, M. (2008) Seven challenges for nanomedicine. *Nature Nanotechnology*, **3**, 242.

7 Sakamoto, J., Annapragada, A., Decuzzi, P. and Ferrari, M. (2007) Antibiological barrier nanovector technology for cancer applications. *Expert Opinion on Drug Delivery*, **4**, 359.

8 Ferrari, M. (2005) Nanovector therapeutics. *Current Opinion in Chemical Biology*, **9**, 343.

9 Kerbel, R.S. (2008) Tumor angiogenesis. *The New England Journal of Medicine*, **358**, 2039.

10 Hobbs, S.K., Monsky, W.L., Yuan, F., Roberts, W.G., Griffith, L., Torchilin, V.P. and Jain, R.K. (1998) Regulation of transport pathways in tumor vessels: role of tumor type and microenvironment. *Proceedings of the National Academy of Sciences of the United States of America*, **95**, 4607.

11 Yuan, F., Dellian, M., Fukumura, D., Leuning, M., Berk, D.D., Yorchilin, P. and Jain, R.K. (1995) *Cancer Research*, **55**, 3752.

12 Links, M. and Brown, R. (1999) Vascular permeability in a human tumor xenograft: molecular size dependence and cutoff size. *Expert Reviews in Molecular Medicine*, **1**, 1.

13 Krishna, R. and Mayer, L.D. (2000) Multidrug resistance (MDR) in cancer. Mechanisms, reversal using modulators of MDR and the role of MDR modulators in influencing the pharmacokinetics of anticancer drugs. *European Journal of Cancer Science*, **11**, 265.

14 Ferrari, M. (2005) Cancer nanotechnology: opportunities and challenges. *Nature Reviews Cancer*, **5**, 161.

15 Goodman, L.S., Wintrobe, M.M., Dameshek, W., Goodman, M.J., Gilman, A. and McLennan, M.T. (1946) Landmark article 21 September 1946: Nitrogen mustard therapy. Use of methyl-bis(beta-chloroethyl)amine hydrochloride and tris(beta-chloroethyl)amine hydrochloride for Hodgkin's disease, lymphosarcoma, leukemia and certain allied and miscellaneous disorders. *Journal of the American Medical Association*, **105**, 475, Reprinted in, *Journal of the American Medical Association*, **1984**, 251, 2255.

16 Gilman, A. (1963) The initial clinical trial of nitrogen mustard. *American Journal of Surgery*, **105**, 574.

17 Baxevanis, C.N., Perez, S.A. and Papamichail, M. (2008) Combinatorial treatments including vaccines, chemotherapy and monoclonal antibodies for cancer therapy. *Cancer Immunology, Immunotherapy*, Epub ahead of print, DOI 10.1007/s00262-008-0576-4.

18 Zitvogel, L., Apetoh, L., Ghiringhelli, F. and Kroemer, G. (2008) Immunological aspects of cancer chemotherapy. *Nature Reviews Immunology*, **8**, 59.

19 National Cancer Institute website (2008) http://www.cancer.gov/ (accessed 12 October 2008).

20 Folkman, J. (1971) Tumor angiogenesis: therapeutic implications. *The New England Journal of Medicine*, **285**, 1182.

21 Folkman, J. (2007) Angiogenesis: an organizing principle for drug discovery? *Nature Reviews Drug Discovery*, **6**, 273.

22 Ferrara, N., Hillan, K.J., Gerber, H.P. and Novotny, W. (2004) Discovery and development of bevacizumab, an anti-VEGF antibody for treating cancer. *Nature Reviews Drug Discovery*, **3**, 391.

23 Hurwitz, H., Fehrenbacher, L., Novotny, W. *et al.* (2004) Bevacizumab plus irinotecan, fluorouracil, and leucovorin for metastatic colorectal cancer. *The New England Journal of Medicine*, **350**, 2335.

24 Sandler, A., Gray, R., Perry, M.C. *et al.* (2006) Paclitaxel-carboplatin alone or with bevacizumab for non-small-cell lung cancer. *The New England Journal of Medicine*, **355**, 2542.

25 Faivre, S., Demetri, G., Sargent, W. and Raymond, E. (2007) Molecular basis for sunitinib efficacy and future clinical development. *Nature Reviews Drug Discovery*, **6**, 734.

26 Motzer, R.J., Michaelson, M.D., Redman, B.G. *et al.* (2006) Activity of SU11248, a multitargeted inhibitor of vascular endothelial growth factor receptor and platelet-derived growth factor receptor, in patients with metastatic renal cell carcinoma. *Journal of Clinical Oncology*, **24**, 16.

27 Escudier, B., Eisen, T., Stadler, W.M. *et al.* (2007) Sorafenib in advanced clear-cell renal-cell carcinoma. *The New England Journal of Medicine*, **356**, 125.

28 Llovet, J., Ricci, S., Mazzaferro, V. *et al.* (2008) Sorafenib in advanced hepatocellular carcinoma. *The New England Journal of Medicine*, **359**, 378.

29 Berenson, A. (2006) (15 February) New York Times.

30 Eskens, F.A. and Verweij, J. (2006) The clinical toxicity profile of vascular endothelial growth factor (VEGF) and vascular endothelial growth factor receptor (VEGFR) targeting angiogenesis inhibitors; a review. *European Journal of Cancer (Oxford, England: 1990)*, **42**, 3127.

31 Verheul, H.M. and Pinedo, H.M. (2007) Possible molecular mechanisms involved in the toxicity of angiogenesis inhibition. *Nature Reviews Cancer*, **7**, 475.

32 Jain, R.K., Duda, D.G., Clark, J.W. and Loeffler, J.S. (2006) Lessons from phase III clinical trials on anti-VEGF therapy for cancer. *Nature Clinical Practice Oncology*, **3**, 24.

33 Kerbel, R.S. (2006) Antiangiogenic therapy: a universal chemosensitization strategy for cancer? *Science*, **312**, 1171.

34 Bergers, G. and Benjamin, L.E. (2003) Tumorigenesis and the angiogenic switch. *Nature Reviews Cancer*, **3**, 401.

35 Pressman, D. and Korngold, L. (1953) The in vivo localization of anti-Wagner-osteogenic-sarcoma antibodies. *Cancer*, **6**, 619.

36 Burnet, F.M. (1967) Immunological aspects of malignant disease. *Lancet*, **1**, 1171.

37 Krzeslak, A., Pomorski, L., Gaj, Z. and Lipinska, A. (2003) Differences in glycosylation of intracellular proteins between benign and malignant thyroid neoplasms. *Cancer Letters*, **196**, 101.

38 Mehl, A.M., Fischer, N., Rowe, M. *et al.* (1998) Isolation and analysis of two strongly transforming isoforms of the Epstein-Barr virus (EBV)-encoded latent membrane protein-1 (LMP1) from a single Hodgkin's lymphoma. *International Journal of Cancer*, **76**, 194.

39 Clark, S.S., McLaughlin, J., Timmons, M. *et al.* (1988) Expression of a distinctive BCR-ABL oncogene in Ph1-positive acute lymphocytic leukemia (ALL). *Science*, **239**, 775.

40 Slamon, D.J., Godolphin, W., Jones, L.A. *et al.* (1989) Studies of the HER2/neu proto-oncogene in human breast and ovarian cancer. *Science*, **244**, 707.

41 Slamon, D.J., Clark, G.M., Wong, S.G., Levin, W.J., Ullrich, A. and McGuire, W.L. (1987) Human breast cancer: Correlation of relapse and survival with amplification of the HER-2/neu oncogene. *Science*, **235**, 177.

42 Dillman, R.O. (2001) Monoclonal antibody therapy for lymphoma: an update. *Cancer Practice*, **9**, 71.

43 Countouriotis, A., Moore, T.B. and Sakamoto, K.M. (2002) Cell surface antigen and molecular targeting in the treatment of hematologic malignancies. *Stem Cells (Dayton, Ohio)*, **20**, 215.

44 Leget, G.A. and Czuczman, M.S. (1998) Use of Rituximab, the new FDA-approved antibody. *Current Opinion in Oncology*, **10**, 548.

45 Khawli, L.A., Miller, G.K. and Epstein, A.L. (1994) Effect of seven new vasoactive immunoconjugates on the enhancement of monoclonal antibody uptake in tumors. *Cancer*, **73**, 824.

46 Goldenberg, D.M. (1988) Targeting of cancer with radiolabeled antibodies. Prospects for imaging and therapy. *Archives of Pathology and Laboratory Medicine*, **112**, 580.

47 Epenetos, A.A., Snook, D., Durbin, H., Johnson, P.M. and Taylor-Papadimitriou, J. (1986) Limitations of radiolabeled monoclonal antibodies for localization of human neoplasms. *Cancer Research*, **46**, 3183.

48 Cobleigh, M.A., Vogel, C.L., Tripathy, D. *et al.* (1999) Multinational study of the efficacy and safety of humanized anti-HER2 monoclonal antibody in women who have HER2-overexpressing metastatic breast cancer that has progressed after chemotherapy for metastatic disease. *Journal of Clinical Oncology*, **17**, 2639.

49 Baselga, J., Tripathy, D., Mendelsohn, J. *et al.* (1996) Phase II study of weekly intravenous recombinant humanized anti-p185HER2 monoclonal antibody in patients with HER2/neu-overexpressing metastatic breast cancer. *Journal of Clinical Oncology*, **14**, 737.

50 Silver, D.A., Pellicer, I., Fair, W.R., Heston, W.D. and Cordon-Cardo, C. (1997) Prostate-specific membrane antigen expression in normal and malignant human tissues. *Clinical Cancer Research*, **3**, 81.

51 Sweat, S.D., Pacelli, A., Murphy, G.P. and Bostwick, D.G. (1998) Prostate-specific membrane antigen expression is greatest in prostate adenocarcinoma and lymph node metastases. *Urology*, **52**, 637.

52 Wright, G.L., Grob, B.M., Haley, C. *et al.* (1996) Upregulation of prostate-specific membrane antigen after androgen-deprivation therapy. *Urology*, **48**, 326.

53 Murphy, G.P., Elgamal, A.A., Su, S.L., Bostwick, D.G. and Holmes, E.H. (1998) Current evaluation of the tissue localization and diagnostic utility of prostate specific membrane antigen. *Cancer*, **83**, 2259.

54 Smith-Jones, P.M., Vallabhajosula, S., Navarro, V., Bastidas, D., Goldsmith, S.J. and Bander, N.H. (2003) Radiolabeled monoclonal antibodies specific to the extracellular domain of prostate-specific membrane antigen: preclinical studies in nude mice bearing LNCaP human prostate tumor. *Journal of Nuclear Medicine*, **44**, 610.

55 McDevitt, M.R., Barendswaard, E., Ma, D. *et al.* (2000) An alpha-particle emitting antibody ([213Bi]J591) for radioimmunotherapy of prostate cancer. *Cancer Research*, **60**, 6095.

56 Bander, N.H., Nanus, D.M., Milowsky, M.I., Kostakoglu, L., Vallabahajosula, S. and Goldsmith, S.J. (2003) Targeted systemic therapy of prostate cancer with a monoclonal antibody to prostate-specific membrane antigen. *Seminars in Oncology*, **30**, 667.

57 Feynman, R. (1960) There's plenty of room at the bottom. *Engineering and Science*, **23**, 22.

58 National Nanotechnology Initiative program . (2008) http://www.nano.gov/NNI_FY09_budget_summary.pdf (accessed 12 October 2008).

59 Theis, T., Parr, D., Binks, P., Ying, J., Drexler, K.E., Schepers, E., Mullis, K., Bai, C., Boland, J.J., Langer, R., Dobson, P., Rao, C.N. and Ferrari, M. (2006) nan' o.tech.nol' o.gy n. *Nature Nanotechnology*, **1**, 8.

60 Heath, J.R. and Davis, M.E. (2008) Nanotechnology and cancer. *Annual Review of Medicine*, **59**, 251.

61 Nie, S., Kim, G.J., Xing, Y. and Simons, J.W. (2007) Nanotechnology applications in cancer. *Annual Review of Biomedical Engineering*, **9**, 257.

62 Riehemann, K., Schneider, S.W., Luger, T.A., Godin, B., Ferrari, M. and Fuchs, H. (2008) Nanomedicine - Developments and perspectives. *Angewandte Chemie - International Edition*, in press, DOI : 10.1002/ange. 200802585.

63 Harris, J.M. and Chess, R.B. (2003) Effect of pegylation on pharmaceuticals. *Nature Reviews Drug Discovery*, **2**, 214.

64 Brannon-Peppas, L. and Blanchette, J.O. (2004) Nanoparticle and targeted systems for cancer therapy. *Advanced Drug Delivery Reviews*, **56**, 1649.

65 Torchilin, V.P. (2007) Targeted pharmaceutical nanocarriers for cancer therapy and imaging. *The APS Journal*, **9**, E128.

66 Saul, J.M., Annapragada, A.V. and Bellamkonda, R.V. (2006) A dual-ligand approach for enhancing targeting selectivity of therapeutic nanocarriers. *Journal of Controlled Release*, **114**, 277.

67 Yang, X., Wang, H., Beasley, D.W. *et al.* (2006) Selection of thioaptamers for diagnostics and therapeutics. *Annals of the New York Academy of Sciences*, **116**, 1082.

68 Souza, G.R., Christianson, D.R., Staquicini, F.I. *et al.* (2006) Networks of gold nanoparticles and bacteriophage as biological sensors and cell-targeting agents. *Proceedings of the National Academy of Sciences of the United States of America*, **103**, 1215.

69 Maeda, H., Wu, J., Sawa, T., Matsumura, Y. and Hori, K. (2000) Tumor vascular permeability and the EPR effect in macromolecular therapeutics: a review. *Journal of Controlled Release*, **65**, 271.

70 Duncan, R. (2006) Polymer conjugates as anticancer nanomedicines. *Nature Reviews Cancer*, **6**, 688.

71 Torchilin, V.P. (2005) Recent advances with liposomes as pharmaceutical carriers. *Nature Reviews Drug Discovery*, **4**, 145.

72 Romberg, B., Hennink, W.E. and Storm, G. (2008) Sheddable coatings for long-circulating nanoparticles. *Pharmaceutical Research*, **25**, 55.

73 Gabizon, A. and Martin, F. (1997) Polyethylene glycol-coated (pegylated) liposomal doxorubicin. Rationale for use in solid tumours. *Drugs*, **54**, 15.

74 Bangham, A.D., Standish, M.M. and Watkins, J.C. (1965) The action of steroids and streptolysin S on the permeability of phospholipid structures to cations. *Journal of Molecular Biology*, **13**, 238.

75 Drummond, D.C., Meyer, O., Hong, K., Kirpotin, D.B. and Papahadjopoulos, D. (1999) Optimizing liposomes for delivery of chemotherapeutic agents to solid tumors. *Pharmacological Reviews*, **51**, 691.

76 Hofheinz, R.D., Gnad-Vogt, S.U., Beyer, U. and Hochhaus, A. (2005) Liposomal encapsulated anti-cancer drugs. *Anti-Cancer Drugs*, **16**, 691.

77 Parveen, S. and Sahoo, S.K. (2006) Nanomedicine: clinical applications of polyethylene glycol conjugated proteins and drugs. *Clinical Pharmacokinetics*, **45**, 965.

78 Zhang, L., Gu, F.X., Chan, J.M., Wang, A.Z., Langer, R.S. and Farokhzad, O.C. (2007) Nanoparticles in medicine: therapeutic applications and developments. *Clinical Pharmacology and Therapeutics*, **83**, 761.

79 Peer, D., Karp, J.M., Hong, S.Y., Farokhzad, O., Margalit, R. and Langer, R.

(2007) Nanocarriers as an emerging platform for cancer therapy. *Nature Nanotechnology*, **2**, 751.

80 Gradishar, W.J., Tjulandin, S., Davidson, N., Shaw, H., Desai, N., Bhar, P., Hawkins, M. and O'Shaughnessy, J. (2005) Phase III trial of nanoparticle albumin-bound paclitaxel compared with polyethylated castor oil-based paclitaxel in women with breast cancer. *Journal of Clinical Oncology*, **23**, 7794.

81 Ringsdorf H. (1975) Structure and properties of pharmacologically active polymers. *Journal of Polymer Science Polymer Symposium*, **51**, 135.

82 Vasey, P.A., Kaye, S.B., Morrison, R., Twelves, C., Wilson, P., Duncan, R., Thomson, A.H., Murray, L.S., Hilditch, T.E. and Murray, T. (1999) Phase I clinical and pharmacokinetic study of PK1 [*N*-(2-hydroxypropyl)methacrylamide copolymer doxorubicin]: first member of a new class of chemotherapeutic agents-drug-polymer conjugates. Cancer Research Campaign Phase I/II Committee. *Clinical Cancer Research*, **5**, 83.

83 Allen, T.M. (2002) Ligand-targeted therapeutics in anticancer therapy. *Nature Reviews Drug Discovery*, **2**, 750.

84 Juweid, M., Neumann, R., Paik, C., Perez-Bacete, M.J., Sato, J., van Osdol, W. and Weinstein, J.N. (1992) Micropharmacology of monoclonal antibodies in solid tumors: direct experimental evidence for a binding site barrier. *Cancer Research*, **52**, 5144.

85 Banerjee, R.K., van Osdol, W., Bungay, P.M., Sung, C. and Dedrick, R.L. (2001) Finite element model of antibody penetration in a prevascular tumor nodule embedded in normal tissue. *Journal of Controlled Release*, **74**, 193.

86 Adams, G.P., Schier, R., McCall, A.M., Simmons, H.H., Horak, E.M., Alpaugh, R.K., Marks, J.D. and Weiner, L.M. (2001) High affinity restricts the localization and tumor penetration of single-chain fv antibody molecules. *Cancer Research*, **61**, 4750.

87 Goren, D., Horowitz, A.T., Zalipsky, S., Woodle, M.C., Yarden, Y. and Gabizon, A. (1996) Targeting of stealth liposomes to erbB-2 (Her/2) receptor: in vitro and in vivo studies. *British Journal of Cancer*, **74**, 1749.

88 Langer, R. (1998) Drug delivery and targeting. *Nature*, **392**, 5.

89 Kang, J., Lee, M.S., Copland, J.A., III, Luxon, B.A. and Gorenstein, D.G. (2008) Combinatorial selection of a single stranded DNA thioaptamer targeting TGF-beta1 protein. *Bioorganic and Medicinal Chemistry Letters*, **18**, 1835.

90 Hajitou, A., Trepel, M., Lilley, C.E. *et al.* (2006) A hybrid vector for ligand-directed tumor targeting and molecular imaging. *Cell*, **125**, 385.

91 Qian, X., Peng, X.H., Ansari, D.O., Yin-Goen, Q., Chen, G.Z., Shin, D.M., Yang, L., Young, A.N., Wang, M.D. and Niel, S. (2008) In vivo tumor targeting and spectroscopic detection with surface-enhanced Raman nanoparticle tags. *Nature Biotechnology*, **26**, 83.

92 Duncan, R. (2003) The dawning era of polymer therapeutics. *Nature Reviews Drug Discovery*, **2**, 347.

93 Hirsch, L.R., Stafford, R.J., Bankson, J.A., Sershen, S.R., Rivera, B., Price, R.E., Hazle, J.D., Halas, N.J. and West, J.L. (2003) Nanoshell-mediated near-infrared thermal therapy of tumors under magnetic resonance guidance. *Proceedings of the National Academy of Sciences of the United States of America*, **100**, 13549.

94 Douziech-Eyrolles, L., Marchais, H., Herve, K., Munnier, E., Souce, M., Linassier, C., Dubois, P. and Chourpa, I. (2007) Nanovectors for anticancer agents based on superparamagnetic iron oxide nanoparticles. *International Journal of Nanomedicine*, **2**, 541.

95 Schroeder, A., Avnir, Y., Weisman, S., Najajreh, Y., Gabizon, A., Talmon, Y., Kost, J. and Barenholz, Y. (2007) Controlling liposomal drug release with low frequency ultrasound: mechanism and feasibility. *Langmuir*, **23**, 4019.

96 Monsky, W.L., Kruskal, J.B., Lukyanov, A.N., Girnun, G.D., Ahmed, M., Gazelle, G.S., Huertas, J.C., Stuart, K.E., Torchilin, V.P. and Goldberg, S.N. (2002) Radio-frequency ablation increases intratumoral liposomal doxorubicin accumulation in a rat breast tumor model. *Radiology*, **224**, 823.

97 Farokhzad, O.C., Cheng, J., Teply, B.A., Sherifi, I., Jon, S., Kantoff, P.W., Richie, J.P. and Langer, R. (2006) Targeted nanoparticle-aptamer bioconjugates for cancer chemotherapy in vivo. *Proceedings of the National Academy of Sciences of the United States of America*, **103**, 6315.

98 Farokhzad, O.C., Karp, J.M. and Langer, R. (2006) Nanoparticle-aptamer bioconjugates for cancer targeting. *Expert Opinion on Drug Delivery*, **3**, 311.

99 Northfelt, D.W., Martin, F.J., Working, P., Volberding, P.A., Russell, J., Newman, M., Amantea, M.A. and Kaplan, L.D. (1996) Doxorubicin encapsulated in liposomes containing surface-bound polyethylene glycol: pharmacokinetics, tumor localization, and safety in patients with AIDS-related Kaposi's sarcoma. *Journal of Clinical Pharmacology*, **36**, 55.

100 Jang, S.H., Wientjes, M.G., Lu, D. and Au, J.L. (2003) Drug delivery and transport to solid tumors. *Pharmaceutical Research*, **20**, 1337.

101 Lankelma, J., Dekker, H., Luque, F.R., Luykx, S., Hoekman, K., van der Valk, P., van Diest, P.J. and Pinedo, H.M. (1999) Doxorubicin gradients in human breast cancer. *Clinical Cancer Research*, **5**, 1703.

102 Tannock, I.F., Lee, C.M., Tunggal, J.K., Cowan, D.S. and Egorin, M.J. (2002) Limited penetration of anticancer drugs through tumor tissue: a potential cause of resistance of solid tumors to chemotherapy. *Clinical Cancer Research*, **8**, 878.

103 Klibanov, A.L., Maruyama, K., Beckerleg, A.M., Torchilin, V.P., and Huang, L. (1991) Activity of amphipathic poly (ethylene glycol) 5000 to prolong the circulation time of liposomes depends on the liposome size and is unfavorable for immunoliposome binding to target. *Biochimica et Biophysica Acta*, **1062**, 142.

104 Souza, G.R., Yonel-Gumruk, E., Fan, D. *et al.* (2008) Bottom-up assembly of hydrogels from bacteriophage and Au nanoparticles: the effect of cis- and trans-acting factors. *PLoS ONE*, **3**, e2242.

105 Sengupta, S., Eavarone, D., Capila, I., Zhao, G., Watson, N., Kiziltepe, T. and Sasisekharan, R. (2005) Temporal targeting of tumour cells and neovasculature with a nanoscale delivery system. *Nature*, **436**, 568.

106 Tasciotti, E., Liu, X., Bhavane, R., Plant, K., Leonard, A.D., Price, B.K., Cheng, M.M., Decuzzi, P., Tour, J.M., Robertson, F.M., and Ferrari, M. (2008) Mesoporous silicon particles as a multistage delivery system for imaging and therapeutic applications. *Nature Nanotechnology*, **3**, 151.

107 Ferrari, M. (2008) Nanogeometry: beyond drug delivery. *Nature Nanotechnology*, **3**, 131.

108 Decuzzi, P., Pasqualini, R., Arap, W. and Ferrari, M. (2008) Intravascular delivery of particulate systems: Does geometry really matter? *Pharmaceutical Research*, 20 August, Epub ahead of print.

109 Goldsmith, H.L. and Spain, S. (1984) Margination of leukocytes in blood flow through small tubes. *Microvascular Research*, **27**, 204.

110 Fahraeus, R. and Lindqvist, T. (1931) The viscosity of the blood in narrow capillary tubes. *The American Journal of Physiology*, **96**, 562.

111 Sharan, M. and Popel, A.S. (2001) A two-phase model for flow of blood in narrow tubes with increased effective viscosity near the wall. *Biorheology*, **38**, 415.

112 Goldman, A.J., Cox, R.G. and Brenner, H. (1967) Slow viscous motion of a sphere parallel to a plane wall. II. Couette flow. *Chemical Engineering Science*, **22**, 653.

113 Decuzzi, P., Lee, S., Bhushan, B. and Ferrari, M. (2005) A theoretical model for the margination of particles within blood vessels. *Annals of Biomedical Engineering*, **33**, 179.

114 Decuzzi, P. and Ferrari, M. (2007) The role of specific and non-specific interactions in receptor-mediated endocytosis of nanoparticles. *Biomaterials*, **28**, 2915–2922.

115 Gavze, E. and Shapiro, M. (1998) Motion of inertial spheroidal particles in a shear flow near a solid wall with special application to aerosol transport in microgravity. *Journal of Fluid Mechanics*, **371**, 59.

116 Filipovic, N., Stojanovic, B., Kojic, N. and Kojic, M. (2008) *Computer Modeling in Bioengineering -Theoretical Background, Examples and Software*. John Wiley & Sons, Chichester, UK.

117 Gentile, F., Chiappini, C., Fine, D., Bhavane, R.C., Peluccio, M.S., Ming-Cheng Cheng, M., Liu, X., Ferrari, M. and Decuzzi, P. (2008) The effect of shape on the margination dynamics of non-neutrally buoyant particles in two-dimensional shear flows. *Journal of Biomechanics*, **41**, 2312.

118 Decuzzi, P. and Ferrari, M. (2006) The adhesive strength of non-spherical particles mediated by specific interactions. *Biomaterials*, **27**, 5307.

119 Gao, H., Shi, W. and Freund, L.B. (2005) Mechanics of receptor-mediated endocytosis. *Proceedings of the National Academy of Sciences of the United States of America*, **102**, 9469.

120 Decuzzi, P. and Ferrari, M. (2008) The receptor-mediated endocytosis of nonspherical particles. *Biophysical Journal*, **94**, 3790.

121 Decuzzi, P. and Ferrari, M. (2008) Design maps for nanoparticles targeting the diseased microvasculature. *Biomaterials*, **29**, 377.

122 The National Cancer Institute, the Nanotechnology Characterization Laboratory (NCL) (2008) http://ncl.cancer.gov (accessed 12 October 2008).

123 Hall, J.B., Dobrovolskaia, M.A., Patri, A.K. and McNeil, S.E. (2007) Characterization of nanoparticles for therapeutics. *Nanomedicine*, **2**, 789.

124 Pelkmans, L., Kartenbeck, J. and Helenius, A. (2001) Caveolar endocytosis of simian virus 40 reveals a new two-step vesicular-transport pathway to the ER. *Nature Cell Biology*, **3**, 473.

125 Serda, R.E., Adolphi, N.L., Bisoffi, M. and Sillerud, L.O. (2007) Targeting and cellular trafficking of magnetic nanoparticles for prostate cancer imaging. *Molecular Imaging*, **6**, 277.

126 Serda, R.E. Gu J, J., Bhavane, R.C., Liu, W., Chiappini, C., Robertson, F., Decuzzi, P. and Ferrari, M. (2008) Microengineering delivery vectors to target inflamed vascular endothelium and reduce RES uptake. (submitted).

127 Tjelle, T.E., Lovdal, T. and Berg, T. (2000) Phagosome dynamics and function. *BioEssays: News and Reviews in Molecular, Cellular and Developmental Biology*, **22**, 255.

128 Davda, J. and Labhasetwar, V. (2002) Characterization of nanoparticle uptake by endothelial cells. *International Journal of Pharmaceutics*, **233**, 51.

129 Dobrovolskaia, M.A. and McNeil, S.E. (2007) Immunological properties of engineered nanomaterials. *Nature Nanotechnology*, **2**, 469.

130 Morimoto, Y. and Tanaka, I. (2008) Effects of nanoparticles on humans. *Sangyo Eiseigaku Zasshi*, **50**, 37.

131 Chen, J., Tan, M., Nemmar, A., Song, W., Dong, M., Zhang, G. and Li, Y. (2006) Quantification of extrapulmonary translocation of intratracheal-instilled particles in vivo in rats: effect of lipopolysaccharide. *Toxicology*, **222**, 195.

Part Three:
Imaging and Probing the Inner World of Cells

Nanotechnology, Volume 5: Nanomedicine. Edited by Viola Vogel

ISBN: 978-3-527-31736-3

4
Electron Cryomicroscopy of Molecular Nanomachines and Cells

Matthew L. Baker, Michael P. Marsh, and Wah Chiu

4.1
Introduction

Genome-sequencing projects continue to provide complete genetic descriptions of an ever-increasing number of model organisms [1–4]. Based on our current knowledge, it has been estimated that life depends on 200–300 core biological processes [5]. Individual gene products rarely function independently; to the contrary, large multicomponent protein assemblies are more often responsible for complex cellular functions. These assemblies are often dynamic and, in many cases, transient. As such, these assemblies are often termed 'molecular nanomachines', capable of carrying out a wide range of functions through often specific and highly intricate interactions [6–10].

Equally as complex as the nanomachines themselves, the individual components can adopt a wide variety of morphologies, functions and interactions. In addressing these complexities, structural genomics seeks to provide a description of all protein folds, where a fold is defined as the three-dimensional (3-D) structure of protein that relates the spatial arrangements and connectivity of secondary structure elements, such as α-helices and β-sheets. As such, the protein fold represents the basic 'building block' of much larger and more complex assemblies that carry out biochemical and cellular processes. To date, more than 51 000 individual protein structures are known [11]; however, far fewer unique folds are recognized.

It is generally accepted that the primary structure of a protein – the amino acid sequence – dictates its 3-D structure, or fold. As such, proteins with a similar primary sequence likely assume similar folds. However, the converse is not necessarily true; proteins with vastly dissimilar sequences can assume similar folds. Nonetheless, the protein fold is ultimately responsible for the necessary 3-D environment for protein function and intermolecular and intramolecular interactions. The description of these folds – and, in particular, their interactions within cellular complexes – is therefore paramount to the understanding of all molecular nanomachines and biological processes.

Nanotechnology, Volume 5: Nanomedicine. Edited by Viola Vogel

ISBN: 978-3-527-31736-3

Electron cryomicroscopy (cryo-EM) is an emerging methodology that is particularly well suited for studying molecular nanomachines at near-native or chemically defined conditions. Cryo-EM can be used to study nanomachines of various sizes, shapes and symmetries, including two-dimensional (2-D) arrays, helical arrays and single particles [12]. With recent advances, cryo-EM can now not only reveal the gross morphology of these nanomachines but also provide highly detailed models of protein folds approaching atomic resolutions [13–17]. In this chapter, we will present the methodology of single-particle cryo-EM, as well as its potential biomedical applications and future prospects.

Complementary to structural studies of nanomachines with cryo-EM, the application of cryo-tomography (cryo-ET) can depict the locations and low-resolution structures of nanomachines in a 3-D cellular environment. The power of cryo-ET comes from its unique ability to observe directly biological nanomachines *in situ*, without the need for isolation and purification. This approach has the potential to capture the structural diversity of nanomachines in their milieu of interacting partners and surrounding cellular context.

4.2
Structure Determination of Nanomachines and Cells

Figure 4.1 shows a series of typical steps in imaging nanomachines using cryo-EM or cryo-ET. The first steps are common to both techniques; biochemical preparation, specimen preservation via rapid freezing; and imaging the frozen, hydrated specimens by low-dose electron microscopy. Although the subsequent steps differ for the two techniques, they both include image processing to generate a 3-D reconstruction, interpreting the 3-D volume density together with other biological data and archiving the density maps and models. In this chapter we will not address how to perform each of the aforementioned mentioned steps, as numerous technical reports and books exist that describe them in detail [12, 18]. Rather, we will briefly summarize these steps and their applications to a few examples of molecular nanomachines and cells.

4.2.1
Experimental Procedures in Cryo-EM and Cryo-ET

In principle, most of these steps are rather straightforward, and the length of time taken to start from a highly purified nanomachine to obtaining a complete structure can range from a few days to months. However, as with any experimental method, various hurdles may be encountered that require further optimization before a reliable structure can be determined.

4.2.1.1 Specimen Preparation for Nanomachines and Cells

Specimen preparation is a critical step in single-particle cryo-EM, which necessarily requires high conformational uniformity while preserving functional activities. In X-ray crystallography, crystallization is a selective process through which only

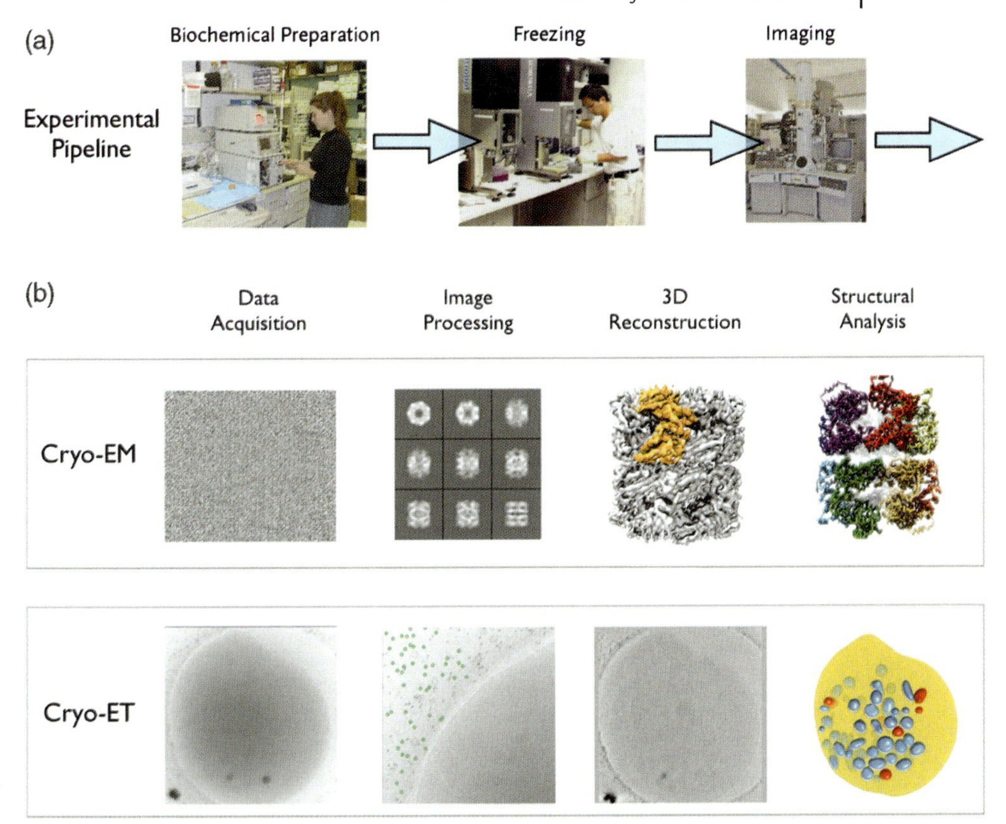

Figure 4.1 The experimental pipeline for cryo-imaging experiments. (a) The first steps of the experiment – specimen preparation, specimen freezing and microcopy – are common to both cryo-EM and cryo-ET; (b) The subsequent steps diverge and differ between the two types of experiments. For cryo-EM, these steps are illustrated with examples of the biological nanomachine GroEL. For cryo-ET, they are illustrated with a cell, the human platelet.

molecules of the same conformations nucleate and crystallize to form a diffracting object. In addition to chemical purification, crystallization also forces the molecules into specific, uniform spatial organization such that diffraction data can be averaged from over billions of molecules in identical conformations. However, cryo-EM experiments image one set of molecules at a time, regardless of their conformations, and thus possibly represents an ensemble of conformations of the molecules in a single micrograph. In order to obtain the highest possible resolution structures by cryo-EM, it is still necessary to computationally average from several hundreds to tens of thousands of a conformationally homogeneous set of particle images recorded in multiple micrographs. Nevertheless, computational methods are being developed to sort out images of particles with different conformations.

The nature of cryo-ET experiments differs substantially from single-particle cryo-EM experiments, and the resolution of the reconstructions is much lower. In contrast to the cryo-EM approach, where images of many conformationally uniform particles are merged to yield a 3-D model, cryo-ET merges many images of the same specimen target, collected at different angles. With this approach, a reconstruction can be computed from the images of a single cell or nanomachine, and so conformational uniformity is not an issue in the most general case. The merging of whole cells or organelles such as single particles is not a reasonable goal, as uniformity can never realistically be expected; however, some subcelluar structures may be sufficiently uniform in conformation to warrant merging and averaging from the 3-D tomogram. Below, we consider such an example when discussing the bacterial flagella motor.

4.2.1.2 **Cryo-Specimen Preservation**

Following biochemical isolation and purification, the first step in a cryo-imaging experiment is to embed a biochemically purified nanomachine or cell under well-defined chemical conditions in ice on a cryo-EM grid [19]. This freezing process is extremely quick in order to prevent the formation of crystalline ice, and thus produces a matrix of vitreous ice in which the water molecules remain relatively unordered. The spread of the nanomachines on the grid should be neither too crowded, such that they would contact each other, nor too dilute as to only have a few nanomachines recorded in each micrograph. For cryo-EM, it is preferable to have the nanomachines situated in random orientations to allow sufficient angular sampling needed for the subsequent 3-D reconstruction procedure. The ideal thickness of the embedding ice is slightly greater than the size of the nanomachine or cell. Excessive ice thickness is detrimental because it diminishes the signal-to-noise ratio (SNR) of the images that can be acquired. Ice that is too shallow can be a problem for cryo-ET experiments, whereby flattening of the specimen can occur. The capillary forces of the solvent, in the fluid phase just prior to vitrification, can compress the sample; this has been reported in vesicles [20] as well as real cells where a 1 μm-thick cell can be reduced to 600 nm [21, 22].

Some specimens are very easy to prepare, while others are more difficult, which necessarily means optimization of the specimen preparation is a trial-and-error process. In general, this step – the preparation of the frozen, hydrated specimens, preserved in vitreous ice with an optimal ice thickness – is often a bottleneck. Analogous to the crystallization process in X-ray crystallography, there is no foolproof recipe for optimal specimen preservation. However, a computer-driven freezing apparatus has made this step more reproducible and tractable in finding optimal conditions for freezing a given specimen [23].

In principle, the frozen, hydrated specimens represent native conformations as they are maintained in an aqueous buffer. Fixation of the nanomachines in a specific orientation can occur prior to freezing. Specimen freezing can also be coordinated with a time-resolved chemical mixing reaction; prototype apparatuses have been built to perform such a time-resolved reaction [24, 25]. It is conceivable that a more sophisticated instrument can be built to allow all sorts of chemical reactions, including those that can be light-activated. Such an approach would allow cryo-EM to follow the structure variations in a chemical process with a temporal resolution of milliseconds [25].

4.2.1.3 Low-Dose Imaging

Once the sample has been frozen, the entire cryo-EM grid can be inserted into the electron cryo-microscope and imaged with electrons ranging from 100 to 400 keV. Electrons at these energies will damage the molecules during imaging [12]. Therefore, low-dose imaging is necessary to minimize the damage to the specimen before the image is recorded. To maintain the frozen, hydrated specimen in vitreous ice inside the electron microscope vacuum, the specimen is kept at low temperature, typically at or below liquid nitrogen temperature. If the specimen temperature is higher than −160 °C, the vitreous ice undergoes a phase transition to crystalline ice and denatures the nanomachines [19].

From a radiation damage perspective, the advantages and disadvantages of keeping biological specimens at different low temperatures have been studied [26–28]. High-quality images have been obtained using liquid helium temperature, and have resulted in high-resolution structures of 2-D crystals [29, 30], helical arrays [31, 32] and single particles [13, 14] where protein backbone traces were feasible. Imaging specimens at liquid nitrogen temperature has also been used successfully for the similar high-resolution structure determination of a broad spectrum of specimens [15, 17, 33, 34]. In the case of cellular cryo-ET, it has been suggested that liquid nitrogen is a preferred temperature [27, 35] because of a significant loss of contrast at liquid helium temperature [36].

4.2.1.4 Image Acquisition

Data collection differs significantly for cryo-EM and cryo-ET. For cryo-EM, images of a field containing multiple, randomly oriented specimens are recorded. Individual particles are recorded only once because of the radiation damage constraints for obtaining the highest possible resolution information. For cryo-ET, a series of images of the same specimen is acquired as the stage is iteratively tilted over an interval spanning approximately 130°. A typical tilt-series might include one image collected every 2° between −65° and +65°. The resolution of the tomographic data is much lower because of the effects of cumulative radiation damage to the specimen throughout the data collection.

For cryo-EM and cryo-ET experiments, images have been traditionally collected on photographic film and subsequently digitized using a high-resolution film scanner. Recent advances in CCD cameras for electron microscopes have made direct digital recording feasible [37–39]. With a modern electron microscope equipped with specialized software for low-dose imaging, data collection is relatively simple and can be either partially or fully automated [40–43].

4.2.2 Computational Procedures in Cryo-EM and Cryo-ET

The recorded image of a nanomachine is essentially a projection (2-D) of its mass density along the path of the irradiating electrons. In order to retrieve its 3-D structure, the particle must be sampled in different angular views [44]. For cryo-ET, this sampling is carried out systematically whereby each image in the tilt-series constitutes a separate angular view. With cryo-EM, the varied orientation of particles in the ice naturally provides an angular distribution of views. The number of views required for the reconstruction is proportional to the diameter of the particle and is inversely proportional to the desired resolution [45]. Because of the noisy nature of the image and the uneven angular distribution (in the case of cryo-EM) of the views, the actual number of the particles used to calculate a reconstruction at a certain resolution is much higher than the theoretical minimum [46, 47]. Ideally, the particles embedded in the vitreous ice are oriented randomly. However in some cases, the particles tend to assume a preferred orientation with respect to the surface of the embedding ice. This can often be overcome by varying the buffer or solvent by adding a small amount of detergent.

4.2.2.1 **Image Processing and Reconstruction**

During the image-processing phase, individual specimen images are aligned with respect to each other and then combined to form a 3-D density map [18, 48, 49]. For cryo-EM studies, the image processing is an iterative process. Several image-processing packages are available for single-particle cryo-EM, such as EMAN [50], SPIDER [51], IMAGIC [52] and Frealign [53]; these are multi-step procedures that can generally be broken into the following steps: (i) identify the locations of each particle; (ii) determine and correct the contrast transfer function and damping function for the particle images; (iii) classify the images according to their conformational identity and orientation parameters; and (iv) average the particle images in each classes and 3-D reconstruct to produce the final 3-D map. The specimen classification, particle averaging and reconstruction of the density map are iterated using the previous iteration as a reference. Iteration of these steps continues until no improvement in the 3-D density maps is made over the previous cycle of refinement. The final resolution of the map is typically assessed by a parameter referred to as Fourier shell correlation (FSC), in which two maps derived from two independent sets of image data are compared [54]. The FSC essentially measures a similarity and reproducibility of two structures in Fourier space; the final resolution is often determined using the 0.5 criterion.

Alignment and reconstruction differ for cryo-ET experiments. For cryo-ET, the nominal angular assignment is known for each image because the tilt-angle of the stage was recorded for each image of the tilt-series. Higher-precision angular assignments must be determined for reconstruction. Alignment processing is frequently simplified by including gold particles in the specimen; these particles have a strong contrast, even under low-dose imaging conditions, and serve as landmarks for registering images with respect to each other [55–57]. Once aligned, the images are then recombined directly by a reconstruction algorithm such as weighted back-projection [58, 59]. Many academically developed processing packages are available that will compute the alignment and reconstruction [60–63]. Although there is no community-accepted convention for assessing the resolution of tomographic reconstructions, a number of statistical approaches have been proposed [64].

4.2.2.2 Structure Analysis and Data Mining

In order to analyze the cryo-EM density maps of large, complex nanomachines, a number of tools have been developed that range from feature detection to domain localization. Perhaps the most well-developed set of tools for the analysis of cryo-EM density maps are those aimed at fitting known crystal or NMR structures to density maps. These tools range from simple rigid-body fitting to complex and dynamic flexible fitting algorithms (for reviews, see Refs [65, 66]). Regardless, each of these tools requires that the structures of a known domain or closely related domain are known.

Recent studies have also shown that cryo-EM density is sufficient for discriminating good models from a gallery of potential structures [67, 68]. In particular, cryo-EM density has been incorporated as a scoring function in a constrained homology modeling approach [69]. As with the aforementioned fitting routines, this approach relies on the availability of a known structure from which a sequence/structure alignment is produced. In the case where a suitable structural template is not known, a constrained *ab initio* modeling approach has also been developed in which the cryo-EM density can be used directly to screen a large gallery of potential models [67]. While no structural template is needed, this approach is restricted to relatively small (<200 amino acids), single-domain proteins.

At subnanometer resolutions, secondary structure elements become visible; α-helices appear as long density rods, while β-sheets appear as thin surfaces [70–73]. By using a variety of feature detection and computational geometry algorithms, secondary structure elements can be reliably identified and quantified [71, 72, 74, 75]. The spatial description of such elements has been used not only to describe protein structure, but also to infer structure and function of individual protein domains [70, 76].

Until recently, the resolution of cryo-EM density necessitated the use of the aforementioned approaches for understanding macromolecular structure and function. Several cryo-EM structures have now achieved resolutions better than 4.5 Å resolution, at which point the pitch of α-helices, separation of β-strands, as well as the densities that connect them, can be seen unambiguously with no reference to crystal structure [13, 14, 17]. In addition to these features, many of the bulky side chains could also be seen. However, it should be noted that these structures still do not have the resolution to utilize standard X-ray crystallographic methods for model construction. However, several *de novo* models have now been constructed directly from these high-resolution cryo-EM density maps using mostly manual assignment and visualization tools [13, 14, 17].

The annotation of cryo-ET maps is a different process because the goals of tomographic imaging are substantially different. Rather than trying to determine high-resolution protein structures, cryo-ET experiments are often focused on how the components of cells or nanomachines are spatially organized. *Segmentation* is a process by which the salient features of the reconstruction of an individual cell or nanomachine are traced. A segmented map hides the noisy data and highlights the structural findings. Tools for segmentation of tomograms include the academic IMOD package [60] and the commercial package Amira (Mercury Systems, Chelmsford, MA, USA). Unfortunately, this is a laborious manual process for which no suitable general-purpose automated routine has been advanced (for a review, see

Ref. [77]). Manual segmentation is applied by annotating one cell or nanomachine at a time. As a research study might require the annotation of many cells or nanomachines, this can drastically increase the time required to annotate the results. In cases where there are uniform nanomachines (e.g. ribosomes) present, these can be computationally identified, extracted and merged to improve the resolution (for a review, see Ref. [78]).

4.2.3 Data Archival

The result of a cryo-EM experiment is typically a 3-D density map with multiple domains and/or models used to annotate the structure and function of the molecular nanomachine. In reaching this model, multiple intermediate data sets and image processing workflows are produced. Databases, such as EMEN [79] and others [80, 81], function on a laboratory scale and can house the final 3-D density maps and model, as well as the original specimen images and all of the intermediate data and processes. The final density map, models and associated metadata can also be deposited in public repositories such as the electron microscopy databank (EMDB) and the protein databank (PDB) [82]. Individual cryo-EM structures are easily retrieved through accession numbers or IDs directly from publicly accessible websites.

4.3 Biological Examples

Cryo-EM is a powerful technique in that it can be used to image a wide variety of specimens under an equally wide array of conditions. Despite the lack of atomic resolution, these structures can provide unprecedented views of the structure and function of molecular nanomachines. In the following sections, we describe two very different samples, and how cryo-EM has provided us with a unique glimpse into their organization. It should also be noted that each of these samples are complex nanomachines that undergo dynamic structural and functional processes in carrying out their intended functions.

The resolution of cryo-EM models is considerably better than that of cryo-ET models, mainly because the heterogeneous cryo-ET particles are rarely suitable for averaging. Even when they are, the average may be the sum of tens to hundreds from extracted tomograms rather than the thousands to millions of asymmetric units in a single-particle reconstruction that contribute to a cryo-EM model. Despite the lower resolution, tremendous insight may still be gained from the cellular context, as evidenced by the two cryo-ET examples presented here.

Both, cryo-EM and cryo-ET offer unique views of nanomachines, and as such integrating the two approaches can generate multiresolution models where a tomogram establishes a low-resolution survey of a cell, and the individual machines in that model are the product of cryo-EM studies. This integrated approached is demonstrated in our last example.

4.3.1
Skeletal Muscle Calcium Release Channel

Ryanodine receptor (RyR1) is a 2.3 MDa homotetramer that regulates the release of Ca^{2+} from the sarcoplasmic reticulum to initiate muscle contraction (for a review, see Ref. [49]). Figure 4.2a shows the 9.6 Å resolution cryo-EM density map of RyR1 reconstructed from ~28 000 particle images (Figure 4.2b) [83]. In this map, the structural organization, including the transmembrane and cytoplasmic regions for each monomer, as well as domains within individual monomers can be clearly seen.

A structural analysis of the RyR1 map using SSEHunter [71] revealed 41 α-helices, 36 in the cytoplasmic region and five in the transmembrane region, as well as seven β-sheets in the cytoplasmic region of a RyR1 monomeric subunit (Figure 4.2c). Interestingly, a kinked inner, pore-lining helix and a pore helix in the transmembrane region bears a remarkable similarity to those of the MthK channel [84]. β-Sheets located in the constricted part that connect the transmembrane and cytoplasmic regions have been seen in the crystal structures of inward rectifier K^+ channels (Kir channels) [85, 86] and a cyclic nucleotide-modulated (HCN2) channel [87]. In Kir channels, this β-sheet has been proposed to form part of the cytoplasmic pore, which is connected to the inner pore. Therefore, this region in the RyR1 may play a role in regulating the ions by interacting with cellular regulators which are yet to be determined.

While there is no crystal structure from any domain or region of RyR1, a homologous domain from the IP3 receptor is known. Using the aforementioned cryo-EM constrained homology modeling approach [69], it was possible to derive three protein folds, based on the ligand-binding suppressor and IP_3-binding core domains from the type 1 IP_3 receptor, for the N-terminal portion (residues 12–565) of the RyR1 primary sequence [88] (Figure 4.2d). Interestingly, these models were localized to a region at the four corners of the RyR1 tetramer, a region that has also been implicated to interact with the dihydropyridine receptor (DHPR) during the

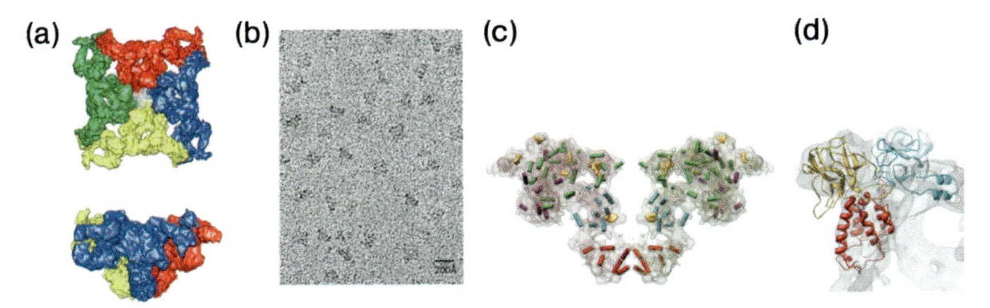

Figure 4.2 RyR1 at 9.6 Å resolution [88]. RyR1 at 9.6 Å resolution. The side and top views shown in (a) were reconstructed from ~28 000 individual particle images. The four subunits are annotated in different colors; (b) A representative view of the particle images; (c) The spatial dispositions of the α-helices (cylinders) and β-sheets (orange planes) in two of the homotetrameric subunits; (d) The three N-terminal models for RyR1 are shown fitted to the cryo-EM density. Model 1 is shown in cyan (residues Q12-S207), model 2 in yellow (residues G216-T407), and model 3 in red (residues A408-Y565). Models 2 and 3 are based on the aforementioned IP_3-binding core domain, while model 1 is based on the ligand-binding suppressor domain.

excitation–contraction coupling of the muscle [89–91]. Also of interest, several disease-related mutations in RyR1 occur within this region. As such, imaging of RyR1 with cryo-EM coupled to structural analysis has resulted into insight into the channels function and role in muscle contraction.

4.3.2 Bacteriophage Epsilon15

Epsilon15 is a 700 Å wide, 22 MDa nanomachine that infects *Salmonella* (Figure 4.3a) [92]. An icosahedral protein shell surrounds its dsDNA genome; at one vertex a large tail assembly protrudes from this shell. Without imposing any symmetry, a reconstruction of the native virion revealed at ~20 Å resolution all the molecular components of the virus, including the portal vertex (Figure 4.3b) [92]. When icosahedral symmetry was imposed during the reconstruction (effectively increasing the number of particle images, as there are 60 asymmetric units per particle), features of the nonicosahedral components such as the portal vertex were averaged out, but the icosahedral position shell proteins could be seen at a finer detail (4.5 Å resolution) [13]. In this high-resolution map, a complete annotation of the capsid components was possible.

While ~4.5 Å resolution is generally insufficient to construct a model by X-ray crystallographic standards, the aforementioned *de novo* model building tools for cryo-EM density maps make it possible with the existence of large α-helices which can be

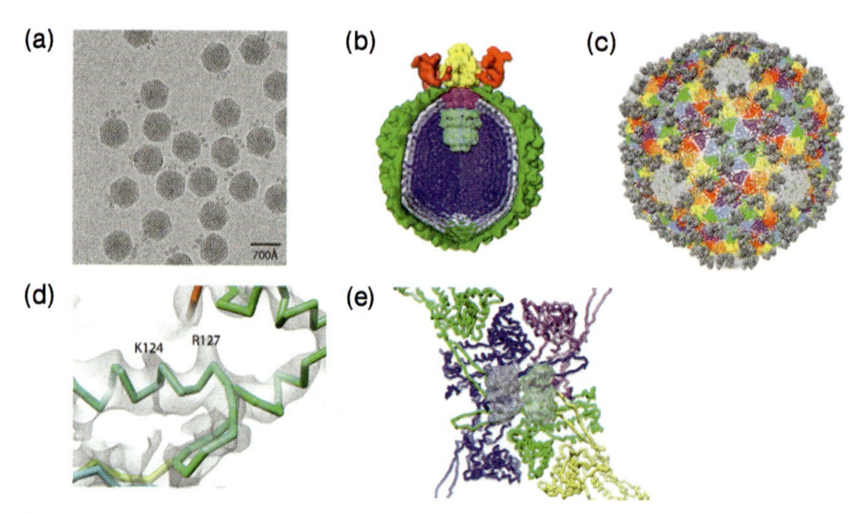

Figure 4.3 Bacteriophage epsilon15. (a) A 300-kV electron image of the phage particles embedded in vitreous ice; (b) A cut-away view of the 20 Å asymmetric reconstruction of epsilon15 [92] shows the molecular components of the portal vertex, the capsid shell protein and the viral DNA. The different molecular components are annotated in different colors; (c) 4.5 Å resolution structure [13], showing the backbone model of the Gp7 and the density of Gp10; (d) Side-chain density in the cryo-EM map; (e) A zoomed-in view of the capsid showing one gp10 spanning across four gp7 molecules, functioning as a 'molecular stapler'.

used to anchor the sequence-to-structure assignment. Gp7, a 420-amino acid protein that makes up the majority of the icosahedral capsid shell, was identified in the reconstruction and shown to have eight α-helices (>2.5 turns) and three β-sheets. Using these features, a complete *de novo* model was constructed (Figure 4.3c) to reveal a structure similar to that of the HK97 major capsid protein [93], despite the lack of detectable sequence similarity. In addition to the detection of a common fold, side-chain density could also be visualized in several regions throughout the map (Figure 4.3d).

Construction of the Gp7 model clearly revealed the presence of a previously undetected capsid protein. Biochemical analysis of the capsid later confirmed this protein to be Gp10 (12 kDa). In analyzing the sequence of this protein, potential structural homology to the PDZ domains [94] was identified. Taken together with the location of this protein of the capsid surface (along the icosahedral twofold symmetry axes), this small protein most likely acts like a molecular staple, bridging four adjacent gp10 molecules and thus assuring stability in the mature capsid (Figure 4.3e).

4.3.3 Bacterial Flagellum

The bacterial flagellum is an intricate biological nanomachine that transduces chemical energy into mechanical energy. The flagellar motor is a complicated assembly of approximately 25 unique polypeptide components that, when assembled correctly and operating under the proper electrochemical gradient, drives the rotary motor at speeds of nearly 300 Hz (for a review, see Ref. [95]). Models of the flagellar motor have been advanced through both single-particle and helical cryo-EM studies of the purified complexes from various spirochetes [96–98]. Cryo-ET has been used to complement these models; the *in situ* perspective derived from tomograms can reveal components of the structure that may be – and if fact, are – lost in the isolation and purification steps.

Recent studies have utilized cryo-ET to examine *in situ* the structure of the motor and the greater flagellar apparatus from the spirochete *Treponema primitia*. In their first report, Jensen and coworkers presented the structural details of the motor [99] in which 20 flagellar motors were computationally isolated from their positions in the cryo-ET reconstructions of 15 intact cells (each cell has two flagella, but not all tomograms captured both). These motors were subsequently merged to yield a ~70 Å resolution model (Figure 4.4a). Like man-made motors, the flagellar motor consists of a rod attached to a rotor that rotates amidst an array of stationary stators; purified flagellar complexes lacked the stators altogether. This tomography-derived model was the first reconstruction to integrate the full motor with the stators, revealing the stators' 16-fold symmetry and their position with respect to the membrane. Besides accounting for the stator density, this model also revealed unexpected density above the stators in the peptidoglycan layer, including a new component termed the P-collar (Figure 4.4b). The P-collar and new findings about stator geometry raise new questions about the motor's mechanistic details.

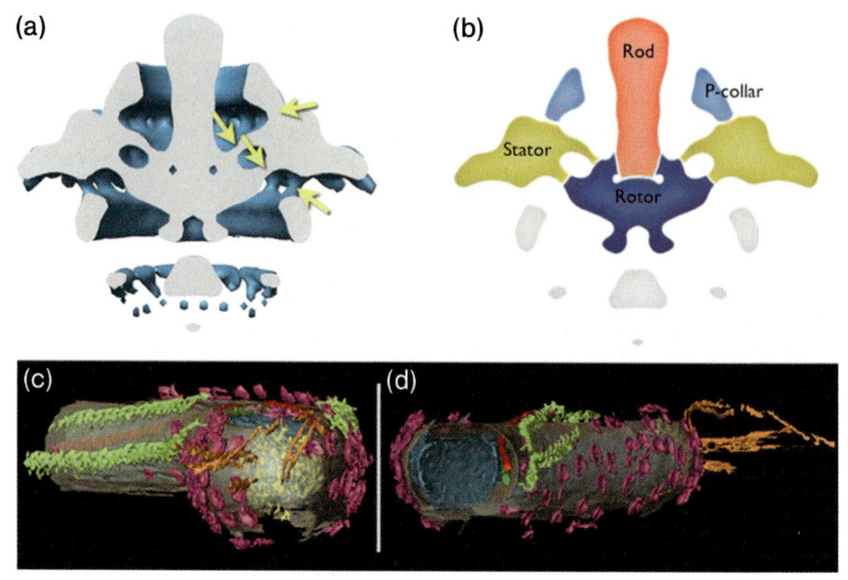

Figure 4.4 Cryo-ET findings on the flagellar motor and cellular features of spirochete *Treponema primitia*. (a) A cutaway surface rendering of the flagellar motor computed by averaging together 20 motors computationally isolated from tomograms. Novel findings included the surprising connectivity between the stator and other components of the motor, marked by yellow arrows; (b) Schematic of the motor organization as revealed through the analysis of the structure shown in (a). The position and geometry of the stator and the existence of the P-collar were missed by previous studies that examined biochemically purified flagellar motors; (c) Surface rendering of a *T. primitia* cell, showing many novel cellular structures, including surface bowls (magenta), the surface hook arcade (yellow) and tip fibrils (orange); the flagella are shown in green and red.

A subsequent cryo-ET study conducted by Jensen and coworkers revealed a plethora of novel structures in the *Treponema* cell, including outer-membrane bowls, polar fibrils, a polar cone and a surface hook arcade that sometimes tracks with the cellular position of flagella (Figure 4.4c and d) [100]. These findings beg new questions as to how, and if, these features relate to flagellar function. More importantly, two distinct periplasmic layers in *T. primitia* were revealed; this observation, when combined with video observations by light microscopy, affirms the rolling cylinder model of motility over the competing gyration model. The importance of this motility mechanism is underscored by the association between motility and pathogenicity in spirochetes such as those that cause Lyme disease and syphilis (for a review, see Ref. [101]).

4.3.4
Proteomic Atlas

Traditional cellular biology studies are frequently limited by carrying out experiments *in vitro* or investigating only fractions of cells. It is an obvious and tremendous

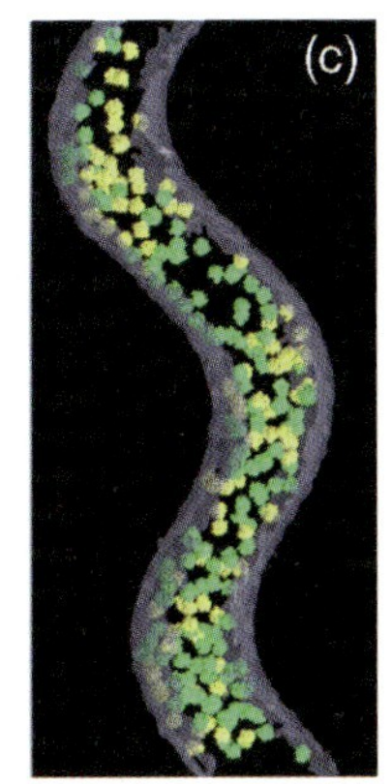

Figure 4.5 A map of ribosome location in the spirochete *Spiroplasma melliferum*. (a) The 0 (projection, an unprocessed cryo-ET image, showing a spirochete cell; (b) A slice of the reconstruction, which shows higher density at positions occupied by ribosomes; (c) A surface rendering of the reconstruction which has been filtered and segmented. High confidence ribosome matches are colored green; intermediate confidence matches are colored yellow.

advantage to integrate the structures and processes of all of the cellular space, enabling investigators to comprehend cells *in toto*. Today, Baumeister and colleagues continue to make strides towards this goal of visualizing a complete cell with all of its major nanomachines. Early proof-of-concept studies have shown that it is possible to identify and differentiate large complexes in the tomograms of synthetic cells [20]. Moreover, recent advances in data processing suggest that even similar assemblies with subtle differences in mass, such as GroEL and GroEL-GroES, can be differentiated [102]. The first application of mapping nanomachines in a cell showed that the total spatial distribution of ribosomes through an entire cell could be directly observed in the spirochete *Spiroplasma melliferum* (Figure 4.5) [103].

The archaebacteria *Thermoplasma acidophilum* is a relatively simple cell with only approximately 1507 open reading frames (ORFs) comprising considerably fewer subcellular assemblies [104]. As such, it is an attractive cryo-ET target for mapping the 3-D position of all major nanomachines – the 'proteomic atlas' – which, ultimately, will reveal unprecedented detail about the 3-D organization of protein–protein networks [105].

4.4
Future Prospects

Today, single-particle cryo-EM has reached the turning point where it is now possible to resolve relatively high-resolution structures of molecular nanomachines under conditions not generally possible with other high-resolution structure determination techniques. Due to the intrinsic nature of the cryo-EM experiment, it can also produce unique and biologically important information, even when a high-resolution structure is already known. Cryo-EM structures of both the ribosome [106] and GroEL [14, 107, 108] have provided significant insight into structural and functional mechanisms, despite being extensively studied using X-ray crystallography.

One obvious challenge for cryo-EM is the pursuit of higher resolution (i.e. close to or better than 3.0 Å), at which point full, all-atom models could be constructed. On the other hand, cryo-EM is not aimed solely at high resolution. Rather, it offers the ability to resolve domains and/or components that are highly flexible at lower resolutions, as well as samples with multiple conformational states [108]. With further developments in the image-processing routines, both high-resolution structure determination and 'computational purification' of samples [108] will further allow for the exploration of complex molecular nanomachines in greater detail.

As with cryo-EM, improvements in data collection and image processing will allow cryo-ET to achieve more accurate and higher-resolution reconstructions of large nanomachines and cells. However – as alluded to in the proteomic atlas – the real strength of cryo-ET is its power to integrate known atomic structures and cryo-EM reconstructions to provide a complete model of *in vivo* protein function [105, 109]. Such integration will ultimately establish a true spatial and temporal view of functional nanomachines within the cell, which can systematically be investigated in either healthy or diseased states. In addition, there is a trend towards the integration of live cell observations made by light microscopy, followed by cryo-ET observations of the same specimens (e.g. [110]). Such hybrid approaches require not only new instrumentation to make sequential observations practical but also the computational tools to integrate the data. These integrated cellular views promise to enhance our understanding of cell structure and function relationships in normal and diseased states at higher spatial and temporal resolutions.

Acknowledgments

These studies were supported by grants from NIH (P41RR02250, 2PN2EY016525, R01GM079429) and NSF (IIS-0705474, IIS-0705644). We thank our collaborators Drs Irina Serysheva, Steve Ludtke, Yao Cong, Maya Topf, Andrej Sali and Susan Hamilton on the RYR1 project; Wen Jiang, Peter Weigele, Jonathan King, Joanita Jakana and Juan Chang on the epsilon15 phage project; and Jose Lopez on human platelet. We also thank Dr Grant Jensen at California Institute of Technology and Drs Wolfgang Baumeister and Julio Ortiz at the Max Planck Institute for providing the artwork for Figures 4.3 and 4.4, respectively.

References

1 Celniker, S.E. and Rubin, G.M. (2003) The *Drosophila melanogaster* genome. *Annual Review of Genomics and Human Genetics*, **4**, 89.

2 Olivier, M., Aggarwal, A., Allen, J., Almendras, A.A., Bajorek, E.S., Beasley, E.M., Brady, S.D., Bushard, J.M., Bustos, V.I., Chu, A., Chung, T.R., De Witte, A.,

Denys, M.E., Dominguez, R., Fang, N.Y., Foster, B.D., Freudenberg, R.W., Hadley, D., Hamilton, L.R., Jeffrey, T.J., Kelly, L., Lazzeroni, L., Levy, M.R., Lewis, S.C., Liu, X., Lopez, F.J., Louie, B., Marquis, J.P., Martinez, R.A., Matsuura, M.K., Misherghi, N.S., Norton, J.A., Olshen, A., Perkins, S.M., Perou, A.J., Piercy, C., Piercy, M., Qin, F., Reif, T., Sheppard, K., Shokoohi, V., Smick, G.A., Sun, W.L., Stewart, E.A., Fernando, J., Tejeda, Tran, N.M., Trejo, T., Vo, N.T., Yan, S.C., Zierten, D.L., Zhao, S., Sachidanandam, R., Trask, B.J., Myers, R.M. and Cox, D.R. (2001) A high-resolution radiation hybrid map of the human genome draft sequence. *Science*, **291**, 1298.

3 Venter, J.C., Adams, M.D., Myers, E.W., Li, P.W., Mural, R.J., Sutton, G.G., Smith, H.O., Yandell, M., Evans, C.A., Holt, R.A., Gocayne, J.D., Amanatides, P., Ballew, R.M., Huson, D.H., Wortman, J.R., Zhang, Q., Kodira, C.D., Zheng, X.H., Chen, L., Skupski, M., Subramanian, G., Thomas, P.D., Zhang, J., Gabor Miklos, G.L., Nelson, C., Broder, S., Clark, A.G., Nadeau, J., McKusick, V.A., Zinder, N., Levine, A.J., Roberts, R.J., Simon, M., Slayman, C., Hunkapiller, M., Bolanos, R., Delcher, A., Dew, I., Fasulo, D., Flanigan, M., Florea, L., Halpern, A., Hannenhalli, S., Kravitz, S., Levy, S., Mobarry, C., Reinert, K., Remington, K., Abu-Threideh, J., Beasley, E., Biddick, K., Bonazzi, V., Brandon, R., Cargill, M., Chandramouliswaran, I., Charlab, R., Chaturvedi, K., Deng, Z., Di Francesco, V., Dunn, P., Eilbeck, K., Evangelista, C., Gabrielian, A.E., Gan, W., Ge, W., Gong, F., Gu, Z., Guan, P., Heiman, T.J., Higgins, M.E., Ji, R.R., Ke, Z., Ketchum, K.A., Lai, Z., Lei, Y., Li, Z., Li, J., Liang, Y., Lin, X., Lu, F., Merkulov, G.V., Milshina, N., Moore, H.M., Naik, A.K., Narayan, V.A., Neelam, B., Nusskern, D., Rusch, D.B., Salzberg, S., Shao, W., Shue, B., Sun, J., Wang, Z., Wang, A., Wang, X., Wang, J., Wei, M., Wides, R., Xiao, C., Yan, C. *et al.* (2001) The sequence of the human genome. *Science*, **291**, 1304.

4 Yu, H., Peters, J.M., King, R.W., Page, A.M., Hieter, P. and Kirschner, M.W. (1998) Identification of a cullin homology region in a subunit of the anaphase-promoting complex. *Science*, **279**, 1219.

5 Martin, A.C. and Drubin, D.G. (2003) Impact of genome-wide functional analyses on cell biology research. *Current Opinion in Cell Biology*, **15**, 6.

6 Alberts, B. (1998) The cell as a collection of protein machines: preparing the next generation of molecular biologists. *Cell*, **92**, 291.

7 Alberts, B. and Miake-Lye, R. (1992) Unscrambling the puzzle of biological machines: the importance of the details. *Cell*, **68**, 415.

8 Levchenko, A. (2001) Computational cell biology in the post-genomic era. *Molecular Biology Reports*, **28**, 83.

9 Sali, A. (2003) NIH workshop on structural proteomics of biological complexes. *Structure*, **11**, 1043.

10 Sali, A. and Chiu, W. (2005) Macromolecular assemblies highlighted. *Structure*, **13**, 339.

11 Berman, H.M., Westbrook, J., Feng, Z., Gilliland, G., Bhat, T.N., Weissig, H., Shindyalov, I.N. and Bourne, P.E. (2000) The Protein Data Bank. *Nucleic Acids Research*, **28**, 235.

12 Glaeser, R.M., Downing, K.H., DeRosier, D.L., Chiu, W. and Frank, J. (2007) *Electron crystallography of biological macromolecules*, Oxford University Press, Oxford, UK. New York.

13 Jiang, W., Baker, M.L., Jakana, J., Weigele, P.R., King, J. and Chiu, W. (2008) Backbone structure of the infectious epsilon15 virus capsid revealed by electron cryomicroscopy. *Nature*, **451**, 1130.

14 Ludtke, S.J., Baker, M.L., Chen, D.H., Song, J.L., Chuang, D.T. and Chiu, W. (2008) De Novo backbone trace of GroEL from single particle electron cryomicroscopy. *Structure*, **16**, 441.

15 Zhang, X., Settembre, E., Xu, C., Dormitzer, P.R., Bellamy, R., Harrison, S.C. and Grigorieff, N. (2008) Near-atomic resolution using electron cryomicroscopy and single-particle reconstruction. *Proceedings of the National Academy of Sciences of the United States of America*, **105**, 1867.

16 Zhou, Z.H. (2008) Towards atomic resolution structural determination by single-particle cryo-electron microscopy. *Current Opinion in Structural Biology*, **18**, 218.

17 Yu, X., Jin, L. and Zhou, Z.H. (2008) 3.88 A structure of cytoplasmic polyhedrosis virus by cryo-electron microscopy. *Nature*, **453**, 415.

18 Frank, J. (2006) *Three-dimensional electron microscopy of macromolecular assemblies: visualization of biological molecules in their native state*, 2nd edn, Oxford University Press, New York.

19 Dubochet, J., Adrian, M., Chang, J.J., Homo, J.C., Lepault, J., McDowall, A.W. and Schultz, P. (1988) Cryo-electron microscopy of vitrified specimens. *Quarterly Reviews of Biophysics*, **21**, 129.

20 Frangakis, A.S., Bohm, J., Forster, F., Nickell, S., Nicastro, D., Typke, D., Hegerl, R. and Baumeister, W. (2002) Identification of macromolecular complexes in cryoelectron tomograms of phantom cells. *Proceedings of the National Academy of Sciences of the United States of America*, **99**, 14153.

21 Grimm, R., Singh, H., Rachel, R., Typke, D., Zillig, W. and Baumeister, W. (1998) Electron tomography of ice-embedded prokaryotic cells. *Biophysical Journal*, **74**, 1031.

22 Nickell, S., Hegerl, R., Baumeister, W. and Rachel, R. (2003) Pyrodictium cannulae enter the periplasmic space but do not enter the cytoplasm, as revealed by cryo-electron tomography. *Journal of Structural Biology*, **141**, 34.

23 Frederik, P.M. and Hubert, D.H. (2005) Cryoelectron microscopy of liposomes. *Methods in Enzymology*, **391**, 431.

24 Berriman, J. and Unwin, N. (1994) Analysis of transient structures by cryo-microscopy combined with rapid mixing of spray droplets. *Ultramicroscopy*, **56**, 241.

25 White, H.D., Walker, M.L. and Trinick, J. (1998) A computer-controlled spraying-freezing apparatus for millisecond time-resolution electron cryomicroscopy. *Journal of Structural Biology*, **121**, 306.

26 Chiu, W., Downing, K.H., Dubochet, J., Glaeser, R.M., Heide, H.G., Knapek, E., Kopf, D.A., Lamvik, M.K., Lepault, J., Robertson, J.D., Zeitler, E. and Zemlin, F. (1986) Cryoprotection in electron microscopy. *Journal of Microscopy*, **141**, 385.

27 Comolli, L.R. and Downing, K.H. (2005) Dose tolerance at helium and nitrogen temperatures for whole cell electron tomography. *Journal of Structural Biology*, **152**, 149.

28 Iancu, C.V., Wright, E.R., Heymann, J.B. and Jensen, G.J. (2006) A comparison of liquid nitrogen and liquid helium as cryogens for electron cryotomography. *Journal of Structural Biology*, **153**, 231.

29 Murata, K., Mitsuoka, K., Hirai, T., Walz, T., Agre, P., Heymann, J.B., Engel, A. and Fujiyoshi, Y. (2000) Structural determinants of water permeation through aquaporin-1. *Nature*, **407**, 599.

30 Henderson, R., Baldwin, J.M., Ceska, T.A., Zemlin, F., Beckmann, E. and Downing, K.H. (1990) An atomic model for the structure of bacteriorhodopsin. *Biochemical Society Transactions*, **18**, 844.

31 Miyazawa, A., Fujiyoshi, Y. and Unwin, N. (2003) Structure and gating mechanism of the acetylcholine receptor pore. *Nature*, **424**, 949.

32 Yonekura, K., Maki-Yonekura, S. and Namba, K. (2003) Complete atomic model of the bacterial flagellar filament by electron cryomicroscopy. *Nature*, **424**, 643.

33 Nogales, E., Wolf, S.G. and Downing, K.H. (1998) Structure of the alpha beta

tubulin dimer by electron crystallography. *Nature*, **391**, 199.

34 Sachse, C., Chen, J.Z., Coureux, P.D., Stroupe, M.E., Fandrich, M. and Grigorieff, N. (2007) High-resolution electron microscopy of helical specimens: a fresh look at tobacco mosaic virus. *Journal of Molecular Biology*, **371**, 812.

35 Iancu, C.V., Tivol, W.F., Schooler, J.B., Dias, D.P., Henderson, G.P., Murphy, G.E., Wright, E.R., Li, Z., Yu, Z., Briegel, A., Gan, L., He, Y. and Jensen, G.J. (2006) Electron cryotomography sample preparation using the Vitrobot. *Nature Protocols*, **1**, 2813.

36 Wright, E.R., Iancu, C.V., Tivol, W.F. and Jensen, G.J. (2006) Observations on the behavior of vitreous ice at approximately 82 and approximately 12K. *Journal of Structural Biology*, **153**, 241.

37 Booth, C.R., Jakana, J. and Chiu, W. (2006) Assessing the capabilities of a 4 k × 4 k CCD camera for electron cryo-microscopy at 300 kV. *Journal of Structural Biology*, **156**, 556.

38 Booth, C.R., Jiang, W., Baker, M.L., Hong Zhou, Z., Ludtke, S.J. and Chiu, W. (2004) A 9 Å single particle reconstruction from CCD captured images on a 200 kV electron cryomicroscope. *Journal of Structural Biology*, **147**, 116.

39 Chen, D.H., Jakana, J., Liu, X., Schmid, M.F. and Chiu, W. (2008) Achievable resolution from images of biological specimens acquired from a 4 k × 4 k CCD camera in a 300-kV electron cryomicroscope. *Journal of Structural Biology*, **163**, 45.

40 Carragher, B., Kisseberth, N., Kriegman, D., Milligan, R.A., Potter, C.S., Pulokas, J. and Reilein, A. (2000) Leginon: an automated system for acquisition of images from vitreous ice specimens. *Journal of Structural Biology*, **132**, 33.

41 Lei, J. and Frank, J. (2005) Automated acquisition of cryo-electron micrographs for single particle reconstruction on an FEI Tecnai electron microscope. *Journal of Structural Biology*, **150**, 69.

42 Zhang, P., Beatty, A., Milne, J.L.S. and Subramaniam, S. (2001) Automated data collection with a Tecnai 12 electron microscope: applications for molecular imaging by cryo-microscopy. *Journal of Structural Biology*, **135**, 251.

43 Marsh, M.P., Chang, J.T., Booth, C.R., Liang, N.L., Schmid, M.F. and Chiu, W. (2007) Modular software platform for low-dose electron microscopy and tomography. *Journal of Microscopy*, **228**, 384.

44 DeRosier, D.L. and Klug, A. (1968) Reconstruction of three-dimensional structures from electron micrographs. *Nature*, **217**, 130.

45 Crowther, R.A., DeRosier, D.J. and Klug, A. (1970) The reconstruction of a three-dimensional structure from projections and its application to electron microscopy. *Proceedings of the Royal Society of London*, **317**, 319.

46 Henderson, R. (1995) The potential and limitations of neutrons, electrons and X-rays for atomic resolution microscopy of unstained biological molecules. *Quarterly Reviews of Biophysics*, **28**, 171.

47 Liu, X., Jiang, W., Jakana, J. and Chiu, W. (2007) Averaging tens to hundreds of icosahedral particle images to resolve protein secondary structure elements using a Multi-Path Simulated Annealing optimization algorithm. *Journal of Structural Biology*, **160**, 11.

48 Jiang, W. and Chiu, W. (2007) Cryoelectron microscopy of icosahedral virus particles. *Methods in Molecular Biology (Clifton, NJ)*, **369**, 345.

49 Serysheva, I.I., Chiu, W. and Ludtke, S.J. (2007) Single-particle electron cryomicroscopy of the ion channels in the excitation-contraction coupling junction. *Methods in Cell Biology*, **79**, 407.

50 Ludtke, S.J., Baldwin, P.R. and Chiu, W. (1999) EMAN: Semi-automated software for high resolution single particle reconstructions. *Journal of Structural Biology*, **128**, 82.

51 Frank, J., Radermacher, M., Penczek, P., Zhu, J., Li, Y., Ladjadj, M. and Leith, A. (1996) SPIDER and WEB: processing and visualization of images in 3D electron microscopy and related fields. *Journal of Structural Biology*, **116**, 190.

52 van Heel, M., Harauz, G., Orlova, E.V., Schmidt, R. and Schatz, M. (1996) A new generation of the IMAGIC image processing system. *Journal of Structural Biology*, **116**, 17.

53 Grigorieff, N. (2007) FREALIGN: high-resolution refinement of single particle structures. *Journal of Structural Biology*, **157**, 117.

54 Saxton, W.O. and Baumeister, W. (1982) The correlation averaging of a regularly arranged bacterial cell envelope protein. *Journal of Microscopy*, **127**, 127.

55 Lawrence, M.C. (1992) *Electron Tomography: Three-dimensional Imaging with the Transmission Electron Microscope*, 1st edn (ed. J. Frank), Springer, p. 197.

56 Luther, P.K., Lawrence, M.C. and Crowther, R.A. (1988) A method for monitoring the collapse of plastic sections as a function of electron dose. *Ultramicroscopy*, **24**, 7.

57 Mastronarde, D.N. (2006) *Electron tomography: Methods for three-dimensional visualization of structures in the cell*, 2nd edn (ed. J. Frank), Springer, p. 187.

58 Radermacher, M. (2006) *Electron tomography: Methods for three-dimensional visualization of structures in the cell*, 2nd edn (ed. J. Frank), Springer, p. 245.

59 Sandberg, K., Mastronarde, D.N. and Beylkin, G. (2003) A fast reconstruction algorithm for electron microscope tomography. *Journal of Structural Biology*, **144**, 61.

60 Kremer, J.R., Mastronarde, D.N. and McIntosh, J.R. (1996) Computer visualization of three-dimensional image data using IMOD. *Journal of Structural Biology*, **116**, 71.

61 Nickell, S., Forster, F., Linaroudis, A., Net, W.D., Beck, F., Hegerl, R., Baumeister, W. and Plitzko, J.M. (2005) TOM software toolbox: acquisition and analysis for electron tomography. *Journal of Structural Biology*, **149**, 227.

62 Winkler, H. (2007) 3D reconstruction and processing of volumetric data in cryo-electron tomography. *Journal of Structural Biology*, **157**, 126.

63 Zheng, S.Q., Keszthelyi, B., Branlund, E., Lyle, J.M., Braunfeld, M.B., Sedat, J.W. and Agard, D.A. (2007) UCSF tomography: an integrated software suite for real-time electron microscopic tomographic data collection, alignment, and reconstruction. *Journal of Structural Biology*, **157**, 138.

64 Cardone, G., Grunewald, K. and Steven, A.C. (2005) A resolution criterion for electron tomography based on cross-validation. *Journal of Structural Biology*, **151**, 117.

65 Rossmann, M.G., Morais, M.C., Leiman, P.G. and Zhang, W. (2005) Combining X-ray crystallography and electron microscopy. *Structure*, **13**, 355.

66 Chiu, W., Baker, M.L., Jiang, W., Dougherty, M. and Schmid, M.F. (2005) Electron cryomicroscopy of biological machines at subnanometer resolution. *Structure*, **13**, 363.

67 Baker, M.L., Jiang, W., Wedemeyer, W.J., Rixon, F.J., Baker, D. and Chiu, W. (2006) Ab initio modeling of the herpesvirus VP26 core domain assessed by CryoEM density. *PLoS Computational Biology*, **2**, e146.

68 Topf, M., Baker, M.L., John, B., Chiu, W. and Sali, A. (2005) Structural characterization of components of protein assemblies by comparative modeling and electron cryo-microscopy. *Journal of Structural Biology*, **149**, 191.

69 Topf, M., Baker, M.L., Marti-Renom, M.A., Chiu, W. and Sali, A. (2006) Refinement of protein structures by Iterative comparative modeling and cryoEM density fitting. *Journal of Molecular Biology*, **357**, 1655.

70 Baker, M.L., Jiang, W., Bowman, B.R., Zhou, Z.H., Quiocho, F.A., Rixon, F.J.

and Chiu, W. (2003) Architecture of the herpes simplex virus major capsid protein derived from structural bioinformatics. *Journal of Molecular Biology*, **331**, 447.

71 Baker, M.L., Ju, T. and Chiu, W. (2007) Identification of secondary structure elements in intermediate-resolution density maps. *Structure*, **15**, 7.

72 Jiang, W., Baker, M.L., Ludtke, S.J. and Chiu, W. (2001) Bridging the information gap: computational tools for intermediate resolution structure interpretation. *Journal of Molecular Biology*, **308**, 1033.

73 Zhou, Z.H., Baker, M.L., Jiang, W., Dougherty, M., Jakana, J., Dong, G., Lu, G. and Chiu, W. (2001) Electron cryomicroscopy and bioinformatics suggest protein fold models for rice dwarf virus. *Nature Structural Biology*, **8**, 868.

74 Kong, Y. and Ma, J. (2003) A structural-informatics approach for mining beta-sheets: locating sheets in intermediate-resolution density maps. *Journal of Molecular Biology*, **332**, 399.

75 Kong, Y., Zhang, X., Baker, T.S. and Ma, J. (2004) A structural-informatics approach for tracing beta-sheets: building pseudo-C (alpha) traces for beta-strands in intermediate-resolution density maps. *Journal of Molecular Biology*, **339**, 117.

76 Baker, M.L., Jiang, W., Rixon, F.J. and Chiu, W. (2005) Common ancestry of herpesviruses and tailed DNA bacteriophages. *Journal of Virology*, **79**, 14967.

77 Sandberg, K. (2007) *Cellular electron microscopy* (ed. J.R. McIntosh), Elsevier, New York, p. 770.

78 Forster, F. and Hegerl, R. (2007) *Cellular electron microscopy* (ed. J.R. McIntosh), Elsevier, New York, p. 742.

79 Ludtke, S.J., Nason, L., Tu, H., Peng, L. and Chiu, W. (2003) Object oriented database and electronic notebook for transmission electron microscopy. *Microscopy and Microanalysis*, **9**, 556.

80 Marabini, R., Vaquerizo, C., Fernandez, J.J., Carazo, J.M., Engel, A. and Frank, J. (1996) Proposal for a new distributed database of macromolecular and subcellular structures from different areas of microscopy. *Journal of Structural Biology*, **116**, 161.

81 Fellmann, D., Pulokas, J., Milligan, R.A., Carragher, B. and Potter, C.S. (2002) A relational database for cryoEM: experience at one year and 50 000 images. *Journal of Structural Biology*, **137**, 273.

82 Fuller, S.D. (2003) Depositing electron microscopy maps. *Structure (Camb.)*, **11**, 11.

83 Ludtke, S.J., Serysheva, I.I., Hamilton, S.L. and Chiu, W. (2005) The pore structure of the closed RyR1 channel. *Structure (Camb.)*, **13**, 1203.

84 Jiang, Y., Lee, A., Chen, J., Cadene, M., Chait, B.T. and MacKinnon, R. (2002) Crystal structure and mechanism of a calcium-gated potassium channel. *Nature*, **417**, 515.

85 Kuo, A., Gulbis, J.M., Antcliff, J.F., Rahman, T., Lowe, E.D., Zimmer, J., Cuthbertson, J., Ashcroft, F.M., Ezaki, T. and Doyle, D.A. (2003) Crystal structure of the potassium channel KirBac1.1 in the closed state. *Science*, **300**, 1922.

86 Nishida, M., Cadene, M., Chait, B.T. and MacKinnon, R. (2007) Crystal structure of a Kir3.1-prokaryotic Kir channel chimera. *The EMBO Journal*, **26**, 4005.

87 Zagotta, W.N., Olivier, N.B., Black, K.D., Young, E.C., Olson, R. and Gouaux, E. (2003) Structural basis for modulation and agonist specificity of HCN pacemaker channels. *Nature*, **425**, 200.

88 Serysheva, I.I., Ludtke, S.J., Baker, M.L., Cong, Y., Topf, M., Eramian, D., Sali, A., Hamilton, S.L. and Chiu, W. (2008) Subnanometer-resolution electron cryomicroscopy-based domain models for the cytoplasmic region of skeletal muscle RyR channel. *Proceedings of the National Academy of Sciences of the United States of America*, **105**, 9610.

89 Wolf, M., Eberhart, A., Glossmann, H., Striessnig, J. and Grigorieff, N. (2003) Visualization of the domain structure of an L-type Ca(2 +) channel using electron cryo-microscopy. *Journal of Molecular Biology*, **332**, 171.

90 Serysheva, I.I., Ludtke, S.J., Baker, M.R., Chiu, W. and Hamilton, S.L. (2002) Structure of the voltage-gated L-type Ca^{2+} channel by electron cryomicroscopy. *Proceedings of the National Academy of Sciences of the United States of America*, **99**, 10370.

91 Block, B.A. and Franzini-Armstrong, C. (1988) The structure of the membrane systems in a novel muscle cell modified for heat production. *The Journal of Cell Biology*, **107**, 1099.

92 Jiang, W., Chang, J., Jakana, J., Weigele, P., King, J. and Chiu, W. (2006) Structure of epsilon15 bacteriophage reveals genome organization and DNA packaging/injection apparatus. *Nature*, **439**, 612.

93 Wikoff, W.R., Liljas, L., Duda, R.L., Tsuruta, H., Hendrix, R.W. and Johnson, J.E. (2000) Topologically linked protein rings in the bacteriophage HK97 capsid. *Science*, **289**, 2129.

94 Jemth, P. and Gianni, S. (2007) PDZ domains: folding and binding. *Biochemistry*, **46**, 8701.

95 Berg, H.C. (2003) The rotary motor of bacterial flagella. *Annual Review of Biochemistry*, **72**, 19.

96 Thomas, D.R., Francis, N.R., Xu, C. and DeRosier, D.J. (2006) The three-dimensional structure of the flagellar rotor from a clockwise-locked mutant of *Salmonella enterica* serovar *Typhimurium*. *Journal of Bacteriology*, **188**, 7039.

97 Suzuki, H., Yonekura, K. and Namba, K. (2004) Structure of the rotor of the bacterial flagellar motor revealed by electron cryomicroscopy and single-particle image analysis. *Journal of Molecular Biology*, **337**, 105.

98 Francis, N.R., Sosinsky, G.E., Thomas, D. and DeRosier, D.J. (1994) Isolation, characterization and structure of bacterial flagellar motors containing the switch complex. *Journal of Molecular Biology*, **235**, 1261.

99 Murphy, G.E., Leadbetter, J.R. and Jensen, G.J. (2006) In situ structure of the complete *Treponema primitia* flagellar motor. *Nature*, **442**, 1062.

100 Murphy, G.E., Matson, E.G., Leadbetter, J.R., Berg, H.C. and Jensen, G.J. (2008) Novel ultrastructures of *Treponema primitia* and their implications for motility. *Molecular Microbiology*, **67**, 1184.

101 Lux, R., Moter, A. and Shi, W. (2000) Chemotaxis in pathogenic spirochetes: directed movement toward targeting tissues? *Journal of Molecular Microbiology and Biotechnology*, **2**, 355.

102 Forster, F., Pruggnaller, S., Seybert, A. and Frangakis, A.S. (2008) Classification of cryo-electron sub-tomograms using constrained correlation. *Journal of Structural Biology*, **161**, 276.

103 Ortiz, J.O., Forster, F., Kurner, J., Linaroudis, A.A. and Baumeister, W. (2006) Mapping 70S ribosomes in intact cells by cryoelectron tomography and pattern recognition. *Journal of Structural Biology*, **156**, 334.

104 Sun, N., Beck, F., Knispel, R.W., Siedler, F., Scheffer, B., Nickell, S., Baumeister, W. and Nagy, I. (2007) Proteomics analysis of *Thermoplasma acidophilum* with a focus on protein complexes. *Molecular and Cellular Proteomics*, **6**, 492.

105 Robinson, C.V., Sali, A. and Baumeister, W. (2007) The molecular sociology of the cell. *Nature*, **450**, 973.

106 Valle, M., Gillet, R., Kaur, S., Henne, A., Ramakrishnan, V. and Frank, J. (2003) Visualizing tmRNA entry into a stalled ribosome. *Science*, **300**, 127.

107 Sewell, B.T., Best, R.B., Chen, S., Roseman, A.M., Farr, G.W., Horwich, A.L. and Saibil, H.R. (2004) A mutant chaperonin with rearranged inter-ring electrostatic contacts and temperature-sensitive dissociation. *Nature Structural & Molecular Biology*, **11**, 1128.

108 Chen, D.H., Song, J.L., Chuang, D.T., Chiu, W. and Ludtke, S.J. (2006) An expanded conformation of single-ring

GroEL-GroES complex encapsulates an 86 kDa substrate. *Structure*, **14**, 1711.

109 Nickell, S., Kofler, C., Leis, A.P. and Baumeister, W. (2006) A visual approach to proteomics. *Nature Reviews. Molecular Cell Biology*, **7**, 225.

110 Sartori, A., Gatz, R., Beck, F., Rigort, A., Baumeister, W. and Plitzko, J.M. (2007) Correlative microscopy: bridging the gap between fluorescence light microscopy and cryo-electron tomography. *Journal of Structural Biology*, **160**, 135.

5
Pushing Optical Microscopy to the Limit: From Single-Molecule Fluorescence Microscopy to Label-Free Detection and Tracking of Biological Nano-Objects

Philipp Kukura, Alois Renn, and Vahid Sandoghdar

5.1
Introduction

The fundamental goal of microscopic imaging is to visualize and identify small objects and to observe their motion. Many techniques based on a wide variety of approaches, including X-ray scattering, scanning probe microscopy, electron diffraction and various optical implementations, have been developed over the past century to improve upon the fundamental limitations of traditional optical microscopy. While each approach has its specific advantages, none by itself can provide a solution to all demands regarding spatial and temporal resolution, sensitivity and *in vivo* imaging capability, as are often desired in biologically motivated studies.

The ultimate resolution would allow one to detect, localize and visualize single molecules, or even atoms. Obtaining structural snapshots at molecular and atomic resolution has become routinely available through X-ray crystallographic techniques. Especially from a biological perspective, such detailed images of molecular and supramolecular structures regularly provide unique insights into their function [1–4]. Comparable resolution is attainable with scanning probe techniques such as scanning tunneling microscopy (STM) or atomic force microscopy (AFM) [5, 6]. While the former is generally limited to nonbiological samples, many biological AFM studies have been reported during the past decades [7, 8]. These experiments have been very informative, both due to their resolution and their ability to apply minute forces on single molecules or nano-objects. Various implementations of electron microscopy have also greatly contributed to biological imaging, due to its ability to provide spatial resolution in the nanometer region, thereby bridging the gap between the ultra high-resolution techniques mentioned above and standard optical imaging approaches [9].

Despite the many successes of these techniques, they all lack an ability to perform real-time, *in vivo* imaging. X-ray scattering experiments require high-quality crystals, which makes studies inside biological media intrinsically impossible. Scanning probe techniques are, by definition, limited to the study of surfaces and are relatively

Nanotechnology, Volume 5: Nanomedicine. Edited by Viola Vogel

ISBN: 978-3-527-31736-3

slow, with acquisition rates rarely exceeding a few Hertz. Electron microscopy requires a vacuum and sometimes metal-coating or cryogenic conditions for high-resolution images, thus making it difficult to perform studies under biologically relevant conditions.

As a consequence, the method of choice for real-time, *in vivo* biological imaging has remained optical microscopy, despite its comparatively low resolution. The optical microscope was invented about 400 years ago, and improvements in resolution to about 200 nm had already taken place by the end of the nineteenth century, through advances in lens design. However, this is a factor of 100 larger than the size of single molecules or proteins that define the ultimately desired resolution, and represents the diffraction limit established by Abbe during the 1880s. Throughout most of the twentieth century, this fundamental limit prevented the optical microscope from opening our eyes to the molecular structure of matter. Nevertheless, numerous recent advances have been made to improve the resolution and, in particular, the contrast of images. In general, these contrast techniques can be divided into two categories, namely linear and nonlinear.

5.1.1
Linear Contrast Mechanisms

The central challenge in visualizing small objects is to distinguish their signals from background scattering, or from reflections that may be caused by other objects or the optics inside the microscope. One possible solution to this problem is provided by dark-field microscopy, which was first reported by Lister in 1830. Here, the design is optimized to reduce all unwanted light to a minimum by preventing the illumination light from reaching the detector [10, 11]. To achieve this, the illumination light is shaped such that it can be removed before detection through the use of an aperture. In general, there are two approaches to achieve dark-field illumination; these are illustrated schematically in Figure 5.1a and b. In Figure 5.1a, the illumination light is modified in such a way as to provide an intense ring of light, for example through the use of a coin-like object blocking all but the outer ring of the incident light. The illumination light can be then removed by an aperture after the sample. In this way, only the light that is scattered by objects of interest, and whose path is thereby deviated, can reach the detector. The other approach, shown in Figure 5.1b, is referred to as total internal reflection microscopy (TIRM) [12]. Here, the illumination is incident on the sample at a steep angle, and is fully reflected at the interface between the glass coverslip and the sample (e.g. water). There, an 'evanescent' region is created, where the light decays exponentially as it enters into the sample. When a particle is placed into this region it scatters energy out of the evanescent part of the beam into the objective. Thus, light will only reach the detector from scatterers that are located within 100 nm of the sample–substrate interface. This technique has been mostly used to detect very weak signals such as those from metallic nanoparticles, or to obtain surface-specific images [13, 14].

Another approach to minimize unwanted light is provided by confocal microscopy, which first appeared during the early 1960s (Figure 5.1c) [15, 16]. In contrast to the

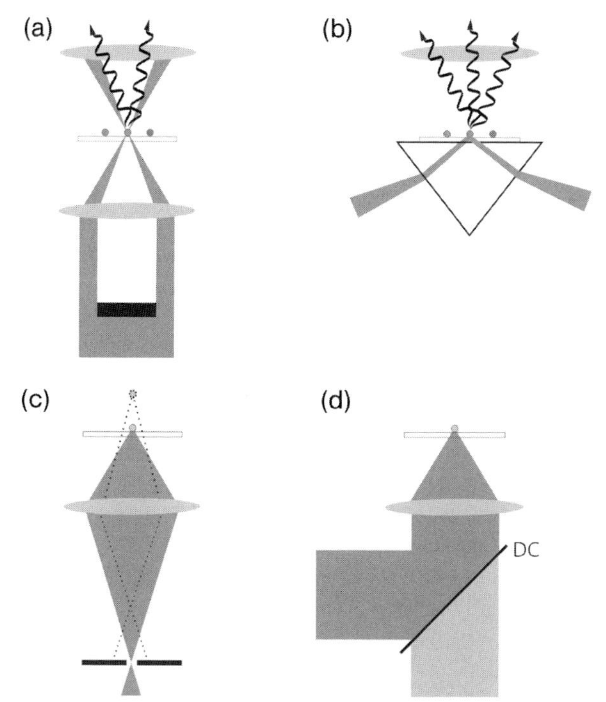

Figure 5.1 Illustration of the major linear contrast mechanisms in optical microscopy. (a) Dark-field illumination; (b) Total internal reflection microscopy (TIRM); (c) Confocal microscopy; (d) Fluorescence microscopy.

previous techniques, where the light from a relatively large sample area is collected ($\sim 20 \times 20\,\mu m$), only light from the focal region of the objective is allowed to reach the detector, thereby considerably reducing the background light. This is achieved through the use of a pinhole that is placed in the confocal plane of the objective. The size of the pinhole is usually chosen so as to match the size of the image of the focal spot. Collecting light in this fashion leads to optical sectioning – that is, the ability to provide three-dimensional (3-D) information by moving the focus in the z-direction through the sample. The major disadvantage of this approach is the need to scan the sample with respect to the illumination, which leads to relatively slow acquisition times.

An alternative for improving the contrast is to exploit the phase of the illumination light. Differential interference contrast (DIC) microscopy, which was introduced during the mid-1950s, takes advantage of slight differences in the refractive index of the specimen to generate additional information about the composition of the sample [17, 18]. For instance, the refractive index of water differs from that of lipids, so areas with a high water content will generate a different signal than those consisting mostly of organic material. The approach splits the illumination light into two slightly displaced beams (<1 μm) at the focus that are recombined before detection. If the two beams travel through material with different indices of refraction, the phase of one relative to the other will change, and this will lead to a small degree of destructive interference after recombination of the two beams. Such areas will thus appear dark, while areas where the sample is homogeneous will appear bright.

So far, we have been only concerned with improving the contrast using scattered light. A different but powerful method is to use *fluorescence* as a contrast mechanism. The first such reports emerged during the early twentieth century, when ultra-violet light was used for the first time in microscopes. However, the breakthrough occurred during the 1950s, when the application of fluorescence labeling for the detection of antigens was demonstrated [19]. Rather than detecting the scattering of incident light, the illumination light is used to excite molecules, causing them to fluoresce. The advantage of this approach is that the excitation light can be reduced by many orders of magnitude by the use of appropriate filters, as the fluorescence is usually red-shifted in energy (Figure 5.1d). One can then observe the species of interest virtually against zero background because the only photons that can reach the detector must be due to fluorescence emission. Such contrast is virtually unachievable with scattering techniques, even when dark-field illumination is employed, because the background scattering cannot be extinguished to such a high degree. The major sources of contrast are biological autofluorescence, specific labeling of the objects of interest with fluorescent dye molecules, or use of the cellular expression system itself to produce fluorescent proteins [20]. Fluorescence can also be used to introduce specificity by spectrally 'coding' the sample. Examples are fluorescence recovery after photobleaching (FRAP) for studying fluidity [21], fluorescence lifetime imaging [22] or simply simultaneous labeling with different fluorophores.

The spectral resolution of fluorescence is, however, rather poor because the emission is usually very broad in energy. Techniques based on vibrational spectroscopy on the other hand, where the resonances are orders of magnitude sharper, provide a much more unique molecular fingerprint, but at the expense of much-reduced signal intensities. In particular, Raman and infrared microscopy yield information about the composition of the sample, without the need for any label [23]. The experimental set-up is very similar to that used for fluorescence microscopy, but is focused on detecting vibrational resonances rather than fluorescence emission. As a matter of fact, Raman experiments can only be successful when the background fluorescence is reduced to a minimum, because the cross-sections for Raman scattering are orders of magnitudes smaller than that of fluorescence. On the upside, every species has a unique Raman spectrum and can thus be identified without the need for any label. Because these experiments must be performed in a confocal arrangement to achieve a sufficiently high photon flux, they provide highly specific and spatially resolved information, even inside cells. The downside is that the intrinsically low Raman cross-sections require large illumination powers, and this might be problematic for live cell imaging, for reasons of phototoxicity.

5.1.2
Nonlinear Contrast Mechanisms

The recent availability of high-power laser sources producing ultrashort pulses on the order of a hundred femtoseconds (10^{-13} s) or less at high pulse powers (>mJ) has opened up completely new areas in microscopy. When such short and intense pulses are focused to a diffraction-limited spot, peak powers of terawatts per cm^2 and above can be achieved. At such high peak powers, nonlinear effects that involve the simultaneous interaction of multiple photons with the sample become observable. The main microscopy-related application that has emerged from this technological jump is that of *two-photon imaging* [24]. Here, rather than exciting a fluorophore with a single resonant photon, for example at 400 nm, two off-resonant photons at 800 nm are used to produce the same excitation. Although nonresonant two-photon cross-sections are negligible compared to their one-photon counterparts (10^{-50} versus 10^{-16} cm^2), the high peak powers coupled with high pulse repetition rates on the order of 100 MHz can compensate for these dramatically lowered cross-sections. The major advantage of this technique is that optical sectioning comes for free, because the high peak intensities are only produced at the focus of the illuminating beam, making confocal pinholes superfluous. Despite the fact that biological tissue is generally transparent in the near-infrared region, sample heating and the rapid destruction of two-photon fluorophores cannot be avoided due to the necessarily high peak intensities used.

Another prominent example of nonlinear microscopy is based on coherent anti-Stokes Raman scattering (CARS) [25], which is the nonlinear equivalent of Raman microscopy. Here, three incident photons are required to produce a single Raman shifted photon. Additional techniques based on second and third harmonic generation microscopies have also appeared [26, 27]. Finally, a very interesting recent development has been discussed in the context of RESOLFT (reversible saturable optically linear fluorescence transitions). Here, a nonlinear process such as stimulated emission is used to deplete the fluorescence to a subdiffraction-limited spot. As a result, the actual volume from which fluorescent photons are emitted is considerably reduced beyond $\lambda/50$ [28].

In this chapter, we will focus our attention on pushing both the sensitivity and resolution limits of state-of-the-art *linear* microscopic techniques. In particular, we will discuss the capabilities and limitations of single-molecule detection in the light of biological applications. After covering the fundamental aspects of resolution, we will outline recent advances in the detection and tracking of nonfluorescent nano-objects. Scattering based labels show much promise in eliminating many of the limitations of fluorescence microscopy. Yet, by going a step further, we will show how these techniques can be used to detect and follow the motion of *unlabeled* biological nanoparticles.

5.2
Single-Molecule Fluorescence Detection: Techniques and Applications

The fundamental question that arises from the previous discussion of current optical imaging methods is: Why is it so difficult, first to 'see' single molecules, and second to achieve molecular resolution with optical microscopes? The former is particularly baffling because the fluorescence of single ions trapped in vacuum *can* be observed with the naked eye, as was shown 20 years ago [29].

5.2.1
Single Molecules: Light Sources with Ticks

To understand the intricacies of detecting single molecules in biological environments, it is useful to ask why it is so easy to observe single ions trapped in a vacuum. The answer is simply – the *absence of any background*. In a vacuum, the emitter is alone, with no other objects or molecules nearby that can either scatter the excitation light or fluoresce upon excitation. In this scenario, single-molecule detection simply becomes an issue of having a good enough detector (curiously, the human eye is one of the best light detectors available, being able to detect single photons with almost 70% efficiency [30]). However, the number of photons that any single molecule can emit via *fluorescence* is strictly limited by its intrinsic photophysics, and cannot be increased at will simply by raising the illumination power.

To understand this concept, it is useful to consider the dynamics that follow the absorption of a single photon by a single molecule. Population of the first excited electronic state is followed by an excited state decay which can take place via two major pathways: nonradiative and radiative decay. The former refers in this simple case to a transition from the excited to the ground state, without the emission of a photon. The excess energy is usually deposited in vibrational degrees of freedom, either of the molecule itself or of the surroundings. In the latter case, the energy is lost through the emission of a photon which is typically lower in energy (red-shifted) than the excitation photon due to the Stokes shift. To generate as many detectable photons per unit time as possible for a given excitation power, one requires: (i) a large absorption cross-section; and (ii) a high fluorescence quantum yield – that is, an efficient conversion of absorbed into emitted photons.

The former requirement brings with it a radiative lifetime on the order of nanoseconds, which can be related to the fundamental considerations of absorption and emission. This limits the total number of emitted photons, irrespective of the total incoming photon flux, because a single absorption–emission cycle takes about 10 ns. Thus, even in ideal circumstances no single molecule can emit more than 10^8 photons per second. The restricted collection properties of objectives, imperfect transmission and reflectivity of optics and limited quantum efficiencies of detectors result in typical effective collection efficiencies of <10%. The corresponding count rates on the order of a few million counts per second can indeed be observed in single-molecule experiments at cryogenic temperatures [31]. Under ambient conditions, which are of relevance for biological investigations, photobleaching puts a limit on the applied excitation intensities. The cause of this bleaching is often the generation and further excitation of triplet states that are accessed through intersystem crossing from the first excited singlet state. Despite the low quantum efficiency of the process (<1%), excitation powers must be chosen at the $kW\,cm^{-2}$ level in order to avoid rapid photobleaching. At these incident light levels, the observed count rates are below 10^5 photons per second [32].

Photobleaching is also the reason why single molecules usually emit a total of 10^5–10^7 photons before turning dark. The most likely cause of this 'sudden death' is triplet–triplet annihilation with molecular oxygen, which is a particular problem in biological environments. The highly reactive singlet oxygen that is generated attacks the dye and oxidizes it, greatly altering its electronic properties and thereby rendering it dark to the excitation photons [33, 34]. To make matters worse, triplet state formation is thought to be the main cause of phototoxicity [35, 36]. This situation can be improved somewhat by deoxygenating the system, or by using oxygen scavengers for *in vitro* experiments in solution [37].

5.2.2
The Signal-to-Noise Ratio Challenge

In addition to the saturation properties of single molecules discussed above, another difficulty arises from the fact that single molecules cannot be excited very efficiently. Even large single-molecule absorption cross-sections are only on the order of $10^{-16}\,cm^2$, compared to focal areas that are no smaller than $\sim 10^{-9}\,cm^2$. Therefore, only one in 10^7 photons that passes through the focus will cause electronic excitation of the molecule. What makes the situation even more problematic, is the fact that a typical focal volume contains on the order of 10^9 molecules. So, even if only one in 10^3 molecules emits a single fluorescence photon per second, the total emission background already matches the maximum fluorescence from the single molecule of about 10^6 photons, even at low temperatures.

As a consequence, initial attempts to detect single molecules used absorption [38], although these were swiftly followed by fluorescence detection [39]. The early studies were performed at cryogenic temperatures, where absorption cross-sections become large so that the saturation regime can be reached at much lower incident powers. At room temperature, however, the task appeared hopeless in the light of the numbers above. One possibility to improve this situation was to develop a technique that is: (i) only surface-sensitive; and (ii) somehow produces a much smaller excitation area. Scanning near-field optical microscopy (SNOM) provides exactly these properties [40, 41]. Here, the light is not focused as in a standard optical microscope, but rather is coupled into a metallized and sharpened tip of an optical fiber equipped with a subwavelength exit hole ($<50\,nm$). The tip is then brought within tens of nanometers of the surface to be studied and scanned laterally. In this way, only molecules that are on the surface are excited by the evanescent field at the tip's aperture, and only in an area that is comparable to the aperture. Therefore, the number of molecules that can contribute to the background signal becomes considerably smaller compared to the confocal arrangement, making the detection of single molecules much more probable. Indeed, the first room-temperature observation of single molecules was achieved using SNOM [42].

The rather difficult experimental set-up necessary for performing SNOM, along with the limitations to study surfaces, motivated the development of far-field single-molecule methods for biological studies. It was quickly realized that far-field methods are capable of generating much larger single-molecule signals than SNOM, with much-reduced experimental demands. Therefore, far-field detection has become the method of choice for detecting single molecules in biological environments [43, 44]. This advance was facilitated by the development of low autofluorescence microscope objectives and immersion oils, as well as improved excitation light rejection through the use of dielectric filters and highly efficient single photon detectors such as avalanche photodiodes. Today, single-molecule detection has become an almost standard technique in biology, chemistry and physics [45]. Single-molecule techniques have been used to directly observe the motion and function of single biological nano-objects such as enzymes, viruses or motor proteins in real time [37, 46, 47].

5.2.3
High-Precision Localization and Tracking of Single Emitters

In the previous section, we discussed the difficulties and current solutions to detecting single molecules. However, we are still faced with the problem, that single molecules are much smaller (~1 nm) than the best possible resolution of an optical microscope (~200 nm), which is linked to the wave nature of light [48]. The crucial point is that the image of a point-like emitter is itself not infinitely small but rather appears as an Airy diffraction image (Figure 5.2a), with the ripples originating from the diffraction at the edges of a circular objective, for example. Because these patterns, which are also known as the point spread functions (PSFs), are caused by the lens, the smaller the aperture of the lens the wider the PSF and the lower the resolution, and vice versa. Here, it is useful to define the term numerical aperture, $NA = n\sin(\varphi)$, where φ is defined as the half collection angle of the objective. Therefore, the larger the NA, the higher the resolution of the microscope. The distance from the central maximum to the first minimum is given by $d_{min} = 0.61$ /NA for a circular aperture, where λ is the wavelength of light and is a common measure of resolution also known as Rayleigh's limit (Figure 5.2b).

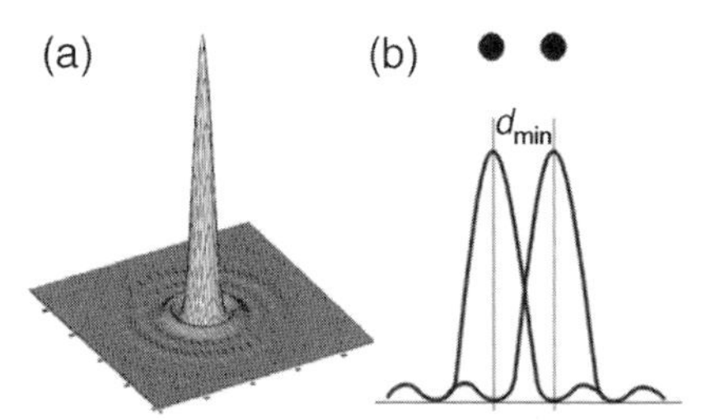

Figure 5.2 Point spread functions and their importance in determining resolution. (a) Surface plot of a typical Airy diffraction pattern; (b) Schematic representation of Rayleigh's criterion. The two graphs represent slices through the 2-D Airy function shown in (a).

Despite the fact that a single molecule smaller than 1 nm yields an image with a diameter of several hundred nanometers, it is possible to determine its location to within a few nanometers. This can be achieved by fitting the data to the theoretical PSF. Thus, the precision of this fit is only limited by the signal-to-noise ratio (SNR) of the acquired Gaussian profile, while the fit tolerance provides the uncertainty in x and y of the emitter's center [49]. To illustrate this, we have acquired confocal images of single rhodamine-labeled GM1 receptors adhered onto an acid-cleaned coverslip (Figure 5.3a). The observed single-step bleaching in Figure 5.3b for the spot highlighted in Figure 5.3a confirms that the image stems from a single molecule. The emission pattern (Figure 5.3c) and the corresponding Gaussian fit (Figure 5.3d) to the highlighted molecule result in a lateral localization accuracy of 10 nm. The high accuracy is due to the fact that all the information in two dimensions can be used for the fit. To illustrate this, three slices and the corresponding fits along the x axis of the spot are shown in Figure 5.3e–g. The center position fluctuates by >20 nm for these three fits due to the limited SNR. A two-dimensional fit, however, provides much higher accuracy, because one effectively fits all slices in every possible direction simultaneously. This approach has been employed in several recent investigations, including the study of lipid diffusion inside supported membrane bilayers [50], and of the mechanism of the molecular motor kinesin stepping along microtubules [51]. While the former study showed a maximum localization accuracy of ~40 nm, the latter state-of-the-art measurements succeeded in realizing molecular resolution (~1.5 nm) with sub-second time resolution.

As can be seen from the previous discussion, the localization accuracy is critically dependent on the SNR with which the PSF can be measured. The limitation arises from the finite number of detectable photons per molecule. The longer the integration time, the higher the accuracy but also the lower the time resolution. As a rule of thumb, a SNR of 10 is required for a localization accuracy of ~10 nm, which translates into roughly a time resolution on the order of several to tens of milliseconds [49, 52]. This makes tracking beyond video rates difficult if the object is to remain visible against the background, especially for *in vivo* imaging [47]. In addition, the total tracking time is limited to a few seconds because of photobleaching. These issues can only be addressed by using labels with no limitations on the number of emitted photons and on photostability. Nevertheless, single molecules have been used successfully to track individual biological nano-objects in real time. An excellent example is given in Figure 5.4, where single adenoviruses were labeled with single dye molecules and then observed before, during and after cell entry [47].

Inorganic quantum dots have become popular as labels in fluorescence microscopy because of their brightness and extreme photostability [53]. Their inherent toxicity is commonly deactivated through the use of protecting layers, and they have been used successfully in intracellular and *in vivo* studies [54]. A major disadvantage of these labels is that, so far, their emission switches off intermittently at unpredictable times and for unknown durations, a process known as *photoblinking*. In addition, once passivated, they can become as large as 15–20 nm.

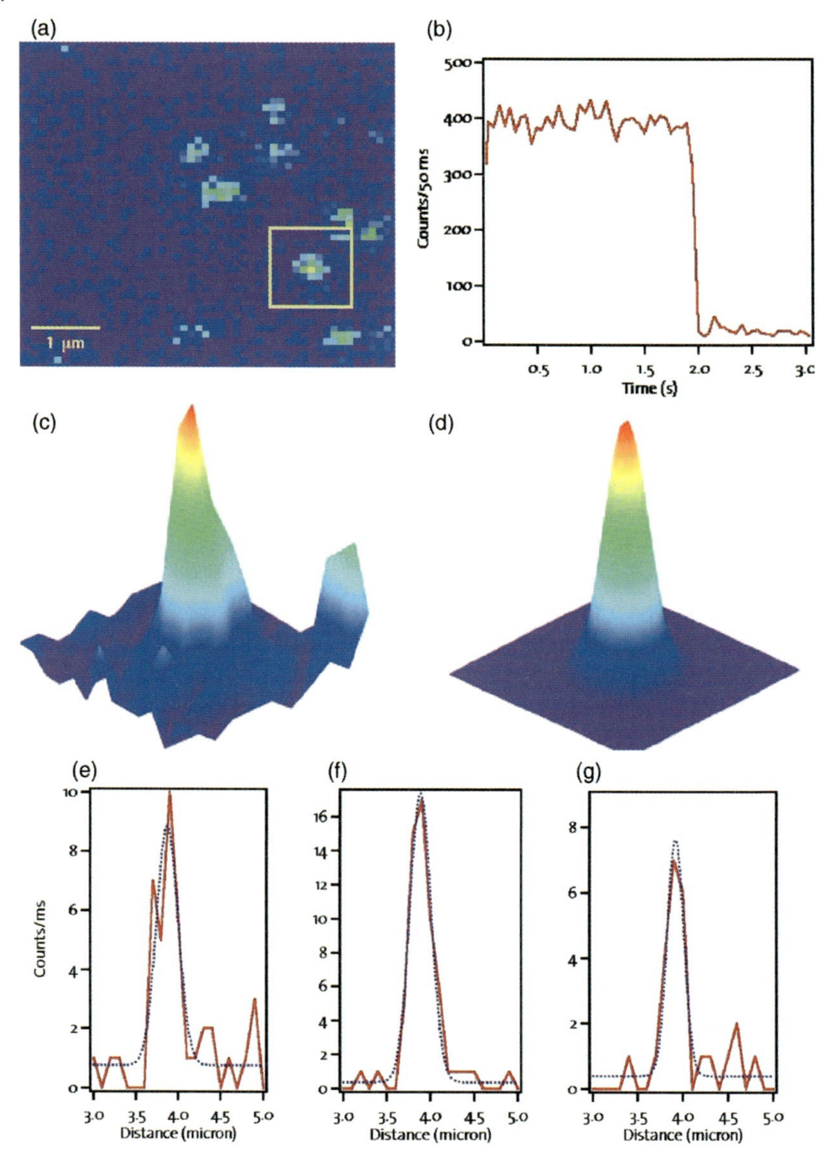

Figure 5.3 High-precision localization of single emitters. (a) Confocal fluorescence scan of a glass coverslip coated with single dye-labeled GM1 receptors. Pixel dwell time: 1 ms; illumination intensity: ~1 kW cm^{-2}; total acquisition time: 10 s; (b) Detector counts as a function of time for the molecule highlighted in (a). The single-step bleaching demonstrates single-molecule sensitivity; (c) Surface plot of the fluorescence collected during the scan from the highlighted molecule; (d) Two-dimensional Gaussian fit to (c); (e–g) Three slices along the *x*-axis of the fluorescence spot (red) accompanied by the corresponding 1-D Gaussian fits (blue, dotted).

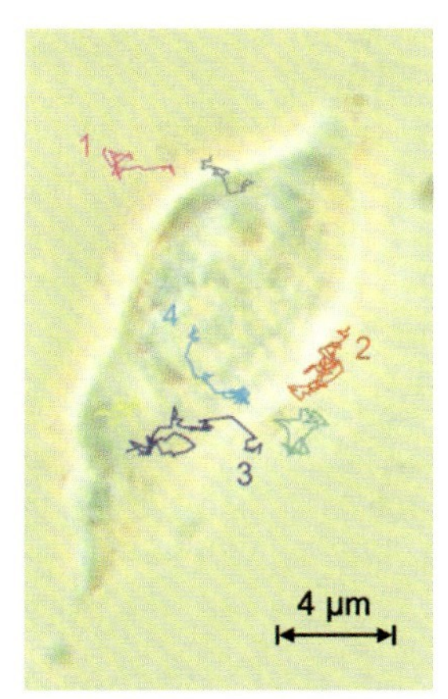

Figure 5.4 Single virus tracking. The various trajectories describe different stages of the infection pathway such as diffusion in solution (1, 2), penetration of the cell membrane (3), diffusion in the cytoplasm (3, 4), penetration of the nuclear envelope (4) and diffusion in the nucleoplasm (5). Adapted from Ref. [47].

5.2.4
Getting Around the Rayleigh Limit: Colocalization of Multiple Emitters

While it may be possible to localize a single emitter with a precision comparable to its size, the fundamental problem defined by Rayleigh's criterion remains if identical objects that are separated by less than half the wavelength of light are to be resolved. This leaves a large gap between the achievable (~200 nm) and the desired molecular resolution (~1 nm). Several approaches have been explored over the past decades, many of which have closed this gap partially and are described in detail [55]. Here, we will focus on concepts that are particularly suited toward the study of single emitters.

One approach is based on the idea that if one emitter could be observed without the other, then the location of each could be pinpointed with high precision by using the methodology outlined above. One way to achieve this task would be to image spectrally orthogonal emitters, as demonstrated by Weiss and coworkers [56], through the use of two inorganic quantum dots fluorescing at different wavelengths. Each of the emitters can be observed independently by separating the emission of the two with a dichroic mirror. Unfortunately, this multicolor colocalization can only be applied to a few particles because fluorescence emission is generally broad, and it is not realistic to use more than two or three emitters simultaneously.

Rather than differentiating between emitters spectrally, another option would be to do this temporally. An early implementation of this approach involved the stepwise photobleaching of individual emitters, as demonstrated by Selvin *et al.* [57]. Here, two emitters (m_1 and m_2) are imaged continuously. The integrated intensity of the emission of both molecules as a function of time shows a two-step behavior due to

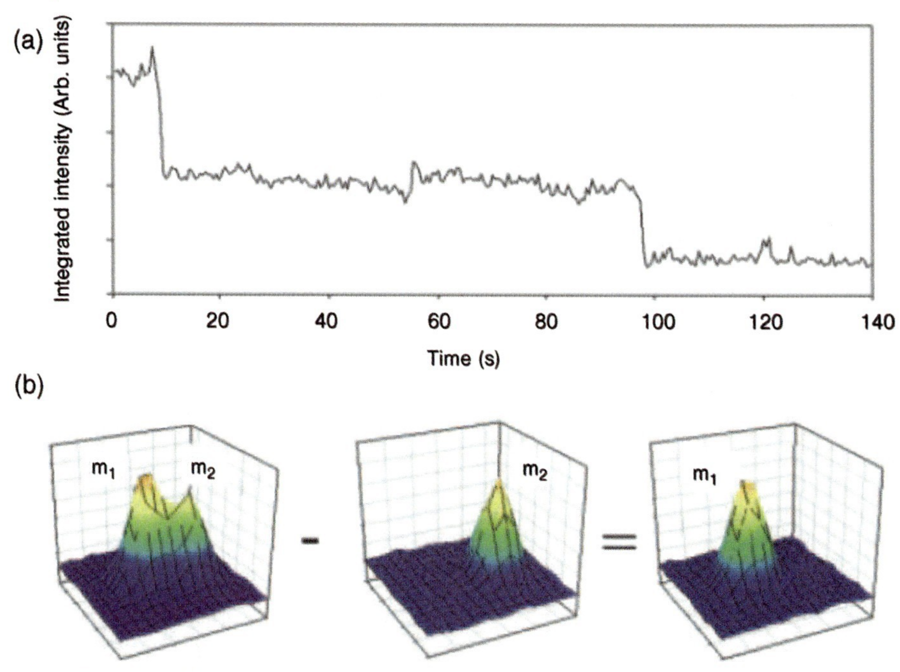

Figure 5.5 Subdiffraction localization through stepwise photobleaching. (a) Summed emission intensity showing clear two-step photobleaching; (b) Schematic representation of the localization procedure. Adapted from Ref. [57].

photobleaching, such as that shown in Figure 5.5a. After the first bleaching event, only m_2 is visible, and this can now be localized with high precision (as described previously). Subtracting the contribution of m_2 from the initial image, where both emitters are present, results in an image representative of m_1 (Figure 5.5b) which can again be localized with high precision. Thus, it is possible to determine the positions of several emitters with near-molecular resolution (down to 1.5 nm). This technique is extremely useful and precise for imaging fairly simple samples containing few fluorophores. However, for general applications such as those required for *in vivo* imaging where many labels are present, it quickly reaches its limit.

This barrier has recently been lifted by a recent approach proposed by Betzig and coworkers [58] and by Zhuang and colleagues [59], based on photoswitchable fluorophores. These methods have been named PALM (photoactivated localization microscopy) and STORM (stochastic optical reconstruction microscopy), respectively. Here, the problem of multiple fluorophores emitting simultaneously is eliminated by initially illuminating the sample in the near-UV (405 nm) at low light levels. This causes a small and stochastically distributed fraction of the total molecules to convert photochemically into an active state. Illumination of the sample in the yellow region (561 nm) then causes the photoactivated molecules to fluoresce.

By observing and fitting the emission pattern from each of these molecules, they can be localized with high precision. Those molecules are subsequently bleached and the process is repeated until the entire sample has been imaged. The resulting images are of spectacular clarity and resolution, especially when compared to standard confocal images, such as in Figure 5.6a–d. The main disadvantage of this approach is currently the low time resolution required by the stochastic activation of a small number of emitters and the reliance on the destruction of the activated signal.

The highest spatial resolution in three dimensions has been achieved at cryogenic temperatures, taking advantage of spectral selectivity [60]. When cooled to a few Kelvin, the absorption and emission from single molecules becomes extremely narrow and highly sensitive to their local environment. Single molecules can then be excited individually and therefore localized with nanometer precision. Whilst, in

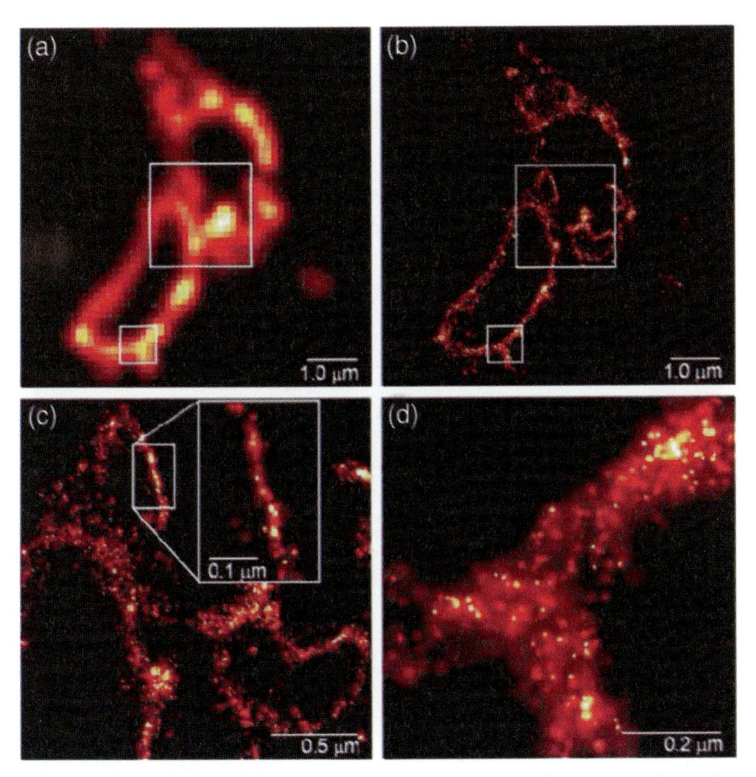

Figure 5.6 Near-molecular resolution using photoactivated localization microscopy (PALM). Comparison between summed molecule TIRF (a) and PALM (b) images from a thin, cryoprepared section of a fixed cell. An enlargement of the large boxed region in (b) reveals smaller associated membranes (c). The inset shows highly localized (10 nm) molecules. An enlargement of the smaller box in (c) shows the distribution of individual molecules within the membrane (d). Adapted from Ref. [58].

principle, this approach should be extendable to biological studies, none has so far been reported due to the high degree of experimental complexity compared to room-temperature studies.

One of the most successful and widespread approaches to achieve spatial information far below the Rayleigh limit involves taking advantage of fluorescence resonance energy transfer (FRET). The principle of this approach is depicted schematically in Figure 5.7. It uses two chromophores, a donor (D) and an acceptor (A), with overlapping absorption and emission bands (Figure 5.7b). When the two emitters are well separated, excitation of the donor will lead to observable emission only from the donor. However, when the two fluorophores are brought into close proximity of each other, the excitation energy originally placed in the donor is efficiently transferred to the acceptor due to the overlapping absorption and emission bands through Förster energy transfer (Figure 5.7c). This indirect excitation of the acceptor leads to fluorescence emission of the acceptor (i.e. far red-shifted compared to that of the donor). Commonly, the emission channels of both the donor and the acceptor are monitored simultaneously by the use of appropriate beam splitters. Thus, a dynamic system where the distance of the two

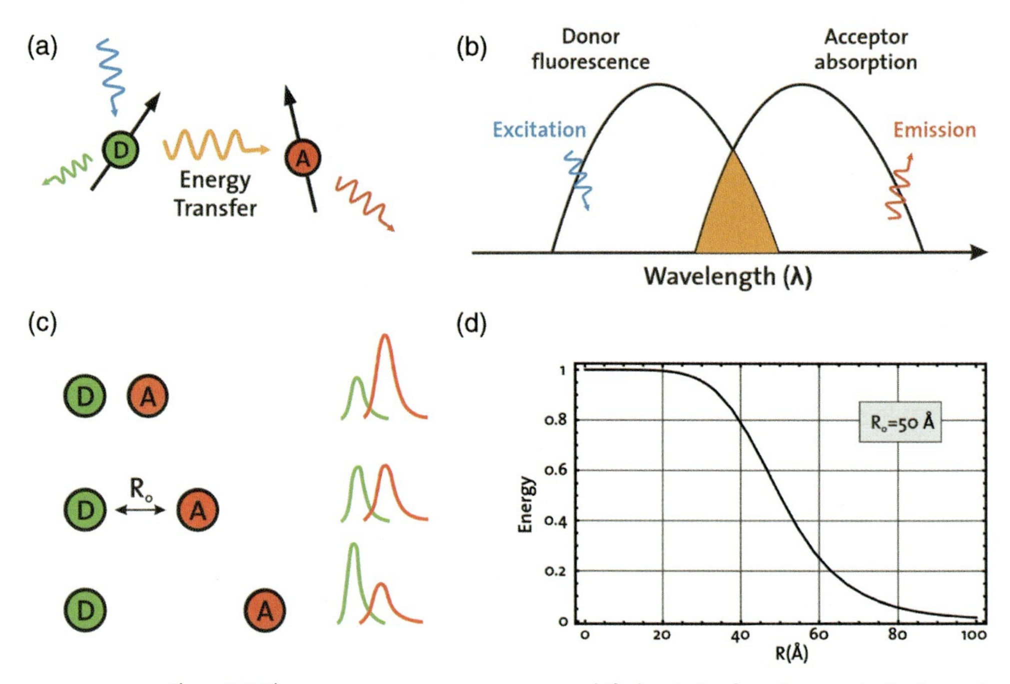

Figure 5.7 Fluorescence resonance energy transfer (FRET). (a) When a donor (D) and an acceptor (A) molecule with overlapping emission and absorption bands (b) are brought into close proximity, energy from the donor is transferred to the acceptor. In this case, red-shifted emission from the acceptor is observed; (c) When the two molecules are separated, donor fluorescence dominates; (d) The distance between the two molecules can be determined with high precision in the 2–8 nm range from the donor to acceptor emission ratio.

labels changes with time, for example due to conformational changes of a protein, will exhibit alternating emission from donor and acceptor [61]. The strong distance-dependence (R^6) of energy transfer efficiency makes FRET an excellent molecular ruler on the sub-10 nm length scale (Figure 5.7d).

5.3 Detection of *Non*-Fluorescent Single Nano-Objects

Despite the amazing capabilities of microscopy using single-molecule labels, two major limitations have become apparent. First, the total number of detectable photons is limited due to photobleaching, thus restricting the total observation times to a few seconds. Second, the saturation properties of single molecules limit the photon count rates to around $10^5\,s^{-1}$ or less. The requirement for a total of $\sim$100 detected photons for reasonable localization [52] results in a maximal time resolution on the order of milliseconds.

Many of the difficulties described above can be eliminated through the use of scattering rather than fluorescence as a contrast mechanism. Here, there is no limit on the number of detected photons because the amount of scattered photons depends only on the incident light level. As a result, an unlimited time resolution is theoretically possible. In addition, because a scattering object acts like a tiny 'mirror', neither bleaching nor blinking is an issue, providing indefinitely long unlimited observation times without dark periods. These advantages have led to the emergence of gold nanoparticles in biological applications with reported time resolutions down to $\sim$25 µs [62]. The downside associated with the use of gold nanoparticles as optical labels is the strong size dependence of the scattering cross-section, which results in a minimum label size on the order of $\sim$30 nm in dark-field detection [14, 63]. The fact that many biological nanoparticles of interest are either comparable (e.g. viruses) or much smaller (proteins) in size than this has restricted the applicability of these labels. In particular, such large labels may strongly perturb the motion of the entity under study.

5.3.1 The Difficulty of Detecting Small Particles Through Light Scattering

The ultimate goal is to detect a single molecule-sized scatterer, as this would combine the advantages of scattering detection and the minimal perturbation of the system. The question becomes: why is it so difficult to detect an object such as a 1 nm gold nanoparticle?

In many ways, the origin of this problem is very similar to that discussed earlier for single molecules. If a single-molecule-sized gold particle could be trapped in a vacuum, the light scattered by the particle could be observed by the naked eye. In a realistic environment, however, background scattering will easily overwhelm the tiny signal generated by the gold particle. For single-molecule fluorescence, one is limited by the maximum photon emission rate, background fluorescence and a small absorption cross-section. For scattering detection, the situation is even worse because *every* object in the focal volume scatters light and the spectral selectivity available in fluorescence detection is lost, making low background measurements a true challenge.There are two possible sol utions to this problem: (i) the number of background scatterers from the focal volume is minimized; or (ii) the scattering signal from the particles of interest is somehow increased. The former has been the traditional approach to detecting small gold nanoparticles through dark-field or total internal reflection microscopy [14]. In this case, the scattered light intensity depends on the square of the polarizability of the particle which, in the electrostatic approximation, can be written as [64]:

$$\alpha(\lambda) = \frac{\pi d^3}{2} \frac{\varepsilon_p - \varepsilon_m}{\varepsilon_p + 2\varepsilon_m}$$

for a spherical particle of diameter d and dielectric constant ε_p inside a medium of dielectric constant ε_m. It easily follows, that the scattering intensity scales to the sixth power with the particle size; a 5 nm particle will therefore scatter light a million times less efficiently than a 50 nm particle! As a consequence, the signal from scatterers smaller than 30 nm drops below the background, even when a dark-field approach is used [14, 63].

Metallic nanoparticles such as those composed of gold or silver exhibit a so-called 'plasmon resonance'. As can be seen in the equation, such a resonance causes the denominator to approach zero in the specific case of $\varepsilon_p(\lambda) \rightarrow -2\varepsilon_m(\lambda)$, and therefore makes the scattering amplitude large. For spherical gold nanoparticles, this resonance occurs conveniently in the visible region of the spectrum at ~530 nm [65]. For biological nanoparticles, the scattering cross-sections are roughly a factor of three smaller than for gold because of the missing plasmon resonance [66].

To circumvent these difficulties, two approaches have emerged recently that allow the detection of gold nanoparticles down to 5 nm. One method involves taking advantage of the absorption of gold nanoparticles, which scales linearly with particle volume and therefore with the third power of the particle diameter. The resulting drop in the sensitivity of signal on the particle size has been utilized both in direct absorption measurements [67] and in the observation of a change in the refractive index through heating of the surrounding medium caused by the absorption of radiation [68]. We will focus here on an alternative method that measures the electric field directly, thereby achieving d^3 sensitivity without the need to heat the sample [69–71]. As we will show, this approach also brings with it the unique advantage of being able to detect biological nanoparticles *without any labels.*

5.3.2
Interferometric Detection of Gold Nanoparticles

To illustrate our approach, it is useful to consider the simple experimental set-up shown in Figure 5.8. Here, the incident laser beam is focused onto the sample, which in the simplest case consists of gold nanoparticles on a glass coverslip. As detection

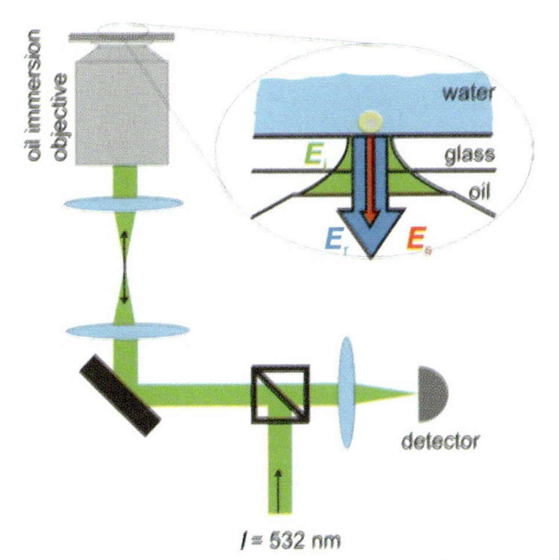

Figure 5.8 Simplified experimental set-up for the interferometric detection of nonfluorescent nanoparticles. The incident light, E_i, is reflected at the glass/water interface (E_r) and collected along with the scattered light from the particle, E_s, by the microscope objective. A portion of this light then passes through the beam splitter and reaches the detector.

occurs in the epi-direction, we are interested in the light returning through the microscope objective. The detector will therefore see the incident light (E_i) reflected at the interface between the glass and the medium. The reflected field reads $E_r = rE_i(\exp(-i\pi/2))$ where r is the field reflectivity and $\pi/2$ denotes the Gouy phase of the reflected focused beam. In addition, light scattered by the particle can be written as $E_s = sE_i = |s|\exp(i\varphi)E_i$ at the detector, where s is proportional to the particle polarizability and therefore to d^3. Here, ϕ signifies the phase change on scattering. The intensity, I_D, measured at the detector is thus

$$I_D = |E_r + E_s|^2 = |E_i|^2(r^2 + s^2 - |r||s| \sin\phi).$$

We can use this equation to illustrate some of the factors discussed above. Dark-field or total internal reflection detection are designed in such a way that $r \to 0$ and therefore only the scattering term, s^2, is detected. This signal drops very rapidly below the background level ($|rE_i|^2$) for particles <30 nm. In this case, the nature of the observed signal depends on the relative magnitudes of the three terms in the above equation. For large particles, the scattering term, s^2, dominates and the particles appear bright against the background (Figure 5.9). As the particle size decreases, s^2 becomes negligible compared to the other two terms, and only the reflection and interference terms contribute to the detected intensity. The particles appear dark against the background due to the destructive interference between the scattered and the reflected beams caused by a $-\pi/2$ Gouy phase shift of the reflected incident beam (Figure 5.10a and b) [71]. The change-over from bright to dark occurs according to the relative magnitudes of the scattering and interference terms, and therefore occurs

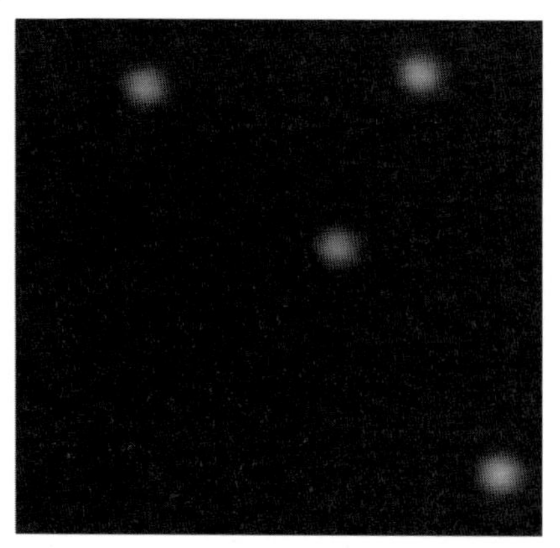

Figure 5.9 Confocal scan of 100 nm gold nanoparticles.

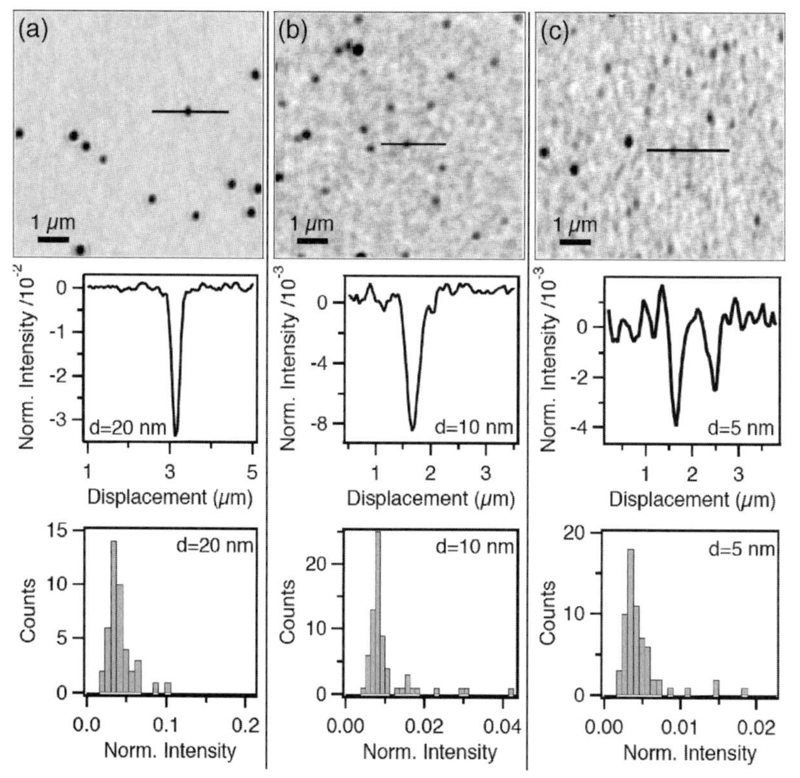

Figure 5.10 Interferometric images of gold nanoparticles spin-coated onto glass coverslips. (a) 20 nm; (b) 10 nm; (c) 5 nm diameter. Representative particle cross-sections, as well as intensity histograms, are provided in each case. Total acquisition time in each case = 10 s; incident power = 2 mW.

earlier for larger r: for example, at 40 nm for air as the surrounding medium, 30 nm for water, and 15 nm for oil that is index-matched fairly well to the glass coverslip.

5.3.2.1 Is it Possible to Detect Molecule-Sized Labels?

To explore the theoretical sensitivity limitations, it is useful to consider the origin of the true noise background that limits the detection sensitivity. This is governed mostly by the noise of the light source itself and the noise of the detector. Both will result in fluctuations in the detected reflected intensity, r, which is the major contributor to the overall detected signal at the detector. Other potential noise sources such as mechanical instabilities of the microscope or beam-pointing instability of the light source are comparatively small and easily corrected for in post-acquisition image processing.

An incident power of 1 mW on the sample will yield $\sim$3 μW of light reaching the detector, taking into account the reflectivity of the glass–water interface and losses due to the limited transmission of optics such as the microscope objective. The ideal detectors for such light intensities are photodiodes, which produce a corresponding photocurrent of 1 μA. The shot noise limit for this photocurrent is on the order of 500 fA $\text{Hz}^{-1/2}$, which is about an order of magnitude above the noise of available amplifiers with 10^7 V/A gain, suggesting that shot noise-limited detection is possible. At this amplification and a realistic detection bandwidth for mechanical scanning of 1 kHz, the electronic shot noise amounts to $\sim 1.5 \times 10^{-5}$ rms, which is a factor of $\sim$300 below the magnitude of the signal observed for 5 nm gold particles. A factor of three reduction in size on the other hand, which would lead to molecular sized labels on the order of 1.3 nm, brings with it a factor of 27 reduction in signal intensity. Thus, such molecular-sized labels should be observable with a SNR of 10 at kHz bandwidths with localization accuracies down to 10 nm!

The previous discussion has shown that neither shot noise nor detector noise limit the detectability of such small labels. One other critical noise source remains: the light source itself. Lasers used in confocal microscopes show optical noise on the order of a small percentage over a wide frequency range. Even state-of-the-art, solid-state, diode-pumped lasers rarely perform better than 0.1%. However, external stabilization using optical fibers, acousto-optic modulators and feedback loops has been shown to reduce laser noise from a small percentage to $\sim 5 \times 10^{-5}$ rms with kHz bandwidth [72], which is comparable to the electronic shot noise calculated above. In addition, the use of single-mode fibers in this stabilization scheme significantly reduces the effects of beam-pointing and mode instabilities, further contributing to the overall stability of the system. Given these simple calculations, it becomes evident that the rapid detection of molecular-sized gold scatterers should be possible. We are currently pursuing such experiments.

All of the images presented in Figure 5.10 have been obtained by scanning the sample across the focus using a piezo translation stage that requires 1–10 s per image. By using scanning mirrors rather than a piezo stage, we have shown previously that it is possible to detect 20 nm particles with up to MHz bandwidths – three orders of magnitude above what is possible using single molecules as labels [73]. In addition, rather than scanning the focus across the surface, the use of a feedback loop enables one to lock the focus to a particle and follow its movements rapidly. The feedback loop is fed by the signal recorded on a four-quadrant detector, where the movement of the particle inside the focus leads to changes in the measured differential voltages. Using the same detector and amplifier combination above, which provide MHz detection bandwidths, the shot noise increases to $\sim 5 \times 10^{-4}$ with mW incident powers. In this way, the tracking of labels as small as 5 nm with MHz bandwidths should be possible.

The sensitivity limitations of the technique are illustrated in Figure 5.10c, which shows a confocal image of 5 nm particles spin-coated on a glass coverslip and covered by water. As can be seen in the image, the particles are visible against the background with a signal contrast on the order of 3×10^{-3}. The distribution width of the signal intensities is in agreement with the manufacturer's specifications with regards to the size of the gold particles, confirming that we are indeed observing single particles.

A close inspection of Figure 5.10b and c reveals the presence of a rather 'noisy' background as the size of the particles and thus their signal magnitude decreases. However, these features are *not* noise, as they are perfectly reproducible in sequential images. Rather, these patterns are due to the surface roughness of the glass coverslips used. Indeed, AFM measurements have shown that the surface roughness amounts to a few nanometers over a few microns. The reproducibility of such nanometer-sized surface roughness demonstrates the excellent sensitivity of this technique to nonmetallic species.

5.3.2.2 The Needle in the Haystack: Finding and Identifying Gold

So far, we have been concerned mostly with detecting and tracking the smallest possible gold nanoparticles. An interesting point to address is how such small labels can be detected in the presence of much larger scatterers, for example in intracellular imaging. Fortunately, gold nanoparticles have a type of 'built-in identification card' in the form of a plasmon resonance in the visible region of the electromagnetic spectrum (Figure 5.11a). As a result, one obtains roughly twice the scattering intensity in the green (532 nm) compared to the blue (488 nm) or red (>560 nm) regions. This wavelength-dependent scattering intensity is in contrast to the constituents of typical biological samples, where the scattering should be roughly identical for both wavelengths.

To demonstrate the possibility of using this interesting feature of gold nanoparticles, we have labeled microtubules with 40 nm gold particles and obtained scattering images simultaneously in the blue and green (Figure 5.11b and c). In the two images, both the nanoparticles and the microtubule are clearly visible. However, when the two images are subtracted from each other (Figure 5.11d), the microtubule disappears while the particles remain. One can thus use this form of differential spectral contrast to ensure that the observed particles are indeed the gold labels of interest and not other scatterers [69].

5.3.3 Combining Scattering and Fluorescence Detection: A Long-Range Nanoscopic Ruler

As yet, we have discussed the advantages and disadvantages of fluorescence and scattering as labels only in biological imaging applications. Now, we present an

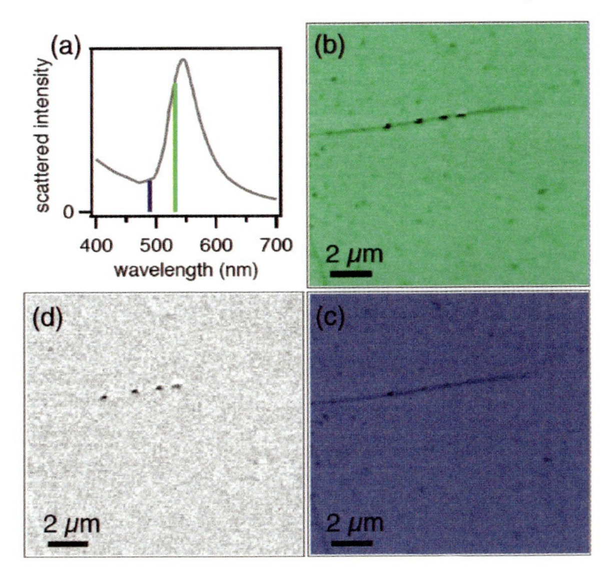

Figure 5.11 Identification of gold scatterers through spectral difference. (a) Plasmon resonance for a 40 nm gold particle; (b) Interferometric image of microtubules labeled with individual 40 nm gold particles acquired at 532 nm; (c) Identical image acquired simultaneously at 488 nm; (d) Subtraction of the blue from the green image. The microtubule with approximately identical scattering cross-sections at the two wavelengths disappears, while the gold particles remain visible.

example where the combination of the two leads to a potentially useful technique. It has been shown previously that nanostructures brought into close vicinity of single emitters can cause enhancement of luminescence and Raman scattering [75]. We have shown recently that a single gold nanoparticle can enhance the fluorescence of a single molecule and the decay rate of its excited state by a factor of 20 [76]. Furthermore, we demonstrated that this strong fluorescence modification is a function of the particle–emitter separation with nanometer sensitivity.

The mechanism of this effect can be intuitively explained as the near-field interaction of the molecular dipole moment, with its image dipole induced in the gold nanoparticle. The dipole–dipole character of this interaction gives rise to a strong distance-dependence much in the same way as in FRET (see also Figure 5.10). However, in this case the interaction range drops much more softly than the $1/r^6$ dependence observed in FRET. Thus, the modification of the fluorescence lifetime close to a nanoparticle can be used as a nanoscopic ruler for distances larger than that of FRET (>10 nm).

To demonstrate this, we have performed studies of systems where a single molecule is linked to a gold nanoparticle with DNA double strands of differing lengths [77]. The techniques of single-molecule detection and microscopy of gold nanoparticles were combined to locate such molecule–particle pairs. The corresponding confocal scans of single functionalized gold nanoparticles of 15 nm

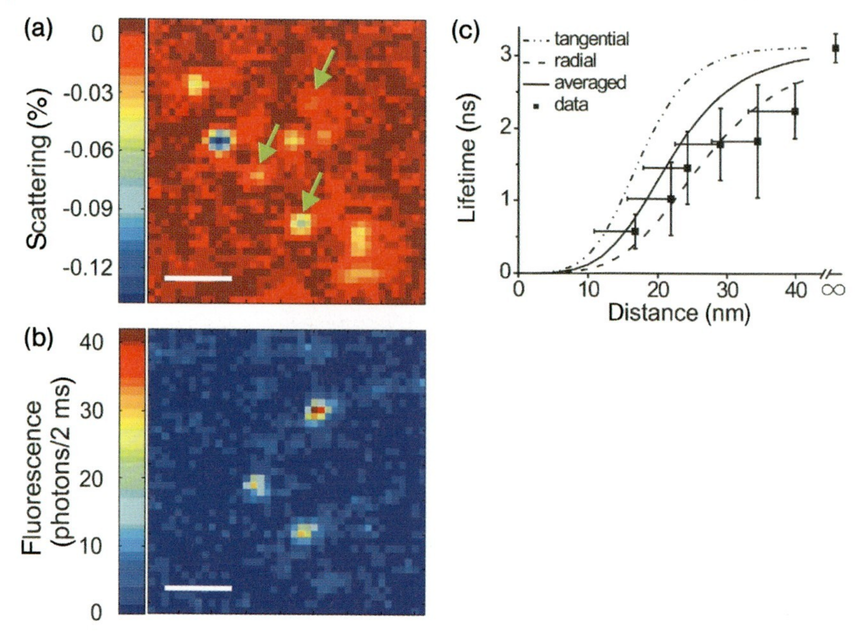

Figure 5.12 Combining scattering and fluorescence detection for a nanoscopic ruler. (a) Scattering image of single 15 nm gold particles functionalized with a single dye molecule via a DNA linker; (b) Simultaneously acquired fluorescence image. The arrows in (a) indicate gold particles that are functionalized with fluorescent markers; (c) Single-molecule fluorescence lifetime dependence on linker length. The dashed and dashed-dotted curves display the calculated fluorescence lifetime for the molecular dipole oriented radially or tangentially with respect to the gold nanoparticle. The solid curve shows the weighted average of the two orientations.

diameter are shown in Figure 5.12a (scattering) and b (fluorescence). As can be seen in the figure, only a fraction of the gold particles contains a fluorescent marker. Figure 5.12c demonstrates the dependence of fluorescence lifetime on linker length. In the absence of a gold nanoparticle, the fluorescence lifetime of the molecule was ~3 ns, but this was reduced to about 0.6 ns for a 15 nm-long DNA linker consisting of 44 base pairs. The interaction length and its slope can be tuned by choosing the particle size and emission wavelength of the dye molecule. The precision of such a nanoscopic ruler is on the order of 1 nm, and is limited by the accuracy with which the fluorescence lifetime can be determined [77].

5.3.4
Label-Free Detection of Biological Nano-Objects

Although we have focused mostly on gold as a scattering label, the previous discussion has also shown that interferometric detection is extremely sensitive to virtually any type of scatterer. This is demonstrated on the one hand by the observation of nanometer surface roughness on glass coverslips, and on the other hand by the visibility of individual microtubules which are hollow shells of only 24 nm diameter. These results suggest that it may be possible to detect biological nano-objects *without the need for any label*. Such detection brings with it the aforementioned advantages of scattering detection, but more importantly eliminates any outside perturbation on the system which may be introduced by labels; moreover, it eliminates the need for labeling chemistry.

To illustrate the capabilities of the technique in this respect, we have obtained scattering images of unlabeled Simian virus 40 (SV40) virions bound to microscope coverslips. SV40 is a small, 45 nm-diameter, tumor-virus consisting of an outer protein shell of 720 copies of the VP1 protein, with a 5000 base-pair DNA-genome in its core (Figure 5.13a) [78]. As can be seen from the image in Figure 5.13b,

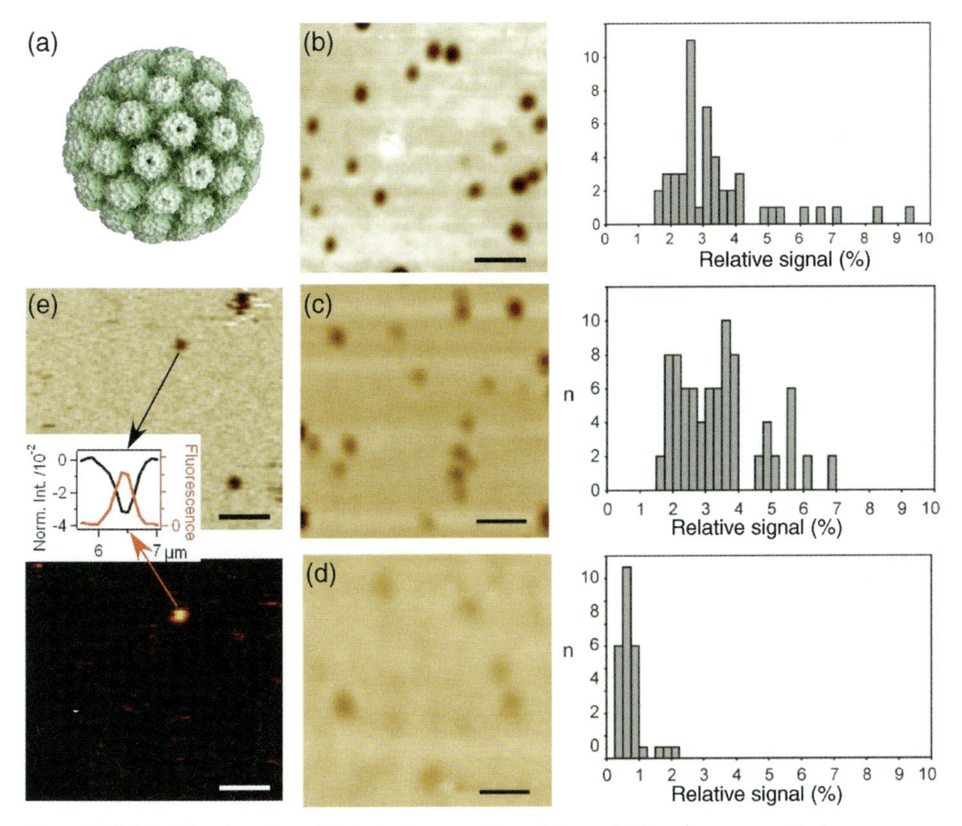

Figure 5.13 Label-free detection of biological nanoparticles. (a) Structure of the SV40 virus as determined from X-ray crystallography; (b) Interferometric images of SV40 adhered to glass and (c) on a supported membrane bilayer; (d) Virus-like particles on a supported membrane bilayer; (e) Simultaneous scattering and fluorescence images of single SV40 viruses labeled with multiple atto-565 fluorophores. The inset shows a cross-section of the particle for both images. Scale bars = 1 µm.

a 45 nm virus shows roughly the same scattering intensity as a 20 nm gold particle ($\sim 3 \times 10^{-2}$). The difference in the observed scattering intensity is due to the lacking plasmon resonance of the virus, and thereby to a reduced polarizability. To test the applicability of this approach in a biologically relevant environment, we have also obtained images of SV40 virions bound to supported lipid bilayer membranes (Figure 5.13c) [66]. Prior to addition of the virus, the membranes showed a homogeneous background signal and few detectable spots caused by unfused vesicles. Within seconds of adding the virus to the solution, Gaussian spots appeared in the image, corresponding to single viruses binding to GM1 pentasaccharide receptors that had been added to the membrane. The signal intensity was comparable to that of viruses bound to the glass coverslips (2.6×10^{-2}). To confirm the validity of the interpretation, we performed two further experiments. First, we acquired images of SV40 virus-like particles, which are identical to the virions, except that the DNA core had been removed. As might be expected, due to the reduced amount of material, the observed signal intensity was lower compared to the viruses at 0.75×10^{-2} (Figure 5.13d). Second, we checked our images by simultaneously acquiring fluorescence and scattering images of fluorescence-labeled SV40 viruses on a supported membrane bilayer (Figure 5.13e). As can be seen in the figure, the two images corresponded perfectly. The inset shows a slice along the virus, demonstrating the complementary nature of the two signals.

In addition, we have performed consecutive confocal scans to investigate diffusion of the virus on the membrane. The trace of the viral motion is superimposed on the image in Figure 5.14a. Computational analysis of the trajectories yielded linear mean square displacement plots, as would be expected for particles undergoing Brownian motion (Figure 5.14b), and exhibited a diffusion constant (D) of 0.0088 ± 0.0004 $\mu m^2 s^{-1}$. These first results demonstrate the power of the interferometric detection of nano-objects, and its potential for the long-term tracking of unlabelled biological entities. The requirement for this is of course a well-defined sample where unwanted scattering has been eliminated, as in the case of the membrane studies presented here.

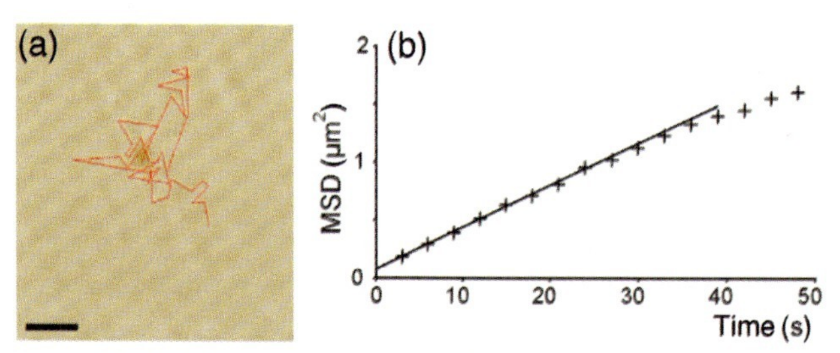

Figure 5.14 Label-free tracking of a single virus diffusing on a supported membrane bilayer. (a) Motion of the virus with 1 s time-resolution acquired over 50 s; (b) Corresponding mean square displacement as deduced from the observed motion in (a), indicating Brownian motion.

5.4
Summary and Outlook

We have discussed the capabilities and limitations of single-molecule fluorescence microscopy, with particular attention being paid to the critical parameters in biological imaging such as the SNR, time resolution and observation time. We have seen that, by detecting and localizing individual fluorescent molecules in a sample, the resolution in optical microscopy can be pushed down to the 1–10 nm regime. The resolution limit in this method is dictated by the noise of the fluorescence signal, and therefore, by the number of photons recorded from each emitter. Currently, problems such as photobleaching and photoblinking prevent an arbitrarily high resolution being achieved in realistic systems. Any efforts to suppress or minimize photobleaching are, therefore, of utmost importance to the future of high-resolution optical microscopy. One interesting possibility is to perform single-molecule detection at cryogenic temperatures where photochemistry is slowed tremendously.

For some imaging and real-time *in vivo* tracking applications, a very promising solution is offered by metallic nanoparticles as alternative labels. The lack of saturation, photobleaching and blinking in light scattering makes such labels ideal candidates to avoid many of the pitfalls of single-molecule spectroscopy. In particular, we have shown how recent advances in the interferometric detection of single gold nanoparticles enable the observation of such labels with sizes down to 5 nm. The ability to illuminate the sample at high power, without saturating the signal, allows a faster integration time and thus a much improved tracking speed, to more than three orders of magnitude above what is possible in single-molecule applications. Finally, we have shown how this interferometric detection technique can be used to observe dielectric objects *without the need for any labels.* Specifically, we have demonstrated label-free detection and the tracking of single SV40 viruses diffusing on artificial lipid bilayer membranes.

The 1990s were witness to a fantastic revival of optical microscopy for high-resolution imaging. Moreover, advances in laser spectroscopy, scanning microscopy, detector technology and photophysics have made it possible to interrogate matter at the single-molecule level, using light *in vitro* and even *in vivo*. In particular, various techniques have shown that the Rayleigh limit can be tackled for specific applications, and have demonstrated optical resolution at the 1–2 nm level. These improvements provide exciting new tools to study a wealth of biological questions at the subcellular level with a spatial resolution more than tenfold higher than can be achieved with conventional confocal microscopy. In other words, the resolution gap between optical microscopy and electron microscopy has been made much smaller. However, what is especially valuable is that this high optical resolution provides a major opportunity to open our eyes to the real-time life processes that occur within a functioning cell.

The road to the optical visualization of every single molecule in the sample, along with its trajectory in the time domain, remains long. However, if the rapid progress made during the past two decades continues then we will have good reason to feel that this chapter will soon become somewhat of an 'antique'. In fact, since the first concept of this chapter, video rate fluorescence imaging with a focal spot of approximately 60 nm in living cells has been achieved using RESOLFT, and this has resulted in some impressive images of synaptic vesicle movement [79]. Furthermore, the initial 2-D studies with PALM and STORM have now been extended to 3-D imaging with a lateral resolution of 20–30 nm and an axial resolution of 50–60 nm [80]. Finally, single-molecule detection has been successfully extended to the investigation of single labels, such as semiconductor quantum dots, in *absorption* at room temperature. This will surely open the way to optical nanoscopy without a need for efficient fluorescent labels [74].

Acknowledgments

The authors thank U. Hakanson, V. Jacobsen, E. Klotzsch, K. Lindfors, A. Schtalheim, J. Seelig and P. Stoller, who each contributed to the development of the interferometric detection technique, and C. Brunner, H. Ewers, A. Helenius, K. Leslie, A. Smith, V. Vogel and C. Weyman for their collaboration on the biological experiments. These studies were performed within the frame of an Integrated Project of the European Union 'Molecular Imaging'. The authors also acknowledge financial support from the ETH Zurich and the Swiss National Foundation (SNF).

References

1 Preston, G.M. and Agre, P. (1991) Isolation of the CDNA for erythrocyte integral membrane-protein of 28-kilodaltons - member of an ancient channel family. *Proceedings of the National Academy of Sciences of the United States of America*, **88**, 11110.

2 Doyle, D.A., Cabral, J.M., Pfuetzner, R.A., Kuo, L., Gulbis, J.M., Cohen, S.L., Chait, B.T. and MacKinnon, R. (1998) The structure of the potassium channel: molecular basis of K+ conduction and selectivity. *Science*, **280**, 69.

3 Palczewski, K., Kumasaka, T., Hori, T., Behnke, C.A., Motoshima, H., Fox, B.A., Le Trong, I., Teller, D.C., Okada, T., Stenkamp, R.E., Yamamoto, M. and Miyano, M. (2000) Crystal structure of rhodopsin: A G protein-coupled receptor. *Science*, **289**, 739.

4 Cramer, P., Bushnell, D.A. and Kornberg, R.D. (2001) Structural basis of transcription: RNA polymerase II at 2.8 Angstrom resolution. *Science*, **292**, 1863.

5 Binnig, G. and Rohrer, H. (1982) Vacuum tunnel microscope. *Helvetica Physica Acta*, **55**, 726.

6 Binnig, G., Quate, C.F. and Gerber, C. (1986) Atomic force microscope. *Physical Review Letters*, **56**, 930.

7 Shao, Z.F., Yang, J. and Somlyo, A.P. (1995) Biological atomic force microscopy: from microns to nanometers and beyond. *Annual Review of Cell and Developmental Biology*, **11**, 241.

8 Hansma, H.G. (2001) Surface biology of DNA by atomic force microscopy. *Annual Review of Physical Chemistry*, **52**, 71.

9 Davis, L.I. (1995) The nuclear-pore complex. *Annual Review of Biochemistry*, **64**, 865.

10 Pluta, M. (1989) *Advanced Light Microscopy*, Elsevier, Amsterdam.

11 Horio, T. and Hotani, H. (1986) Visualization of the dynamic instability of individual microtubules by dark-field microscopy. *Nature*, **321**, 605.

12 Prieve, D.C., Luo, F. and Lanni, F. (1987) Brownian-motion of a hydrosol particle in a colloidal force-field. *Faraday Discussions*, 297.

13 Joos, U., Biskup, T., Ernst, O., Westphal, I., Gherasim, C., Schmidt, R., Edinger, K., Pilarczyk, G. and Duschl, C. (2006) Investigation of cell adhesion to structured surfaces using total internal reflection fluorescence and confocal laser scanning microscopy. *European Journal of Cell Biology*, **85**, 225.

14 Sonnichsen, C., Geier, S., Hecker, N.E., von Plessen, G., Feldmann, J., Ditlbacher, H., Lamprecht, B., Krenn, J.R., Aussenegg, F.R., Chan, V.Z.H., Spatz, J.P. and Moller, M. (2000) Spectroscopy of single metallic nanoparticles using total internal reflection microscopy. *Applied Physics Letters*, **77**, 2949.

15 Minsky, M. (1988) Memoir on inventing the confocal scanning microscope. *Scanning*, **10**, 128.

16 Wilson, T. (1990) *Confocal Microscopy*, Academic Press, London.

17 Nomarski, G. (1955) Nouveau dispositif pour l'observation en contraste de phase differentiel. *Journal de Physique et le Radium*, **16**, S88.

18 Murphy, D.B. (2001) *Fundamentals of Light Microscopy and Electronic Imaging*, Wiley-Liss, New York.

19 Coons, A.H. and Kaplan, M.H. (1950) Localization of antigen in tissue cells 2. Improvements in a method for the detection of antigen by means of fluorescent antibody. *The Journal of Experimental Medicine*, **91**, 1.

20 Chalfie, M., Tu, Y., Euskirchen, G., Ward, W.W. and Prasher, D.C. (1994) Green fluorescent protein as a marker for gene-expression. *Science*, **263**, 802.

21 Axelrod, D., Koppel, D.E., Schlessinger, J., Elson, E. and Webb, W.W. (1976) Mobility measurement by analysis of fluorescence photobleaching recovery kinetics. *Biophysical Journal*, **16**, 1055.

22 Bastiaens, P.I.H. and Squire, A. (1999) Fluorescence lifetime imaging microscopy: spatial resolution of biochemical processes in the cell. *Trends In Cell Biology*, **9**, 48.

23 Richards-Kortum, R. and Sevick-Muraca, E. (1996) Quantitative optical spectroscopy for tissue diagnosis. *Annual Review of Physical Chemistry*, **47**, 555.

24 Konig, K. (2000) Multiphoton microscopy in life sciences. *Journal of Microscopy*, **200**, 83.

25 Cheng, J.X., Jia, Y.K., Zheng, G.F. and Xie, X.S. (2002) Laser-scanning coherent anti-Stokes Raman scattering microscopy and applications to cell biology. *Biophysical Journal*, **83**, 502.

26 Campagnola, P.J., Wei, M.D., Lewis, A. and Loew, L.M. (1999) High-resolution nonlinear optical imaging of live cells by second harmonic generation. *Biophysical Journal*, **77**, 3341.

27 Barad, Y., Eisenberg, H., Horowitz, M. and Silberberg, Y. (1997) Nonlinear scanning laser microscopy by third harmonic generation. *Applied Physics Letters*, **70**, 922.

28 Hell, S.W. (2007) *Science*, **316**, 1153.

29 Nagourney, W., Janik, G. and Dehmelt, H. (1983) Linewidth of single laser-cooled (Mg-24) + ion in radiofrequency trap. *Proceedings of the National Academy of Sciences of the United States of America*, **80**, 643.

30 Kim, J.E., Tauber, M.J. and Mathies, R.A. (2001) Wavelength dependent cis-trans isomerization in vision. *Biochemistry*, **40**, 13774.

31 Moerner, W.E. and Orrit, M. (1999) Illuminating single molecules in condensed matter. *Science*, **283**, 1670.

32 Xie, X.S. and Trautman, J.K. (1998) Optical studies of single molecules at room temperature. *Annual Review of Physical Chemistry*, **49**, 441.

33 Eggeling, C., Widengren, J., Rigler, R. and Seidel, C.A.M. (1998) Photobleaching of fluorescent dyes under conditions used for

single-molecule detection: evidence of two-step photolysis. *Analytical Chemistry*, **70**, 2651.

34 Renn, A., Seelig, J. and Sandoghdar, V. (2006) Oxygen-dependent photochemistry of fluorescent dyes studied at the single molecule level. *Molecular Physics*, **104**, 409.

35 Foyer, C.H., Lelandais, M. and Kunert, K.J. (1994) Photooxidative stress in plants. *Physiologia Plantarum*, **92**, 696.

36 Hoebe, R.A., Van Oven, C.H., Gadella, T.W.J., Dhonukshe, P.B., Van Noorden, C.J.F. and Manders, E.M.M. (2007) Controlled light-exposure microscopy reduces photobleaching and phototoxicity in fluorescence live-cell imaging. *Nature Biotechnology*, **25**, 249.

37 Yildiz, A., Forkey, J.N., McKinney, S.A., Ha, T., Goldman, Y.E. and Selvin, P.R. (2003) Myosin V walks hand-over-hand: single fluorophore imaging with 1.5-nm localization. *Science*, **300**, 2061.

38 Moerner, W.E. and Kador, L. (1989) Optical-detection and spectroscopy of single molecules in a solid. *Physical Review Letters*, **62**, 2535.

39 Orrit, M. and Bernard, J. (1990) Single pentacene molecules detected by fluorescence excitation in a para-terphenyl crystal. *Physical Review Letters*, **65**, 2716.

40 Pohl, D.W., Denk, W. and Lanz, M. (1984) Optical stethoscopy - image recording with resolution lambda/20. *Applied Physics Letters*, **44**, 651.

41 Lewis, A., Isaacson, M., Harootunian, A. and Muray, A. (1984) Development of a 500-A spatial-resolution light-microscope 1. Light is efficiently transmitted through gamma-16 diameter apertures. *Ultramicroscopy*, **13**, 227.

42 Betzig, E. and Chichester, R.J. (1993) Single molecules observed by near-field scanning optical microscopy. *Science*, **262**, 1422.

43 Nie, S.M., Chiu, D.T. and Zare, R.N. (1994) Probing individual molecules with confocal fluorescence microscopy. *Science*, **266**, 1018.

44 Macklin, J.J., Trautman, J.K., Harris, T.D. and Brus, L.E. (1996) Imaging and time-resolved spectroscopy of single molecules at an interface. *Science*, **272**, 255.

45 Special Issue on Single Molecules (1999) *Science*, **283**, 1667.

46 Vale, R.D., Funatsu, T., Pierce, D.W., Romberg, L., Harada, Y. and Yanagida, T. (1996) Direct observation of single kinesin molecules moving along microtubules. *Nature*, **380**, 451.

47 Seisenberger, G., Ried, M.U., Endress, T., Buning, H., Hallek, M. and Brauchle, C. (2001) Real-time single-molecule imaging of the infection pathway of an adeno-associated virus. *Science*, **294**, 1929.

48 Hecht, E. (2001) *Optics*, Addison Wesley, New York.

49 Thompson, R.E., Larson, D.R. and Webb, W.W. (2002) Precise nanometer localization analysis for individual fluorescent probes. *Biophysical Journal*, **82**, 2775.

50 Schmidt, T., Schutz, G.J., Baumgartner, W., Gruber, H.J. and Schindler, H. (1995) Characterization of photophysics and mobility of single molecules in a fluid lipid-membrane. *The Journal of Physical Chemistry*, **99**, 17662.

51 Yildiz, A., Tomishige, M., Vale, R.D. and Selvin, P.R. (2004) Kinesin walks hand-over-hand. *Science*, **303**, 676.

52 Ober, R.J., Ram, S. and Ward, E.S. (2004) Localization accuracy in single-molecule microscopy. *Biophysical Journal*, **86**, 1185.

53 Alivisatos, A.P., Gu, W.W. and Larabell, C. (2005) Quantum dots as cellular probes. *Annual Review of Biomedical Engineering*, **7**, 55.

54 Michalet, X., Pinaud, F.F., Bentolila, L.A., Tsay, J.M., Doose, S., Li, J.J., Sundaresan, G., Wu, A.M., Gambhir, S.S. and Weiss, S. (2005) Quantum dots for live cells, in vivo imaging, and diagnostics. *Science*, **307**, 538.

55 Pawley, J.B. (2006) *Handbook of Biological Confocal Microscopy*, 3rd edn, Springer Verlag.

56 Lacoste, T.D., Michalet, X., Pinaud, F., Chemla, D.S., Alivisatos, A.P. and Weiss, S. (2000) Ultrahigh-resolution multicolor colocalization of single fluorescent probes. *Proceedings of the National Academy of Sciences of the United States of America*, **97**, 9461.

57 Gordon, M.P., Ha, T. and Selvin, P.R. (2004) Single-molecule high-resolution imaging with photobleaching. *Proceedings of the National Academy of Sciences of the United States of America*, **101**, 6462.

58 Betzig, E., Patterson, G.H., Sougrat, R., Lindwasser, O.W., Olenych, S., Bonifacino, J.S., Davidson, M.W., Lippincott-Schwartz, J. and Hess, H.F. (2006) Imaging intracellular fluorescent proteins at nanometer resolution. *Science*, **313**, 1642.

59 Rust, M.J., Bates, M. and Zhuang, X.W. (2006) Sub-diffraction-limit imaging by stochastic optical reconstruction microscopy (STORM). *Nature Methods*, **3**, 793.

60 Hettich, C., Schmitt, C., Zitzmann, J., Kühn, S., Gerhardt, I. and Sandoghdar, V. (2002) Nanometer resolution and coherent optical dipole coupling of two individual molecules. *Science*, **298**, 385.

61 Ha, T., Enderle, T., Ogletree, D.F., Chemla, D.S., Selvin, P.R. and Weiss, S. (1996) Probing the interaction between two single molecules: fluorescence resonance energy transfer between a single donor and a single acceptor. *Proceedings of the National Academy of Sciences of the United States of America*, **93**, 6264.

62 Kusumi, A., Nakada, C., Ritchie, K., Murase, K., Suzuki, K., Murakoshi, H., Kasai, R.S., Kondo, J. and Fujiwara, T. (2005) Paradigm shift of the plasma membrane concept from the two-dimensional continuum fluid to the partitioned fluid: high-speed single-molecule tracking of membrane molecules. *Annual Review of Biophysics and Biomolecular Structure*, **34**, 351.

63 Schultz, S., Smith, D.R., Mock, J.J. and Schultz, D.A. (2000) Single-target molecule detection with nonbleaching multicolor optical immunolabels. *Proceedings of the National Academy of Sciences of the United States of America*, **97**, 996.

64 Bohren, C.F. and Huffman, D.R. (1983) *Absorption and Scattering of Light by Small Particles*, John Wiley and Sons, New York.

65 Kelly, K.L., Coronado, E., Zhao, L.L. and Schatz, G.C. (2003) The optical properties of metal nanoparticles: the influence of size, shape, and dielectric environment. *Journal of Physical Chemistry B*, **107**, 668.

66 Ewers, H., Jacobsen, V., Klotzsch, E., Smith, A., Helenius, A. and Sandoghdar, V. (2007) Label-free optical detection and tracking of single virions bound to their receptors in supported membrane bilayers. *Nano Letters*, **7**, 2263.

67 Arbouet, A., Christofilos, D., Del Fatti, N., Vallee, F., Huntzinger, J.R., Arnaud, L., Billaud, P. and Broyer, M. (2004) Direct measurement of the single-metal-cluster optical absorption. *Physical Review Letters*, **93**, 127401.

68 Boyer, D., Tamarat, P., Maali, A., Lounis, B. and Orrit, M. (2002) Photothermal imaging of nanometer-sized metal particles among scatterers. *Science*, **297**, 1160.

69 Jacobsen, V., Stoller, P., Brunner, C., Vogel, V. and Sandoghdar, V. (2006) Interferometric optical detection and tracking of very small gold nanoparticles at a water-glass interface. *Optics Express*, **14**, 405.

70 Ignatovich, F.V. and Novotny, L. (2006) Real-time and background-free detection of nanoscale particles. *Physical Review Letters*, 96.

71 Lindfors, K., Kalkbrenner, T., Stoller, P. and Sandoghdar, V. (2004) Detection and spectroscopy of gold nanoparticles using supercontinuum white light confocal microscopy. *Physical Review Letters*, **93**, 037401.

72 Carter, A.R., King, G.M., Ulrich, T.A., Halsey, W., Alchenberger, D. and Perkins, T.T. (2007) Stabilization of an optical microscope to 0.1 nm in three dimensions. *Applied Optics*, **46**, 421.

73 Jacobsen, V., Klotzsch, E. and Sandoghdar, V. (2007) *Nano Biophotonics*, Vol. **3** (eds H. Masuhura, S. Kawata and F. Tokunaga), Elsevier, Amsterdam, p. 143.

74 Kukura, P., Celebrano, M., Renn, A. and Sandoghdar, V. (2008) Seeing a single quantum emitter when it is dark. *Nano Letters*, doi 10.1021/nl801735y.

75 Moskovits, M. (1985) Surface-enhanced spectroscopy. *Reviews of Modern Physics*, **57**, 783.

76 Kühn, S., Hakanson, U., Rogobete, L. and Sandoghdar, V. (2006) Enhancement of single-molecule fluorescence using a gold nanoparticle as an optical nanoantenna. *Physical Review Letters*, **97**, 017402.

77 Seelig, J., Leslie, K., Renn, A., Kühn, S., Jacobsen, V., van de Corput, M., Wyman, C. and Sandoghdar, V. (2007) Nanoparticle-induced fluorescence lifetime modification as nanoscopic ruler: demonstration at the single molecule level. *Nano Letters*, **7**, 685.

78 Liddington, R.C., Yan, Y., Moulai, J., Sahli, R., Benjamin, T.L. and Harrison, S.C. (1991) Structure of Simian virus-40 at 3.8-A resolution. *Nature*, **354**, 278.

79 Westphal, V., Rizzoli, S.O., Lauterbach, M.A., Kamin, D., Jahn, R. and Hell, S.W. (2008) Video-rate far-field optical nanoscopy dissects synaptic vesicle movement. *Science*, **320**, 246.

80 Huang, B., Wang, W., Bates, M. and Zhuang, X. (2008) Three-dimensional super-resolution imaging by stochastic optical reconstruction microscopy. *Science*, **319**, 810.

6
Nanostructured Probes for *In Vivo* Gene Detection

Gang Bao, Phillip Santangelo, Nitin Nitin, and Won Jong Rhee

6.1
Introduction

The ability to image specific RNAs in living cells in real time can provide essential information on RNA synthesis, processing, transport and localization, as well as on the dynamics of RNA expression and localization in response to external stimuli. Such an ability will also offer unprecedented opportunities for advancement in molecular biology, disease pathophysiology, drug discovery and medical diagnostics. Over the past decade or so, an increasing amount of evidence has come to light suggesting that RNA molecules have a wide range of functions in living cells, from physically conveying and interpreting genetic information, to essential catalytic roles, to providing structural support for molecular machines, and to gene silencing. These functions are realized through control of the expression level and stability, both temporally and spatially, of specific RNAs in a cell. Therefore, determining the dynamics and localization of RNA molecules in living cells will significantly impact on the molecular biology and medicine.

Many *in vitro* methods have been developed to provide a relative (mostly semi-quantitative) measure of gene expression level within a cell population, by using purified DNA or RNA obtained from cell lysates. These methods include the polymerase chain reaction (PCR) [1], Northern hybridization (or Northern blotting) [2], expressed sequence tag (EST) [3], serial analysis of gene expression (SAGE) [4], differential display [5] and DNA microarrays [6]. These technologies, combined with the rapidly increasing availability of genomic data for numerous biological entities, present exciting possibilities for the understanding of human health and disease. For example, pathogenic and carcinogenic sequences are increasingly being used as clinical markers for diseased states. However, the use of *in vitro* methods to detect and identify foreign or mutated nucleic acids is often difficult in a clinical setting, due to the low abundance of diseased cells in blood, sputum and stool samples. Further, these methods cannot reveal the spatial and temporal variation of RNA within a single cell.

Nanotechnology, Volume 5: Nanomedicine. Edited by Viola Vogel

ISBN: 978-3-527-31736-3

Labeled linear oligonucleotide (ODN) probes have been used to study intracellular mRNA via *in situ* hybridization (ISH) [7], in which cells are fixed and permeabilized to increase the probe delivery efficiency. Unbound probes are removed by washing to reduce the background and achieve specificity [8]. In order to enhance the signal level, multiple probes targeting the same mRNA can be used [7], although fixation agents and other supporting chemicals can have a considerable effect on the signal level [9] and possibly also on the integrity of certain organelles, such as mitochondria. Thus, the fixation of cells (by using either crosslinking or denaturing agents) and the use of proteases in ISH assays may prevent an accurate description of intracellular mRNA localization from being obtained. It is also difficult to obtain a dynamic picture of gene expression in cells using ISH methods.

Of particular interest is the fluorescence imaging of specific messenger RNAs (mRNAs) – in terms of both their expression level and subcellular localization – in living cells. As shown schematically in Figure 6.1, for eukaryotic cells a pre-mRNA molecule is synthesized in the cell nucleus. After processing (including splicing and polyadenylation), the mature mRNAs are transported from the cell nucleus to the cytoplasm, and often are localized at specific sites. The mRNAs are then translated by

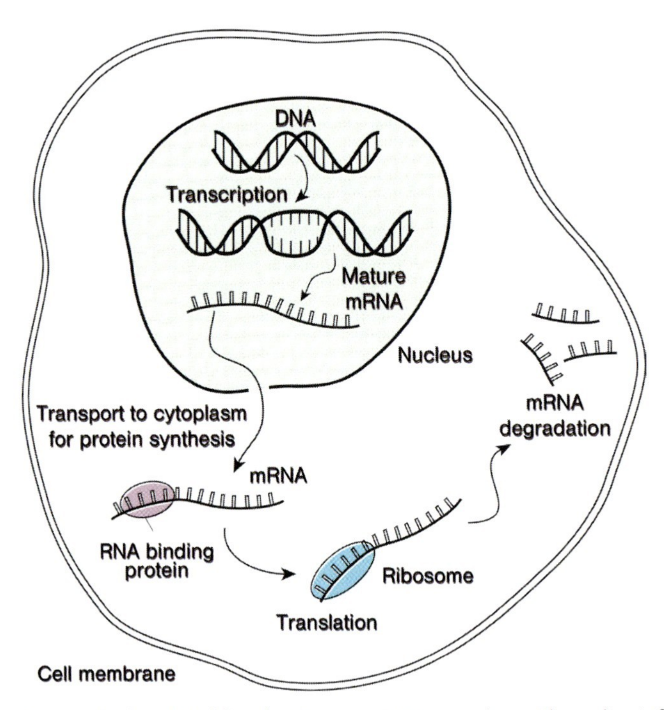

Figure 6.1 The mRNA life cycle. Messenger RNA (mRNA) encoding the chemical 'blueprint' for a protein is synthesized (transcribed) from a DNA template, and the pre-mRNA is processed (spliced) to produce a mature mRNA; this is then transported to specific locations in the cell cytoplasm. The coding information carried by mRNA is used by the ribosomes to produce proteins (translation). After a certain time the message is degraded. mRNAs are almost always complexed with RNA-binding proteins to form ribonucleoprotein (RNP) molecules.

ribosomes to produce specific proteins, and then degraded by RNases after a certain period of time. The limited lifetime of mRNA enables a cell to alter its protein synthesis rapidly, and in response to its changing needs. During the entire life cycle of an mRNA, it is always complexed with RNA-binding proteins to form a ribonucleoprotein (RNP). This has significant implications for the live-cell imaging of mRNAs (as discussed below).

To detect RNA molecules in living cells, with not only high specificity but also high sensitivity and signal-to-background ratio, it is important that the probes recognize RNA targets with high specificity, convert target recognition *directly* into a measurable signal, and differentiate between true and false-positive signals. This is especially important for low-abundance genes and clinical samples containing only a small number of diseased cells. It is also important for the probes to quantify low gene expression levels with great accuracy, and have fast kinetics in tracking alterations in gene expression in real time. For detecting genetic alterations such as mutations, insertions and deletions, the ability to recognize single nucleotide polymorphisms (SNPs) is essential. In order to achieve this optimal performance, it is necessary to have a good understanding of the structure–function relationship of the probes, the probe stability and the RNA target accessibility in living cells. It is also necessary to achieve an efficient cellular delivery of probes, with minimal probe degradation.

In the following sections we will review the fluorescent probes that are most often used for RNA detection, and discuss the critical issues in live-cell RNA detection, including probe design, target accessibility, the cellular delivery of probes, as well as detection sensitivity, specificity and signal-to-background ratio. Emphasis is placed on the design and application of molecular beacons, although some of the issues are common to other oligonucleotide probes.

6.2 Fluorescent Probes for Live-Cell RNA Detection

Several classes of molecular probes have been developed for RNA detection in living cells, including: (i) tagged linear ODN probes; (ii) oligonucleotide hairpin probes; and (iii) probes using fluorescent proteins as reporter. Although probes composed of full-length RNAs (mRNA or nuclear RNA) tagged with a fluorescent or radioactive reporter have been used to study the intracellular localization of RNA [10–12], they are not discussed here as they cannot be used to measure the expression level of specific RNAs in living cells.

6.2.1 Tagged Linear ODN Probes

Single fluorescently labeled linear oligonucleotide probes have been developed for RNA tracking and localization studies in living cells [13–15]. Although these probes may recognize specific endogenous RNA transcripts in living cells via Watson–Crick base pairing, and thus reveal subcellular RNA localization, this approach lacks the ability to distinguish background from true signal, as both bound probes (i.e. those hybridized to RNA target) and unbound probes give a fluorescence signal. Such an approach might also lack detection specificity, as a partial match between the probe and target sequences could induce probe hybridization to RNA molecules of multiple genes. A novel way to increase the signal-to-noise ratio (SNR) and improve detection specificity is to use two linear probes with a fluorescence resonance energy transfer (FRET) pair of (donor and acceptor) fluorophores [13]. However, the dual-linear probe approach may still have a high background signal due to direct excitation of the acceptor and emission detection of the donor fluorescence. Further, it is difficult for linear probes to distinguish targets that differ by a few bases as the difference in free energy of the two hybrids (with and without mismatch) is typically rather small. This limits the application of linear ODN probes in biological and disease studies.

6.2.2 ODN Hairpin Probes

Hairpin nucleic acid probes have the potential to be highly sensitive and specific in live-cell RNA detection. As shown in Figure 6.2a and b, one class of such probes is that of '*molecular beacons*'; these are dual-labeled oligonucleotide probes with a fluorophore at one end and a quencher at the other end [16]. They are designed to form a stem–loop hairpin structure in the absence of a complementary target, so that the fluorescence of the fluorophore is quenched. Hybridization with the target nucleic acid opens the hairpin and physically separates the fluorophore from quencher, allowing a fluorescence signal to be emitted upon excitation (Figure 6.2b). Under optimal conditions, the fluorescence intensity of molecular beacons can increase more than 200-fold upon binding to their targets [16], and this enables them to function as sensitive probes with a high signal-to-background ratio. The stem–loop hairpin structure provides an adjustable energy penalty for hairpin opening which improves probe specificity [17, 18]. The ability to transduce target recognition *directly* into a fluorescence signal with a high signal-to-background ratio, coupled with an improved specificity, has allowed molecular beacons to enjoy a wide range of biological and biomedical applications. These include multiple analyte detection, real-time enzymatic cleavage assaying, cancer cell detection, real-time monitoring of PCR, genotyping and mutation detection, viral infection studies and mRNA detection in living cells [14, 19–32].

As illustrated in Figure 6.2a, a *conventional molecular beacon* has four essential components: loop, stem, fluorophore (dye) and quencher. The loop usually consists of 15–25 nucleotides and is selected to have a unique target sequence and proper melting temperature. The stem, which is formed by two complementary short-arm sequences, is typically four to six bases long and chosen to be independent of the target sequence (Figure 6.2a).

A novel design of hairpin probes is the *wavelength-shifting molecular beacon*, which can fluoresce in a variety of different colors [33]. As shown in Figure 6.2c, in this design, a molecular beacon contains two fluorophores (dyes): a first fluorophore that absorbs

(a)

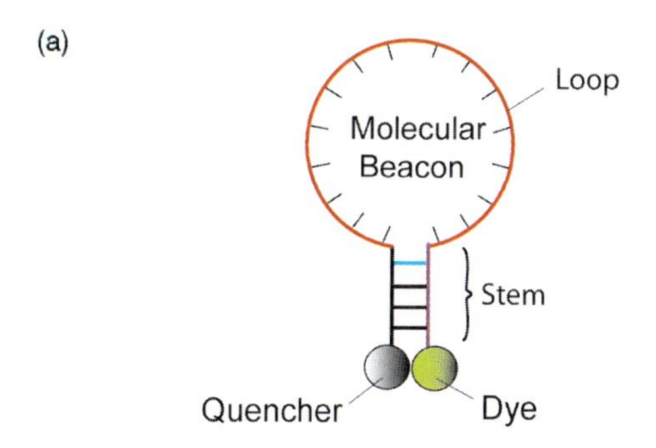

(b)

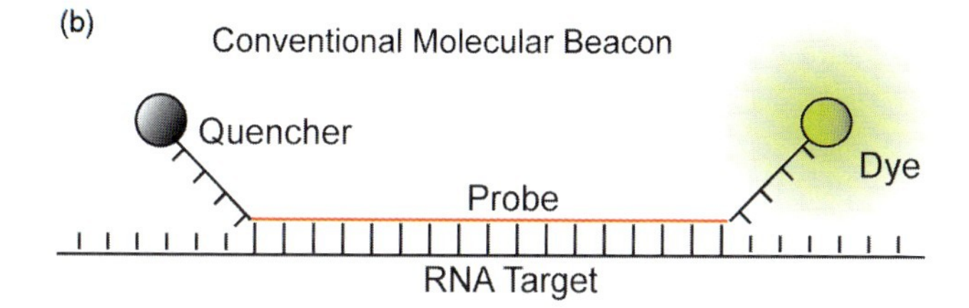

(c)

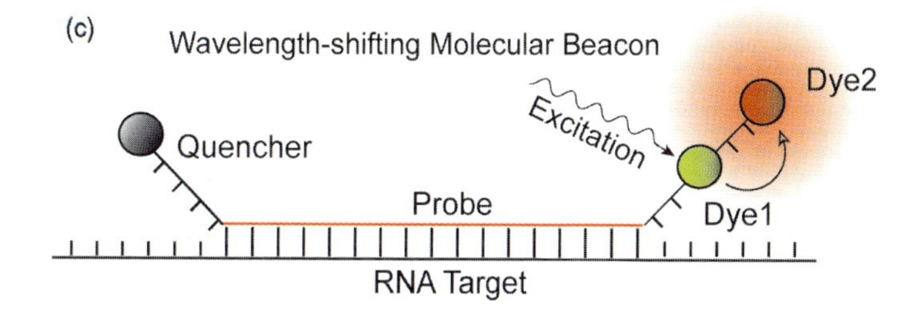

(d)

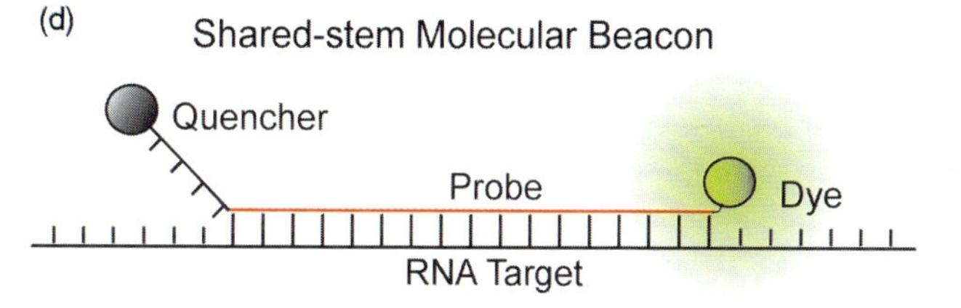

Figure 6.2 Illustrations of molecular beacon designs. (a) Molecular beacons are stem–loop hairpin oligonucleotide probes labeled with a reporter fluorophore at one end and a quencher molecule at the other end; (b) Conventional molecular beacons are designed such that the short complementary arms of the stem are independent of the target sequence; (c) Wavelength-shifting molecular beacons contain two fluorophores: one absorbs in the wavelength range of the monochromatic light source, and the other emits light at the desired emission wavelength due to FRET; (d) Shared-stem molecular beacons are designed such that one arm of the stem participates in both stem formation and target hybridization.

strongly in the wavelength range of the monochromatic light source, and a second fluorophore that emits at the desired emission wavelength due to fluorescence resonance energy transfer from the first fluorophore to the second fluorophore. It has been shown that wavelength-shifting molecular beacons are substantially brighter than conventional molecular beacons, which contain a fluorophore that cannot efficiently absorb energy from the available monochromatic light source.

One major advantage of the stem–loop hairpin probes is that they can recognize their targets with higher specificity than can linear ODN probes. The results of solution studies [17, 18] have suggested that, by using molecular beacons it is possible to discriminate between targets that differ by a single nucleotide. In contrast to current techniques for detecting SNPs – which are often labor-intensive and time-consuming – molecular beacons may provide a simple and promising tool for detecting SNPs in disease diagnosis.

The basic features of molecular beacon versus fluorescence *in situ* hybridization (FISH) are compared in Figure 6.3. Specifically, molecular beacons are dual-labeled hairpin probes of 15–25 nt, while FISH probes are dye-labeled linear oligonucleotides of 40–50 nt. The molecular beacon-based approach has the advantage of detecting RNA in live cells, without the need for cell fixation and washing. However, it does requires the cellular delivery of probes and has a low target accessibility (this is discussed below). The advantage of FISH assays is the ease of probe design due to a better target accessibility. Although FISH assays can be used to image the localization of mRNA in fixed cells, they rely on stringent washing to achieve signal specificity, and do not have the ability to image the dynamics of gene expression in living cells.

In the conventional molecular beacon design, the stem sequence is typically independent of the target sequence (see Figure 6.2b), although sometimes two end bases of the probe sequence, each adjacent to one arm sequence of the stem, could be complementary with each other, thus forming part of the stem (the light blue base of the stem shown in Figure 6.2a). Molecular beacons can also be designed such that all the bases of one arm of the stem (to which a fluorophore is conjugated) are complementary to the target sequence, thus participating in both stem formation and target hybridization (shared-stem molecular beacons) [34] (Figure 6.2d). The advantage of this shared-stem design is to help fix the position of the fluorophore that is attached to the stem arm, limiting its degree-of-freedom of motion, and increasing the FRET in the dual-FRET molecular beacon design (as discussed below).

A dual-FRET molecular beacon approach was developed [26–28] to overcome the difficult that, in live-cell RNA detection, molecular beacons are often degraded by nucleases or open due to nonspecific interaction with hairpin-binding proteins, causing a significant amount of false-positive signal. In this dual-probe design, a pair of molecular beacons labeled with a donor and an acceptor fluorophore, respectively are employed (Figure 6.4). The probe sequences are chosen such that this pair of molecular beacons hybridizes to adjacent regions on a single RNA target (Figure 6.4). As FRET is very sensitive to the distance between donor and acceptor fluorophores, and typically occurs when the donor and acceptor fluorophores are

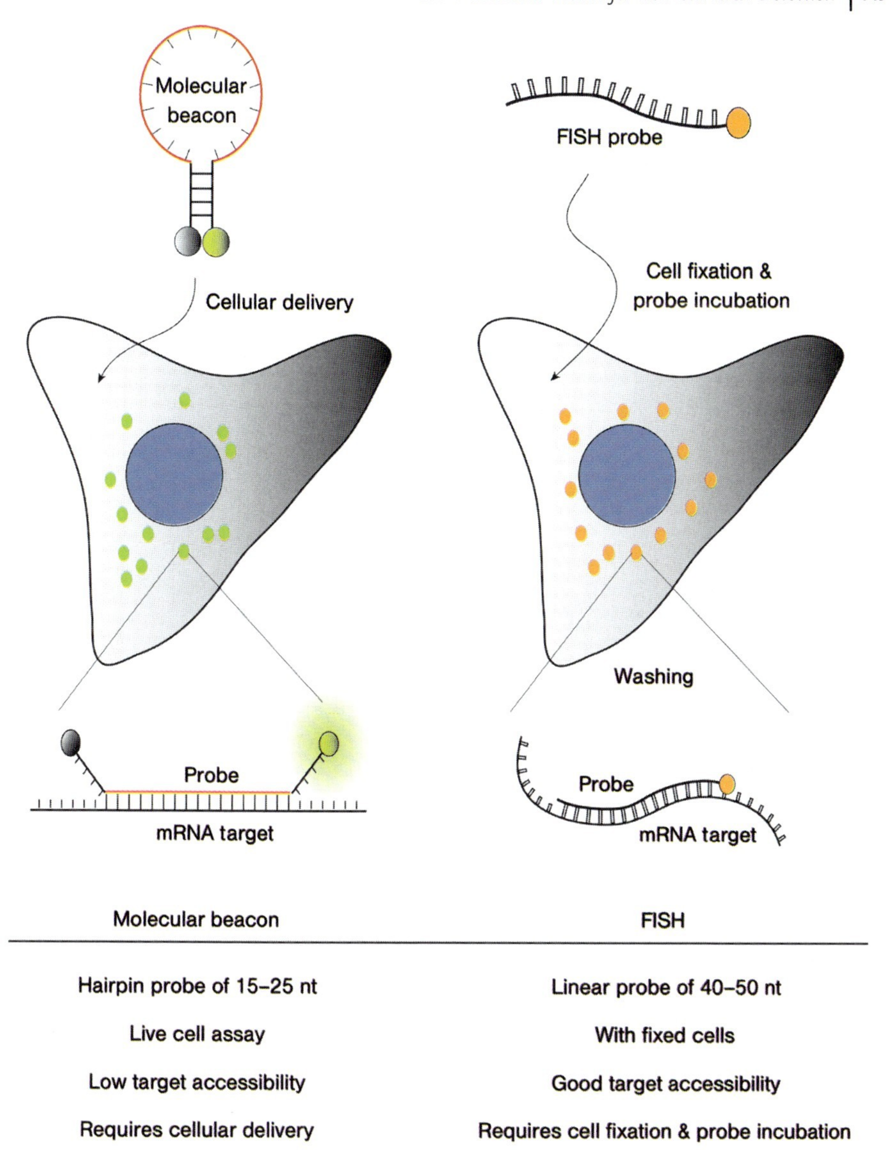

Figure 6.3 Comparison of molecular beacon and FISH approaches.

within ~10 nm, the FRET signal is generated by the donor and acceptor beacons only if both probes are bound to the same RNA target. Thus, the sensitized emission of the acceptor fluorophore upon donor excitation serves as a positive signal in the FRET-based detection assay; this can be differentiated from non-FRET false-positive signals due to probe degradation and nonspecific probe opening. This approach combines the low background signal and high specificity of molecular

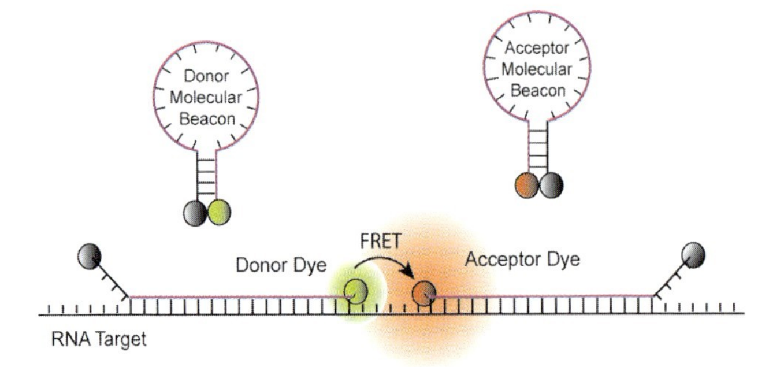

Figure 6.4 A schematic showing the concept of dual-FRET molecular beacons. Hybridization of the donor and acceptor molecular beacons to adjacent regions on the same mRNA target results in FRET between donor and acceptor fluorophores upon donor excitation. By detecting the FRET signal, fluorescence signals due to probe/target binding can be readily distinguished from that due to molecular beacon degradation and nonspecific interactions.

beacons with the ability of FRET assays in differentiating between true target recognition and false-positive signals, leading to an enhanced ability to quantify RNA expression in living cells [28].

6.2.3 Fluorescent Protein-Based Probes

In addition to oligonucleotide probes, tagged RNA-binding proteins such as those with green fluorescent protein (GFP) tags have been used to detect mRNA in live cells [35]. One limitation here is that it requires the identification of a unique protein, which only binds to the specific mRNA of interest. To address this issue, a coat protein of the RNA bacteriophage MS2 was tagged with GFP, after which a RNA sequence corresponding to several MS2 binding sites was introduced to the mRNA of interest. This allowed for the specific targeting of the *nanos* mRNA in live *Drosophila* eggs [36]. The GFP-MS2 approach has been used to track the localization and dynamics of RNA in living cells with single-molecule sensitivity [37, 38]. However, as unbound GFP-tagged MS2 proteins also produce a fluorescence signal, the background signal in the GFP-MS2 approach could be high, leading to a low signal-to-background ratio in live-cell imaging of RNA.

An interesting fluorescent protein-based approach that overcomes this problem is to utilize the fluorescent protein complementation [39, 40]. In this method (split-GFP), a RNA-binding protein is dissected into two fragments, which are respectively fused to the split fragments of a fluorescent protein. Binding of the two tagged fragments of the RNA-binding protein to adjacent sites on the same mRNA molecule (or two parts of an aptamer sequence inserted to the mRNA sequence) brings the two halves of the fluorescent protein together, thus reconstituting the fluorescent protein and restoring fluorescence [40]. Alternatively, two RNA-binding proteins that bind specifically to adjacent sites on the same mRNA molecule can be tagged with the split fragments of a fluorescent protein, such that their binding to the target mRNA results in the restoration of fluorescence [39]. The advantage of this novel approach is that the background signal is low; there is no fluorescence signal unless the RNA-binding proteins (or protein fragments) are bound to the target mRNA. The split-GFP method, however, may have difficulties in tracking the dynamics of RNA expression in real time, as reconstitution of the fluorescent protein from the split fragments typically takes 2–4 h, during which time the RNA expression level may change. Transfection efficiency may also be a major concern in the GFP-based approaches, in that usually only a small percentage of the cells express the fluorescent proteins following transfection. This limits the application of the split-GFP methods in detecting diseased cells using mRNA as a biomarker for the disease.

6.3 Probe Design and Structure–Function Relationships

6.3.1 Target Specificity

There are three major design issues of molecular beacons: probe sequence; hairpin structure; and fluorophore/quencher selection. In general, the probe sequence is selected to ensure specificity, and to have good target accessibility. The hairpin structure, as well as the probe and stem sequences, are determined to have the proper melting temperature, while the fluorophore–quencher pair should produce a high signal-to-background ratio. To ensure specificity, for each gene to target, it is possible to use the NCBI BLAST [41] or similar software to select multiple target sequences of 15–25 bases that are unique to the target RNA. As the melting temperature of the molecular beacons affects both the signal-to-background ratio and detection specificity (especially for mutation detection), it is often necessary to select the target sequence with a balanced G-C content, and to adjust the loop and stem lengths and the stem sequence of the molecular beacon to realize the optimal melting temperature. In particular, it is necessary to understand the effect of molecular beacon design on melting temperature so that, at 37 °C, single-base mismatches in target mRNAs can be differentiated. This is also a general issue for detection specificity in that, for any specific probe sequence selected, there might be multiple genes in the mammalian genome that have sequences which differ from the probe sequence by only a few bases. Therefore, it is important to design the molecular beacons so that only the specific target RNA would produce a strong signal.

Several approaches can be taken to validate the signal specificity. For example, one could either upregulate or downregulate the expression level of a specific RNA, quantify the level using RT-PCR, and then compare the PCR result with that of molecular beacon-based imaging of the same RNA in living cells. However, complications may arise when the approach used to change the RNA expression level in living cells has an effect on multiple genes, as this would lead to some ambiguity, even when the PCR and beacon results match. It is possible that the best way to down-regulate the level of a specific mRNA in live cells is to use small interfering RNA (siRNA) treatment, which typically leads to a >80% reduction of the specific mRNA level. As the effect of siRNA treatment varies depending on the specific probe used, the siRNA delivery method, cell type and optimization of the protocol (i.e. probe design and delivery method/conditions) is often needed.

6.3.2
Molecular Beacon Structure–Function Relationships

The loop, stem lengths and sequences are critical design parameters for molecular beacons, since at any given temperature they largely control the fraction of molecular beacons that are bound to the target [17, 18]. In many applications, the choices of the probe sequence are limited by target-specific considerations, such as the sequence surrounding a single nucleotide polymorphism (SNP) of interest. However, the probe and stem lengths, and stem sequence, can be adjusted to optimize the performance (i.e. specificity, hybridization rate and signal-to-background ratio) of a molecular beacon for a specific application [17, 34].

In order to demonstrate the effect of molecular beacon structure on its melting behavior, the melting temperature for molecular beacons with various stem–loop structures is displayed in Figure 6.5a. In general, the melting temperature was found to increase with probe length, but appeared to plateau at a length of ~20 nucleotides. The stem length of the molecular beacon was also found to have a major influence on the melting temperature of the molecular beacon–target duplexes.

While both the stability of the hairpin probe and its ability to discriminate targets over a wider range of temperatures increase with increasing stem length, it is accompanied by a decrease in the hybridization on-rate constant (see Figure 6.5b).

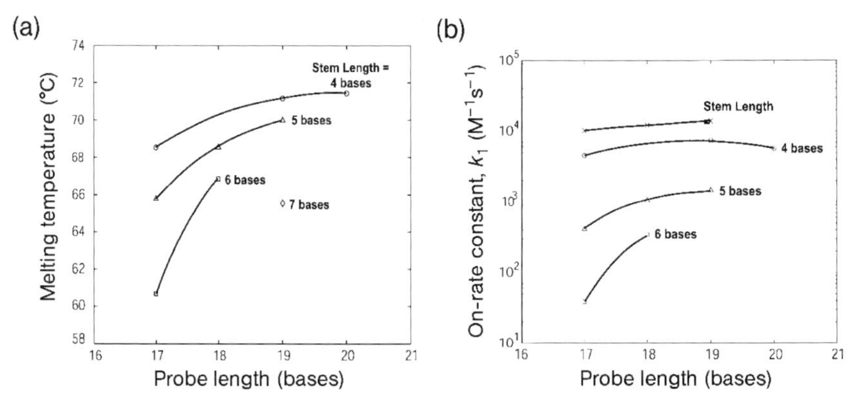

Figure 6.5 Structure–function relationships of molecular beacons. (a) Melting temperatures for molecular beacons with different structures in the presence of target; (b) The rate constant of hybridization k_1 (on-rate constant) for molecular beacons with various probe and stem lengths hybridized to their complementary targets.

For example, molecular beacons with a four-base stem had an on-rate constant up to 100-fold higher than did molecular beacons with a six-base stem. Changing the probe length of a molecular beacon may also influence the rate of hybridization, as shown in Figure 6.5b.

The results of thermodynamic and kinetic studies showed that, if the stem length was too large then it would be difficult for the beacon to open on hybridization. But, if the stem length was too small, then a large fraction of beacons might open due to the thermal force. Likewise, and relative to the stem length, whilst a longer probe might lead to a lower dissociation constant, it might also reduce the specificity, as the relative free energy change due to a one base mismatch would be smaller. A long probe length may also lead to coiled conformations of the beacons, resulting in reduced kinetic rates. Consequently, the stem and probe lengths must be carefully chosen in order to optimize both hybridization kinetics and molecular beacon specificity [17, 34]. In general, molecular beacons with longer stem lengths have an improved ability to discriminate between wild-type and mutant targets in solution, over a broader range of temperatures. This effect can be attributed to the enhanced stability of the molecular beacon stem–loop structure and the resulting smaller free energy difference between closed (unbound) molecular beacons and molecular beacon–target duplexes, which generates a condition where a single-base mismatch reduces the energetic preference of probe–target binding. Longer stem lengths, however, are accompanied by a reduced probe–target hybridization kinetic rate. On a similar note, molecular beacons with short stems have faster hybridization kinetics but suffer from lower signal-to-background ratios compared to molecular beacons with longer stems.

6.3.3
Target Accessibility

One critical issue in molecular beacon design is target accessibility, as is the case for most oligonucleotide probes for live-cell RNA detection. It is well known that a functional mRNA molecule in a living cell is always associated with RNA-binding proteins, thus forming a RNP. An mRNA molecule also often has double-stranded portions and forms secondary (folded) structures (Figure 6.6). Therefore, when designing a molecular beacon it is necessary to avoid targeting mRNA sequences that are double-stranded, or occupied by RNA-binding proteins, for otherwise the probe will have to penetrate into the RNA double strand or compete with the RNA-binding protein in order to hybridize to the target. In fact, molecular beacons designed to target a specific mRNA often show no signal when delivered to living cells. One difficulty in molecular beacon design is that, although predictions of mRNA secondary structure can be made using software such as *Beacon Designer* (www.premierbiosoft.com) and *mfold* (http://www.bioinfo.rpi.edu/applications/mfold/old/dna/), they may be inaccurate due to limitations of the biophysical models used, and the limited understanding of protein–RNA interaction. Therefore, for each gene to be targeted it may be necessary to select multiple unique sequences along the target RNA, and then to design, synthesize and test the corresponding molecular beacons in living cells in order to select the best target sequence.

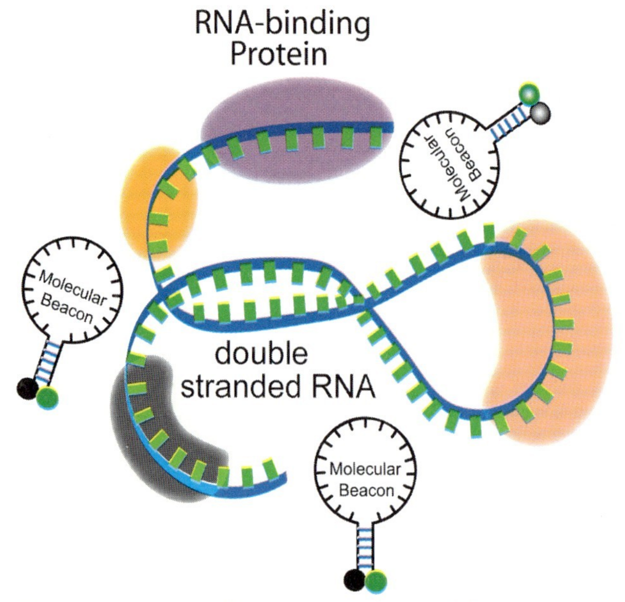

Figure 6.6 A schematic illustration of a segment of the target mRNA with a double-stranded portion and RNA-binding proteins. A molecular beacon must penetrate into the mRNA double strand or compete with the RNA-binding protein(s) in order to hybridize to the target.

In aiming to reveal the possible molecular beacon design rules, the accessibility of BMP-4 mRNA was studied using different beacon designs [42]. Specifically, molecular beacons were designed to target the start codon and termination codon regions, the siRNA and anti-sense oligonucleotide probe sites (which were identified previously) and also the sites that were chosen at random. All of the target sequences are unique to BMP-4 mRNA. Of the eight molecular beacons designed to target BMP-4 mRNA, only two were found to produce a strong signal: one which targeted the start codon region, and one which targeted the termination codon region. It was also found that, even for a molecular beacon which functioned well, shifting its targeting sequence by only a few bases towards the 3′ or 5′ ends caused a significant reduction in the fluorescence signal from beacons in a live-cell assay. This indicated that the target accessibility was quite sensitive to the location of the targeting sequence. These results, together with molecular beacons validated previously, suggest that the start and termination codon regions and the exon–exon junctions are more accessible than other locations in an mRNA.

6.3.4
Fluorophores and Quenchers

With a correct backbone synthesis and fluorophore/quencher conjugation, a molecular beacon can – in theory – be labeled with any desired reporter–quencher pair.

However, the correct selection of the reporter and quencher could also improve the signal-to-background ratio and multiplexing capabilities. The selection of a fluorophore label for a molecular beacon as reporter is normally less critical than for the hairpin probe design, as many conventional dyes can yield satisfactory results. However, the correct selection may yield additional benefits such as an improved signal-to-background ratio and multiplexing capabilities. As each molecular beacon utilizes only one fluorophore, it is possible to use multiple molecular beacons in the same assay, assuming that the fluorophores are chosen with minimal emission overlap [19]. Molecular beacons can even be labeled simultaneously with two fluorophores – that is with 'wavelength shifting' reporter dyes (see Figure 6.2c), allowing multiple reporter dye sets to be excited by the same monochromatic light source but to fluoresce in a variety of colors [33]. Clearly, multicolor fluorescence detection of different beacon/target duplexes may in time become a powerful tool for the simultaneous detection of multiple genes.

For dual-FRET molecular beacons (see Figure 6.4), the donor fluorophores typically emit at shorter wavelengths compared with the acceptor. Energy transfer then occurs as a result of long-range dipole–dipole interactions between the donor and acceptor. The efficiency of such energy transfer depends on the extent of the spectral overlap of the emission spectrum of the donor with the absorption spectrum of the acceptor, the quantum yield of the donor, the relative orientation of the donor and acceptor transition dipoles [43], and the distance between the donor and acceptor molecules (usually four to five bases). In selecting the donor and acceptor fluorophores so as to create a high signal-to-background ratio, it is important to optimize the above parameters, and to avoid direct excitation of the acceptor fluorophore at the donor excitation wavelength. It is also important to minimize donor emission detection at the acceptor emission detection wavelength. Examples of FRET dye pairs include Cy3 (donor) with Cy5 (acceptor), TMR (donor) with Texas Red (acceptor), and fluorescein (FAM) (donor) with Cy3 (acceptor).

By contrast, it is relatively straightforward to select the quencher molecules. Organic quencher molecules such as dabcyl, BHQ2 (blackhole quencher II) (Biosearch Tech), BHQ3 (Biosearch Tech) and Iowa Black (IDT) can all effectively quench a wide range of fluorophores by both FRET and the formation of an exciton complex between the fluorophore and the quencher [44].

6.4 Cellular Delivery of Nanoprobes

One of the most critical aspects of measuring the intracellular level of RNA molecules using synthetic probes is the ability to deliver the probes into cells via the plasma membrane, which itself is quite lipophilic and restricts the transport of large, charged molecules. Thus, the plasma membrane serves as a very robust barrier to polyanionic molecules such as hairpin oligonucleotides. Further, even if the probes enter the cells successfully, the efficiency of delivery in an imaging assay should be defined not only by how many probes enter the cell, or how many cells have probes internalized, but also by how many probes remain functioning inside the cells. This is a different situation from both antisense and gene delivery applications, where the reduction in level of protein expression is the final metric used to define efficiency or success. For measuring RNA molecules (including mRNA and rRNA) in the cytoplasm, a large amount of the probe should remain in the cytoplasm.

Existing cellular delivery techniques can be divided into two categories, namely *endocytic* and *nonendocytic*. Endocytic delivery typically employs cationic and polycationic molecules such as liposomes and dendrimers, whereas nonendocytic methods include microinjection and the use of cell-penetrating peptides (CPPs) or streptolysin *O* (SLO). Probe delivery via the endocytic pathway typically takes 2–4 h. It has been reported that ODN probes internalized via endocytosis are predominantly trapped inside endosomes and lysosomes, where they are degraded by the action of cytoplasmic nucleases [45]. Consequently, only 0.01% to 10% of the probes remain functioning after having escaped from endosomes and lysosomes [46].

Oligonucleotide probes (including molecular beacons) have been delivered into cells via microinjection [47]. In most cases, the ODNs were accumulated rapidly in the cell nucleus and prevented the probes from targeting mRNAs in the cell cytoplasm. The depletion of intracellular ATP or lowering the temperature from 37 to 4 °C did not have any significant effect on ODN nuclear accumulation, thus ruling out any active, motor protein-driven transport [47]. It is unclear if the rapid transport of ODN probes to the nucleus is due to electrostatic interaction, or is driven by a microinjection-induced flow, or the triggering of a signaling pathway. There is no fundamental biological reason why ODN probes should accumulate in the cell nucleus, but to prevent such accumulation streptavidin (60 kDa) molecules were conjugated to linear ODN probes via biotin [13]. After being microinjected into the cells, the dual-FRET linear probes could hybridize to the same mRNA target in the cytoplasm, resulting in a FRET signal. More recently, it was shown that when transfer RNA (tRNA) transcripts were attached to molecular beacons with a 2′-*O*-methyl backbone and injected into the nucleus of HeLa cells, the probes were exported into the cytoplasm. Yet, when these constructs were introduced into the cytoplasm, they remained cytoplasmic [48]. However, even without the problem of unwanted nuclear accumulation, microinjection is an inefficient process for delivering probes into a large number of cells.

Another nonendocytic delivery method is that of *toxin-based cell membrane permeabilization*. For example, SLO is a pore-forming bacterial toxin that has been used as a simple and rapid means of introducing oligonucleotides into eukaryotic cells [49–51]. SLO binds as a monomer to cholesterol and oligomerizes into a ring-shaped structure to form pores of approximately 25–30 nm in diameter, allowing the influx of both ions and macromolecules. It was found that SLO-based permeabilization could achieve an intracellular concentration of ODNs which was approximately 10-fold that achieved with electroporation or liposomal-based delivery. As cholesterol composition varies with cell type, however, the permeabilization protocol must be optimized for each cell type by varying the temperature, incubation time, cell number and SLO concentration. One essential feature of toxin-based permeabilization is that it is reversible. This can be achieved by introducing oligonucleotides with SLO under serum-free conditions and then removing the mixture and adding normal media with the serum [50, 52].

Cell-penetrating peptides have also been used to introduce proteins, nucleic acids and other biomolecules into living cells [53–55]. Included among the family of peptides with membrane-translocating activity are antennapedia, HSV-1 VP22 and the HIV-1 Tat peptide. To date, the most widely used peptides are HIV-1 Tat peptide and its derivatives, due to their small sizes and high delivery efficiencies. The Tat peptide is rich in cationic amino acids (especially arginine, which is very common in many CPPs); however, the exact mechanism of CPP-induced membrane translocation remains elusive.

A wide variety of cargos have been delivered to living cells, both in cell culture and in tissues, using CPPs [56, 57]. For example, Allinquant *et al.* [58] linked the antennapedia peptide to the 5′ end of DNA oligonucleotides (with biotin on the 3′ end) and incubated both peptide-linked ODNs and ODNs alone (as control) with cells. By detecting biotin via a streptavidin–alkaline phosphatase amplification, the peptide-linked ODNs were shown to be internalized very efficiently into all cell compartments compared to control ODNs. Moreover, no indication of endocytosis was found. Similar results were obtained by Troy *et al.* [59], with a 100-fold increase in antisense delivery efficiency when the ODNs were linked to antennapedia peptides. Recently, Tat peptides were conjugated to molecular beacons using different linkages (Figure 6.7); the resultant peptide-linked molecular beacons were delivered into living cells to target glyceraldehyde phosphate dehydrogenase (GAPDH) and survivin mRNAs [29]. It was shown that, at relatively low concentrations, peptide-linked molecular beacons were internalized into living cells within 30 min, with near-100% efficiency. Further, peptide-based delivery did not interfere with either specific targeting by, or hybridization-induced florescence of, the probes. In addition, the peptide-linked molecular beacons were seen to possess self-delivery, targeting and reporting functions. In contrast, the liposome-based (Oligofectamine) or dendrimer-based (Superfect) delivery of molecular beacons required 3–4 h and resulted in a punctate fluorescence signal in the cytoplasmic vesicles and a high background in

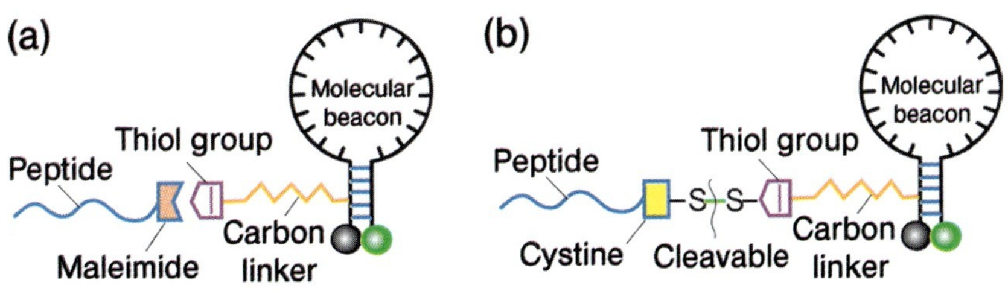

Figure 6.7 A schematic of peptide-linked molecular beacons. (a) A peptide-linked molecular beacon using the thiol–maleimide linkage in which the quencher arm of the molecular beacon stem is modified by adding a thiol group which can react with a maleimide group placed to the C terminus of the peptide to form a direct, stable linkage; (b) A peptide-linked molecular beacon with a cleavable disulfide bridge in which the peptide is modified by adding a cysteine residue at the C terminus; the cysteine then forms a disulfide bridge with the thiol-modified molecular beacon. This disulfide bridge design allows the peptide to be cleaved from the molecular beacon by the reducing environment of the cytoplasm.

both the cytoplasm and nucleus of cells [29]. These results showed clearly that the cellular delivery of molecular beacons using a peptide-based approach is far more effective than conventional transfection methods.

6.5 Living Cell RNA Detection Using Nanostructured Probes

Sensitive gene detection in living cells presents a significant challenge. In addition to issues of target accessibility, detection specificity and probe delivery (as discussed above), the achievement of a high detection sensitivity and a high signal-to-background ratio requires not only careful design of the probes and advanced fluorescence microscopy imaging, but also a better understanding of RNA biology and probe–target interactions. It is likely that different applications have different requirements on the properties of probes. For example, the rapid determination of RNA expression level and localization requires fast probe/target hybridization kinetics, whereas the long-term monitoring of gene expression dynamics requires probes with a high intracellular stability.

To demonstrate the capability of molecular beacons in the sensitive detection of specific endogenous mRNAs in living cells, dual-FRET molecular beacons were designed to detect K-ras and survivin mRNAs in HDF and MIAPaCa-2 cells, respectively [28]. K-*ras* is one of the most frequently mutated genes in human cancers [60]. A member of the G-protein family, K-ras is involved in transducing growth-promoting signals from the cell surface. Survivin, one of the inhibitor of apoptosis proteins (IAPs), is normally expressed during fetal development but not in most normal adult tissues [61], and thus can be used as a tumor biomarker for several types of cancer. Each FRET probe pair consisted of two molecular beacons – one labeled with a donor fluorophore (Cy3, donor beacon) and a second labeled with an acceptor fluorophore (Cy5, acceptor beacon). These molecular beacons were designed to hybridize to adjacent regions on an mRNA target so that the two fluorophores lay within the FRET range (~6 nm) when probe/target hybridization occurred for both beacons. BHQ-2 and BHQ-3 were used as quenchers for the donor and acceptor molecular beacons, respectively. One pair of molecular beacons targets a segment of the wild-type K-*ras* gene, the codon 12 mutations of which are involved in the pathogenesis of many cancers. A negative control dual-FRETmolecular beacon pair was also designed ('random beacon pair'), the specific 16-base target sequence of which was selected using random walking, and thus had no exact match in the mammalian genome. It was found that detection of the FRET signal significantly reduced false-positives, leading to sensitive imaging of K-*ras* and survivin mRNAs in live HDF and MIAPaCa-2 cells. For example, FRET detection gave a ratio of 2.25 of K-*ras* mRNA expression in stimulated versus unstimulated HDF cells, which was comparable to a ratio of 1.95 using RT-PCR but contrasted to the single-beacon result of 1.2. The detection of survivin mRNA also indicated that, compared to the single-beacon approach, dual-FRET molecular beacons gave a lower background signal, which in turn led to a higher signal-to-background ratio [28].

6.5.1
Biological Significance

An intriguing discovery in detecting K-*ras* and survivin mRNAs using dual-FRET molecular beacons is the clear and detailed mRNA localization in living cells [28]. To demonstrate this point, a fluorescence image of K-*ras* mRNA in stimulated HDF cells is shown in Figure 6.8a, indicating an intriguing filamentous localization pattern. The localization pattern of K-*ras* mRNA was further studied and found to be colocalized with mitochondria inside live HDF cells [62]. As K-ras proteins interact with proteins such as Bcl-2 in the mitochondria to mediate both anti-apoptotic and pro-apoptotic pathways, it seems that cells localize certain mRNAs where the corresponding proteins can easily bind to their partners.

The survivin mRNA, however, is localized in MIAPaCa-2 cell very differently. As shown in Figure 6.8b, in which the fluorescence image was superimposed with a white-light image of the cells, the survivin mRNAs seemed to localize in a nonsymmetrical pattern within MIAPaCa-2 cells, often to one side of the nucleus of the cell. These mRNA localization patterns raise many interesting biological questions. For example, how are mRNAs transported to their destination, and how is the destination recognized? Also, to which subcellular organelle might the mRNAs be colocalized? And what is the biological implication of mRNA localization? Although mRNA localization in living cells is believed to be closely related to the post-transcriptional regulation of gene expression, much remains to be seen if such localization indeed targets a protein to its site of function by producing the protein 'right on the spot'.

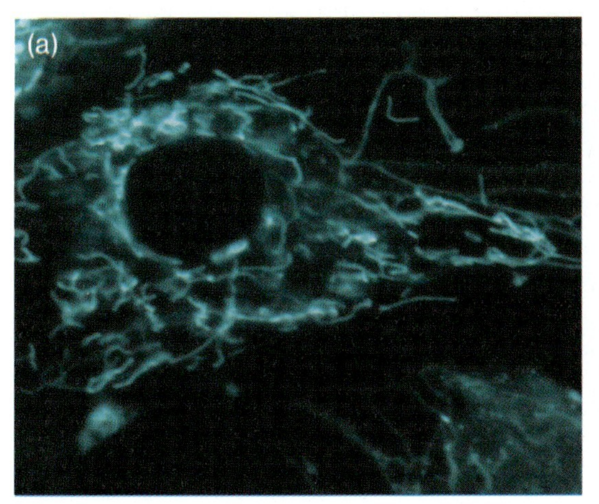

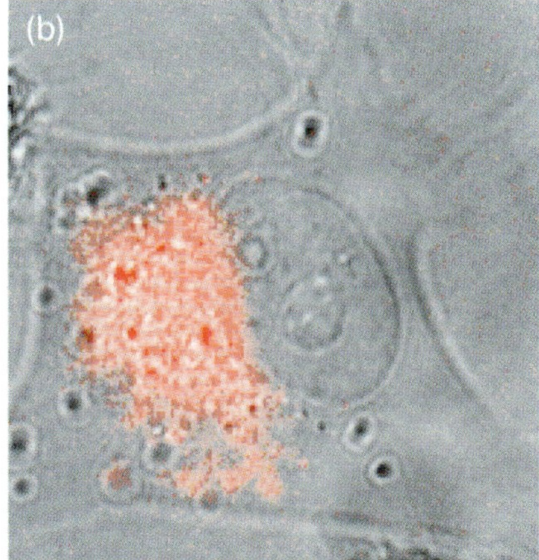

Figure 6.8 mRNA localization in HDF and MIAPaCa-2 cells. (a) Fluorescence images of K-*ras* mRNA in stimulated HDF cells. Note the filamentous K-*ras* mRNA localization pattern; (b) A fluorescence image of survivin mRNA localization in MIAPaCa-2 cells. Note that survivin mRNAs are often localized to one side of the nucleus of the MIAPaCa-2 cells.

The transport and localization of oskar mRNA in *Drosophila melanogaster* oocytes has also been visualized [26]. In these studies, molecular beacons with a 2′-*O*-methyl backbone were delivered into cells using microinjection, and the migration of oskar mRNA was tracked in real time, from the nurse cells where it is produced to the posterior cortex of the oocyte where it is localized. Clearly, the direct visualization of specific mRNAs in living cells with molecular beacons will provide important insights into the intracellular trafficking and localization of RNA molecules.

As another example of targeting specific genes in living cells, molecular beacons were used to detect the viral genome and characterize the spreading of bovine respiratory syncytial virus (bRSV) in living cells [63]. It was found that a molecular beacon signal could be detected in single living cells infected by bRSV with high detection sensitivity, and the signal revealed a connected, highly three-dimensional, amorphous inclusion-body structure not seen in fixed cells. Figure 6.9 shows the molecular beacon signal indicating the spreading of viral infection at days 1, 3, 5 and

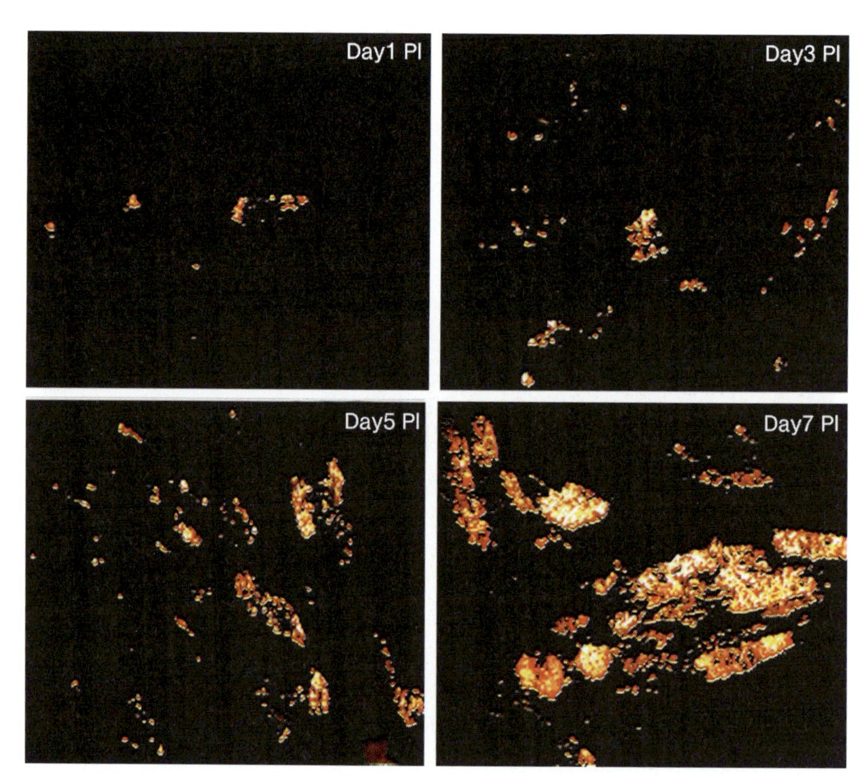

Figure 6.9 Live-cell fluorescence imaging of the genome of bovine respiratory syncytial virus (bRSV) using molecular beacons, showing the spreading of infection in host cells at days 1, 3, 5 and 7 post-infection (PI). Primary bovine turbinate cells were infected by a clinical isolate of bRSV, CA-1, with a viral titer of $2 \times 10^{3.6}$ $TCID_{50}$ ml^{-1}. Molecular beacons were designed to target several repeated sequences of the gene-end-intergenic-gene-start signal within the bRSV genome, with a SNR of 50–200.

7 post-infection, and demonstrates the ability of molecular beacons to monitor and quantify – in real time – the viral infection process. Molecular beacons were also used to image the viral genomic RNA (vRNA) of human RSV (hRSV) in live Vero cells, revealing the dynamics of filamentous virion egress, and providing an insight as to how viral filaments bud from the plasma membrane of the host cell [64].

6.6 Engineering Challenges in New Probe Development

Nanostructured molecular probes such as molecular beacons have the potential to enjoy a wide range of applications that require the sensitive detection of genomic sequences. For example, molecular beacons can be used as a tool for the detection of single-stranded nucleic acids in homogeneous *in vitro* assays [65, 66]. Surface-immobilized molecular beacons used in microarray assays allow for the high-throughput parallel detection of nucleic acid targets, while avoiding the difficulties associated with PCR-based labeling [65, 67]. Another novel application of molecular beacons is the detection of double-stranded DNA targets using PNA 'openers' that form triplexes with the DNA strands [68]. Further, proteins can be detected by synthesizing an 'aptamer molecular beacon' [69, 70] which, upon binding to a protein, undergoes a conformational change that results in the restoration of fluorescence.

The most exciting application of nanostructured oligonucleotide probes, however, is that of living cell *gene detection*. As demonstrated, molecular beacons can detect endogenous mRNA in living cells with high specificity, sensitivity and signal-to-background ratio, and thus have the potential to provide a powerful tool for both laboratory and clinical studies of gene expression *in vivo*. For example, molecular beacons can be used in high-throughput cell-based assays to quantify and monitor the dose-dependent changes of specific mRNA expression in response to different drug leads. The ability of molecular beacons to detect and quantify the expression of specific genes in living cells will also facilitate disease studies, such as viral infection detection and cancer diagnosis.

A number of challenges exist in the detection and quantification of RNA expression in living cells. In addition to the issues of probe design and target accessibility, quantifying gene expression in living cells in terms of mRNA copy-number per cell poses a significant challenge. For example, it is necessary to distinguish between true and background signals, to determine the fraction of mRNA molecules hybridized with probes, and to quantify the possible self-quenching effect of the reporter, especially when mRNA is highly localized. As the fluorescence intensity of the reporter may be altered by the intracellular environment, it is also necessary to create an internal control by, for example, injecting fluorescently labeled oligonucleotides with known quantity into the same cells and obtaining the corresponding fluorescence intensity. Furthermore, unlike RT-PCR studies – where the mRNA expression is averaged over a large number of cells (usually $>10^6$) – in the optical imaging of mRNA expression in living cells only a relatively small number of cells (typically

<1000) are observed. Therefore, the average copy number per cell may change with the total number of cells observed due to the (often large) cell-to-cell variation of mRNA expression.

Another issue in living cell gene detection using hairpin ODN probes is the possible effect of probes on normal cell function, including protein expression. As has been revealed in antisense therapy research, the complementary pairing of a short segment of an exogenous oligonucleotide to mRNA can have a profound impact on protein expression levels, and even cell fate. For example, tight binding of the probe to the translation start site may block mRNA translation. Binding of a DNA probe to mRNA can also trigger RNase H-mediated mRNA degradation. However, the probability of eliciting antisense effects with hairpin probes may be very low when low concentrations of probes (<200 n*M*) are used for mRNA detection, in contrast to the high concentrations (typically 20 μ*M*; [51]) employed in antisense experiments. Further, it generally takes 4 h before any noticeable antisense effect occurs, whereas the visualization of mRNA with hairpin probes requires less than 2 h after delivery. However, it is important to carry out a systematic study of the possible antisense effects, especially for molecular beacons with a 2′-*O*-methyl backbone, which may also trigger unwanted RNA interference.

As a new approach for *in vivo* gene detection, nanostructured probes can be further developed to have an enhanced sensitivity and a wider range of applications. For example, it is likely that hairpin ODN probes with quantum dot as the fluorophore will have a better ability to track the transport of individual mRNAs from the cell nucleus to the cytoplasm. Hairpin ODN probes with a near-infrared (NIR) dye as the reporter, combined with peptide-based delivery, have the potential to detect specific RNAs in tissue samples, animals or even humans. It is also possible to use lanthanide chelate as the donor in a dual-FRET probe assay and to perform time-resolved measurements to dramatically increase the SNR, thus achieving high sensitivity while detecting low-abundance genes. Although very challenging, the development of these and other nanostructured ODN probes will significantly enhance our ability to image, track and quantify gene expression *in vivo*, and provide a powerful tool for basic and clinical studies of human health and disease.

There are many possibilities for nanostructured probes to become clinical tools for disease detection and diagnosis. For example, molecular beacons could be used to perform cell-based early cancer detection using clinical samples such as blood, saliva and other body fluids. In this case, cells in the clinical sample are separated, while the molecular beacons designed to target specific cancer genes are delivered to the cell cytoplasm for detecting mRNAs of the cancer biomarker genes. Cancer cells having a high level of the target mRNAs (e.g. survivin) or mRNAs with specific mutations that cause cancer (e.g. K-*ras* codon 12 mutations) would show high levels of fluorescence signal, whereas normal cells would show just a low background signal. This would allow cancer cells to be distinguished from normal cells. When using this approach, the target mRNAs would not be diluted compared to approaches using a cell lysate, such as PCR. Thus, molecular beacon-based assays have the potential for the positive identification of cancer cells in a clinical sample, with high specificity and sensitivity. It might also be possible to detect cancer cells *in vivo* by using NIR-dye-labeled molecular beacons in combination with endoscopy. Nanostructured probes could also be used for the cell-based detection of other diseases. As illustrated above, well-designed molecular beacons can rapidly detect viral infection in living cells, with high specificity and sensitivity. Another possibility might be to analyze the vulnerability of atherosclerotic plaques by designing nanostructured probes to image biomarkers (mRNAs or proteins) of vulnerable plaques in blood samples. Although there remain significant challenges, imaging methods using nanostructured probes possess a truly great potential to become a powerful clinical tool for disease detection and diagnosis.

Acknowledgments

These studies were supported by the National Heart Lung and Blood Institute of the NIH as a Program of Excellence in Nanotechnology (HL80711), by the National Cancer Institute of the NIH as a Center of Cancer Nanotechnology Excellence (CA119338), and by the NIH Roadmap Initiative in Nanomedicine through a Nanomedicine Development Center award (PN2EY018244).

References

1 Saiki, R.K., Scharf, S., Faloona, F., Mullis, K.B., Horn, G.T., Erlich, H.A. and Arnheim, N. (1985) *Science*, **230**, 1350.

2 Alwine, J.C., Kemp, D.J., Parker, B.A., Reiser, J., Renart, J., Stark, G.R. and Wahl, G.M. (1979) *Methods in Enzymology*, **68**, 220.

3 Adams, M.D., Dubnick, M., Kerlavage, A.R., Moreno, R., Kelley, J.M., Utterback, T.R., Nagle, J.W., Fields, C. and Venter, J.C. (1992) *Nature*, **355**, 632.

4 Velculescu, V.E., Zhang, L., Vogelstein, B. and Kinzler, K.W. (1995) *Science*, **270**, 484.

5 Liang, P. and Pardee, A.B. (1992) *Science*, **257**, 967.

6 Schena, M., Shalon, D., Davis, R.W. and Brown, P.O. (1995) *Science*, **270**, 467.

7 Bassell, G.J., Powers, C.M., Taneja, K.L. and Singer, R.H. (1994) *The Journal of Cell Biology*, **126**, 863.

8 Buongiorno-Nardelli, M. and Amaldi, F. (1970) *Nature*, **225**, 946.

9 Behrens, S., Fuchs, B.M., Mueller, F. and Amann, R. (2003) *Applied and Environmental Microbiology*, **69**, 4935.

10 Huang, Q. and Pederson, T. (1999) *Nucleic Acids Research*, **27**, 1025.

11 Glotzer, J.B., Saffrich, R., Glotzer, M. and Ephrussi, A. (1997) *Current Biology*, **7**, 326.

12 Jacobson, M.R. and Pederson, T. (1998) *Proceedings of the National Academy of Sciences of the United States of America*, **95**, 7981.

13 Tsuji, A., Koshimoto, H., Sato, Y., Hirano, M., Sei-Iida, Y., Kondo, S. and Ishibashi, K. (2000) *Biophysical Journal*, **78**, 3260.

14 Dirks, R.W., Molenaar, C. and Tanke, H.J. (2001) *Histochemistry and Cell Biology*, **115**, 3.

15 Molenaar, C., Abdulle, A., Gena, A., Tanke, H.J. and Dirks, R.W. (2004) *The Journal of Cell Biology*, **165**, 191.

16 Tyagi, S. and Kramer, F.R. (1996) *Nature Biotechnology*, **14**, 303.

17 Tsourkas, A., Behlke, M.A., Rose, S.D. and Bao, G. (2003) *Nucleic Acids Research*, **31**, 1319.

18 Bonnet, G., Tyagi, S., Libchaber, A. and Kramer, F.R. (1999) *Proceedings of the*

National Academy of Sciences of the United States of America, **96**, 6171.

19 Tyagi, S., Bratu, D.P. and Kramer, F.R. (1998) *Nature Biotechnology*, **16**, 49.

20 Li, J.J., Geyer, R. and Tan, W. (2000) *Nucleic Acids Research*, **28**, E52.

21 Molenaar, C., Marras, S.A., Slats, J.C., Truffert, J.C., Lemaitre, M., Raap, A.K., Dirks, R.W. and Tanke, H.J. (2001) *Nucleic Acids Research*, **29**, E89.

22 Sokol, D.L., Zhang, X., Lu, P. and Gewirtz, A.M. (1998) *Proceedings of the National Academy of Sciences of the United States of America*, **95**, 11538.

23 Vet, J.A., Majithia, A.R., Marras, S.A., Tyagi, S., Dube, S., Poiesz, B.J. and Kramer, F.R. (1999) *Proceedings of the National Academy of Sciences of the United States of America*, **96**, 6394.

24 Kostrikis, L.G., Tyagi, S., Mhlanga, M.M., Ho, D.D. and Kramer, F.R. (1998) *Science*, **279**, 1228.

25 Piatek, A.S., Tyagi, S., Pol, A.C., Telenti, A., Miller, L.P., Kramer, F.R. and Alland, D. (1998) *Nature Biotechnology*, **16**, 359.

26 Bratu, D.P., Cha, B.J., Mhlanga, M.M., Kramer, F.R. and Tyagi, S. (2003) *Proceedings of the National Academy of Sciences of the United States of America*, **100**, 13308.

27 Tsourkas, A., Behlke, M.A., Xu, Y. and Bao, G. (2003) *Analytical Chemistry*, **75**, 3697.

28 Santangelo, P.J., Nix, B., Tsourkas, A. and Bao, G. (2004) *Nucleic Acids Research*, **32**, e57.

29 Nitin, N., Santangelo, P.J., Kim, G., Nie, S. and Bao, G. (2004) *Nucleic Acids Research*, **32**, e58.

30 Tyagi, S. and Alsmadi, O. (2004) *Biophysical Journal*, **87**, 4153.

31 Peng, X.H., Cao, Z.H., Xia, J.T., Carlson, G.W., Lewis, M.M., Wood, W.C. and Yang, L. (2005) *Cancer Research*, **65**, 1909.

32 Medley, C.D., Drake, T.J., Tomasini, J.M., Rogers, R.J. and Tan, W. (2005) *Analytical Chemistry*, **77**, 4713.

33 Tyagi, S., Marras, S.A. and Kramer, F.R. (2000) *Nature Biotechnology*, **18**, 1191.

34 Tsourkas, A., Behlke, M.A. and Bao, G. (2002) *Nucleic Acids Research*, **30**, 4208.

35 Brodsky, A.S. and Silver, P.A. (2002) *Methods (San Diego, Calif.)*, **26**, 151.

36 Forrest, K.M. and Gavis, E.R. (2003) *Current Biology*, **13**, 1159.

37 Shav-Tal, Y., Darzacq, X., Shenoy, S.M., Fusco, D., Janicki, S.M., Spector, D.L. and Singer, R.H. (2004) *Science*, **304**, 1797.

38 Haim, L., Zipor, G., Aronov, S. and Gerst, J.E. (2007) *Nature Methods*, **4**, 409.

39 Ozawa, T., Natori, Y., Sato, M. and Umezawa, Y. (2007) *Nature Methods*, **4**, 413.

40 Valencia-Burton, M., McCullough, R.M., Cantor, C.R. and Broude, N.E. (2007) *Nature Methods*, **4**, 421.

41 States, D.J., Gish, W. and Altschul, S.F. (1991) *Methods (San Diego, Calif.)*, **3**, 66.

42 Rhee, W.J., Santangelo, P.J., Jo, H. and Bao, G. (2007) *Nucleic Acids Research*, **36**, e30.

43 Lakowicz, J.R. (1999) *Principles of Fluorescence Spectroscopy*, 2nd edn, Plenum Press, New York.

44 Marras, S.A., Kramer, F.R. and Tyagi, S. (2002) *Nucleic Acids Research*, **30**, e122.

45 Price, N.C. and Stevens, L. (1999) *Fundamentals of Enzymology: The Cell and Molecular Biology of Catalytic Proteins*, 3rd edn, Oxford University Press, New York.

46 Dokka, S. and Rojanasakul, Y. (2000) *Advanced Drug Delivery Reviews*, **44**, 35.

47 Leonetti, J.P., Mechti, N., Degols, G., Gagnor, C. and Lebleu, B. (1991) *Proceedings of the National Academy of Sciences of the United States of America*, **88**, 2702.

48 Mhlanga, M.M., Vargas, D.Y., Fung, C.W., Kramer, F.R. and Tyagi, S. (2005) *Nucleic Acids Research*, **33**, 1902.

49 Giles, R.V., Ruddell, C.J., Spiller, D.G., Green, J.A. and Tidd, D.M. (1995) *Nucleic Acids Research*, **23**, 954.

50 Barry, M.A. and Eastman, A. (1993) *Archives of Biochemistry and Biophysics*, **300**, 440.

51 Giles, R.V., Spiller, D.G., Grzybowski, J., Clark, R.E., Nicklin, P., Tidd, D.M. (1998) *Nucleic Acids Research*, **26**, 1567.

52 Walev, I., Bhakdi, S.C., Hofmann, F., Djonder, N., Valeva, A., Aktories, K. and Bhakdi, S. (2001) *Proceedings of the National Academy of Sciences of the United States of America*, **98**, 3185.

53 Snyder, E.L. and Dowdy, S.F. (2001) *Current Opinion in Molecular Therapeutics*, **3**, 147.

54 Wadia, J.S. and Dowdy, S.F. (2002) *Current Opinion in Biotechnology*, **13**, 52.

55 Becker-Hapak, M., McAllister, S.S. and Dowdy, S.F. (2001) *Methods (San Diego, Calif.)*, **24**, 247.

56 Wadia, J.S. and Dowdy, S.F. (2005) *Advanced Drug Delivery Reviews*, **57**, 579.

57 Brooks, H., Lebleu, B. and Vives, E. (2005) *Advanced Drug Delivery Reviews*, **57**, 559.

58 Allinquant, B., Hantraye, P., Mailleux, P., Moya, K., Bouillot, C. and Prochiantz, A. (1995) *The Journal of Cell Biology*, **128**, 919.

59 Troy, C.M., Derossi, D., Prochiantz, A., Greene, L.A. and Shelanski, M.L. (1996) *The Journal of Neuroscience*, **16**, 253.

60 Minamoto, T., Mai, M. and Ronai, Z. (2000) *Cancer Detection and Prevention*, **24**, 1.

61 Altieri, D.C. and Marchisio, P.C. (1999) *Laboratory Investigation; A Journal of Technical Methods and Pathology*, **79**, 1327.

62 Santangelo, P.J., Nitin, N. and Bao, G. (2005) *Journal of Biomedical Optics*, **10**, 44025.

63 Santangelo, P., Nitin, N., LaConte, L., Woolums, A. and Bao, G. (2006) *Journal of Virology*, **80**, 682.

64 Santangelo, P.J. and Bao, G. (2007) *Nucleic Acids Research*, **35**, 3602.

65 Liu, X. and Tan, W. (1999) *Analytical Chemistry*, **71**, 5054.

66 Kambhampati, D., Nielsen, P.E. and Knoll, W. (2001) *Biosensors and Bioelectronics*, **16**, 1109.

67 Steemers, F.J., Ferguson, J.A. and Walt, D.R. (2000) *Nature Biotechnology*, **18**, 91.

68 Kuhn, H., Demidov, V.V., Coull, J.M., Fiandaca, M.J., Gildea, B.D. and Frank-Kamenetskii, M.D. (2002) *Journal of the American Chemical Society*, **124**, 1097.

69 Hamaguchi, N., Ellington, A. and Stanton, M. (2001) *Analytical Biochemistry*, **294**, 126.

70 Yamamoto, R., Baba, T. and Kumar, P.K. (2000) *Genes to Cells: Devoted to Molecular & Cellular Mechanisms*, **5**, 389.

7
High-Content Analysis of Cytoskeleton Functions by Fluorescent Speckle Microscopy

Kathryn T. Applegate, Ge Yang, and Gaudenz Danuser

7.1
Introduction

In 1949, Linus Pauling observed that hemoglobin in patients with sickle cell anemia is structurally different from that in healthy individuals [1]. This seminal discovery of a 'molecular disease' overturned a century-old notion that all diseases were caused by structural problems at the cellular level. Today, we know that disease can arise from aberrations in the expression, regulation or structure of a single molecule. Frequently, such aberrations interfere with one or more of the cell's basic morphological activities, including cell division, morphogenesis and maintenance in different tissue environments, or cell migration.

The advent of molecular pathology precipitated the rise of molecular biology and genomics, which in turn jump-started other large-scale '-omics' fields. In parallel, sophisticated imaging, quantitative image analysis and bioinformatics approaches were developed. These methods have enabled a quantum leap in our knowledge base about the molecular underpinnings of life, and what goes wrong during disease. Much has already been translated to the clinic. For example, mutation and gene expression profiles can be used to prescribe targeted drugs to breast cancer patients [2], and in the US many states have adopted metabolic screening programs to test newborns for a growing number of disorders [3].

Yet on the whole, the genomic era has failed to yield the 'goldmine of personalized interventions' that it first promised. Drug development pipelines rely heavily on high-throughput screens to identify compounds that have a desired effect on the biochemical activity of a particular drug target. These screens, however, cannot resolve whether a 'hit' will be active in living cells and specific to the pathway of interest. To avoid this limitation, high-content screens – also called 'phenotypic' or 'imaging' screens – use automated image analysis methods to detect desired changes in cells photographed under the light microscope [4]. Changes in gross cell morphology or the spatial activity of a protein of interest become apparent when analyzing a large population of cells. Although the identification of a molecular target causing a

Nanotechnology, Volume 5: Nanomedicine. Edited by Viola Vogel

ISBN: 978-3-527-31736-3

phenotypic change can be rate-limiting, assays can often be designed with the drug target already in mind. The larger challenge is to derive meaningful, quantitative phenotypic information from images [4].

The problem of extracting phenotypic information is compounded by the fact that many phenotypic differences can only be resolved in *time*; the dynamics of molecules, not just their concentrations and localization, are important in disease development. In other words, a cell may look healthy in a still image, but an analysis of the underlying dynamics of cell-adhesion proteins, for example, may reveal that the cell has metastatic potential. Current phenotypic screens can only distinguish between coarse, spatially oriented phenotypes [5], while many diseases exhibit extremely subtle, yet significant, phenotypes. New methods are needed to extract and correlate dynamic descriptors if we are to design drugs and other nanomedical intervention strategies with minimal side effects.

In this chapter, we review quantitative fluorescent speckle microscopy (qFSM), a relatively young imaging technology that has been used to characterize the dynamic infrastructure of the cell. qFSM has the potential to become a unique assay for live-cell phenotypic screening that will guide the development of drugs and other nanomedical strategies based on the dynamics of subcellular structures. We begin by summarizing how regulation of the cytoskeleton contributes to important cell morphological processes that go awry in disease. We then describe how fluorescent speckles form to mark the dynamics of subcellular structures. Next, we illustrate critical biological insights that have been gleaned from qFSM experiments. We conclude with new applications and an outlook on the future of qFSM.

7.2 Cell Morphological Activities and Disease

The filamentous actin (F-actin), intermediate filament (IF) and microtubule (MT) cytoskeleton systems are key mediators of cell morphology (Figure 7.1). Each filament system is unique in its physical properties and extensive subset of associated proteins [6]. Endogenous and exogenous chemical and mechanical signals control the precise arrangement of these dynamic polymers, and defects in their regulation are seen in a wide variety of diseases [7].

7.2.1 Cell Migration

One of the most fundamental cell morphological functions is migration. Many cell types in the body are motile, including fibroblasts, epithelial cells, neurons, leukocytes and stem cells. Failure to migrate, or migration to the wrong location in the body, can lead to congenital heart or brain defects, atherosclerosis, chronic inflammation, neurodegenerative disease, compromised immune response, defects in wound healing and tumor metastasis [8].

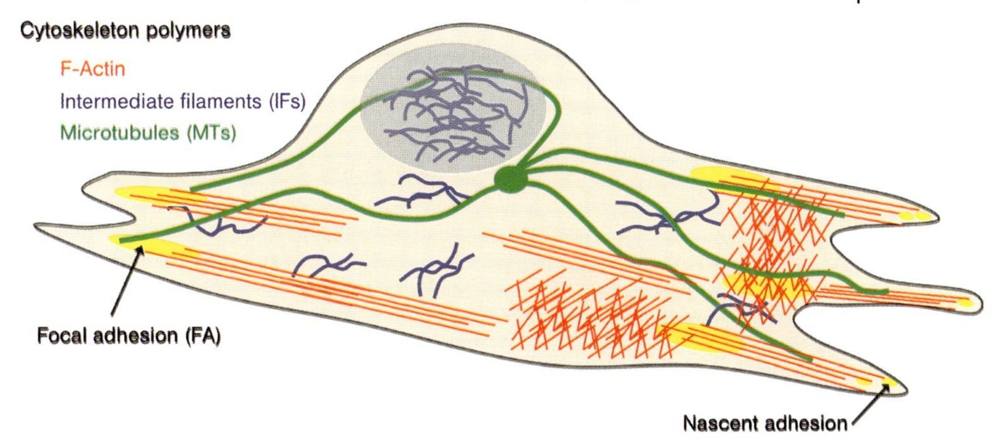

Figure 7.1 The F-actin, microtubule and intermediate filament cytoskeletons and adhesions in a migrating cell. A mesh-like F-actin network (red) at the leading edge drives the plasma membrane forward. Contractile F-actin bundles (red) linked to strong adhesions (yellow) in the front and weak adhesions in the back promote tail retraction. MTs (green) are implicated in cell polarization and adhesion regulation. The IFs (blue) are a heterogeneous group of filamentous proteins which help maintain cell structural integrity.

Cell migration is a remarkably complex behavior at the molecular level. Cell crawling involves three basic steps [9]: (i) leading edge protrusion; (ii) contraction of the F-actin polymer network; and (iii) tail retraction. Net movement of the cell cannot occur unless the F-actin network is anchored to the substrate. Otherwise, the protruding forces generated by F-actin polymerization at the leading edge plasma membrane and the contraction of the network by myosin molecular motors would simply deform the cell, without creating traction. Cells accomplish this via adhesion organelles located on the cell's ventral surface (Figure 7.1). In the case of uniform cell adhesion, actin network contraction would pull equally on the front and rear of the cell, resulting in zero movement. Directional movement thus requires an adhesion *gradient*, which is established by adhesions assembling at the protruding edge and weakening as they mature towards the cell rear [10]. The control of such a spatial gradient is extremely complex and involves many molecular components [11], the interactions of which are still poorly understood.

Like F-actin and adhesions, MTs are also important for migration in many cell types. They are implicated in polarizing the cell by delivering signaling molecules and regulating the turnover of adhesions [12]. In addition, mechanical interactions between MTs and F-actin may also contribute to the control of cell movement and morphology [13]. Tight spatial and temporal coordination between the F-actin, MT and adhesion systems is critical for cell migration.

7.2.2
Cell Division

Division is another cellular function that depends on a complex and highly regulated series of molecular events. Division is essential during embryogenesis and development, and also occurs constantly in tissues of the adult body. In the intestines alone, approximately 10^{10} old cells are shed and replaced every day [14]. When DNA replication is complete, the pairs of chromosomes must be pulled apart symmetrically and segregated to opposite ends of the cell. Segregation is accomplished by a dynamic structure called the spindle, which is composed of MTs and motor proteins. In the final step of cell division – cytokinesis – the spindle elongates and a contractile actin structure develops to pinch off the membrane, partitioning organelles and cytoplasm into two daughter cells. While these processes progress with remarkable fidelity in healthy individuals, unchecked and faulty cell division are hallmarks of oncogenesis [15] and age-related disorders [16]. Cell cycling of adult neurons may be implicated in Alzheimer's disease [17]. Analysis of the architectural dynamics of the F-actin and MT structures involved in these steps is critical if we are to intervene with abnormal events in cell division associated with disease development.

7.2.3
Response to Environmental Changes

Cells also must be able to respond to physical changes in the environment. Almost 15 years ago, Wang *et al.* reported that applying a mechanical stimulus to integrin transmembrane receptors in adhesions caused the cytoskeleton to stiffen in proportion to the load [18]. The increased stiffness required the presence of intact F-actin, MTs and IFs. Such adaptation is central to the formation of multicellular tissues and functional organelles [19]. Recent studies have also provided evidence that mechanical cues relayed through the cytoskeleton systems dictate stem cell fate. Naïve mesenchymal stem cells cultured on a soft, brain-like matrix differentiate into neurons, while those cultured on a more rigid, bone-like matrix develop into osteoblasts [20]. When F-actin contraction by myosin motors is inhibited, lineage specification by elasticity is blocked. Mechanistic analyses of these differentiation-defining processes will require a quantitative analysis of the underlying cytoskeleton dynamics.

7.2.4
Cell–Cell Communication

Many diseases are directly linked to abnormalities in cell–cell or intracellular communication. For example, normal epithelial cells grow into organized, confluent monolayers because their growth is constantly monitored and controlled by cell–cell contacts. Carcinoma cells, on the other hand, lose this cue and grow into a mass. In addition to cancer, the loss of control over cell–cell contacts can result in embryonic death, severe developmental defects, neuropathy, skin-blistering diseases, diabetes, autoimmune disorders and atherosclerosis [21]. Besides the communication between cells in direct attachment, cells also communicate via a number of pathways, from the endo- or exocytosis of a few receptor-bound molecules to the uptake of large particles via phagocytosis [6]. Pathogens have managed to hijack these pathways to enter and exit cells [22]. In all of these communication pathways, the three cytoskeleton systems are on center stage. Analogously, the import of drugs or nanomedical devices into defective cells requires the specific activation of one of these pathways in the right place, at the right time. Thus, our ability to understand disease and to precisely manipulate cells depends on our ability to analyze the cytoskeleton structure and dynamics *in situ* in living cells. We will now introduce qFSM as one of the emerging tools to achieve this goal.

7.3
Principles of Fluorescent Speckle Microscopy (FSM)

As reviewed by Danuser and Waterman-Storer [23], FSM enables quantitative analysis of the dynamics of subcellular structures *in vitro* and *in vivo*. Like many innovations in science, FSM was discovered by accident when Clare Waterman-Storer noticed that the microinjection of a small amount of X-rhodamine-labeled tubulin into cells gave rise to MTs with a speckled appearance [24]. Over the past ten years, sample preparation, speckle imaging and computational image analysis have been developed and improved to yield robust measurements of intracellular flow and assembly/disassembly dynamics for both MT and F-actin structures. Very recently, FSM has also been applied to measure interaction dynamics between molecular assemblies, such as transient F-actin and adhesion coupling in migrating epithelial cells [25] and F-actin and MT comovement in epithelial cells [26] and neurons [27]. Moreover, computational post-processing of FSM data has yielded indirect information, such as intracellular forces [28]. Thus, FSM is a prime imaging mode for interrogating the cytoskeleton's many roles in cell physiology and disease.

FSM is a twist on conventional live-cell fluorescence microscopy, where structures are visualized either by the expression of a fluorescent protein fused to the protein of interest, or by the microinjection of a covalently labeled protein into the cell. Typically in fluorescence microscopy, high expression levels or large amounts of injected fluorescent protein are necessary to achieve a high signal-to-noise ratio (SNR). This approach reveals protein localization and, to an extent, the movement of molecular structures in cells. However, the ability to report protein dynamics is limited due to the inherently high background fluorescence from out-of-focus incorporated and diffusing, unincorporated fluorescent protein. In addition, it is impossible to detect the movement or turnover of protein subunits within a larger, uniformly labeled structure. Laser photobleaching and photoactivation can help by marking structures in defined regions of the cell and allowing the measurement of recovery or movement in the local region at steady state [29–34]. Similar to these techniques is the ratiometric method of fluorescence localization after photobleaching (FLAP), which shows the diffusive and convective transport of unincorporated protein [35, 36].

In contrast to conventional fluorescence microscopy, FSM probes the dynamics of protein assemblies which contain very few (<1%) fluorescent subunits among a vast majority of unlabeled subunits. When imaged by high-resolution, diffraction-limited optics, the scattered distribution of fluorescence yields a punctate (speckled) pattern that reveals the motion of the entire structure, the reorganization of subunits within the structure, and the association and dissociation of subunits. Thus, FSM provides the same information as the aforementioned photomarking techniques, but does so across much or all of the cell simultaneously. As FSM does not require active marking, it allows the continuous detection of nonsteady-state dynamics within protein assemblies, and reveals spatial and temporal relationships between these dynamic events at submicron and second resolution. FSM also reduces out-of-focus fluorescence and improves the visibility of fluorescently labeled structures and their dynamics in three-dimensional (3-D) polymer arrays, such as the mitotic spindle [37–39].

In the early years of its development, FSM used wide-field epifluorescence light microscopy and digital imaging with a sensitive, low-noise, cooled charge-coupled-device (CCD) camera [24]. Since then, FSM has been transferred to confocal and total internal reflection fluorescence (TIRF) microscopes [37, 40–42]. The development of fully automated, computer-based tracking and the statistical analysis of speckle behavior proved to be critical steps in establishing FSM as a routine method for measuring cytoskeleton architectural dynamics. Thus, FSM is an integrated technology in which sample preparation, imaging and image analysis are optimized to achieve detailed information about polymer dynamics.

7.4 Speckle Image Formation

7.4.1 Speckle Formation in Microtubules (MTs): Stochastic Clustering of Labeled Tubulin Dimers in the MT Lattice

MTs exhibit a variation in fluorescence intensity along their lattices when cells are injected with a small amount of labeled tubulin dimers, leading to a speckled appearance (Figure 7.2a and b). Several possibilities exist for how speckles arise in this situation:

- The fluorescent tubulin could form oligomers or aggregates on the MT.
- Cellular organelles or MT-associated proteins (MAPs) could be bound to the MT and conceal or quench the fluorescence in some regions.
- Random variation of the number of fluorescent tubulin subunits in each diffraction-limited image region along the MT could occur as the MT assembles from a pool of labeled and unlabeled dimers [43].

The first hypothesis was discounted by showing that labeled tubulin dimers sediment similarly to unlabeled dimers in an analytical ultracentrifugation assay. Next, it was shown that MTs assembled from purified tubulin *in vitro* exhibited similar speckle patterns to MTs in cells, where MAPs and organelles are present [43]. Thus, the most plausible explanation for MT speckle formation is that variations exist in the number of fluorescent tubulin subunits in each resolution-limited image region along the MT.

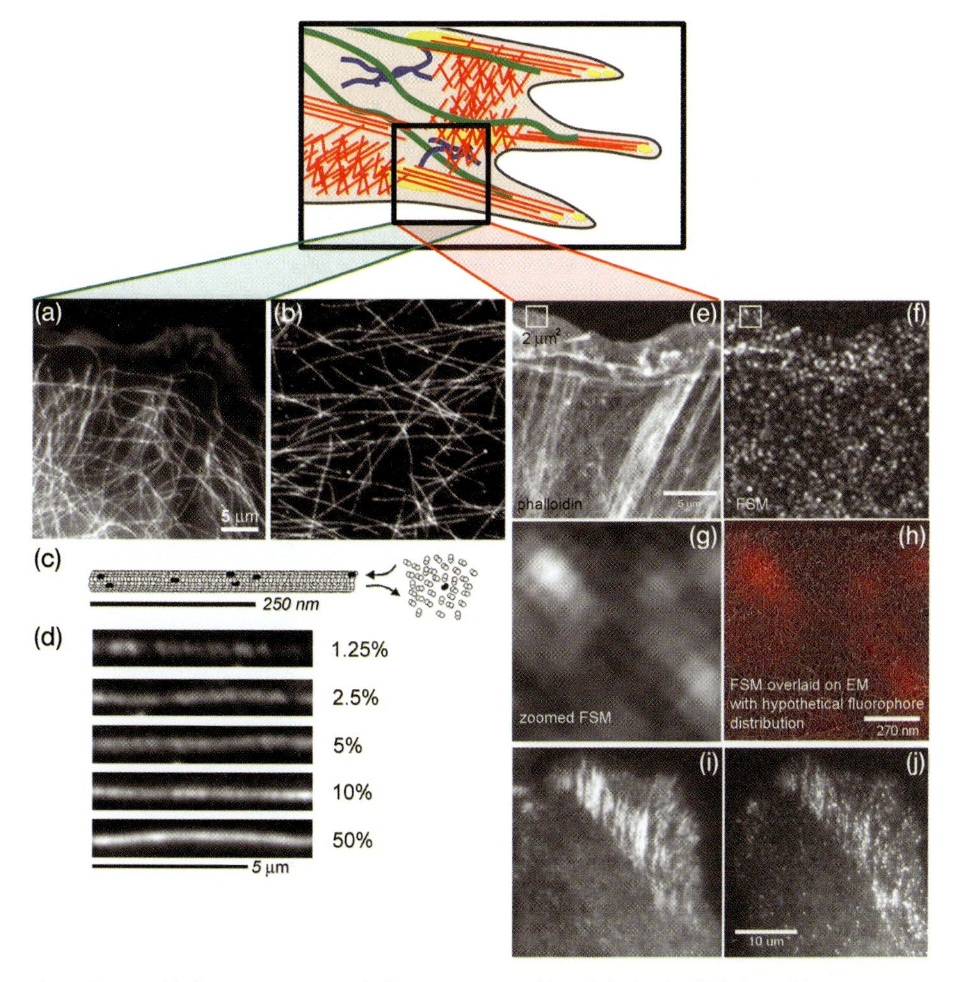

Figure 7.2 Speckle formation in microtubules, F-actin and FAs. The cartoon is an inset of Figure 7.1, showing that MTs (a) and F-actin (e, f) in cells can be imaged in separate channels. (a, b) Comparison of random speckle pattern of fluorescence along MTs for (a) a living epithelial cell microinjected with X-rhodamine-labeled tubulin and for (b) MTs assembled *in vitro* from 5% X-rhodamine-labeled tubulin; (c) Model for fluorescent speckle pattern formation in a MT grown from a tubulin pool containing a small fraction of labeled dimers; (d) Dependence of speckle contrast on the fraction of labeled tubulin dimers. (e, f) Speckle formation in actin filament networks. An epithelial cell was microinjected with a low level of X-rhodamine-labeled actin, fixed, and stained to show structure with Alexa-488 phalloidin. (e) Phalloidin image showing the organization of actin filaments in amorphous filament networks and bundles; (f) In the single FSM image, much of the structural information is lost, but time-lapse FSM series contain dynamic information of filament transport and turnover not accessible with higher-level labeling of the cytoskeleton; (g) Close-up of $2 \times 2\,\mu m$ window in panels e and f; (h) Colorized speckle signal overlaid onto a quick-freeze deep etch image of the same-sized region of the actin cytoskeleton in the leading edge of a fibroblast (kindly provided by Tatyana Svitkina). The hypothetical fluorophore distribution (red) could give rise to such a speckle pattern, indicating the scale of speckles in comparison with the polymer network ultrastructure. It also illustrates that a small proportion of the total actin fluoresces and that fluorophores from different filaments contribute to the same speckle; (i, j) Low-level expression of the GFP-tagged FA protein vinculin results in speckled FAs in TIRF images. A cell expressing GFP-vinculin was fixed and immunofluorescently stained with antibodies to (i) vinculin to reveal the position of FAs, which in the (j) GFP channel appear speckled because of the low level of incorporation of GFP-vinculin. Figure reproduced with permission from Ref. [23].

To understand how speckles originate, consider the incorporation of fluorescent subunits into the helical MT lattice, which consists of 1625 α/β tubulin dimers per micron (Figure 7.2c) [44]. The image of an individual MT under the light microscope results from a convolution of the fluorophore distribution along the MT with the point-spread function (PSF) of the microscope. Ignoring the vertical dimension, the in-focus slice of the PSF is given by the Airy disk. Given the emission wavelength λ of the fluorophore and the numerical aperture (NA) of the microscope objective, the first ring of the Airy disk with zero intensity has a radius $r = 0.61\,\lambda/\mathrm{NA}$ [45]. Objects separated by less than r cannot be resolved. For X-rhodamine-labeled MTs ($\lambda = 620$ nm) at a high NA (1.4), the Airy disk radius $r = 270$ nm, which corresponds to 440 tubulin subunits (270 nm $\times$ 1.625 subunits nm^{-1}). A given fraction of fluorescent dimers, f, produces a mean number of fluorescent dimers $n = 440{\cdot}f$ per PSF. Variations in the number of fluorescent dimers per PSF relative to this mean produce the speckle pattern along the MT. The speckle pattern contrast in this example can be approximated by the ratio of the standard deviation and the mean of a binomial distribution with 440 elements: $c = \sqrt{440 \cdot f \cdot (1-f)}/(440 \cdot f)$. Accordingly, the contrast c can be increased by decreasing f (Figure 7.2d) or by making the Airy disk smaller – that is, effectively lowering the number of tubulin subunits per Airy disk. The latter is accomplished by using optics with the highest NA possible. Experiments have shown that fractions in the range of 0.5 to 2%, where speckles consisting of three to eight fluorophores are optimal for the speckle imaging of individual microtubules [46].

7.4.2
Speckle Formation in Other Systems: The Platform Model

Conventional fluorescence microscopy of the F-actin network reveals a variety of actin-based structures, such as stress fibers and filopodia, but this approach is not conducive to the observation of structural dynamics (see Section 7.3). When fluorescently labeled actin is injected into a cell at a low level relative to the amount of endogenous, unlabeled actin, actin-rich structures appear relatively evenly speckled (Figure 7.2e and f) [26, 47–50]. Importantly, while the speckle pattern reveals the local dynamics of actin structures the overall architectural organization of the cytoskeleton is no longer visible. Speckle formation can also be found when expressing green fluorescent protein (GFP)-fused actin at a very low level [48, 51] or by injecting trace amounts of the labeled actin-binding molecule phalloidin [52, 53]. In contrast to speckle formation in isolated MTs, labeled actin subunits bind within a highly cross-linked 3-D network of F-actin filaments [54–56]. The mesh size of an F-actin network in living cells is nearly always below the resolution limit of the light microscope (Figure 7.2g and h). Consequently, unless f is kept extremely low so that only one fluorophore falls into the PSF volume [48], most fluorescent speckles arise from subunits on multiple actin filaments.

The same concept applies to speckle formation within adhesion sites [40]. GFP-fusions to adhesion proteins, including vinculin, talin, paxilin, α-actinin, zyxin and $\alpha_v$integrin [57], have been expressed in epithelial cells from crippled promoters to achieve very low expression levels. Labeled proteins assembling with endogenous, unlabeled proteins give the adhesions a speckled appearance (Figure 7.2i and j). As with F-actin networks, the speckles represent randomly distributed fluorescent adhesion proteins that are temporarily clustered in the adhesion complex within the volume of one PSF.

In summary, a speckle is defined as a diffraction-limited region that is significantly higher in fluorophore concentration (i.e. higher in fluorescence intensity) than its local neighborhood. For a speckle signal to be detected in an image, the contributing fluorescent molecules must be associated with a molecular scaffold, or 'speckle platform', during the 0.1–2.0 s exposure time required by most digital cameras to acquire the dim FSM image. Conversely, unbound, diffusible fluorescent molecules visit many pixels during the exposure and yield an evenly distributed background signal, instead of speckles [48]. The same idea was illustrated for the MT MAP ensconsin [58] and for the MT kinesin motor Eg5 [59]. Thus, association with the platform can occur when labeled subunits either *become* part of the platform, as with tubulin or actin, or simply *bind* to it, as in the case of cytoskeleton-associated proteins or adhesion molecules.

7.5 Interpretation of Speckle Appearance and Disappearance

7.5.1 Naïve Interpretation of Speckle Dynamics

Following the platform model, one would expect the appearance of a speckle to correspond to the local association of subunits with the platform. Conversely, the disappearance of a speckle would mark the local dissociation of subunits. In other words, FSM allows – in principle – the direct kinetic measurement of subunit turnover in space and time via speckle lifetime analysis. In addition, once a speckle is formed, it may undergo motion that indicates the coordinated movement of labeled subunits on the platform and/or the movement of the platform itself.

7.5.2 Computational Models of Speckle Dynamics

The interpretation of speckle dynamics becomes significantly more complicated when individual speckles arise from fluorophores distributed over multiple polymers. To examine how speckle appearance and disappearance relate to the rates of assembly and disassembly of F-actin, we performed Monte Carlo simulations of fluorophore incorporation into growing and shrinking filaments in dense, branched networks, and generated synthetic FSM time-lapse sequences [60]. The first lesson learned from this modeling was that the speckle density is independent of whether the network assembles or disassembles; it depends only on how many Airy disks can be resolved per square micron. With NA $= 1.4/100\times$ optics, this amounts to $\sim$4 (approximately 2×2), as confirmed by Figure 7.3a. The graph displays the mean

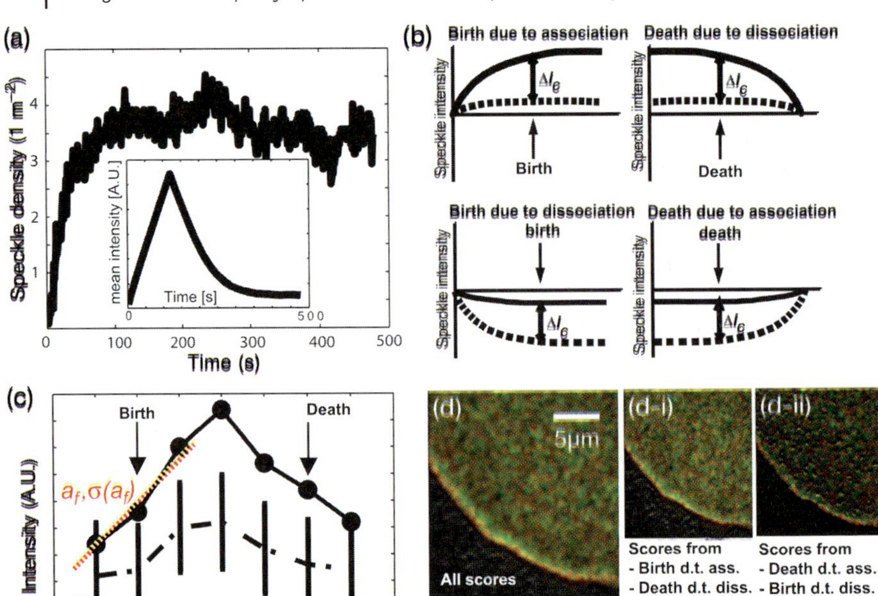

Figure 7.3 Relationship between speckle appearance (birth) and disappearance (death) and the turnover in the underlying macromolecular assembly. (a) Simulated speckle density in an actin filament network assembling for 120 s and disassembling for 360 s. Inset: Mean intensity indicating the overall change in bound fluorophore over time; (b) Classification of speckle birth and death due to monomer association and dissociation with the network. A speckle appears when the difference between foreground (solid line) and background (dashed line) is greater than a threshold ΔI_C, which is a function of the camera noise and the shot noise of the signal [60]; (c) Measurement of intensity changes in foreground (solid line) and background (dashed line) during a speckle birth. The entire lifetime of the speckle is shown (40 s). Dash-dotted line: Mean between foreground and background; error bars: ΔI_C computed in every time point. Birth and death are defined as the time points at which the intensity difference exceeds ΔI_C for the first and the last time. Red dotted line: Regression line to the foreground intensity values before, at and after birth. Blue dotted line: Regression line to the background intensity values. The cause of speckle birth is inferred by statistical classification of the two slopes and their standard deviations (see text). The statistically more dominant of the two slopes, if also significant relative to image noise, defines the score of the event (foreground slope in the example given); (d) Spatial averaging of scores accumulated over a defined time window of an FSM time-lapse sequence yields maps of net polymerization (red) and depolymerization (green). Scores from birth due to association and death due to dissociation (d-i) or from birth due to dissociation and death due to association (d-ii) reveal the same spatial distribution of polymerization and depolymerization. Figure in parts reproduced with permission from Refs [60, 83].

speckle density from five simulations of a network that starts with no fluorophores, assembles for 120 s (inset: mean fluorescence intensity increases), and disassembles for 360 s (inset: mean fluorescence intensity decreases) at equal rates. The density does not change after saturation at 100 s and remains constant, despite the further addition of fluorophores for another 20 s. This suggests that monomer association can cause an equal number of speckle appearances and disappearances. The same holds true in the opposite sense during network disassembly.

Whereas, the NA of the optics defines the maximum number of resolvable speckles per unit area, the labeling ratio influences the speckle density indirectly. For multi-fluorophore speckles the ratio f is the main determinant of speckle contrast. When increasing this ratio, the speckle density drops because the difference between the peak intensity of a speckle and its surroundings is no longer distinguishable from intensity fluctuations due to noise. Similarly, at labeling ratios where speckles represent the image of single fluorophores ($f < 0.1\%$), a further decrease in f reduces the speckle density proportionally. Across the optimal range of ratios for multi-fluorophore speckles ($0.5\% < f < 3\%$) the density is almost constant. These model predictions were largely confirmed experimentally by Adams *et al.* [40].

The origin of constant speckle density in the range of $0.5\% < f < 3\%$ is illustrated in Figure 7.3b. A speckle represents a local image intensity maximum significantly above the surrounding background. The critical intensity difference ΔI_c depends on both the camera noise and the shot noise. The shot noise is by itself a function of the speckle intensity [60]. Speckles may appear (speckle birth) for two reasons: (i) the intensity of a local maximum becomes brighter because of the association of fluorescent subunits; or (ii) the intensity of the surrounding background becomes dimmer because of subunit dissociation in the neighborhood. In both cases, a speckle birth is detected when the peak-to-background intensity difference exceeds ΔI_c. Analogously, speckles may disappear (speckle death) either because of subunit dissociation in the location of a speckle or because of subunit association in the neighborhood.

7.5.3
Statistical Analysis of Speckle Dynamics

With the classification scheme in Figure 7.3b, speckles become time-specific, diffraction-limited probes of turnover of subcellular structures. The change in foreground or background intensity that causes the birth or death of a speckle is, on average, proportional to the net number of subunits Δm added to or removed from the PSF volume between two frames. This defines an algorithm for the local measurement of network assembly or disassembly kinetics [60]:

- Calculation of changes in foreground and background intensities. After detection of a speckle birth/death event, regression lines are fitted to the foreground and background intensities for one time point before, during and after the event (Figure 7.3c). Intensity values before birth and after death are extrapolated [60]. The line fits provide two estimates a_f and a_b of the slopes of foreground and background intensity variation. They also yield the standard deviations σa_f and σa_b of the slopes, which are derived from the residuals of the intensity values to the regression line. Noisy data, poorly represented by the regression model, generate large values of σ; intensity values in perfect match with the model result in small values of σ.

- Each of the two slopes is tested for statistical significance. Insignificant intensity changes are discarded.

- If both foreground and background slopes are significant, the one with the higher significance (lower *p*-value) is selected as the cause of the event. In the example in Figure 7.3c the foreground slope has the higher significance. The magnitude of the more significant slope is recorded as the score of the birth/death event. If neither foreground nor background slope is statistically significant, no score is generated.

Score values represent instantiations of a random variable with an expectation value $\mu = \alpha \Delta m \cdot f$ and variance $\sigma^2 = \alpha^2 \Delta m^2 \cdot f(1-f)$, where α denotes the unknown intensity of one fluorophore. In addition, the scores are perturbed by noise. However, assuming that the net rate Δm remains constant for a small probing window, the intrinsic score variation and noise are approximately eliminated by averaging all scores falling into the window. The choice of the window size depends on the density of significant scores and the demand for spatial or temporal resolution. The more scores averaged by time integration, the less spatial averaging is required, and *vice versa*.

Figure 7.3d displays rates of actin assembly (red) and disassembly (green) of the F-actin network at the edge of an epithelial cell. Score values were averaged over 10 min, reflecting the steady-state turnover. The two smaller panels indicate the rate distributions calculated from scores extracted from speckle births due to monomer association and from speckle deaths due to monomer dissociation only (Figure 7.3d-i), and from births due to monomer dissociation and from deaths due to monomer association only (Figure 7.3d-ii). Both panels display the same distribution of loci of strong assembly (for example, the cell edge) and disassembly but at different event densities. Figure 7.3d-i thus corresponds to the naïve interpretation of speckle appearance and disappearance. These events contribute ~70% of all scores; the other ~30% of significant scores is related to the counterintuitive cases of speckle birth and death, and neglecting these would significantly reduce the sample size. How many intuitive versus counterintuitive cases occur depends on the fraction of labeled monomers. The lower the fraction, the fewer counterintuitive cases are observed, with a lower boundary defined by the single-fluorophore speckle regime, where all speckle appearances are due to monomer association and all disappearances are due to monomer dissociation.

The processing of only short time intervals around speckle birth and death events focuses the analysis on image events that are more likely to have originated from monomer exchange rather than from intensity fluctuations due to image noise, bleaching and in- and out-of-focus speckle motion. In addition, the algorithm rejects ~60% of all speckle birth and death events as insignificant [61]. That is, these events are not classifiable as induced by monomer exchange with the certainty the user chooses as the confidence level for the analysis. Bleaching affects all speckle scores, and thus can be corrected based on global drifts in the image signal [60]. It has also been shown that, with a NA = 1.4 objective lens, focus drifts smaller than 100 nm over three frames (e.g. 30 nm per 1–5 s) have no effect on the mapping of network turnover. Thus, the statistical model described in this section provides a robust method for calculating spatiotemporal maps of assembly and disassembly of subcellular structures such as F-actin networks.

7.5.4
Single- and Multi-Fluorophore Speckles Reveal Different Aspects of the Architectural Dynamics of Cytoskeleton Structures

Intuitively, it seems that FSM would be most powerful if implemented as a single-molecule imaging method, where speckle appearances and disappearances unambiguously signal association and dissociation of fluorescent subunits to the platform [48]. However, the much simpler signal analysis of single-molecule images is counterweighed by several disadvantages not encountered when using multifluorophore speckles. First, establishing that an image contains only single-fluorophore speckles can be challenging, especially when the signal of one fluorophore is close to the noise floor of the imaging system. Especially in 3-D structures a large number of speckles have residual contributions from at least one other fluorophore, and those mixtures must be eliminated from the statistics. Second, the imaging of single-fluorophore speckles is practically more demanding than multifluorophore FSM and requires longer camera exposures to capture the very dim signals, thus reducing the temporal resolution. Third, in addition to the lower temporal resolution, single-fluorophore FSM offers lower spatial resolution because the density of speckles drops significantly with the extremely low labeling ratio required for single-molecule-imaging conditions. In dense, crosslinked structures such as the F-actin network the intensity variation during appearance and disappearance events of multifluorophore speckles can distinguish between fast and slow turnover. In contrast, single-fluorophore speckles deliver on/off information only. Thus, in order to measure rates of the turnover of molecular structures on a continuous scale, single-fluorophore speckle analysis must rely on spatial and temporal averaging, which further decreases the resolution, while multifluorophore speckles provide this information at a finer spatial and shorter temporal scale.

On the other hand, multifluorophore speckles cannot resolve the dynamics of closely apposed individual units with subcellular structures. If the dynamics of the individual building blocks of a structure is of interest, and the lower density of spatial and temporal samples can be afforded, single-fluorophore speckles are adequate probes. If information about individual units *and* the ensemble of units is needed, single-fluorophore and multifluorophore speckle imaging can be combined in two spectrally distinct channels. This has been demonstrated in an analysis of the *Xenopus* spindle [62], where combined single- and multifluorophore qFSM revealed the overall dynamics of MTs in the spindle, as well as the dynamics and length distribution of individual MTs within densely packed bundles inside the spindle (see Section 7.9.2).

In summary, FSM can probe different aspects of the architectural dynamics of subcellular structures at different spatial and temporal scales via modulation of the ratio of labeled to unlabeled subunits. Currently, the exact labeling ratio is difficult to control in a given experiment. Statistical clustering analysis of the resulting speckle intensity is required to identify the distribution of the numbers of fluorophores within speckles. In the near future, these mathematical methods will be complemented with sophisticated molecular biology that will allow relatively precise titration of the labeled subunits. Together, these approaches will be invaluable to a systematic mapping of the heterogeneous dynamics of complex subcellular structures such as the cytoskeleton.

7.6 Imaging Requirements for FSM

Time-lapse FSM requires the imaging of high-resolution, diffraction-limited regions containing one to ten fluorophores and the inhibiting of fluorescence photobleaching. This requires a sensitive imaging system with little extraneous background fluorescence, efficient light collection, a camera with low noise, high quantum efficiency, high dynamic range, high resolution, and the suppression of fluorescence photobleaching with illumination shutters and/or oxygen scavengers [49, 63, 64]. In addition, all fluorescently labeled molecules must be functionally competent to bind their platform; otherwise, they will contribute to diffusible background and reduce the speckle contrast [65]. The reader is referred to a review by Gupton and Waterman-Storer [66] for an in-depth discussion of the hardware requirements for obtaining FSM images.

Because FSM is achieved by the level of fluorescent protein in the sample, it is adaptable to various modes of high-resolution fluorescence microscopy, such that the specific advantages of each mode can be exploited in combination with the quantitative capabilities of FSM. For example, we have performed FSM on both spinning-disk confocal microscopy [41] and total internal reflection fluorescence microscopy (TIRFM) [40] systems to gain speckle data in two spectral channels with the specific image advantages of confocal and TIRFM. A comparison of FSM images of the actin cytoskeleton in migrating epithelial cells acquired by wide-field epifluorescence, spinning-disk confocal microscopy and TIRFM is shown in Figure 7.4a–c. Clearly, speckle contrast is improved by reducing out-of-focus fluorescence with either of the latter techniques. Contrast in TIRFM images is further improved over the spinning-disk confocal image because the evanescent field excitation depth is reduced to 100–200 nm into the specimen. When quantified, the effect of the reduced effective imaging volume on modulation and detectability of actin and focal adhesion (FA) speckles showed that TIRF-FSM indeed affords major improvements in these parameters over wide-field epifluorescence for imaging macromolecular assemblies at the ventral surface of living cells, both in thin peripheral and thick central cell regions [40]. Importantly, to date FSM has proved to be incompatible with all commercial laser-scanning confocal microscope systems. This is because these instruments use photomultipliers as detectors that are noisy and have a limited dynamic range compared to the

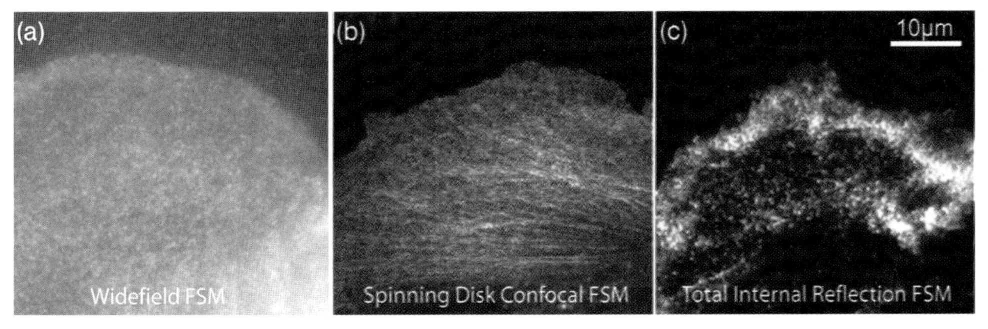

Figure 7.4 Comparison of X-rhodamine–actin FSM images of the edge of migrating Ptk1 epithelial cells using (a) wide-field epifluorescence, (b) spinning-disk confocal microscopy and (c) TIRF microscopy. Panels (a) and (b) were acquired using a Nikon 100 × 1.4 NA Plan Apo phase contrast objective lens and a 14 bit Hamamatsu Orca II camera with 6.7 micron pixels. Panel (c) was acquired with a Nikon 100 × 1.45 NA Plan Apo TIRF objective lens and a 14 bit Hamamatsu Orca II ER with 6.4 micron pixels. Note that speckle contrast and the ability to detect speckles in more central cell regions increases from panels (a) to (c). Note, however, in the TIRF image that speckles are very bright a few microns back from the edge, most likely where the cell is in closer contact with the substrate. Figure reproduced with permission from Ref. [23].

low-noise, high dynamic range CCDs used with spinning-disk confocal microscope systems.

7.7 Analysis of Speckle Motion

7.7.1 Tracking Speckle Flow: Early and Recent Developments

In addition to revealing the kinetics of association and dissociation of subunits to and from the molecular platform, speckles also show the movement of subunits within the platform and of the platform itself. In early applications of FSM, speckle motion was quantified by hand-tracking a few speckles – a tedious, error-prone and incomplete way of analyzing the wealth of information contained by these images [26, 27, 48]. Alternatively, kymographs provided average estimates of speckle velocities [37, 38, 49, 50, 59, 67–70].

Initial attempts to automate the extraction of more complete speckle flow maps from FSM time-lapse sequences of F-actin networks relied on correlation-based tracking. The speckled area of a source frame in the movie was divided into small probing windows, with each window being displaced until the normalized cross-correlation of the window with the signal of the next frame in the movie was maximized. This approach reported the average motion of all the speckles falling into the window. The window size pitted robustness in correlation against spatial resolution: the larger the window, the more unique was the speckle pattern to be recognized in the target frame. On the other hand, larger windows increased the averaging of distinct speckle motions within the window.

Underlying the method of cross-correlation tracking is the assumption that the signal of a probing window, although translocated in space, does not change between source and target frame. In practice, this assumption is always violated because of noise. However, the cross-correlation of two image signals appears to be tolerant toward spatially uncorrelated noise, making it a prime objective function in computer vision tracking [71–73]. The many speckle appearances and disappearances in F-actin networks, however, introduce signal perturbations that cannot be tackled by the cross-correlation function [74]. Instead, we have developed a particle flow method, in which the movement of each speckle was tracked individually [74]. Speckles were linked between consecutive frames by nearest-neighbor assignment in a distance graph, in which conflicts between multiple speckles in the source frame competing for the same speckle in the target frame were resolved by global optimization [75]. An extension of the graph to linking speckles in three consecutive frames allowed enforcement of smooth and unidirectional trajectories, so that speckles moving in antiparallel flow fields could be tracked [74].

Surprisingly, cross-correlation-based tracking was successful in measuring average tubulin flux in meiotic spindles [76]. Simulated time-lapse sequences showed that if a significant subpopulation of speckles in the probing window moves jointly, then the coherent component of the flow can be estimated even when the rest of the speckles move randomly or, as in the case of the spindle apparatus, a smaller population moves coherently in opposite directions. However, the tracking result will be ambiguous if the window contains multiple, coherently moving speckle subpopulations of equal size. Miyamoto *et al.* [76] carefully chose windows in the central region of a half-spindle, where the motion of speckles towards the nearer of the two poles dominated speckle motion in the opposite direction and random components. The approach was aided further by several features of the spindle system: tubulin flux in a spindle is quasi-stationary; speckle appearances and disappearances are concentrated at the spindle midzone and in the pole regions, which were both excluded from the probing window; and the flow fields were approximately parallel inside the probing window.

Encouraged by these results, we returned to cross-correlation tracking of speckle flow in F-actin networks [77]. The advantage of cross-correlation tracking over particle flow tracking is that there is no requirement to detect the same speckle in at least two consecutive frames. Hence, speckle flows can be tracked in movies with high noise levels and weak speckle contrast [77]. In order to avoid trading correlation stability for spatial resolution, we capitalized on the fact that cytoskeleton transport is often stationary on the timescale of minutes. Thus, although the correlation of a single pair of probing windows in source and target frames is ambiguous (Figure 7.5b-i), rendering the tracking of speckle flow impossible (Figure 7.5c-i), time-integration of the correlation function over *multiple* frame pairs yields robust displacement estimates for probing windows as small as the Airy disk area (Figure 7.5b-ii, c-ii). Figure 7.5d presents a complete high-resolution speckle flow map extracted by integration over 20 frames ($\sim$3 min).

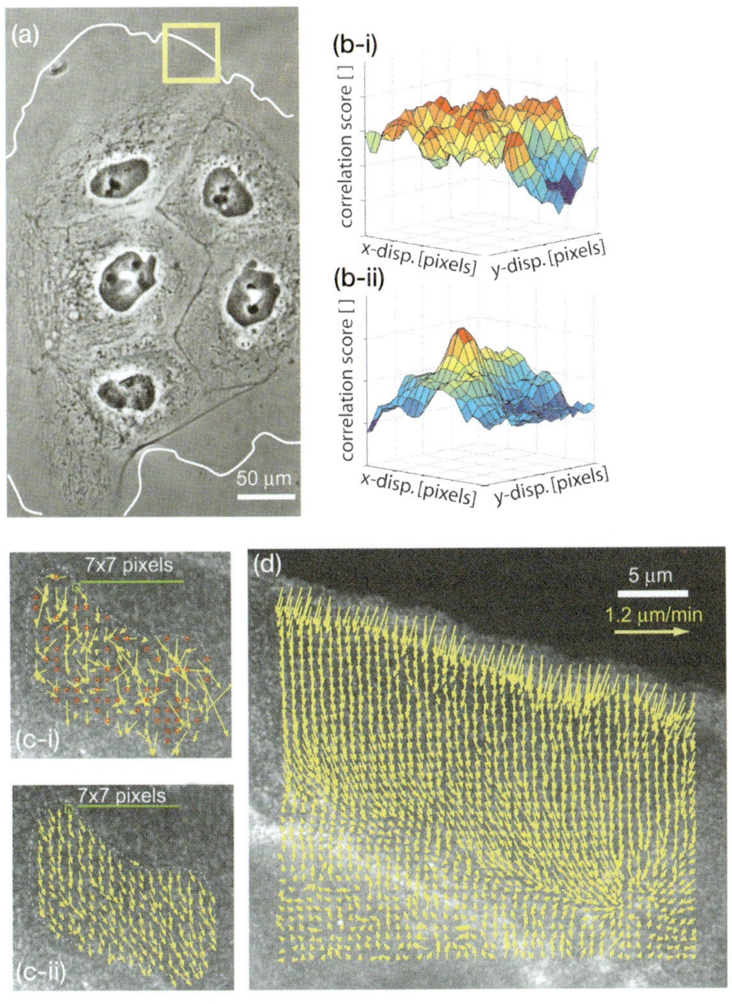

Figure 7.5 Tracking quasi-stationary speckle flow using multiframe correlation. (a) Island of epithelial cells; (b) Cross-correlation for a single frame pair (b-i) and integrated for 20 frame pairs (b-ii); (c) Region of a speckled actin network tracked with a probing window of 7×7 pixels (400×400 nm) using a single frame pair (c-i) and 20 frame pairs (c-ii); (d) Speckle flow map corresponding to the inset in (a) extracted by integration of the correlation score over 20 frame pairs. Speckle flow in this movie is almost stationary, justifying the time integration. Figure reproduced with permission from Ref. [77].

7.7.2
Tracking Single-Speckle Trajectories

The extraction of kinetic data according to Figure 7.3 requires the accurate localization of speckle birth and death events. For this, it was necessary to devise methods capable of tracking full trajectories at the single-speckle level. The large number

(>100 000) of dense speckles poses a significant challenge. Details of the current implementation of single-particle tracking of speckles are described by Ponti *et al.* [78]. Our approach follows the framework of most particle-tracking methods – that is, the detection of speckles as particles on a frame-by-frame basis, and the subsequent assignment of corresponding particles in consecutive frames. Assignment is iterated to close gaps in the trajectories created by the short-term instability of the speckle signal. Our implementation of this framework included two algorithms that address particularities of the speckle signal:

- Speckles are detected in an iterative statistical framework, which accounts for signal overlap between proximal speckles.
- Speckle assignments between consecutive frames are executed in a hybrid approach combining speckle flow and single-speckle tracking.

Speckle flow fields are extracted iteratively from previous solutions of single-speckle trajectories [78], or by initial correlation-based tracking [77]. The fields are then employed to propagate speckle motion from the source to the target frame, prior to establishing the correspondence between the projected speckle position and the effective speckle position in the target frame by global nearest-neighbor assignment [79, 80].

Motion propagation allows us to cope with two problems of FSM data. First, in many cases the magnitude of speckle displacements between two frames significantly exceeds half the distance between speckles. Hence, no solution to the correspondence problem exists without prediction of future speckle locations. Second, speckles undergo sharp spatial gradients in speed and direction of motion. A global propagation scheme discarding regional variations will thus fail, whereas an iterative extraction of the flow field permits a gradually refined trajectory reconstruction in these areas.

Figure 7.6a displays the single-speckle trajectories for speckles initiated in the first 20 frames of the same movie for which speckle flow computation is demon-

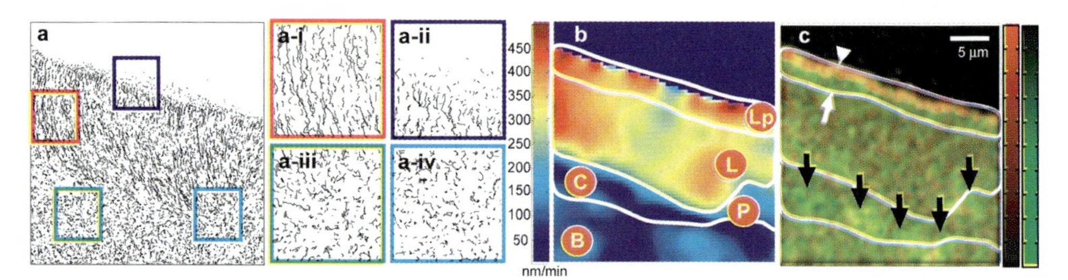

Figure 7.6 Tracking single-speckle trajectories. (a) Trajectories of speckles initiated in the first 20 frames of an actin FSM movie. (a-i to a-iv) Close-ups in different areas indicating regional variation in directional persistence, velocity and lifetime of the trajectories; (b) Speed distribution averaged over all 220 frames of the movie; (c) Distribution of polymerization (red channel) and depolymerization (green channel) calculated from scores averaged over 220 frames. Four regions of the actin network with distinct kinematic (motion) and kinetic (turnover) properties can be segmented (see text). Figure reproduced with permission from Ref. [23].

strated in Figure 7.5. The color-framed close-ups indicate regional differences between trajectories. Window (*a-i*) contains mostly straight trajectories with an average lifetime of 88 s; the trajectories in window (*a-ii*) are also straight, with an average lifetime of 60 s. In contrast, trajectories in windows (*a-iii*) and (*a-iv*) exhibit less directional persistence and have average lifetimes of 65 s and 59 s, respectively. It was concluded from such data that the F-actin cytoskeleton is regulated in a regionally variable fashion.

Figure 7.6b and c present the steady-state speed of actin network transport and turnover extracted from ~100 000 trajectories. Three different patterns of turnover are recognized that correspond to regions with different average speeds. At the cell edge, a ~1 μm-wide band of network assembly (red color; white arrowhead) abuts a ~1 μm-wide band of disassembly (green color; white arrow). The yellow shade in the assembly band indicates that filament polymerization and depolymerization significantly overlap. This 2 μm-wide cell border, which is referred to as the lamellipodium (Lp), exhibits on average the fastest F-actin retrograde flow. Predominant disassembly is found ~10 μm from the cell edge (black arrows), where the speed of F-actin flow is minimal. Here, the retrograde flow of the cell front encounters the anterograde flow of the cell body (B); this region is thus called the 'convergence zone' (C). Between the lamellipodium and the convergence zone is a region called the lamella (L), where assembly and disassembly alternate in a random pattern, accompanied by relatively coherent retrograde flow of moderate speed. The same pattern of network turnover is observed in the cell body.

7.7.3 Mapping Polymer Turnover Without Speckle Trajectories

It frequently occurs that a lower speckle contrast or a high image noise does not allow the precise identification of single-speckle trajectory endpoints. However, the trackable subsections of the trajectories are usually sufficient to extract the overall structure of speckle flow. In this case, an alternative scheme relying on the continuity of the optical density of the speckle field permits the mapping of turnover at lower resolution [81], as indicated in Figure 7.7 for the example of a crawling fish keratocyte, a cell system where the generation of clear fluorescent speckle patterns has proven difficult [52].

7.8 Applications of FSM for Studying Protein Dynamics *In Vitro* and *In Vivo*

Applications of FSM have thus far focused mostly on the study of F-actin and MT cytoskeleton systems, although other systems have also been analyzed in this way. A

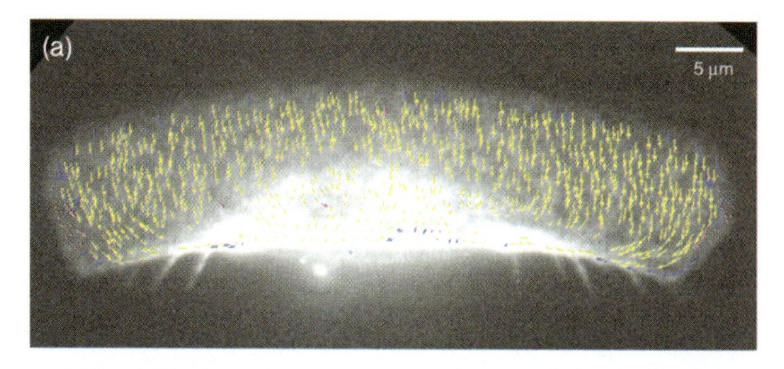

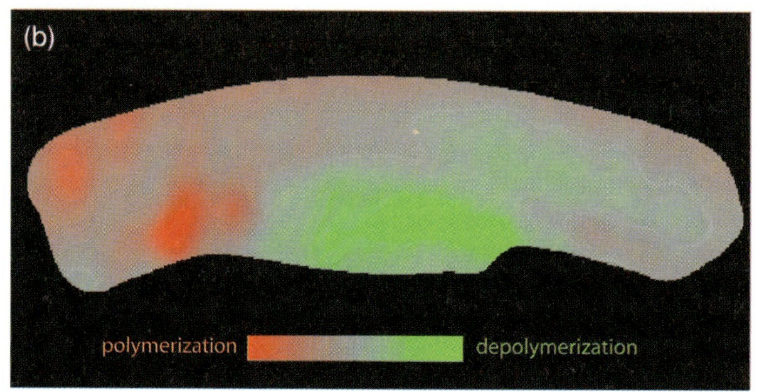

Figure 7.7 Reconstruction of F-actin network turnover from speckle flow in a fish keratocyte with poor speckle contrast. (a) Flow is calculated using multiframe correlation (as in Figure 7.5); (b) The calculated turnover map is based on conservation of image intensity. Without explicit identification of speckle births and deaths, essential details in the fine structure of the network turnover are lost.

summary of the FSM literature can be found in Table 7.1, where the major biological findings and technical advances in FSM made to date are listed. Most of the FSM data analysis has been limited to kymograph measurements of average speckle flow (see Section 7.7) and to the manual tracking of a few hundred speckles to extract lifetime information [48] and selected trajectories of cytoskeleton structures [26, 27, 69]. A systematic analysis of the full spatiotemporal information offered by FSM regarding transport and turnover in molecular assemblies is far from complete, although significant progress has already been made. In the following section, we describe comprehensive qFSM analyses both of F-actin cytoskeleton dynamics in migrating epithelial cells, and of the architectural dynamics of the spindle. Some of the most interesting results of these studies, showcasing the technical possibilities of qFSM, are also summarized.

Table 7.1 Selection of fluorescent speckle microscopy (FSM) general references and biological findings by subject.

Previous FSM Reviews and Methods Chapters

Previous reviews:

- Speckle image formation and the first FSM applications studying microtubule and F-actin dynamics *in vitro* and *in vivo*[a,b]
- Introduction to the 'platform concept' of FSM, where labeled subunit association with and dissociation from a macromolecular structure produces a stochastic signal indicative of turnover and movement of the structures[c]
- Algorithms development for quantitative FSM (through 2003)[d]
- FSM imaging of cytoskeleton dynamics in neurons[e]
- Comparison of the FSM signal in wide-field epifluorescence and total internal reflection fluorescence (TIRF) microscopy[f]
- FSM imaging and signal analysis, history of biological questions, and corresponding methodological advances[g]

Methods:

- Methods for microscope set-up and cell preparations for FSM imaging in living cells[h,i,j,k,j]
- Method of intracellular force reconstruction by FSM analysis[m]

FSM Analysis of Microtubule (MT) Motion and Assembly Dynamics

MTs assembled from pure tubulin (in vitro):

- MTs assembled from labeled and unlabeled pure tubulin *in vitro* exhibit fluorescent speckles, showing that cellular factors or organelles do not contribute to the speckle pattern[n]
- Speckles containing one fluorophore can be detected using conventional wide-field epifluorescence[o]
- For MTs *in vitro*, treadmilling is not unidirectional, suggesting that it is powered by differences in dynamic instability between plus and minus ends[p]
- The visualization of individual speckled MTs in a pool of tubulin, even when labeled at ratios <2%, requires a reduction of out-of-focus fluorescence by spinning-disk confocal microscopy[p]
- The KinI subfamily of kinesin-related proteins mediates depolymerization of MTs at both ends *in vitro*[q]

MT flux in meiotic spindles assembled from Xenopus laevis *egg extracts (in vitro):*

- FSM allows detailed analysis of dense polymers like the spindle[a]
- Spinning disk FSM reveals MT bundles, whereas in wide-field FSM those bundles are not detectable[r]
- MTs both polymerize and depolymerize at the kinetochores[r]
- Flux rates are different for kinetochore and nonkinetochore MT bundles[r,s]
- Monopoles do not exhibit MT flux during spontaneous bipolarization, and the onset of flux is correlated with the onset of bipolarity. This suggests that arrays of antiparallel MTs are required for flux generation[t]
- Disruption of depolymerization factors at the poles yields kinesin Eg5-dependent elongation of the metaphase spindle[u,v]
- MT flux is predominantly driven by ensembles of processive kinesin Eg5 motors[w]
- FSM and cross-correlation analysis reveal MT minus ends throughout the spindle, with more at the poles than at the spindle equator[x]
- MTs within bundles move at heterogeneous speeds, and FSM can be used to determine the length distribution of individual MTs in the spindle[y]

(Continued)

Table 7.1 (*Continued*)

MT flux in mitotic spindles in tissue culture cells (in vivo):
- The majority of poleward flux of kinetochore MTs in mammalian PtK1 epithelial cells is driven by a polar pulling-in mechanism, whereas Eg5, which plays a dominant role in *Xenopus* egg extracts, makes a minor contribution[z]

MT flux in spindles of other cell systems:
- MT flux makes a significant contribution to poleward chromosome movement during anaphase A in *Drosophila melanogaster* embryos[aa,bb]
- Three mitotic motors exhibit different roles in anaphase B in *Drosophila* embryos[cc]
- Two functionally distinct MT-destabilizing *KinI* enzymes are responsible for normal chromatid-to-pole motion in *Drosophila*[dd]
- MT flux in crane-fly spermatocytes increases from metaphase to anaphase and is faster than chromosome poleward motion, suggesting that MT plus ends are still polymerizing[ee]

MTs in interphase tissue cells:
- In living cells, optimal speckle contrast occurs at fractions of labeled tubulin in the 0.1–0.5% range, where the fluorescence of each speckle corresponds to one to seven fluorophores per resolvable unit[n,o]
- Cytoplasmic dynein, a MT-associated motor, promotes the formation and growth of immobile MTs in organized astral arrays, as opposed to organizing the array by powering the motion of pre-existing polymers[ff]
- Overexpression of Ncd, a kinesin-14, in mammalian fibroblasts results in generation of sliding forces between adjacent MTs in bundles[gg]

MTs in neurons:
- In the axon shaft proximal to the cell body, individual MTs are stationary, suggesting that tubulin dimers are transported down the axon to promote axonal growth and branching[hh]
- In regions of axon growth (i.e. growth cones and interstitial branches), short segments of MTs move, suggesting that exploratory behavior of neurons is promoted by MT transport[ii]
- Measurement of MT growth and transport in growth cones reveals they grow towards the periphery while being transported towards the axon[jj,kk]

MT dynamics in S. cerevisiae *and* S. pombe*:*
- During mating of S. *cerevisiae*, the nucleus and spindle pole body are oriented and tethered to the shmoo tip by a MT-dependent search and capture mechanism, where MT growth and shrinkage are localized mostly to the shmoo tip[ll]
- Kinetochore MTs grow and shrink only at the plus ends and do not exhibit poleward flux[mm]
- Astral MT plus end growth and shortening at the cell cortex plays an important role in positioning the nucleus during interphase and the spindle during mitosis[nn]

MT dynamics in pathogens:
- Dynamics of MTs in the fungal pathogen *Ustilago maydis* determines cell polarity[oo]
- MT dynamics through the cell cycle alter morphogenesis in the fungal pathogen *Candida albicans*[pp]

FSM Analysis of Microtubule-Associated Protein (MAP) Dynamics

MT plus-end binding proteins:
- CLIP-170 binds at the growing MT end, stays stationary relative to the MT, and dissociates after some time[qq]
- MT assembly in meiotic *Xenopus* egg extract spindles is visualized by localization of speckle-like EB1 comets[rr]

MT motors:
- The motor protein Eg5 stays stationary in spindles despite the flux of MTs, suggesting that Eg5 may be bound to a putative non-MT spindle matrix[ss]

(*Continued*)

Table 7.1 (*Continued*)

Other MAPs:

- Co-imaging of full-length ensconsin or its MT-binding domain (EMTB) conjugated to GFP with fluorescent MTs suggests that dynamics of MAP:MT interactions is at least as rapid as tubulin: MT dynamics in the polymerization reaction[tt]
- Binding and unbinding of ensconsin generates a speckle pattern along the MT, the dynamics of which can be evaluated to study the phosphorylation-dependent regulation of the turnover[uu]
- Multi-GFP tandems on MAPs significantly increase the speckle contrast and stability[uu]

FSM Analysis of F-actin Dynamics in Migrating and Non-migrating Cells

Actin network dynamics in migrating tissue cells:

- F-actin in polarized cells is organized in four distinct zones: a lamellipodium with rapid retrograde flow and constant polymerization; a lamella with slower retrograde flow; a contraction zone with no flow; and a zone of anterograde flow. The spatial transition from retrograde to anterograde flow suggests the presence of a contractile belt powered by myosin II which may drive cell migration[vv]
- Single-fluorophore speckles can reveal F-actin turnover, as known from speckle analysis in MTs, despite the complex filamentous structure of the F-actin meshwork[ww]
- Statistical clustering analysis of single speckle dynamics reveals two kinetically and kinematically distinct, yet spatially overlapping, actin networks that mediate cell protrusion[xx]
- Spatiotemporal correlation of F-actin assembly maps and GFP-Arp2/3 clustering indicate that, in the lamellipodium, actin assembly is mediated by Arp2/3, while lamellar assembly is independent of Arp2/3 activity[yy]
- Arp2/3- and cofilin-regulated assembly of the lamellipodia is not required for epithelial cell protrusion[zz]
- Molecular kinetics of Arp2/3 and capping protein can be measured by single-molecule FSM[aaa]
- The dynamics of actin-binding proteins (capping protein, Arp2/3, tropomyosin) exhibit spatial differentiation in the lamellipodium and lamella of *Drosophila* S2 cells[bbb]

Actin network dynamics in neurons:

- Steady-state retrograde flow in neuronal growth cones depends on both myosin II contractility and actin-network treadmilling[ccc]

Actin networks in keratocytes and keratocyte fragments:

- Mechanical stimulation of keratocyte fragments activates acto-myosin contraction and causes directional motility[ddd]
- Keratocytes exhibit F-actin retrograde flow relative to the substrate[eee] in a biphasic relationship between flow magnitude and adhesiveness[fff]
- Actin and myosin undergo polarized assembly, suggesting force generation occurs at the lamellipodium/cell body transition zone[ggg]
- Directed motility is initiated by symmetry breaking actin-myosin network reorganization and contractility at the cell rear[hhh]

Actin in contact-inhibited epithelial cells:

- Unlike migrating cells, cortical actin in contact-inhibited cells is spatially stationary but undergoes rapid turnover[iii, jjj]
- The spatiotemporal mapping of F-actin network turnover from speckle signal analysis during appearance and disappearance events can be carried out at high resolution in contact-inhibited cells[jjj]

Actin dynamics in S. cerevisiae, *using GFP-tubulin:*

- FSM can be used to visualize bud-associated assembly and motion of F-actin cables in budding yeast[kkk]

Multi-Spectral FSM Analysis of F-actin and Other Macromolecular Structures

(*Continued*)

Table 7.1 (*Continued*)

Co-motion of the F-actin cytoskeleton and other structures, using spectrally distinct fluorescent analogues:

- FSM of MTs and F-actin in cytoplasmic extracts of *Xenopus* eggs confirms two basic types of interaction between the polymers: a cross-linking activity and a motor-mediated interaction[lll]
- Dynamic interactions between MTs and actin filaments are required for axon branching and directed axon outgrowth[mmm]
- Direction of MT growth is guided by the tight association of MTs with F-actin bundles[nnn]
- F-actin contraction may be involved in the breaking of MTs[vv,ooo]
- Rho and Rho effectors have differential effects on F-actin and MT dynamics during growth cone motility[ppp]
- In migrating epithelial cells, the dynamics of MTs and F-actin is coordinated by signaling pathways downstream of Rac1[qqq]
- FSM of F-actin and several focal adhesion (FA) proteins reveals a differential transmission of F-actin network motion through the adhesion structure to the extracellular matrix[rrr]

FSM Analysis of Protein Turnover in Focal Adhesions (FAs)

- FSM of low-level GFP-fusion protein expression, in combination with TIRF microscopy, allows quantification of molecular dynamics within FA protein assemblies at the ventral surface of living cells[f,rrr]

[a] Waterman-Storer, C.M., Desai, A., Bulinski, J.C. and Salmon E.D. (1998) *Curr. Biol.*, **8**, 1227.
[b] Keating, T.J. and Borisy, G.G. (2000) *Curr Biol.*, **10**, R22.
[c] Waterman-Storer, C.M. and Danuser, G. (2002) *Curr. Biol.*, **12**, R633.
[d] Danuser, G. and Waterman-Storer, C.M. (2003) *J. Microsc.*, **211**, 191.
[e] Dent, E.W. and Kalil, K. (2003) in *Methods in Enzymology*, Vol. 361, Academic Press, San Diego, pp. 390.
[f] Adams, M., Matov, A., Yarar, D., Gupton, S., Danuser, G. and Waterman-Storer, C.M. (2004) *J. Microsc.*, **216**, 138.
[g] Danuser, G. and Waterman-Storer, C.M. (2006) *Annu. Rev. Biophys. Biomol. Struct.*, **35**, 361.
[h] Adams, M.C., Salmon, W.C., Gupton, S.L., Cohan, C.S., Wittmann, T., Prigozhina, N. and Waterman-Storer, C.M. (2003) *Methods*, **29**, 29.
[i] Waterman-Storer, C.M. (2002) in *Current Protocols in Cell Biology* (eds J.S. Bonifacino, M. Dasso, J.B. Harford, J. Lippincott-Schwartz and K.M. Yamada), Wiley, New York.
[j] Maddox, P.S., Moree, B., Canman, J.C. and Salmon, E.D. (2003) *Methods in Enzymology*, **360**, 597.
[k] Gupton, S.L. and Waterman-Storer, C.M. (2006) in *Cell Biology: A Laboratory Handbook*, Vol. 3, 3 edn. (eds J. Celis, N. Carter, K. Simons, J.V. Small, T. Hunter and D. Shotton), Academic Press, San Diego, pp. 137.
[l] Waterman-Storer, C., Desai, A. and Salmon, E.D. (1999) in *Methods in Cell Biology*, Vol. 61, p. 155.
[m] Ji, L., Loerke, D., Gardel, M. and Danuser, G. (2007) in *Methods in Cell Biology*, Vol. 83.
[n] Waterman-Storer, C.M. and Salmon, E.D. (1998) *Biophys. J.*, **75**, 2059.
[o] Waterman-Storer, C.M. and Salmon, E.D. (1999) *FASEB J*, **13**, 225.
[p] Grego, S., Cantillana, V. and Salmon, E.D. (2001) *Biophys. J.*, **81**, 66.
[q] Hunter, A.W., Caplow, M., Coy, D.L., Hancock, W.O., Diez, S., Wordeman, L. and Howard, J. (2003) *Molecular Cell*, **11**, 445.
[r] Maddox, P., Straight, A., Coughlin, P., Mitchison, T.J. and Salmon, E.D. (2003) *J. Cell Biol.*, **162**, 377.
[s] Vallotton, P., Ponti, A., Waterman-Storer, C.M. Salmon, E.D. and Danuser, G. (2003) *Biophys. J.*, **85**, 1289.
[t] Mitchison, T.J., Maddox, P., Groen, A., Cameron, L., Perlman, Z., Ohi, R., Desai, A., Salmon, E.D. and Kapoor, T.M. (2004) *Mol. Biol. Cell*, **15**, 5603.
[u] Shirasu-Hiza, M., Perlman, Z.E., Wittmann, T., Karsenti, E. and Mitchison, T.J. (2004) *Curr. Biol.*, **14**, 1941.
[v] Gaetz, J. and Kapoor, T.M. (2004) *J. Cell Biol.*, **166**, 465.
[w] Miyamoto, D.T., Perlman, Z.E., Burbank, K.S., Groen, A.C. and Mitchison, T.J. (2004) *J. Cell Biol.*, **167**, 813.
[x] Burbank, K.S., Groen, A.C., Perlman, Z.E., Fisher, D.D. and Mitchison, T.J. (2006) *J. Cell Biol.*, **175**, 369.

[yy]Yang, G., Houghtaling, B.R., Gaetz, J., Liu, J.Z., Danuser, G. and Kapoor, T.M. (2007) *Nat. Cell Biol.*, **9**, 1233.
[z]Cameron, L.A., Yang, G., Cimini, D., Canman, J.C., Evgenieva, O.K., Khodjakov, A., Danuser, G. and Salmon, E.D. (2006) *J. Cell Biol.*, **173**, 173.
[aa]Maddox, P., Desai, A., Oegema, K., Mitchison, T.J. and Salmon, E.D. (2002) *Curr. Biol.*, **12**, 1670.
[bb]Brust-Mascher, I. and Scholey, J.M. (2002) *Mol. Biol. Cell*, **13**, 3967.
[cc]Brust-Mascher, I., Civelekoglu-Scholey, G., Kwon, M., Mogilner, A. and Scholey, J.M. (2004) *Proc. Natl Acad. Sci. USA*, **101**, 15938.
[dd]Rogers, G.C., Rogers, S.L., Schwimmer, T.A., Ems-McClung, S.C., Walczak, C., Vale, R.D., Scholey, J.M. and Sharp, D.J. (2004) *Nature*, **427**, 364.
[ee]LaFountain, J.R. Jr., Cohan, C.S., Siegel, A.J. and LaFountain, D.J. (2004) *Mol. Biol. Cell*, **15**, 5724.
[ff]Vorobiev, I., Malikov, V. and Rodionov, V. (2001) *Proc. Natl Acad. Sci. USA*, **98**, 10160.
[gg]Oladipo, A., Cowan, A. and Rodionov, V. (2007) *Mol. Biol. Cell*, **18**, 3601.
[hh]Chang, S., Svitkina, T.M., Borisy, G.G. and Popov, S.V. (1999) *Nat. Cell Biol.*, **1**, 399.
[ii]Dent, E.W., Callaway, J.L., Szebenyi, G., Baas, P.W. and Kalil, K. (1999) *J. Neurosci.*, **19**, 8894.
[jj]Kabir, N., Schaefer, A.W., Nakhost, A., Sossin, W.S. and Forscher, P. (2001) *J. Cell Biol.*, **5**, 1033.
[kk]Zhou, F.-Q., Waterman-Storer, C.M. and Cohan, C.S. (2002) *J. Cell Biol.* **2002**, **157**, 839.
[ll]Maddox, P., Chin, E., Mallavarapu, A., Yeh, E., Salmon, E.D. and Bloom, K. (1999) *J. Cell Biol.*, **144**, 977.
[mm]Maddox, P.S., Bloom, K.S. and Salmon, E.D. (2000) *Nat. Cell Biol.*, **2**, 36.
[nn]Tran, P.T., Marsh, L., Doye, V., Inoue, S. and Chang, F. (2001) *J. Cell Biol.*, **153**, 397.
[oo]Steinberg, G., Wedlich-Soldner, R., Brill, M. and Schulz, I. (2001) *J. Cell Sci.*, **114**, 609.
[pp]Finley, K.R. and Berman, J. (2005) *Eukaryotic Cell*, **4**, 1697.
[qq]Perez, F., Diamantopoulos, G.S., Stalder, R. and Kreis, T.E. (1999) *Cell*, **96**, 517.
[rr]Tirnauer, J.S., Salmon, E.D. and Mitchison, T.J. (2004) *Mol. Biol. Cell*, **15**, 1776.
[ss]Kapoor, T.M. and Mitchison, T.J. (2001) *J. Cell Biol.*, **154**, 1125.
[tt]Faire, K., Waterman-Storer, C.M., Gruber, D., Masson, D., Salmon, E.D. and Bulinski, J.C. (1999) *J. Cell Sci.*, **112**, 4243.
[uu]Bulinski, J.C., Odde, D.J., Howell, B.J., Salmon, T.D. and Waterman-Storer, C.M. (2001) *J. Cell Sci.*, **114**, 3885.
[vv]Salmon, W.C., Adams, M.C. and Waterman-Storer, C.M. (2002) *J. Cell Biol.*, **158**, 31.
[ww]Watanabe, Y. and Mitchison, T.J. (2002) *Science* **2002**, **295**, 1083.
[xx]Ponti, A., Machacek, M., Gupton, S.L., Waterman-Storer, C.M. and Danuser, G. (2004) *Science*, **305**, 1782.
[yy]Ponti, A., Matov, A., Adams, M., Gupton, S. Waterman-Storer, C.M. and Danuser, G. (2005) *Biophys. J.*, **89**, 3456.
[zz]Gupton, S.L., Anderson, K.L., Kole, T.P., Fischer, R.S., Ponti, A., Hitchcock-DeGregori, S.E., Danuser, G., Fowler, V.M., Wirtz, D., Hanein, D. and Waterman-Storer, C.M. (2005) *J. Cell Biol.*, **168**, 619.
[aaa]Miyoshi, T., Tsuji, T., Higashida, C., Hertzog, M., Fujita, A., Narumiya, S., Scita, G. and Watanabe, N. (2006) *J. Cell Biol.*, **175**, 947.
[bbb]Iwasa, J.H. and Mullins, R.D. (2007) *Curr. Biol.*, **17**, 395.
[ccc]Medeiros, N.A., Burnette, D.T. and Forscher, P. (2006) *Nat. Cell Biol.*, **8**, 215.
[ddd]Verkhovsky, A.B., Svitkina, T.M., Borisy, G.G. (1999) *Curr. Biol.*, **9**, 11.
[eee]Vallotton, P., Danuser, G., Bohnet, S., Meister, J.J. and Verkhovsky, A. (2005) *Mol. Biol. Cell*, **16**, 1223.
[fff]Jurado, C., Haserick, J.R. and Lee, J. (2005) *Mol. Biol. Cell*, **16**, 507.
[ggg]Schaub, S., Bohnet, S., Laurent, V.M., Meister, J.-J. and Verkhovsky, A.B. (2007) *Mol. Biol. Cell*, E06.
[hhh]Yam, P.T., Wilson, C.A., Ji, L., Hebert, B. Barnhart, E.L., Dye, N.A., Wiseman, P.W., Danuser, G. and Theriot, J.A. (2007) *J. Cell Biol.*, **178**, 1207.
[iii]Waterman-Storer, C.M., Salmon, W.C. and Salmon, E.D. (2000) *Mol. Biol. Cell* **2000**, **11**, 2471.
[jjj]Ponti, A., Vallotton, P., Salmon, W.C., Waterman-Storer, C.M. and Danuser, G. (2003) *Biophys. J.*, **84**, 3336.
[kkk]Yang, H.-C. and Pon, L.A. (2002) *Proc. Natl Acad. Sci. USA*, **99**, 751.
[lll]Waterman-Storer, C., Duey, D.Y., Weber, K.L., Keech, J., Cheney, R.E., Salmon, E.D. and Bement, W. M. (2000) *J. Cell Biol.*, **150**, 361.
[mmm]Dent, E.W. and Kalil, K. (2001) *J. Neurosci.*, **15**, 9757.
[nnn]Schaefer, A.W., Kabir, N. and Forscher, P. (2002) *J. Cell Biol.*, **158**, 139.

[ooo]Gupton, S.L., Salmon, W.C. and Waterman-Storer, C.M. (2002) *Curr. Biol.*, **12**, 1891.
[ppp]Zhang, X.-F., Schaefer, A.W., Burnette, D.T., Schoonderwoert, V.T. and Forscher, P. (2003) *Neuron*, **40**, 931.
[qqq]Wittmann, T., Bokoch, G.M. and Waterman-Storer, C.M. (2003) *J. Cell Biol.*, **161**, 845.
[rrr]Hu, K., Ji, L., Applegate, K., Danuser, G. and Waterman-Storer, C.M. (2007) *Science*, **315**, 111.

7.9 Results from Studying Cytoskeleton Dynamics

7.9.1 F-Actin in Cell Migration

Over the past few years, qFSM has critically driven our understanding of actin cytoskeleton dynamics in cell migration. It has provided unprecedented details of the spatial organization of F-actin turnover and the transport and deformation of F-actin networks *in vivo*. In the following sections, we review a few key discoveries, enabled by qFSM, that have defined a new paradigm for the functional linkage between actin cytoskeleton regulation and epithelial cell migration.

7.9.1.1 F-Actin in Epithelial Cells is Organized Into Four Dynamically Distinct Regions

Figures 7.5 and 7.6 indicate the steady-state organization of the F-actin cytoskeleton in four kinematically and kinetically distinct zones:

- The lamellipodium, characterized by fast retrograde flow and two narrow bands of assembly and disassembly resulting from the fast treadmilling of actin between its polymeric and monomeric states [54, 82].
- The lamella, characterized by reduced retrograde flow and assembly and disassembly in random punctate patterns.
- The cell body, characterized by anterograde flow and turnover patterns similar to those of the lamella.
- The convergence zone, where the flows of the lamella and cell body meet and where strong depolymerization suggests that the lamella and cell body are materially separate structures.

qFSM also delivers nonsteady-state measurements of flow and turnover, revealing distinct variations in the periodicity of turnover between these regions [78]. In combination with pharmacological perturbation, qFSM was used to dissect the mechanisms of retrograde flow. Thus, lamellipodium flow was found to be independent of myosin motor contraction, whereas lamella flow was blocked by the specific inhibition of myosin II activity [83]. The lamellipodium and the lamella also exhibited different sensitivity to the disruption of filament assembly, disassembly and severing, which suggested that these regional differences might be associated with differential molecular regulation [83]. This hypothesis has thus far been confirmed by immunostaining studies [83, 84] and by the expression of constitutively active and dominant-negative constructs of regulatory proteins [84, 85]. Here, qFSM provides a critical insight into cytoskeleton dynamic responses to shifted activation of regulatory factors.

In summary, these data demonstrate how qFSM can be used to quantify spatiotemporal modulations of the kinetics and kinematics of molecular assemblies and to identify dynamically distinct structural modules, even when they are composed of the same base protein.

7.9.1.2 Actin Disassembly and Contraction are Coupled in the Convergence Zone

A similar spatiotemporal correlation analysis was performed to examine the relationship of F-actin network depolymerization and contraction in the convergence zone [81]. It was first established that transient increases in speckle flow convergence are coupled to transient increases in disassembly. This begged the question whether the rate of speckle flow convergence increases because disassembly boosts the efficiency of myosin II motors in contracting a more compliant network, or because motor contraction mediates network disassembly. To address this question, we transiently perfused cells with Calyculin A, a type II phosphatase inhibitor that increases myosin II activity. Unexpectedly, we reproducibly measured a strong burst of disassembly long before flow convergence was affected. This evidence suggested that myosin II contraction can actively promote the depolymerization of F-actin, for example, by breaking filaments. The link between F-actin contractility and turnover has since been confirmed by fluorescence recovery after photobleaching measurements in the contractile ring required for cytokinesis [86]. In summary, these data demonstrate the correlation of two qFSM parameters to decipher the relationship between deformation and plasticity of polymer networks inside cells.

7.9.1.3 Two Distinct F-Actin Structures Overlap at the Leading Edge

The transition between the lamellipodium and the lamella is characterized by a narrow band of strong disassembly adjacent to a region of mixed assembly and disassembly and a sharp decrease in retrograde flow velocity (see Figure 7.6). Together, these features defined a unique mathematical signature for tracking the boundary between the two regions over time (Figure 7.8a). In view of the different speckle velocities and lifetimes between the two regions, it was speculated that the same boundary could be tracked by spatial clustering of speckle properties. It was predicted that fast, short-living speckles (class 1) would preferentially localize in the lamellipodium, whereas slow, longer-living speckles (class 2) would be dominant in the lamella. To test this hypothesis, we solved a multiobjective optimization problem in which the thresholds of velocity ν_{th} and lifetime τ_{th} separating the two classes, as well as the boundary ∂Lp between lamellipodium and lamella, were determined subject to the rule $\{\partial Lp, \tau_{th}, \nu_{th}\} = \max(N_1/(N_1 + N_2) \in Lp) \& \min(N_1/(N_1 + N_2) \in La)$ (Figure 7.8b and c), where N_1 and N_2 denote the number of speckles in classes 1 and 2, respectively. The prediction was confirmed in the lamella, where class 1 speckles occupied a statistically insignificant fraction. However, class 2 speckles made up 30–40% of the lamellipodium, indicating that in this region speckles with different kinetic and kinematic behavior colocalize. This information was previously lost in the averaged analysis of single-speckle trajectories. When mapping the scores of class 1 and class 2 speckles separately (Figure 7.8di-dii), it was discovered that class 1 speckles define the bands of polymerization and depolymerization characteristic

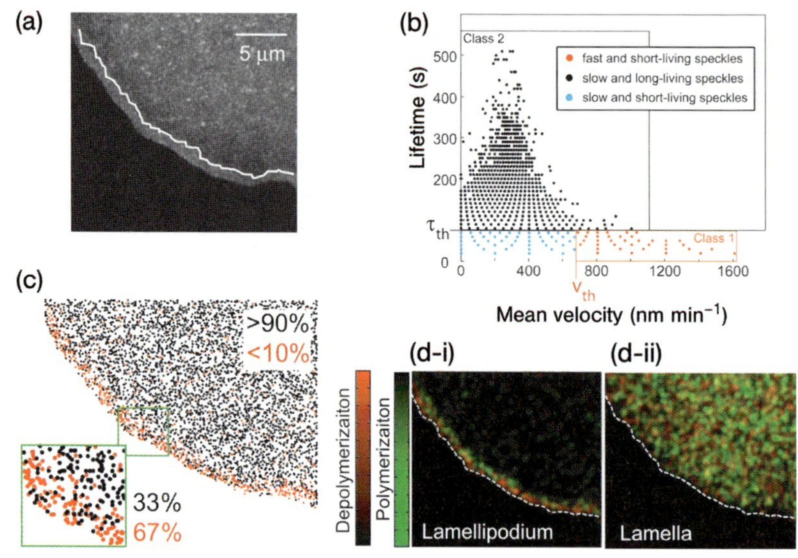

Figure 7.8 Distinction of two spatially overlapping actin networks based on heterogeneity of single-speckle properties. (a) Raw FSM image overlaid with the boundary between lamellipodium and lamella computed from spatial gradients in F-actin turnover and flow velocity; (b,c) Cluster analysis of speckle lifetime and velocity (see text); (d) Class 1 speckles constitute the rapidly treadmilling lamellipodium. Class 2 speckles constitute the lamella with a punctate pattern of random actin turnover. Both networks spatially overlap in the first 2 μm from the cell edge. Figure reproduced in parts with permission from Refs [83] and [23].

of the lamellipodium, and that class 2 speckles define the puncta of assembly and disassembly characteristic of the lamella, which reaches all the way to the leading edge. Subsequent experiments specifically disrupting actin treadmilling in the lamellipodium confirmed the finding that the lamellipodium and lamella form two spatially overlapping, yet kinetically, kinematically and molecularly different, F-actin networks [83, 84].

7.9.2
Architecture of *Xenopus laevis* Egg Extract Meiotic Spindles

During cell division, MTs form a spindle, which maintains stable bipolar attachment to chromosomes over tens of minutes. A sophisticated checkpoint system senses the status of attachment and generates a signal to progress with symmetric segregation of the replicated sister chromatids into the newly forming mother and daughter cells [6]. The minus ends of polar MTs are preferentially located at the spindle poles, whereas the plus ends continually switch between growth and shrinkage, a process known as 'MT dynamic instability' [87]. Strikingly, dynamic instability in vertebrate spindles occurs within a few tens of seconds, a time scale at least an order of magnitude shorter than the existence of the spindle [88]. In addition to MT dynamic instability at the MT plus end, individual MTs are transported toward the spindle poles, a behavior known as 'poleward flux'. Poleward flux has only been observed in higher eukaryotic spindles, including the *Xenopus laevis* extract spindle system [89]. How the overall stability of spindle architecture is maintained under the much faster dynamics of its building blocks is largely unknown. qFSM has made several critical contributions to the mechanistic analysis of spindle architecture (Table 7.1). For example, it has revealed detailed maps of the organization of heterogeneous MT poleward flux (Figure 7.9a–c) [74, 90], and it was also used to show that MTs form distinct types of bundles with different flux dynamics, depending on whether they are attached to chromosomes (kinetochore fibers) or form a scaffold of overlapping fibers emanating from opposite poles (interpolar MT fibers) [38]. Together, these data have indicated an enormous architectural complexity, which requires fine regulation of the dynamics of each MT. However, the high MT density in the spindle has precluded measurement of the dynamics of individual MTs within bundles. Speckles generally consist of multiple fluorophores distributed over many different MTs.

Recently, this difficulty has been overcome by single-fluorophore speckle imaging of *Xenopus laevis* extract spindles [62]. As a cell-free spindle model, the extract spindle allows convenient control over fluorescent tubulin levels to achieve sparse labeling of MTs (Figure 7.9d). For low labeling ratios *f*, speckle intensities cluster in multiples of ~500 AU, indicating that speckles are composed of a discrete low number (e.g. one, two, three or four) of fluorophores (Figure 7.9e and f) [62]. At the lowest concentrations of labeled tubulin, only one intensity cluster with a mean value of ~500 AU was found. Furthermore, the average intensity of a detectable speckle remained constant over time at ~500 AU (Figure 7.9g), although the speckle number decreased due to photobleaching, Together, the cluster analysis of speckle intensities and the photobleaching analysis confirmed that >98% of the speckles reflected the image of a single fluorophore.

7.9.2.1 **Individual MTs within the Same Bundle Move at Different Speeds**

Single-fluorophore speckles were then used to investigate how individual MTs in close proximity move relative to one another. In order to avoid any *a priori* assumptions, the spatial organization of spindle MTs was mapped using the dense flow field measured in a spectrally distinct channel displaying multifluorophore speckles. Path integration of the flow field allowed the construction of equally spaced bands of uniform width (480 nm), which reflects the average position of MT bundles within the spindle (Figure 7.9h). The band width was chosen to match the diffraction limit of the microscope, and is consistent with electron microscopy studies which showed that MTs form bundles typically a few hundred nanometers wide, with individual MTs 20–50 nm apart [91, 92]. Next, the pairwise difference between the velocities of speckles located in the same band showed that MTs spaced at a distance comparable to the width of MT bundles exhibit remarkably heterogeneous movement (Figure 7.9i). Thus, individual MTs appear to slide past one another over very short distances, suggesting that the spindle is a MT scaffold that is continuously restructured at the scale of tens of seconds.

7.9.2.2 **The Mean Length of Spindle MTs is 40% of the Total Spindle Length**

Despite the heterogeneous movement of the majority of speckles within one band, a small percentage (~1% of all speckles) moved in synchronized pairs: not only did

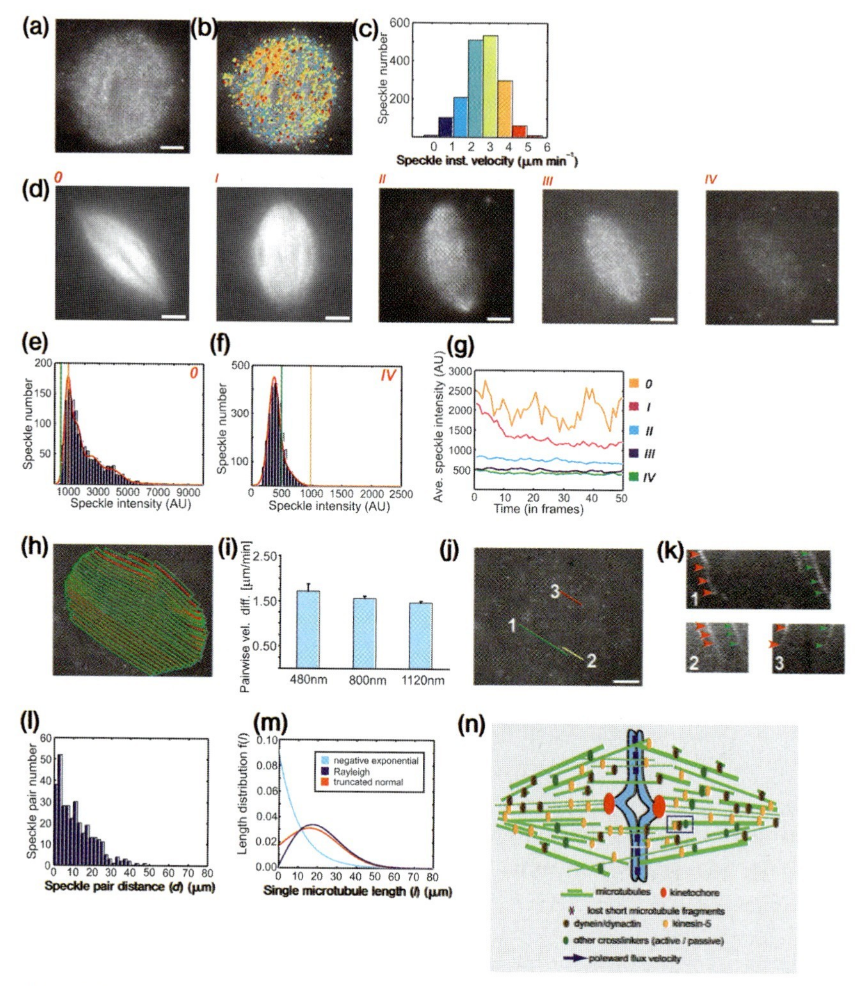

Figure 7.9 Analyzing the dynamic MT architecture of the *Xenopus laevis* egg extract spindle using single-fluorophore imaging. (a) Spindle imaged using multifluorophore tubulin speckles. Scale bar = 10 μm; (b,c) Speckle trajectories (b) overlaid onto the spindle from (a) were recovered using particle tracking. Color-coding by velocity range is specified in the velocity histogram (c). Average velocity mean ± SD = 2.68 ± 0.95 μm min^{-1} ($n = 1699$ tracks); (*d*) Single-fluorophore imaging conditions were determined by sequentially reducing labeled tubulin concentrations (***0***, 3.3 n*M*; ***I***, 1.1 n*M*; ***II***, 0.33 n*M*; ***III***, 0.11 n*M*; ***IV***, 0.033 n*M*; ***V***, 0.011 n*M*). Scale bars = 10 μm; (e, f) Speckle intensity distributions of spindles in (d) (concentrations ***0*** and ***V***), each fitted by a mixture of normal distributions (red lines) calculated from cluster analysis [103]. Lines mark 500 AU (green) and 1000 AU (brown); (g) Changes of average speckle intensity over time due to photobleaching within spindles in (d); (h) Based on two-color speckle imaging and path integration of the MT poleward flux vector field, uniform bands were generated at equal distances to trace MTs within the spindle [62]; (i) Difference in instantaneous velocities between pairs of speckles within bands constructed as in (h). Error bars represent standard errors of the mean (480 nm: $n = 78$ pairs; 800 nm: $n = 659$ pairs; 1120 nm: 1605 pairs); (j) Examples of synchronously moving speckle pairs identified within bands constructed as in (h). Each pair resided in the

they stay within the same band and move in the same direction at the same time (Figure 7.9j and k), but they also concurrently changed velocities (Figure 7.9k, 1–3). Clearly, these single-fluorophore speckle pairs must reside on the same MT. By applying stringent detection criteria on spatial colocalization, temporal overlap, relative distance change and relative velocity change, a total of 328 synchronous speckle pairs was identified from 13 spindles. Interestingly, 90% of the speckle distances were less than half the length of the spindle (Figure 7.9l). To estimate the lengths of spindle MTs from the measured distances between synchronously moving speckle pairs, a mathematical model was developed of the stochastic incorporation of labeled tubulin into a population of MTs with an *a priori* unknown length distribution $f(l)$ [62]. Assuming a hypothetical function $f(l)$, the model defined the expected cumulative distribution P of distances d between two speckles, given the event A that the speckles reside on the same MT:

$$P(D<d|A) = \frac{\int_0^d l^2 e^{-c \cdot l \cdot r} f(l)dl + \int_d^{+\infty} (2dl - d^2) e^{-c \cdot l \cdot r} f(l)dl}{\int_0^{+\infty} u^2 e^{-c \cdot u \cdot r} f(u)du}.$$

Here, l denotes the steady-state length of individual MTs in microns, c is the number of tubulin dimers per micron (1625), and r is the fraction of labeled tubulin ($\sim 2.86 \times 10^{-6}$ for 0.066–0.033 nM labeled tubulin). Fitting the above formula for the expected cumulative distribution to the measured cumulative histogram of distances between speckle pairs made it possible to estimate parameters of $f(l)$ (Figure 7.9m). For instance, it was estimated that the ratio between the mean length of MTs and the spindle length is $\sim$0.4.

In summary, by integrating single-fluorophore imaging with computational image analysis, it was found that spindle MTs in close proximity move at highly heterogeneous velocities, and that the majority have a length shorter than the spindle pole-to-metaphase plate distance. These results, along with molecular perturbation data (not shown), suggest that MTs in the vertebrate meiotic spindle are dynamically organized as a crosslinked 'tiled-array' in a way similar to how the actin network is organized in motile cells (Figure 7.9n) [62]. This model challenges longstanding textbook models, which assume that the majority of MTs emanate from the two poles. The mechanical stability of the tiled-array is maintained by dynamic crosslinks. Thus, a structure can be formed where the stability of the ensemble is much higher than the stability of its individual building blocks. It is speculated that the design of cytoskeleton structures

same band, coexisted over a time interval of at least 10 s, and varied synchronously in flux velocity. Scale bar = 10 μm; (k) Kymograph representation of the synchronous movement of the speckle pairs shown in (j); (l) Histogram of the measured distances between speckle pairs (328 pairs from $n = 13$ spindles); (m) Estimated length distributions under different models: exponential distribution (mean ± SD: 11.75 ± 11.75 μm) (light blue), Rayleigh distribution (mean ± SD: 22.00 ± 11.50 μm) (blue), truncated normal distribution (TND; mean ± SD: 20.11 ± 12.23 μm). TND was selected based on its minimal fitting error and statistically validated [62]. Spindle length: mean ± SD: 49.0 ± 5.0 μm ($n = 13$ spindles); (n) A tile-array architectural model of the *Xenopus* extract spindle. Figure reproduced with permission from Ref. [62].

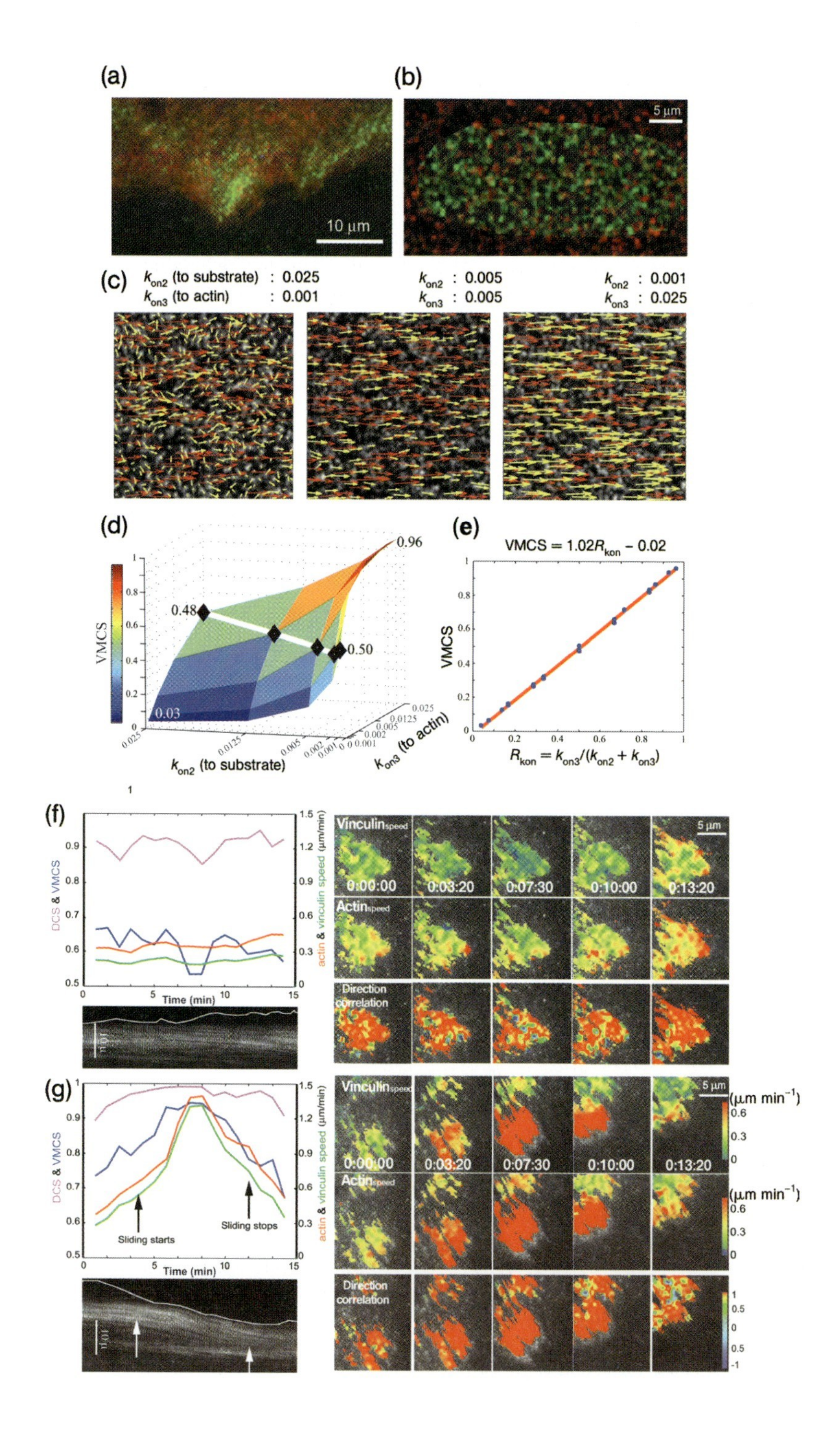
(a)
10 μm
(b)
5 μm
(c)
k_{on2} (to substrate) : 0.025
k_{on3} (to actin) : 0.001
k_{on2} : 0.005
k_{on3} : 0.005
k_{on2} : 0.001
k_{on3} : 0.025
(d)
VMCS
0.96
0.48
0.50
0.03
k_{on2} (to substrate)
k_{on3} (to actin)
(e)
VMCS = 1.02R_{kon} − 0.02
VMCS
$R_{kon} = k_{on3}/(k_{on2} + k_{on3})$
(f)
DCS & VMCS
actin & vinculin speed (μm/min)
Time (min)
Vinculin speed
Actin speed
Direction correlation
0:00:00
0:03:20
0:07:30
0:10:00
0:13:20
5 μm
(g)
DCS & VMCS
actin & vinculin speed (μm/min)
Sliding starts
Sliding stops
Time (min)
Vinculin speed
Actin speed
Direction correlation
(μm min⁻¹)

follows the general principle of coupling many short and dynamic components into larger, longlasting ensembles to achieve both the flexibility and stability needed for cellular life under constantly changing conditions.

7.9.3 Hierarchical Transmission of F-Actin Motion Through Focal Adhesions

Cell migration requires a delicate spatial balance of cell adherence to the substrate. Dynamic structures called focal complexes assemble next to the leading edge and mature over time into FAs, macromolecular assemblies of more than 100 different proteins. Focal adhesions tether the F-actin network to integrin receptors, which in turn bind to the substrate. Forces generated by F-actin polymerization and/or contraction are transmitted to the extracellular matrix (ECM) via the coupling of F-actin to FAs. It has long been known that F-actin and FA proteins are coupled; many FA proteins bind directly or indirectly to F-actin [93–95] or integrin receptors [96–98], and the ends of contractile actin bundles often appear to be embedded in FAs [57, 99].

However, despite many years of intensive research aimed at identifying the molecular parts list of FAs, it has been impossible to determine the hierarchy of interactions between specific FA proteins and the F-actin cytoskeleton in living cells. Hu *et al.* used two-color total internal reflection fluorescent speckle microscopy (TIR-FSM) to simultaneously image X-rhodamine actin and various GFP-tagged FA proteins (Figure 7.10a) [25]. As expected, F-actin retrograde flow slowed down directly

Figure 7.10 Measuring the coupling between F-actin and FA proteins. (a) F-actin (red) and FA protein vinculin (green) in a live cell; (b) Simulated F-actin (red) and FA protein vinculin (green) from a Monte Carlo simulation. (c–e) Varying the association and dissociation rate constants of FA proteins in Monte Carlo simulations affects the speed of FA speckle motion and the coupling to F-actin. In these simulations, FA proteins switched between unbound (1), FA platform-bound (2) and F-actin-bound (3) states to allow coupling to F-actin flow. The VMCS and DCS were calculated using a simulation of F-actin flowing from left to right at $\nu = 0.25\,\mu m\,min^{-1}$; (c) Tracked motion of FA speckles (yellow vectors) for three representative FSM movies out of 25. VMCS increased from left to right as FA protein speckle flow became more aligned with F-actin speckle flow and more FA proteins were bound to the F-actin network; (d) Surface plot of VMCS determined by tracking 25 simulated FSM movies with the same dissociation rate constant ($k_{off} = k_{off21} = k_{off31} = 0.005$) and variable association rate constants to the FA platform (k_{on2}) and to F-actin (k_{on3}). For conditions $k_{on2} = k_{on3}$ (black diamonds), the VMCS was $\sim$0.5, in agreement with the notion that half the FA proteins are stationary while the other half move at velocity ν; (e) Scatter plot of VMCS versus $R_{kon} = k_{on3}/(k_{on2} + k_{on3})$. R_{kon} is a measure of the fraction of total FA proteins bound to F-actin. (f,g) Temporal variation of F-actin and vinculin speckle speeds, DCS, and VMCS within a stable (f) and a sliding (g) FA. The top left panels show graphs of average speeds of F-actin (red) and vinculin (green) speckles, vinculin-actin VMCS (blue) and vinculin-actin DCS (pink). The bottom left panels show kymographs of GFP-vinculin taken in the direction parallel to actin retrograde flow. The position of the cell edge (white) shows that the FA remains stationary in (f), whereas in (g) the FA initiates sliding at $\sim$4 min (left arrow) and stops at $\sim$12 min (right arrow). Right panels show maps of vinculin and actin speckle speeds and DCS. During retraction and FA sliding, vinculin alters its binding to F-actin. Times are shown as h:min:s. Figure reproduced with permission from Ref. [25].

over the FAs, suggesting that the latter may dampen flow by engaging F-actin to the ECM. Furthermore, when the motions of three classes of FA proteins were compared with F-actin, major differences in the speeds and visual coherence of the flow fields were observed. The ECM-binding α_V integrin exhibited slow, incoherent retrograde flow compared to actin, while the FA 'core' proteins paxilin, zyxin and focal-adhesion kinase (FAK), which do not bind F-actin or the ECM directly but have structural or signaling roles, moved slightly faster and more coherently. The third class, composed of the actin-binding proteins α-actinin, vinculin and talin, moved significantly faster (close to the speed of actin) and with the highest coherence.

The next step was to estimate coupling by quantifying the degree of correlated motion between FA and F-actin speckles. Correlated motion would strongly indicate that FA proteins help to transmit the force generated during actin polymerization and myosin II-mediated contraction to the ECM. To quantify this coupling, both speckle flow maps were interpolated to a common grid and two parameters were calculated: the direction coupling score (DCS) and the velocity magnitude coupling score (VMCS). The DCS measures the level of directional similarity between F-actin and FA speckle motion, while the VMCS measures the component of FA motion in the direction of actin, thus taking into account both direction and speed. The quantitative interpretation of these parameters, in terms of the kinetics of molecular interactions between F-actin and FA components, required mathematical modeling. Monte Carlo simulations were used to generate synthetic two-color movies of speckles associated with two transiently coupled protein structures (Figure 7.10b–d). Relating the binding/unbinding events at the molecular level to the relative movement of speckles in the two structures, along with an analysis of the noise characteristics of such movies, revealed that the VMCS is a linear reporter of the ratio between the time that the FA component is bound to F-actin alone and the time it is bound to both F-actin and the substrate (Figure 7.10e). Thus, the relative movement of two speckle fields directly reflects the degree of interaction between two protein structures in living cells.

When the DCS and VMCS were calculated for the three classes of FA proteins, integrin had the lowest coupling to actin, and F-actin-binding proteins the highest. The core proteins showed intermediate coupling. Such quantitative analysis provides even more insight when scores are compared over time. For example, coupling scores stayed constant for stationary FAs in a protrusive area of the cell edge, but F-actin–vinculin coupling increased in FAs that slid backwards in a retracting area of the cell edge (Figure 7.10f and g). Multicolor qFSM analysis therefore suggests a hierarchical molecular clutch model of force transmission, in which the efficiency of force transmission depends on the make-up of the FAs [25].

7.10
Outlook: Speckle Fluctuation Analysis to Probe Material Properties

Speckle trajectories probe different dynamic phenomena at different spatial and temporal scales. So far, by using the long-range directed components of speckle trajectories, qFSM has been used to measure the flow and deformation of F-actin and

MT networks. These movements are induced by molecular forces coordinated over several microns (e.g. by the activity of a large number of molecular motors or the concerted polymerization of many filaments). On a shorter spatiotemporal scale, speckle trajectories contain components associated with the microscopic deformations of polymer scaffolds that are induced by less-coordinated local actions of individual motors and thermal forces. The positional fluctuations of speckles can also be attributed to the sliding of locally decoupled filaments, to filament bending inside the network, and to photometric shifts of the speckle centroids due to local fluorophore exchange. These fluctuations occur at a length scale shorter than the mesh size of the polymer scaffold, and are independent between speckles. When calculating the cross-correlation of trajectories of two speckles separated by a distance greater than the mesh size, these fluctuations cancel out. However, even after directional components are eliminated, the cross-correlation between two-speckle trajectories decays with $1/r$, where r denotes the distance between them. The magnitude of the correlation indicates how much of the fluctuations are spatially transmitted through the material. Soft materials have a higher rate of transmission, and hence a higher correlation magnitude, than stiff materials. This is shown in Figure 7.11a for the example of a soft and a stiff F-actin network

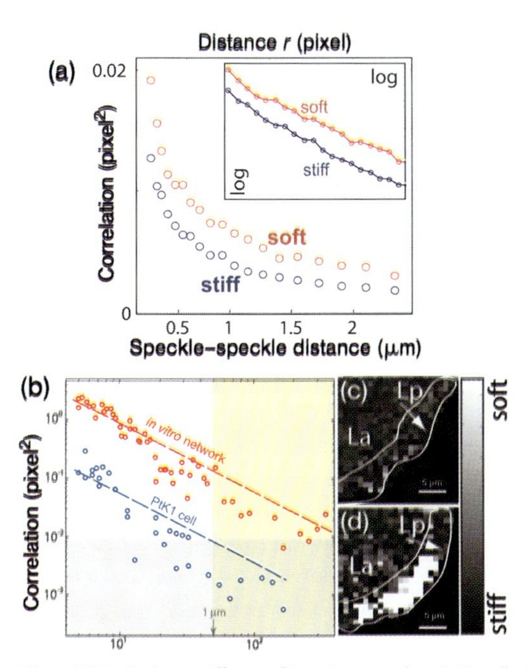

Figure 7.11 Probing stiffness of F-actin networks inside cells. (a) Correlation of random motion of two speckles as a function of their interspeckle distance r. The curves follow a $1/r$ decay (see inset), as predicted for a viscoeleastic medium; (b) Log-log plot of the correlation of random speckle motion as a function of r. *In vitro* networks are at least one order of magnitude softer than plated PtK1 cells. The gray area indicates the noise floor. Yellow area: for $r > 1\,\mu m$ the correlation is insignificant. Compliance of F-actin networks in a control cell (c) and in a cell expressing constitutively active cofilin (S3A) (d) which softens the lamellipodium network, most likely due to its selective severing activity on Lp filaments. Panel (a) reproduced with permission from Ref. [23].

measured *in vitro*. The $1/r$ decay of the fluctuation correlation is known from two-point microrheology, in which embedded beads instead of speckles are used to track thermal fluctuations in polymer networks [100, 101]. Thus, spatially correlated yet undirected components of speckle motion could be used to probe material properties of polymer networks inside a cell at the scale of the interspeckle distance. Figure 7.11b compares the stiffness between an *in vitro* network of entangled actin filaments and a cortical F-actin network in an epithelial cell. The marked difference originates in the dense crosslinking of *in vivo* networks, both intracellularly and extracellularly.

The stiffness of cellular networks is so high that correlations above noise are measurable only for speckles at a distance $<1\,\mu m$ (Figure 7.11b, gray zone). The possibility to extract meaningful information from speckle fluctuations over these short distances relies on recent enhancements of speckle tracking to an accuracy of approximately one-tenth of a pixel, even when speckles overlap. A module was also implemented that performs correlation analysis in small windows to map out the spatial modulation of material properties. Figure 7.11c and d compare the stiffness maps of a control epithelial cell and a cell expressing constitutively active cofilin(S3A). While the control cell has minimal spatial variation, the cell with cofilin(S3A) has a much softer lamellipodium (Lp) network but an unchanged lamella (La). These data show that cofilin – a severing factor and promoter of F-actin depolymerization – acts selectively in the lamellipodium network. High cofilin activity eventually eliminates the crosslink between the lamellipodium and lamella, resulting in a substantial softening of the lamellipodium network structure [85]. This example illustrates the potential of qFSM to derive spatial maps of the mechanical properties of cytoskeleton structures from speckle fluctuations.

7.11
Conclusions

Over the past few years, FSM has become a versatile tool for simultaneously probing the motion, deformation, turnover and materials properties of macromolecular assemblies. Despite the many exciting discoveries already made using FSM (see Table 7.1), it is a technology still in its infancy. In a next step, FSM measurements will be combined with correlational analyses to establish how assemblies operate as dynamic and plastic structures, enabling a broad variety of cell functions. In parallel, FSM will continue to go multispectral, so that these parameters can be correlated among different macromolecular structures. This requires major modifications to the current qFSM software to cope with the explosion of combinatorial data in two or more simultaneously imaged speckle channels.

With regards to future applications, FSM has the potential to uncover new biology outside the cytoskeleton field, and the analyses of FA dynamics have made some initial steps in this direction. Projects are also under way to apply qFSM to studies of the dynamic interaction of clathrin, dynamin and actin structures during endocytosis; of individual interphase MTs, MT-associated proteins and F-actin; and of DNA repair [102].

In addition to advancing basic research, FSM will hopefully become an important tool in drug discovery, particularly in the area of cancer. Already, by measuring MT dynamics in FSM-amenable cell lines transfected with patient-derived mutations in tumor suppressor genes, it has been observed that differential disease phenotypes are reproducibly replicated by different phenotypes of MT dynamics in nondividing cells (unpublished data). These subtle – but statistically highly significant – shifts in MT dynamics, that are not resolvable in static images of fixed cells, may disrupt the balance of the MT dynamics-mediated organization of signals within the cell and/or cell morphological functions. At the scale of multicellular tissues, these defects may result in detrimental responses that trigger tumor formation and metastatic behavior. Thus, FSM could enable the development of a screen for tumor-specific and patient-specific cancer drugs that would reverse the differences between cancer and control cells in terms of cytoskeleton dynamics. Such specific diagnostic tools at the subcellular scale may, at an early stage, allow the identification of efficient compounds and compound combinations with less harsh side effects than the current antimitotic chemotherapies.

Acknowledgments

These studies were supported by NIH through grants R01 GM67230 and NIH R01 GM60678 to the Danuser laboratory. Fellowship support from the Burroughs-Wellcome LJIS program (G.Y.) and the National Science Foundation (K.A.) is also acknowledged. We thank our collaborators, Clare Waterman-Storer, Edward Salmon, Tarun Kapoor, Julie Theriot, Paul Forscher and their laboratory members, for image data and uncountable discussions, without which the development of qFSM would not have been possible. We also thank James Lim and Dinah Loerke for sharing their unpublished data.

References

1 Pauling, L., Itano, H., Singer, S.J. and Wells, I. (1949) *Science*, **110**, 543.
2 Nahta, R. and Esteva, F.J. (2003) *Clinical Cancer Research*, **9**, 5078.
3 Garg, U. and Dasouki, M. (2006) *Clinical Biochemistry*, **39**, 315.
4 Eggert, U.S. and Mitchison, T.J. (2006) *Current Opinion in Chemical Biology*, **10**, 232.
5 Dorn, J.F., Danuser, G. and Yang, G. (2008) in *Fluorescent Proteins*, 2nd edn, Academic Press, Elsevier, Vol. 85, pp. 497.
6 Alberts, B., Johnson, A., Lewis, J., Raff, M., Roberts, K. and Walter, P. (2002) *Molecular Biology of the Cell*, 4th edn, Garland Science, New York.
7 Ramaekers, F.C. and Bosman, F.T. (2004) *The Journal of Pathology*, **204**, 351.
8 http://www.cellmigration.org (2007) (accessed 4 October 2007).
9 Abercrombie, M. (1978) *Proceedings of the Royal Society of London. Series, B, Biological Sciences*, **207**, 129.
10 Lauffenburger, D.A. and Horwitz, A.F. (1996) *Cell*, **84**, 359.

11 Ridley, A.J., Schwartz, M.A., Burridge, K., Firtel, R.A., Ginsberg, M.H., Borisy, G.B., Parsons, J.T. and Horwitz, A.R. (2003) *Science*, **302**, 1704.

12 Small, J.V., Geiger, B., Kaverina, I. and Bershadsky, A. (2002) *Nature Reviews Molecular Cell Biology*, **3**, 957.

13 Rodriguez, O.C., Schaefer, A.W., Mandato, C.A., Forscher, P., Bement, W.M. and Waterman-Storer, C.M. (2003) *Nature Cell Biology*, **5**, 599.

14 Lewis, E.M. (1991) *Journal of Surgical Oncology*, **47**, 243.

15 Ganem, N.J., Storchova, Z. and Pellman, D. (2007) *Current Opinion in Genetics and Development*, **17**, 157.

16 Skop, A.R., Liu, H., Yates, J. III, Meyer, B.J. and Heald, R. (2004) *Science*, **305**, 61.

17 Yang, Y., Varvel, N.H., Lamb, B.T. and Herrup, K. (2006) *The Journal of Neuroscience*, **26**, 775.

18 Wang, N., Butler, J.P. and Ingber, D.E. (1993) *Science*, **260**, 1124.

19 Chen, C.S., Mrksich, M., Huang, S., Whitesides, G.M. and Ingber, D.E. (1997) *Science*, **276**, 1425.

20 Engler, A.J., Sen, S., Sweeney, H.L. and Discher, D.E. (2006) *Cell*, **126**, 677.

21 Ruch, R.J. (2002) *Toxicological Sciences*, **68**, 265.

22 Stevens, J.M., Galyov, E.E. and Stevens, M.P. (2006) *Nature Reviews Microbiology*, **4**, 91.

23 Danuser, G. and Waterman-Storer, C.M. (2006) *Annual Review of Biophysics and Biomolecular Structure*, **35**, 361.

24 Waterman-Storer, C.M. and Salmon, E.D. (1997) *The Journal of Cell Biology*, **139**, 417.

25 Hu, K., Ji, L., Applegate, K., Danuser, G. and Waterman-Storer, C.M. (2007) *Science*, **315**, 111.

26 Salmon, W.C., Adams, M.C. and Waterman-Storer, C.M. (2002) *The Journal of Cell Biology*, **158**, 31.

27 Schaefer, A.W., Kabir, N. and Forscher, P. (2002) *The Journal of Cell Biology*, **158**, 139.

28 Ji, L., Loerke, D., Gardel, M. and Danuser, G. (2007) *Methods in Cell Biology*, **83**, 199–235.

29 Lippincott-Schwartz, J. and Patterson, G.H. (2003) *Science*, **300**, 87.

30 Mitchison, T.J. (1989) *The Journal of Cell Biology*, **109**, 637.

31 Theriot, J.A. and Mitchison, T.J. (1991) *Nature*, **352**, 126.

32 Wadsworth, P. and Salmon, E. (1986) *The Journal of Cell Biology*, **102**, 1032.

33 Wang, Y. (1985) *The Journal of Cell Biology*, **101**, 597.

34 Wolf, D.E. (1989) *Methods in Cell Biology*, **30**, 271.

35 Dunn, G.A., Dobbie, I.M., Monypenny, J., Holt, M.R. and Zicha, D. (2002) *Journal of Microscopy, Oxford*, **205**, 109.

36 Zicha, D., Dobbie, I.M., Holt, M.R., Monypenny, J., Soong, D.Y.H., Gray, C. and Dunn, G.A. (2003) *Science*, **300**, 142.

37 Maddox, P., Desai, A., Oegema, K., Mitchison, T.J. and Salmon, E.D. (2002) *Current Biology*, **12**, 1670.

38 Maddox, P., Straight, A., Coughlin, P., Mitchison, T.J. and Salmon, E.D. (2003) *The Journal of Cell Biology*, **162**, 377.

39 Waterman-Storer, C. Desai, A. and Salmon, E.D. (1999) *Methods in Cell Biology*, **61**, 155.

40 Adams, M., Matov, A., Yarar, D., Gupton, S., Danuser, G. and Waterman-Storer, C.M. (2004) *Journal of Microscopy*, **216**, 138.

41 Adams, M.C., Salmon, W.C., Gupton, S.L., Cohan, C.S., Wittmann, T., Prigozhina, N. and Waterman-Storer, C.M. (2003) *Methods (San Diego, Calif.)*, **29**, 29.

42 Grego, S., Cantillana, V. and Salmon, E.D. (2001) *Biophysical Journal*, **81**, 66.

43 Waterman-Storer, C.M. and Salmon, E.D. (1998) *Biophysical Journal*, **75**, 2059.

44 Desai, A. and Mitchison, T.J. (1997) *Annual Review of Cell and Developmental Biology*, **13**, 83.

45 Inoue, S. and Spring, K.R. (1997) *Video Microscopy: The Fundamentals*, 2nd edn, Plenum, New York and London.

46 Danuser, G. and Waterman-Storer, C.M. (2003) *Journal of Microscopy*, **211**, 191.

47 Verkhovsky, A.B., Svitkina, T.M. and Borisy, G.G. (1999) *Current Biology*, **9**, 11.

48 Watanabe, Y. and Mitchison, T.J. (2002) *Science*, **295**, 1083.

49 Waterman-Storer, C.M., Desai, A., Bulinski, J.C. and Salmon, E.D. (1998) *Current Biology*, **8**, 1227.

50 Waterman-Storer, C.M. Salmon, W.C. and Salmon, E.D. (2000) *Molecular Biology of the Cell*, **11**, 2471.

51 Jurado, C., Haserick, J.R. and Lee, J. (2005) *Molecular Biology of the Cell*, **16**, 507.

52 Vallotton, P., Danuser, G., Bohnet, S., Meister, J.J. and Verkhovsky, A. (2005) *Molecular Biology of the Cell*, **16**, 1223.

53 Zhang, X.-F., Schaefer, A.W., Burnette, D.T., Schoonderwoert, V.T. and Forscher, P. (2003) *Neuron*, **40**, 931.

54 Pollard, T.D., Blanchoin, L. and Mullins, R.D. (2000) *Annual Review of Biophysics and Biomolecular Structure*, **29**, 545.

55 Small, V. (1981) *The Journal of Cell Biology*, **91**, 695.

56 Svitkina, T.M., Verkhovsky, A.B., McQuade, K.M. and Borisy, G.G. (1997) *The Journal of Cell Biology*, **139**, 397.

57 Geiger, B., Bershadsky, A., Pankov, R. and Yamada, K.M. (2001) *Nature Reviews Molecular Cell Biology*, **2**, 793.

58 Bulinski, J.C., Odde, D.J., Howell, B.J., Salmon, T.D. and Waterman-Storer, C.M. (2001) *Journal of Cell Science*, **114**, 3885.

59 Kapoor, T.M. and Mitchison, T.J. (2001) *The Journal of Cell Biology*, **154**, 1125.

60 Ponti, A., Vallotton, P., Salmon, W.C., Waterman-Storer, C.M. and Danuser, G. (2003) *Biophysical Journal*, **84**, 3336.

61 Ponti, A. (2004) High-resolution analysis of F-actin meshwork kinetics and kinematics using computational fluorescent speckle microscopy. Dissertation No. 15286, *ETH Zurich (Zurich)*.

62 Yang, G., Houghtaling, B.R., Gaetz, J., Liu, J.Z., Danuser, G. and Kapoor, T.M. (2007) *Nature Cell Biology*, **9**, 1233.

63 Mikhailov, A.V. and Gundersen, G.G. (1995) *Cell Motility and the Cytoskeleton*, **32**, 173.

64 Waterman-Storer, C.M. Sanger, J. and Sanger, J. (1993) *Cell Motility and the Cytoskeleton*, **26**, 19.

65 Waterman-Storer, C.M. (2002) *Current Protocols in Cell Biology* (eds J.S. Bonifacino, M. Dasso, J.B. Harford, J. Lippincott-Schwartz and K.M. Yamada), John Wiley & Sons, New York.

66 Gupton, S.L. and Waterman-Storer, C.M. (2006) *Cell Biology: A Laboratory Handbook*, 3rd edn, Vol. **3**, (eds J. Celis, N. Carter, K. Simons, J.V. Small, T. Hunter and D. Shotton), Academic Press, San Diego, pp. 137.

67 Brust-Mascher, I. and Scholey, J.M. (2002) *Molecular Biology of the Cell*, **13**, 3967.

68 Gaetz, J. and Kapoor, T.M. (2004) *The Journal of Cell Biology*, **166**, 465.

69 Gupton, S.L., Salmon, W.C. and Waterman-Storer, C.M. (2002) *Current Biology*, **12**, 1891.

70 Wittmann, T., Bokoch, G.M. and Waterman-Storer, C.M. (2003) *The Journal of Cell Biology*, **161**, 845.

71 Jepson, A.D., Fleet, D.J. and El-Maraghi, T.F. (2003) *IEEE Transactions on Pattern Analysis and Machine Intelligence*, **25**, 1296.

72 Micheli, E.D., Torre, V. and Uras, S. (1993) *IEEE Transactions on Pattern Analysis and Machine Intelligence*, **15**, 434.

73 Ye, M., Haralick, R.M. and Shapiro, L.G. (2003) *IEEE Transactions on Pattern Analysis and Machine Intelligence*, **25**, 1625.

74 Vallotton, P., Ponti, A., Waterman-Storer, C.M., Salmon, E.D. and Danuser, G. (2003) *Biophysical Journal*, **85**, 1289.

75 Ahuja, R.K., Magnanti, T.M. and Orlin, J.B. (1993) *Network Flows: Theory, Algorithms and Optimization*, Prentice-Hall, Inc., New Jersey.

76 Miyamoto, D.T., Perlman, Z.E., Burbank, K.S., Groen, A.C. and Mitchison, T.J. (2004) *The Journal of Cell Biology*, **167**, 813.

77 Ji, L. and Danuser, G. (2005) *Journal of Microscopy*, **220**, 150.

78 Ponti, A., Matov, A., Adams, M., Gupton, S., Waterman-Storer, C.M. and Danuser, G. (2005) *Biophysical Journal*, **89**, 3456.

79 Blackman, S.S. and Popoli, R. (1999) *Design and Analysis of Modern Tracking Systems*, Artech House, Norwood, MA.

80 Burkard, K.E. and Cela, E. (1999) in *Handbook of Combinatorial Optimization*, Vol. Supp. A, (eds D.Z. Du and P.M. Pardalos), Kluwer Academic Publishers, Dordrecht, NL, p. 75.

81 Vallotton, P., Gupton, S.L., Waterman-Storer, C.M. and Danuser, G. (2004) *Proceedings of the National Academy of Sciences of the United States of America*, **101**, 9660.

82 Pollard, T.D. and Borisy, G.B. (2003) *Cell*, **112**, 453.

83 Ponti, A., Machacek, M., Gupton, S.L., Waterman-Storer, C.M. and Danuser, G. (2004) *Science*, **305**, 1782.

84 Gupton, S.L., Anderson, K.L., Kole, T.P., Fischer, R.S., Ponti, A., Hitchcock-DeGregori, S.E., Danuser, G., Fowler, V.M., Wirtz, D., Hanein, D. and Waterman-Storer, C.M. (2005) *The Journal of Cell Biology*, **168**, 619.

85 Delorme, V., Machacek, M., DerMardirossian, C., Andersen, K.L., Wittmann, T., Hanein, D., Waterman-Storer, C.M., Danuser, G. and Bokoch, G. (2007) *Developmental Cell*, **13** (5), 646–662.

86 Murthy, K. and Wadsworth, P. (2005) *Current Biology*, **15**, 724.

87 Mitchison, T. and Kirschner, M. (1984) *Nature*, **312**, 237.

88 Kinoshita, K., Arnal, I., Desai, A., Drechsel, D.N. and Hyman, A.A. (2001) *Science*, **294**, 1340.

89 Sawin, K.E. and Mitchison, T.J. (1991) *The Journal of Cell Biology*, **112**, 941.

90 Burbank, K.S., Groen, A.C., Perlman, Z.E., Fisher, D.D. and Mitchison, T.J. (2006) *The Journal of Cell Biology*, **175**, 369.

91 Mastronarde, D.N., McDonald, K.L., Ding, R. and McIntosh, J.R. (1993) *The Journal of Cell Biology*, **123**, 1475.

92 Mitchison, T.J., Maddox, P., Groen, A., Cameron, L., Perlman, Z., Ohi, R., Desai, A., Salmon, E.D. and Kapoor, T.M. (2004) *Molecular Biology of the Cell*, **15**, 5603.

93 Maruyama, K. and Ebashi, S. (1965) *Journal of Biochemistry*, **58**, 13.

94 Muguruma, M., Matsumura, S. and Fukazawa, T. (1990) *Biochemical and Biophysical Research Communications*, **171**, 1217.

95 Johnson, R.P. and Craig, S.W. (1995) *Nature*, **373**, 261.

96 Tanaka, T., Yamaguchi, R., Sabe, H., Sekiguchi, K. and Healy, J.M. (1996) *FEBS Letters*, **399**, 53.

97 Calderwood, D.A., Zent, R., Grant, R., Rees, D.J.G., Hynes, R.O. and Ginsberg, M.H. (1999) *The Journal of Biological Chemistry*, **274**, 28071.

98 Burridge, K. and Mangeat, P. (1984) *Nature*, **308**, 744.

99 Burridge, K. and Chrzanowska-Wodnicka, M. (1996) *Annual Review of Cell and Developmental Biology*, **12**, 463.

100 Crocker, J.C., Valentine, M.T., Weeks, E.R., Gisler, T., Kaplan, P.D., Yodh, A.G. and Weitz, D.A. (2000) *Physical Review Letters*, **85**, 888.

101 Gardel, M.L., Shin, J.H., MacKintosh, F.C., Mahadevan, L., Matsudaira, P. and Weitz, D.A. (2004) *Science*, **304**, 1301.

102 Soutoglou, E., Dorn, J.F., Sengupta, K., Jasin, M., Nussenzweig, A., Ried, T., Danuser, G. and Misteli, T. (2007) *Nature Cell Biology*, **9**, 675.

103 Fraley, C. and Raftery, A.E. (2002) *Journal of the American Statistical Association*, **97**, 611.

8
Harnessing Biological Motors to Engineer Systems for Nanoscale Transport and Assembly*

Anita Goel and Viola Vogel

By considering how the biological machinery of our cells carries out many different functions with a high level of specificity, we can identify a number of engineering principles that can be used to harness these sophisticated molecular machines for applications outside their usual environments. Here, we focus on two broad classes of nanomotors that burn chemical energy to move along linear tracks: assembly nanomotors and transport nanomotors.

8.1
Sequential Assembly and Polymerization

The molecular machinery found in our cells is responsible for the sequential assembly of complex biopolymers from their component building blocks (monomers): polymerases make DNA and RNA from nucleic acids, and ribosomes construct proteins from amino acids. These assembly nanomotors operate in conjunction with a master DNA or RNA template that defines the order in which individual building blocks must be incorporated into a new biopolymer. In addition to recognizing and binding the correct substrates (from a pool of many different ones), the motors must also catalyze the chemical reaction that joins them into a growing polymer chain. Moreover, both types of motors have evolved highly sophisticated mechanisms so that they are able not only to discriminate the correct monomers from the wrong ones, but also to detect and repair mistakes as they occur [1].

Molecular assembly machines or nanomotors (Figure 8.1a) must effectively discriminate between substrate monomers that are structurally very similar. Polymerases must be able to distinguish between different nucleosides, and ribosomes need to recognize particular transfer RNAs (tRNAs) that carry a specific amino acid. These well-engineered biological nanomotors achieve this by pairing complementary Watson–Crick base pairs and comparing the geometrical fit of the monomers to their respective polymeric templates. This molecular discrimination makes use of the

*Reprinted by Permission from Macmillan Publishers Ltd: nature nanotechnology, Vol 2, August 2008.

Nanotechnology, Volume 5: Nanomedicine. Edited by Viola Vogel

ISBN: 978-3-527-31736-3

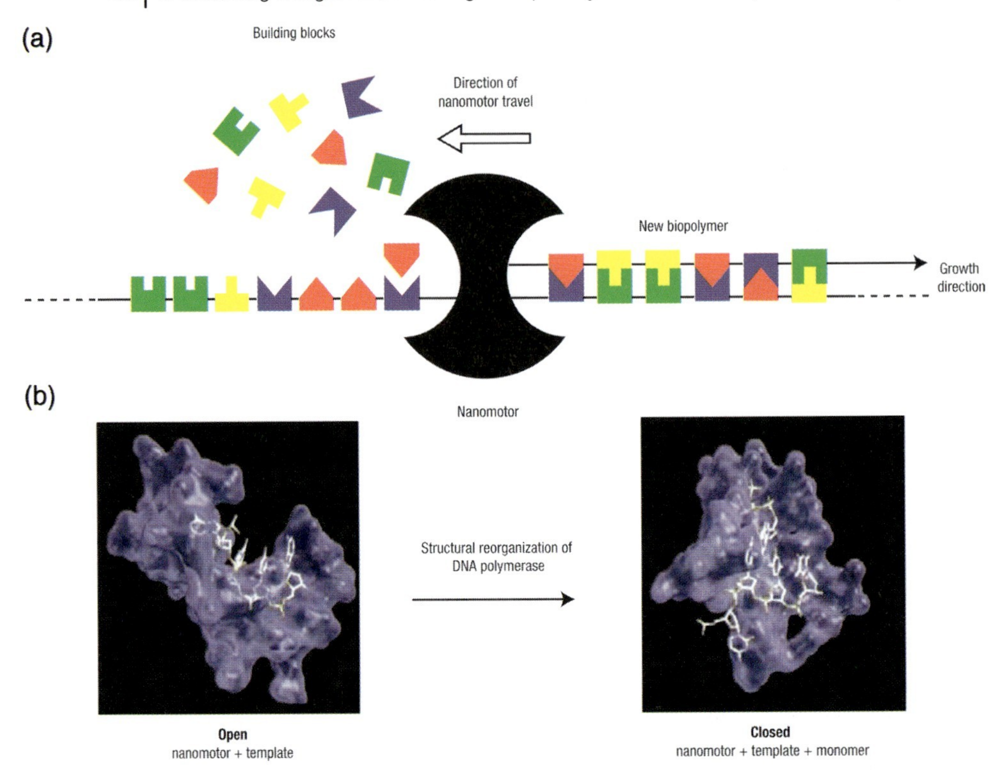

Figure 8.1 Molecular discrimination during sequential assembly. (a), The polymerase nanomotor discriminates between four different building blocks as it assembles a DNA or RNA strand complementary to its template sequence. Molecular discrimination between substrate monomers that are structurally very similar is achieved by comparing the geometrical fit of the monomers to their respective polymeric templates; (b), The T7 DNA polymerase motor undergoes an internal structural transition from an open state (when the active site samples different nucleotides) to a closed state (when the correct nucleotide is incorporated into the nascent DNA strand). Nucleotides are added to the nascent strand one at a time. This structural transition is the rate-limiting step in the replication cycle and is thought to be dependent on the mechanical tension in the template strand [2, 9, 107, 116, 121, 127, 128, 131]. Figure adapted from Ref. [127]; © 2001 PNAS.

differential binding strengths of correctly matched and mismatched substrates, which is determined by the complementarity of the base-pairing between them.

Figure 8.1b illustrates the assembly process used by the DNA polymerase nanomotor. A template of single-stranded DNA binds to the nanomotor with angstrom-level precision, forming an open complex. The open complex can 'sample' the free nucleosides available. Binding of the correct nucleoside induces a conformational change in the nanomotor, which then allows the new nucleoside to be added to the growing DNA strand [1]. The tight-fitting complementarity of shapes between the polymerase binding site and the properly paired base pair guarantees a 'geometric selection' for the correct nucleotide [2]. A similar mechanism is seen in *Escherichia coli* RNA polymerase, where the binding of an incorrect monomer inhibits the conformational change in the motor from an 'open' (inactive) to a 'closed' (active) conformation [3].

Ribosome motors carry out tasks much more complex than polymerases. Instead of the four nucleotide building blocks used by polymerases to assemble DNA or RNA, ribosomes must recognize and selectively arrange 20 amino acids to synthesize a protein. This fact alone increases the chance of errors. Nevertheless, ribosomes obviously work (and do so along the same principles of geometric fit and conformational change as do polymerases) and are able to build amino acid polymers that are subsequently folded into functional proteins. But ribosomal motors can be tricked, much more easily than DNA motors, into building the 'incorrect' sequences when supplied with synthetic amino acids that resemble real ones [4].

8.1.1
Engineering Principle No. 1: discrimination of similar building blocks

> *Nanomotors used in the sequential assembly of biopolymers can discriminate efficiently between similar building blocks.*

The structure of molecular machines can be visualized with angstrom-level resolution using X-ray crystallography, and the sequential assembly processes they drive can be probed in real time using single-molecule techniques [5–9]. By elucidating nanomotor kinetics under load, such nanoscale techniques provide detailed insights into the single-molecule dynamics of nanomotor-driven assembly processes. Techniques such as optical and magnetic tweezers, for example, have further elucidated the polymer properties of DNA [7, 10–12] and the force-dependent kinetics of molecular motors [13–18]. Single-molecule fluorescence methods such as fluorescence energy transfer, in conjunction with such biomechanical tools, are illuminating the internal conformational dynamics of these nanomotors [19–21].

As the underlying design principles of assembly nanomotors are revealed, it will become increasingly possible to use these biomachines for *ex vivo* tasks. Sequencing and PCR are two such techniques that already harness polymerase nanomotors for the *ex vivo* replication of nucleic acids. The polymerase chain reaction, or PCR, is a landmark, Nobel prize-winning technique [22] invented in the 1980s that harnessed polymerase nanomotors to amplify a very small starting sample of DNA to billions of molecules. Likewise, there are many conceivable future applications that either use assembly nanomotors *ex vivo* or mimic some of their design principles. Efforts are already under way to control these nanomotors better, and thus to improve such *ex vivo* sequential assembly processes for industrial use (see, for example, the websites www.cambrios.com; www.helicosbio.com; www.nanobiosym.com; www.pacificbiosciences.com).

In contrast, current *ex vivo* methods to synthesize block copolymers rely primarily on random collisions, resulting in a wide range of length distributions and much less control over the final sequence [23]. Sequential assembly without the use of nanomotors remains limited to the synthesis of comparatively short peptides, oligonucleotides and oligosaccharides [24–26]. Common synthesizers still lack both the precision of monomer selection and the inbuilt proofreading machinery for monomer repair that nanomotors have. Building such copolymers with polymerase nanomotors *ex vivo* would yield much more homogeneous products of the correct sequence and precise length. Natural (e.g., nanomotor-enabled) designs could inspire new technologies to synthesize custom biopolymers precisely from a given blueprint.

Ribosome motors have likewise been harnessed *ex vivo* to drive the assembly of new bioinorganic heterostructures [27] and peptide nanowires [28, 29] with gold-modified amino acids inserted into a polypeptide chain. These ribosomes are forced to use inorganically modified tRNAs to sequentially assemble a hybrid protein containing gold nanoparticles wherever the amino acid cysteine was specified by the messenger RNA template. Such hybrid gold-containing proteins can then attach themselves selectively to materials used in electronics, such as gallium arsenide [28]. This application illustrates how biomotors could be harnessed to synthesize and assemble even nonbiological constructs such as nanoelectronic components (see www.cambrios.com).

Assembly nanomotors achieve such high precision in sequential assembly by making use of three key features: (i) geometric shape-fitting selection of their building blocks (e.g., nucleotides); (ii) motion along a polymeric template coupled to consumption of an energy source (e.g., hydrolysis of ATP molecules); and (iii) intricate proofreading machinery to correct errors as they occur. Furthermore, nanomotor-driven assembly processes allow much more stable, precise and complex nanostructures to be engineered than can be achieved by thermally driven self-assembly techniques alone [30–32].

We should also ask whether some of these principles, which work so well at the nanoscale, could be realized at the micrometer scale as well. Whitesides and coworkers, for example, have used simple molecular self-assembly strategies, driven by the interplay of hydrophobic and hydrophilic interactions, to assemble micro-fabricated objects at the mesoscale [33, 34]. Perhaps the design principles used by nanomotors to improve precision and correct errors could also be harnessed to engineer future *ex vivo* systems at the nanoscale, as well as on other length scales. Learning how to engineer systems that mimic the precision and control of nano-motor-driven assembly processes may ultimately lead to efficient fabrication of complex nanoscopic and mesoscopic structures.

8.2 Cargo Transport

Cells routinely use another set of nanomotors (i.e., transport nanomotors) to recognize, sort, shuttle and deliver intracellular cargo along filamentous freeways to well-defined destinations, allowing molecules and organelles to become highly organized (for reviews, see Refs. [35–44]). This is essential for many life processes. Motor proteins transport cargo along cytoskeletal filaments to precise targets, concentrating molecules in desired locations. In intracellular transport, myosin motors are guided by actin

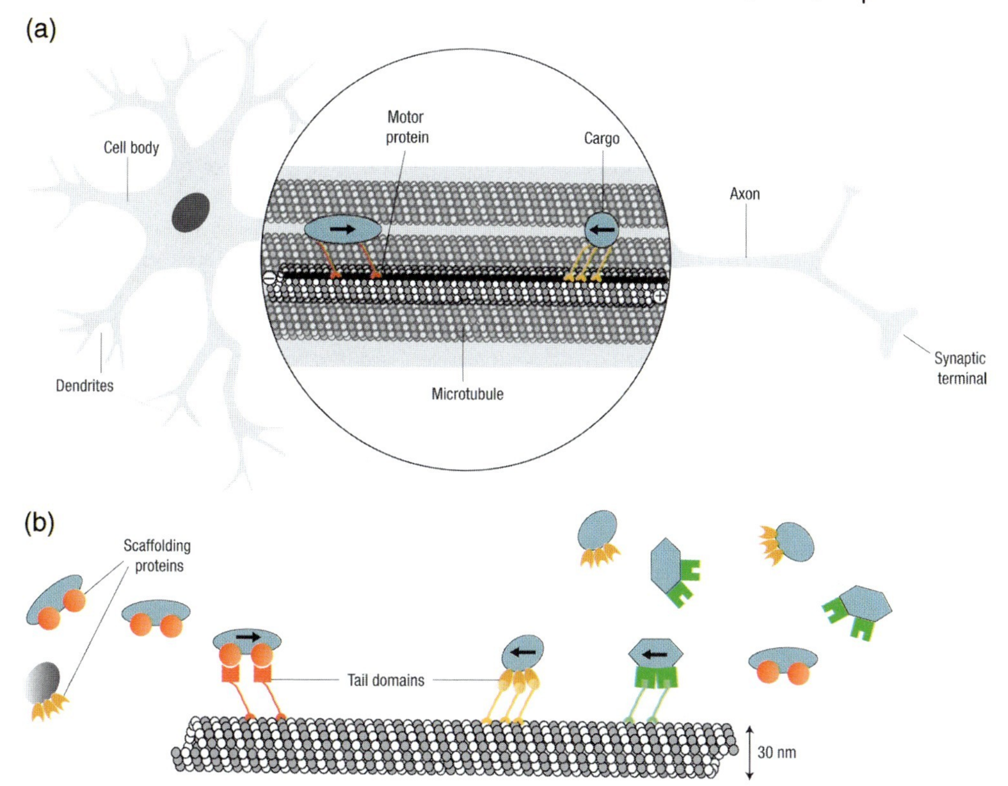

Figure 8.2 Motor-specific cargo transport in neurons. (a), The axon of neurons consists of a bundle of highly aligned microtubules along which cargo is trafficked from the cell body to the synapse and *vice versa*. Most members of the large kinesin family (red) transport cargo towards the periphery, while other motors, including dyneins (yellow), transport cargo in the opposite direction. Motors preferentially move along a protofilament rather then side-stepping (one randomly selected protofilament is shown in dark gray). Protofilaments are assembled from the dimeric protein tubulin (white and gray spheres) which gives microtubules their structural polarity. The protofilaments then form the hollow microtubule rod. When encountering each other on the same protofilament, the much more tightly bound kinesin has the 'right of way', perhaps even forcing the dynein to step sidewise to a neighboring protofilament [52–55]. (b), Each member of a motor family selects its own cargo (blue shapes) through specific binding by scaffolding proteins (colored symbols) or directly by the cargo's tail domains.

filaments, whereas dynein and kinesin motors move along rodlike microtubules. Figure 8.2a illustrates how conventional kinesins transport molecular cargo along nerve axons towards the periphery, efficiently transporting material from the cell body to the synaptic region [45]. Dyneins, in contrast, move cargo in the opposite direction, so that there is active communication and recycling between both ends (see reviews [42, 46]). In fact, the blockage of such bidirectional cargo transport along nerve axons can give rise to substantial neural disorders [47–50].

The long-range guidance of cargo is made possible by motors pulling their cargo along filamentous rods. Microtubules, for example, are polymerized from the dimeric tubulin into protofilaments that assemble into rigid rods around 30 nm in diameter [36]. These polymeric rods are inherently unstable: they polymerize at one end (plus) while depolymerizing from the other (minus) end, giving rise to a structural polarity. The biological advantage of using transient tracks is that they can be rapidly reconfigured on demand and in response to changing cellular needs, or to various external stimuli. Highly efficient unidirectional cargo transport is realized in cells by bundling microtubules into transport highways where all microtubules are oriented in the same direction. Excessively tight bundling of microtubules, however, can greatly impair the efficiency of cargo transport, by blocking the access of motors and cargo to the microtubules in the bundle interior. Instead, microtubule-associated proteins are thought to act as repulsive polymer-brushes, thereby regulating the proximity and interactions between neighboring microtubules [51].

Traffic control is an issue when using the filaments as tracks on which kinesin and dynein motors move in opposite directions. Although different cargoes can be selectively recognized by different members of the motor protein families and shuttled to different destinations, what happens if motors moving in opposite directions encounter each other on the same protofilament (Figure 8.2b)? If two of these motors happen to run into each other, kinesin seems to have the 'right of way'. As kinesin binds the microtubule much more strongly, it is thought to force dynein to step sideways to a neighboring protofilament [52]. Dynein shows greater lateral movement between protofilaments than kinesin [52–54] as there is a strong diffusional component to its steps [55]. When a microtubule becomes overcrowded with only kinesins, the runs of individual kinesin motors are minimally affected. But when a microtubule becomes overloaded with a mutant kinesin that is unable to step efficiently, the average speed of wild-type kinesin is reduced, whereas its processivity is hardly changed. This suggests that kinesin remains tightly bound to the microtubule when encountering an obstacle and waits until the obstacle unbinds and frees the binding site for kinesin's next step [56].

8.2.1
Engineering Principle No. 2: various track designs

Various track designs enable motors to pull their cargo along filamentous tracks, whereas others allow motors bound to micro- or nanofabricated tracks to propel the filaments which can then serve as carriers.

It is not a trivial task to engineer transport highways *ex vivo*, particularly in versatile geometries with intersections and complex shapes. Individual filaments typically allow only one-dimensional transport, as the motor-linked cargo drops off once the end of the filament is reached. Furthermore, conventional kinesin makes only a few hundred 8 nm-sized steps before dissociating from the microtubule [57, 58], further limiting the use of such a system for *ex vivo* applications.

Instead of having the motors transport their cargo along filaments, motors have been immobilized on surfaces in an inverted geometry that enables the filaments to

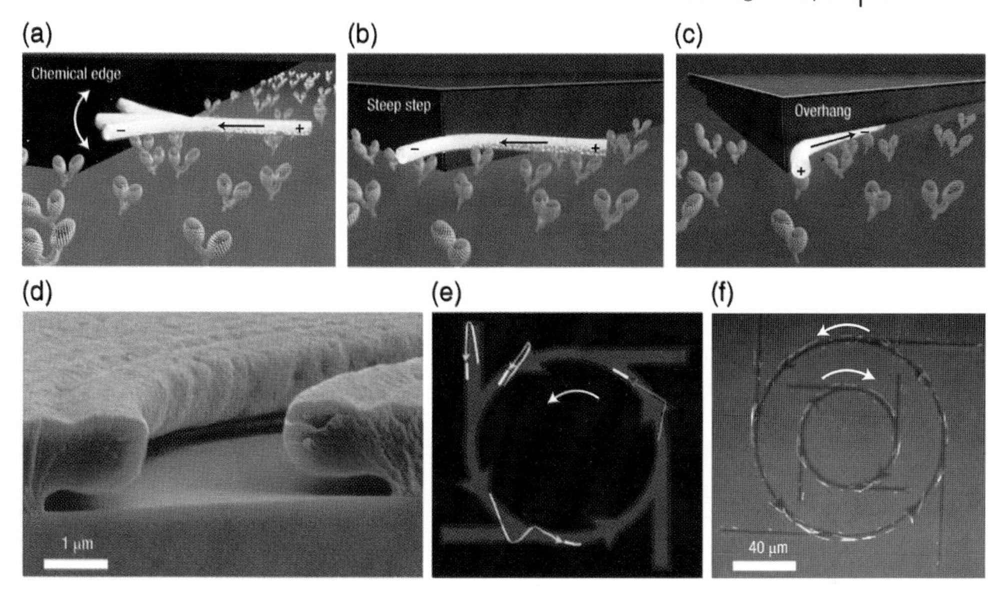

Figure 8.3 Track designs to guide nanomotor-driven filaments *ex vivo*. A variety of track designs have been used. (a), A chemical edge (adhesive stripes coated with kinesin surrounded by nonadhesive areas). The filament crosses the chemical edge and ultimately falls off as it does not find kinesins on the nonadhesive areas [61]; (b), Steep channel walls keep the microtubule on the desired path as they are forced to bend [61,65]; (c), Overhanging walls have been shown to have the highest guidance efficiency [64]; (d), Electron micrograph of a microfabricated open channel with overhanging walls [64]; (e), Breaking the symmetry of micropatterns can promote directional sorting of filament movement [63, 65, 69, 138]. The trajectories of four microtubules are shown: movement into reflector arms causes the tubule to turn around (yellow), an arrow-shaped direction rectifier allows those travelling in the desired direction to continue (red) and forces others to turn around (blue). At intersections, tubules preferentially continue straight on (green); (f), The complex microfabricated circuit analysed in (e) with open channels and overhanging walls, demonstrating unidirectional movement of microtubules.

be collectively propelled forward [45]. The head domains of the kinesin and myosin motors can rotate and swivel with respect to their feet domains, which are typically bound in random orientations to the surface. These motor heads detect the structural anisotropy of the microtubules and coherently work together to propel a filament forward [59, 60].

Various examples of such inverted designs for motor tracks have been engineered to guide filaments efficiently. Some of these are illustrated in Figure 8.3. Inverted motility assays can be created, for example, by laying down tracks of motor proteins in microscopic stripes of chemical adhesive on an otherwise flat, protein-repellent surface, surrounded by nonadhesive surface areas. Such chemical patterns (Figure 8.3a) have been explored to guide actin filaments or microtubules. The loss rate of guiding filaments increases exponentially with the angle at which they approach an adhesive/nonadhesive contact line [61]. The passage of the contact line by filaments at nongrazing angles, followed by their drop off, can be prevented by using much narrower lanes whose size is of the order of the diameter of the moving object. Such nanoscale kinesin tracks provide good guidance and have been fabricated by nanotemplating [62].

Alternatively, considerably improved guidance has been accomplished by topographic surface features (Figure 8.3b). Microtubules hitting a wall are forced to bend along this obstacle and will continue to move along the wall [63–66]. The rigidity of the polymeric filaments used as shuttles thus greatly affects how tracks should be designed for optimal guidance. Whereas microtubules with a persistence length of a few millimeters can be effectively guided in channels a few micrometers wide as they are too stiff to turn around [61], the much more flexible actin filaments require channel widths in the submicrometer range [67, 68]. Finally, the best long-distance guidance of microtubules has been obtained so far with overhanging walls [64, 69] (Figure 8.3c). The concept of topographic guidance in fact works so well that swarms of kinesin-driven microtubules have been used as independently moving probes to image unknown surface topographies. After averaging all their trajectories in the focal plane for an extended time period, the image grayscale is determined by the probability of a surface pixel being visited by a microtubule in a given time frame [70].

But how can tracks be engineered to produce *uni*directional cargo transport? All the motor-propelled filaments must move in the same direction to achieve effective long-distance transport. When polar filaments land from solution onto a motor-covered surface, however, their orientations and initial directions of movement are often randomly distributed. Initially, various physical means, such as flow fields [71], have been introduced to promote their alignment. Strong flows eventually either force gliding microtubules to move along with the flow, or force microtubules, if either their plus or minus end is immobilized on a surface [72], to rotate around the anchoring point and along with the flow. The most universal way to control the local direction in which the filamentous shuttles are guided is to make use of asymmetric channel features. Figure 8.3d–f illustrates how filaments can be actively sorted according to their direction of motion by breaking the symmetry of the engineered tracks. This 'local directional sorting' has been demonstrated on surfaces patterned with open-channel geometries, where asymmetric intersections are followed by dead-ended channels (that is, reflector arms), or where channels are broadened into arrow heads. Both of these topographical features not only selectively pass filaments moving in the desired direction, but can also force filaments moving in the opposite direction to turn around [65, 69, 73, 74]. Once directional sorting has been accomplished, electric fields have been used to steer the movement of individual microtubules as they pass through engineered intersections [75, 76].

In addition to using isolated nanomotors, hybrid biodevices and systems that harness self-propelling microbes could be used to drive transport processes along engineered tracks. Flagellated bacteria, for example, have been used to generate both translational and rotational motion of microscopic objects [77]. These bacteria can be attached head-on to solid surfaces, either via polystyrene beads or polydimethylsiloxane, thereby enabling the cell bodies to form a densely packed monolayer, while their flagella continue to rotate freely. In fact, a microrotary motor, fuelled by glucose and comprising a 20 μm-diameter silicon dioxide rotor, can be driven along a silicon track by the gliding bacterium *Mycoplasma* [78]. Depending on the specific application and the length scale on which transport needs to be achieved, integrating bacteria into such biohybrid devices (that work under physiological conditions) might ultimately prove more robust than relying solely on individual nanomotors.

8.3 Cargo Selection

To maintain intracellular contents in an inhomogeneous distribution far from equilibrium, the intracellular transport system must deliver molecular cargo and organelles on demand to precise destinations. This tight spatiotemporal control of molecular deliveries is critical for adequate cell function and survival. Molecular cargo or organelles are typically barcoded so that they can be recognized by their specific motor protein (Figure 8.4). Within cells, motors recognize cargo either from the cargo's tail domains directly, or via scaffolding proteins that link cargo to their tail domain [43].

8.3.1 Engineering Principle No. 3: barcoding

> *Engineered molecular recognition sites enable cargo to be selectively bonded to moving shuttles.*

Although most cargo shuttled around by motors can be barcoded using the existing repertoire of biological scaffolding proteins, synthetic approaches are needed for all those *ex vivo* applications where the cargo has to be specifically linked to moving filaments. The loading and transport of biomedically relevant or engineered cargo has already been demonstrated (Figure 8.4) [79–83]. Typical approaches are to tag the cargo

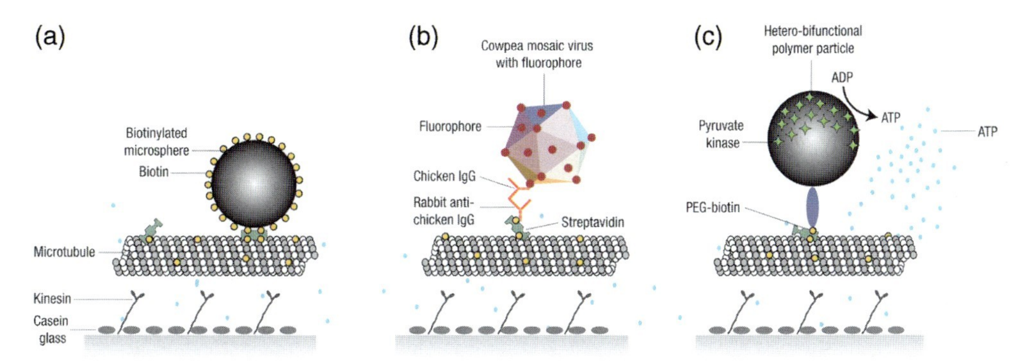

Figure 8.4 Selecting specific cargo by molecular recognition. A versatile toolbox exists by which synthetic and biological cargo can be coupled to microtubules. (a) Biotinylated objects are coupled via avidin or streptavidin to biotinylated microtubules. (b) Biological molecules, viruses [79, 81] or cells can be coupled by antibody recognition. (c) Backpacks of chemically or biologically active reagents can be shuttled around, including bioprobes [80] or tiny ATP factories [93] as shown here.

with antibodies or to biotinylate microtubules and coat the cargo with avidin or streptavidin (Figure 8.4) (for reviews, see Refs. [74, 79]), as done for polymeric and magnetic beads [84, 85] (Figure 8.4a), gold nanoparticles [86–88], DNA [87, 89, 90] and viruses [79, 81] (Figure 8.4b), and finally mobile bioprobes and sensors [80, 81, 91] (Figure 8.4c). However, if too much cargo is loaded onto the moving filaments and access of the propelling motors is even partially blocked, the transport velocity can be significantly impaired [92]. Finally, the binding of cargo to a moving shuttle can be used to regulate its performance. In fact, microtubules have recently been furnished with a backpack that selfsupplies the energy source ATP. Cargo particles bearing pyruvate kinase have been tethered to the microtubules to provide a local ATP source [93] (Figure 8.4c). The coupling of multiple motors to cargo or other scaffold materials can affect the motor performance. If single-headed instead of double-headed kinesins are used, cooperative interactions between the monomeric motors attached to protein scaffolds increase hydrolysis activity and microtubule gliding velocity [59].

At the next level of complexity – successful cargo tagging – sorting and delivery will depend on the engineering of integrated networks of cargo loading, cargo transport and cargo delivery zones. Although the construction of integrated transport circuits is still in its infancy, microfabricated loading stations have been built [88] (Figure 8.5). The challenge here is to immobilize cargo on loading stations such that it is not easily detached by thermal motion, yet to allow for rapid cargo transfer to passing microtubules. By properly tuning bond strength and multivalency, and most importantly by taking advantage of the fact that mechanical strain weakens bonds, cargo can be efficiently stored on micropatches and transferred after colliding with a microtu-

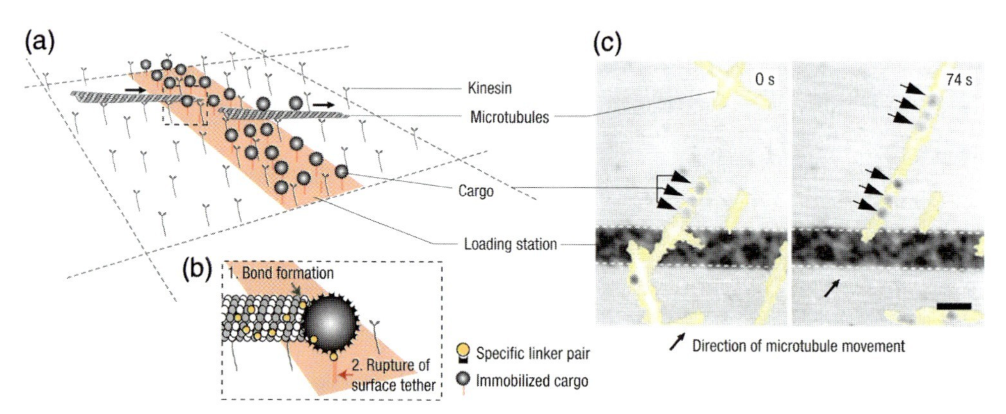

Figure 8.5 Cargo loading stations [93]; (a) Stripes of immobilized cargo are fabricated by binding thiolated oligonucleotides to micropatterned lines of gold. Hybridization with complementary strands exposing antibodies at their terminal ends allows them to immobilize a versatile range of cargos that carry antibodies on their surfaces. (b) The challenge is to tune the bond strength and valency to prevent thermal activation during cargo storage on the loading station. On collision with the shuttle (microtubule), the cargo must rapidly break off the bond it has formed with the station88. Fortunately, however, tensile mechanical force acting on a noncovalent bond shortens its lifetime; (c, d) These concepts are used in the design of the loading stations shown here, where a microtubule moves through a stripe of immobilized gold cargo and picks up a few beads.

bule [88]. Considerable fine-tuning of bond strength can be accomplished by using DNA oligomers hybridized such that the bonds are either broken by force all at once (a strong bond) or in sequence (a weak bond) [94].

As discussed above, filaments are most commonly used to shuttle molecular cargo in most emerging devices that harness linear motors for active transport. Alternatively, if the filamentous tracks could be engineered in versatile geometries, the motors themselves could be used to drag cargo coupled to the molecular recognition sites of their tail domains as in the native systems. We could thus make use of the full biological toolbox of already known or engineered scaffolding proteins that link specific motors to their respective cargoes [40, 43]. So far, assemblies of microtubules organized into complex, three-dimensional patterns such as asters, vortices and networks of interconnected poles [95, 96] have been successfully created in solution, and mesoscopic needles and rotating spools of microtubule bundles held together by noncovalent interactions have been engineered on surfaces [31]. All of these mesoscopic structures are uniquely related to active motor-driven motion, and would not have formed purely by self-assembly without access to an energy source.

To increase the complexity of microtubule track networks, densely packed arrays of microtubules have been grown in confined spaces, consisting of open microfabricated channels with user-defined geometrical patterns [97]. The key to achieving directed transport, however, is for all microtubules within each bundle or array to be oriented in the same direction. This has been accomplished by making use of directed motility in combination with sequential assembly procedures (Figure 8.6). First, microtubule seedlings have been oriented in open microfabricated and kinesin-

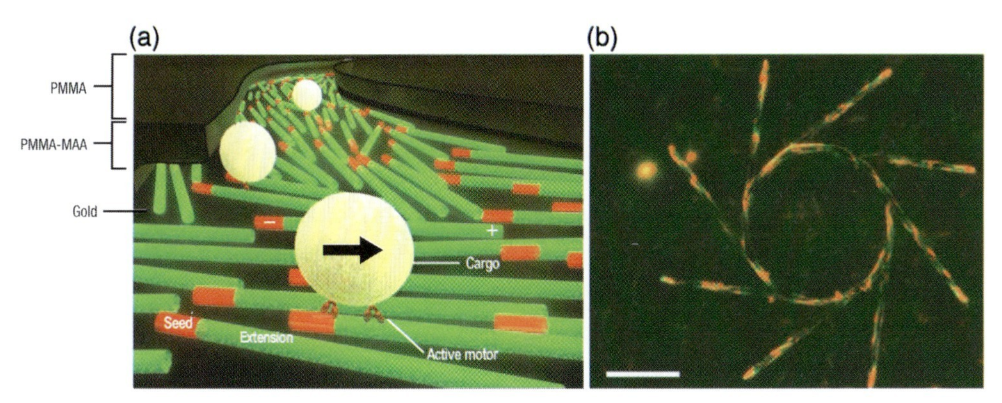

Figure 8.6 Filament tracks made from engineered bundles of microtubules [97]. Active transport is used to produce bundles of microtubules and confine them to user-defined geometries. (a) Sequential assembly procedure: first, microtubule seedlings (labelled in red) are allowed to orient themselves in open kinesin-coated microfabricated channels that contained reflector arms. Second, and after mild fixation, the oriented seedlings are polymerized into mature microtubules through the addition of tubulin into the solution (labelled green) which preferentially binds to the plus-end (polymerizing end) of the microtubules. (b) Fluorescence image of microtubules that have been grown in the confined space provided by the open channels until the channels were filled with dense networks of microtubules all oriented in the same direction [97]. Scale bar-40 μm.

coated channels that contain reflector arms. Once oriented by self-propelled motion, the seedlings were polymerized into mature microtubules that were confined to grow in the open channels until the channels were filled with dense networks of microtubules all oriented in the same direction [97]. Single kinesins take only a few hundred steps before they fall off, but the walking distance can be greatly increased if the cargo is pulled by more than one motor [98]. Such approaches to fabricating networks of microtubule bundles could be further expanded to engineer future devices that use either the full toolbox of native scaffolding proteins or new scaffolding proteins that target both biological and synthetic cargo.

Nanoengineers would not be the first to harness biological motors to transport their cargo. Various pathogens are known to hijack microtubule or actin-based transport systems within host cells (reviewed in Ref. [99]). *Listeria monocytogenes*, for example, propels itself through the host cell cytoplasm by means of a fast-polymerizing actin filament tail [100]. Likewise, the vaccinia virus, a close relative of smallpox, uses actin polymerization to enhance its cell-to-cell spreading [101], and the alpha herpesvirus hijacks kinesins to achieve long-distance transport along the microtubules of neuronal axons [102]. Signaling molecules and pathogens that cannot alter cell function and behavior by simply passing the outer cell membrane can thus hijack the cytoskeletal highways to get transported from the cell periphery to the nucleus.

8.3.2
Engineering Principle No. 4: active transport of tailored drugs and gene carriers

By taking advantage of the existing cytoskeleton, tailored drugs and gene carriers can be actively transported to the cell nucleus.

Indeed, many viruses [37, 103, 104] as well as nonviral therapeutic gene carriers, such as polyethylenimine/DNA or other polymer-based gene transfer systems (i.e., polyplexes) [105, 106] take advantage of nanomotor-driven transport along microtubule filaments to accelerate their way through the cytoplasm towards the nucleus. Nanomotor-driven transport to the nucleus leads to a much more efficient nuclear localization than could ever be achieved by slow random diffusion through the viscous cytoplasm. Active gene carrier transport can lead to more efficient perinuclear accumulation within minutes [37, 105, 106]. In contrast, nonviral gene carriers that depend solely on random diffusion through the cytoplasm move much more slowly and thus have considerably reduced transfection efficiencies. Understanding how to 'hijack' molecular and cellular transport systems, instead of letting a molecule become a target for endosomal degradation [37, 91], will ultimately allow the design of more efficient drug and gene carrier systems.

8.4
Quality Control

Nanomanufacturing processes, much like macroscopic assembly lines, urgently need procedures that offer precise control over the quality of the product, including the ability to recognize and repair defects. Living systems use numerous quality control procedures to detect and repair defects occurring during the synthesis and assembly of biological nanostructures. As yet, this has not been possible in synthetic nanosystems. Many cellular mechanisms for damage surveillance and error correction rely on nanomotors. Such damage control can occur at two different levels as follows.

8.4.1 Engineering Principle No. 5: error recognition and repair at the molecular level

> *Certain motor proteins recognize assembly mistakes and repair them at the molecular level.*

DNA replication represents one of the most complex sequential assembly processes in a cell. Here, the genetic information stored in the four-base code must be copied with ultra-high precision. Errors generated during replication can have disastrous biological consequences. Figure 8.7 illustrates the built-in mechanism used by the polymerase (DNAp) motor to repair mistakes made during the process of DNA replication [107]. When the DNAp motor misincorporates a base while replicating the template DNA strand, it slows down and switches gears from the polymerase to the exonuclease cycle. Once in exonuclease mode, it will excise the mismatched base pair and then rapidly switch back to the polymerase cycle to resume forward replication. Similar error correction mechanisms, known as 'kinetic proofreading', are conjectured to occur in RNA polymerases and ribosomal machineries [1, 13, 108–113].

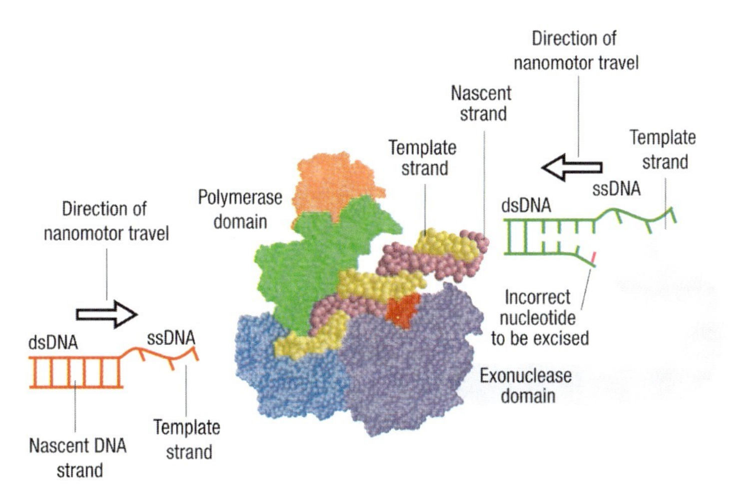

Figure 8.7 Quality control procedures for damage recognition and molecular repair. The DNA polymerase motor (DNAp) contains two active sites. It switches from polymerase (copying) to exonuclease (error correction) activity when it encounters a mismatched base. Mismatched bases are detected as they have weaker bonding interactions—the 'melting' temperature is lower—and this increases the chance of switching from the polymerase to the exonuclease active site [107]. In the exonuclease mode, the motor excises the incorrect base from the nascent DNA strand.

8.4.2
Engineering Principle No. 6: error recognition and repair at the system level

Integrated systems of motors and signaling molecules are needed to recognize and repair damage at the supramolecular level.

Nerve cells have evolved a highly regulated axonal transport system that contains an integrated damage surveillance system [114]. The traffic regulation of motors moving in opposite directions on a microtubule typically occurs in special 'turnaround' zones at the base and tip of an axon [43], but a zone for switching the organelle's direction can also be created when axonal transport is blocked at the site of nerve injury [46] (see Figure 8.2). When irreparable, such blockages are often signatures of neurodegenerative diseases. For example, amyloid precursor protein [47] or tau [115] can give rise to the accumulation of protein aggregates that inhibit anterograde axonal transport, a mechanism potentially implicated in Alzheimer's disease.

At present, there are no synthetic materials that can, in a self-regulated manner, recognize and repair defects at either the molecular or supramolecular level. Molecular recognition and repair is typically attributed to a tightly fitted stereochemical complementarity between binding partners. Nanoscale tools applied to the study of molecular recognition and repair are also elucidating the functional roles of the different structural conformations (and hence three-dimensional shapes) of the motors. For instance, the DNAp motor is in one particular conformation when it binds DNA in its copying (i.e., polymerization) mode and in an entirely different conformation (i.e., the exonuclease mode) when it binds DNA to proofread or excise a mistaken base from the replicated DNA strand [107]. In contrast, damage control at the supramolecular level (e.g., during axonal transport) is achieved by the trafficking of signaling molecules. Deciphering the underlying engineering design principles of damage surveillance and error correction mechanisms in biological systems will inevitably allow better quality-control procedures to be integrated into nanoengineered systems.

8.5
External Control

8.5.1
Engineering Principle No. 7: performance regulation on demand

As with macroscopic engines, external controls can regulate the performance of nanomotors on demand.

Learning how to control and manipulate the performance of nanomotors externally is another critical hurdle in harnessing nanomotors for *ex vivo* applications. By finding or engineering appropriate external knobs in the motor or its environment, its nanoscale movement can be tightly regulated, switched on and off, or otherwise manipulated on demand.

To achieve external control over the nanoscale movement of biological motors, it is important to identify the correct external parameters that can be used to control their dynamics. These external modulators of motor function ('handles') can be either naturally occurring or somehow artificially engineered into the motor to make it susceptible to a particular external control knob or regulator. Because the motion of nanomotors is typically driven by a series of conformational changes in the protein, mechanical load or strain on the motor molecule can also affect the dynamics of the motor. Nanomotors apply mechanical strain to their filaments or substrates as they go through various internal conformational changes. This mechanical strain is intimately related to their dynamics along the substrate and hence their functional performance. Certain interstate transition rates can depend, for example [107], on the amount of intramolecular strain in the motor protein. Applying a mechanical load to a motor perturbs key mechanical transitions in the motor's kinetic pathway, and can thereby affect rates of nucleotide binding, ATP hydrolysis and product release. Single-molecule techniques are beginning to elucidate how mechanical strain on a motor protein might be used to regulate its biological functions (e.g., nanoscale assembly or transport) [13, 55, 107, 116–120].

The single-molecule dynamics of the DNAp motor, as it converts single-stranded (ss) DNA to double-stranded (ds) DNA, has been probed, for example, through the differential elasticity of ssDNA and dsDNA (see Figure 8.8). The T7 DNA polymerase motor replicates DNA at rates of more than 100 bases per second, and this rate steadily decreases with mechanical tension greater than about 5 pN on the DNA template [9]. The motor can work against a maximum of about 34 pN of template tension [9]. The replication rates for the Klenow and Sequenase DNA polymerases also decrease when the ssDNA template tension exceeds 4 pN, and completely ceases at tensions greater than 20 pN [121]. Likewise, single-molecule techniques have allowed direct observation of the RNA polymerase (RNAp) motor moving one base at a time [122], and occasionally pausing and even backtracking [123]. Although RNAp motors are typically five- to tenfold slower than DNAp motors, the effects of DNA template tension on their dynamics are still being investigated [6]. Similarly, ribosome motors, which translate messenger RNA (mRNA) into amino acids at roughly 10 codons per second, have been found to generate about 26.5 ± 1 pN of force [124]. The underlying design principles by which these nanomotors operate are being further elucidated by theoretical models [107, 116, 125–128] that describe nanomachines at a level commensurate with single-molecule data. Furthermore, these molecular assembly machines can be actively directed, driven and controlled by environmental signals [107].

Consequently, an external load or force applied to the substrate or to the motor itself can be used to slow down a motor's action or stall its movement. The stalling forces of kinesin and dynein are 6 and 1 pN, respectively [58, 129]. For example, the binding of two kinesin domains to a microtubule track creates an internal strain in the motor that prevents ATP from binding to the leading motor head. In this way, the two motor domains remain out-of-phase for many mechanochemical cycles and thereby provide an efficient, adaptable mechanism for achieving highly processive movement [130]. Beyond stalling the movement of motors by a mechanical load, other types of perturbations can also influence the dynamics of molecular motors,

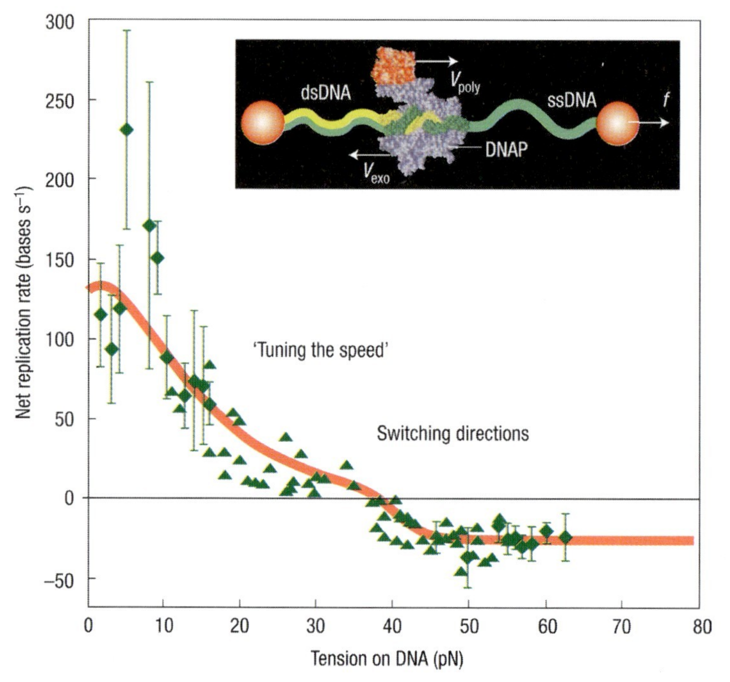

Figure 8.8 Precision control of nanomotors with external control 'knobs'. The net replication rate of a DN Ap motor can be controlled by the mechanical tension on the DNA template strand. Single-molecule data for the motor's force-dependent velocity (two sets of data—diamonds and triangles—are shown, relating to constant force and constant extension measurements) can be described by a network model (red curve) as shown here. The change in net replication rate shows how external controls can change the dynamics of the nanomotor. This model illustrates how environmental control knobs can tune the dynamics of the nanomotor by altering the rate constants associated with its various internal transitions [106]. Tensions between 0 and 35 pN control the net replication rate, whereas tensions above 35 pN actually reverse the velocity of the nanomotor. Inset, experimental setup: a single DNA molecule is stretched between two plastic beads as the motor catalyses the conversion of singlestranded to double-stranded DNA. Figure adapted from Ref. [106].

including the stretching of substrate molecules like DNA [13]. Although this external control over nanomotors has been demonstrated in a few different contexts *ex vivo*, a rich detailed mechanistic understanding of how such external control knobs can modulate the dynamics of the molecular motor is emerging from recent work on the DNA polymerase motor [9, 107, 116, 121, 127, 128, 131].

Remote-controlling the local ATP concentration by the photoactivated release of caged ATP can allow a nanomotor-driven transport system to be accelerated or stopped on demand [84]. External control knobs or regulators can also be engineered into the motors. For instance, point mutations can be introduced into the gene encoding the motor protein, such that it is engineered to respond to light, temperature, pH or other stimuli [43, 85]. Engineering light-sensitive switches into nanomotors enables the rate of ATPase [43, 132] to be regulated, thereby providing an alternate handle for tuning the motor's speed, even while the ATP concentration is kept constant and high. When additional ATP-consuming enzymes are present in solution, the rate of ATP depletion regulates the distance the shuttles move after being activated by a light pulse and before again coming to a halt [84].

Future applications could require that, instead of all the shuttles being moved at the same time, only those in precisely defined locations be activated, on demand. Some of the highly conserved residues within motors help to determine the motor's ATPase rate [43]. Introducing chemical switches near those locations might provide a handle for chemical manipulation of the motor's speed. In fact, this has already been realized for a rotary motor [132] as well as for a linear kinesin motor, where the insertion of a Ca^{2+}-dependent chemical switch makes the ATPase activity steeply dependent on Ca^{2+} concentrations [133]. In addition to caged ATP, caged peptides that block binding sites could be used to regulate the motility of such systems. Caged peptides derived from the kinesin C-terminus domain have already been used to achieve photo control of kinesin-microtubule motility [134]. Instead of modulating the rate of ATP hydrolysis, the access of microtubules to the motor's head domain can also be blocked in an environmentally controlled manner. In fact, temperature has already been shown to regulate the number of kinesins that are accessible while embedded in a surface-bound film of thermoresponsive polymers [135].

The nanomotor-driven assembly of DNA by the DNA polymerase motor provides an excellent example of how precision control over the nanomotor can be achieved by various external knobs in the motor's environment [107, 116, 127, 128]. The DNAp motor moves along the DNA template by cycling through a given sequence of geometric shape changes. The sequence of shapes or internal states of the nanomachine can be denoted by nodes on a simple network [107, 116, 127, 128]. As illustrated in Figure 8.8, this approach elucidates how mechanical tension on a DNA molecule can precisely control (or 'tune') the nanoscale dynamics of the polymerase motor along the DNA track by coupling into key conformational changes of the motor [107].

Macroscopic knobs to precision-control the motor's movement along DNA tracks can be identified by probing how the motor's dynamics vary with each external control knob (varied one at a time). Efforts are currently under way to control even more precisely the movement of these nanomotors along DNA tracks by tightly controlling the parameters in the motor's environment (see www.nanobiosym.com). Concepts of fine-tuning and robustness could also be extended to describe the sensitivity of other nanomotors (modelled as simple biochemical networks) to various external control parameters [107]. Furthermore, such a network approach [107] provides experimentally testable predictions that could aid the design of future molecular-scale manufacturing methods that integrate nanomotor-driven assembly schemes. External control of these nanomotors will be critical in harnessing them for nanoscale manufacturing applications.

8.6
Concluding Remarks

We have reviewed several key engineering design principles that enable nanomotors moving along linear templates to perform a myriad of tasks. Equally complex biomimetic tasks have not yet been mastered *ex vivo*, either by harnessing biological motors or via synthetic analogues. Engineering insights into how such tasks are carried out by the biological nanosystems will inspire new technologies that harness nanomotor-driven processes to build new systems for nanoscale transport and assembly.

Sequential assembly and nanoscale transport, combined with features currently attributed only to biological materials, such as self-repair and healing, might one day become an integral part of future materials and biohybrid devices. In the near term, molecular biology techniques could be used to synthesize and assemble nanoelectronic components with more control (www.cambrios.com; see also Ref. [29]). Numerous proof-of-concept experiments using nanomotors integrated into synthetic microdevices have already been demonstrated (for reviews, see Refs. [74, 136]). Among many others, these applications include stretching surface-bound molecules by moving microtubules [87, 90]; probing the lifetime of a single receptor–ligand interaction via a cantilevered microtubule that acts as a piconewton force sensor [85]; topographic surface imaging by self-propelled probes [70]; and cargo pick-up from loading stations [88] as illustrated in Figure 8.5.

Although much progress is being made in the synthesis of artificial motors (see Ref. [137]), it has been difficult, in practice, to synthesize artificial motors that come even close in performance to their natural counterparts (see Ref. [39]). Harnessing biological motors to perform nanoscale manufacturing tasks might thus be the best near-term strategy. Although many individual nanoparts can be easily manufactured, the high-throughput assembly of these nanocomponents into complex structures is still nontrivial. At present, no *ex vivo* technology exists that can actively guide such nanoscale assembly processes. Despite advances in deciphering the underlying engineering design principles of nanomotors, many hurdles still impede harnessing them for *ex vivo* transport and sequential assembly in nanosystems. Although the use of biological nanomotors puts intrinsic constraints on the conditions under which they can be assembled and used in biohybrid devices, many of their sophisticated tasks are still poorly mimicked by synthetic analogues. Understanding the details of how these little nanomachines convert chemical energy into controlled movements will nevertheless inspire new approaches to engineer synthetic counterparts that might some day be used under harsher conditions, operate at more extreme temperatures, or simply have longer shelf lives.

Certain stages of the materials production process might one day be replaced by nanomotor-driven sequential self-assembly, allowing much more control at the molecular level. Biological motors are already being used to drive the efficient fabrication of complex nanoscopic and mesoscopic structures, such as nanowires [31] and supramolecular assemblies. Techniques for precision control of nanomotors that read DNA are also being used to engineer integrated systems for rapid DNA detection and analysis (www.nanobiosym.com). The specificity and control of assembly and transport shown by biological systems offers many opportunities to those interested in assembly of complex nanosystems. Most importantly, the intricate schemes of proofreading and damage repair—features that have not yet been realized in any manmade nanosystems—should provide inspiration for those interested in producing synthetic systems capable of similarly complex tasks.

Acknowledgments

We thank Sheila Luna, Christian Brunner and Jennifer Wilson for the artwork, and all of our collaborators who contributed thoughts and experiments. At the same time, we apologize to all authors whose work we could not cite owing to space limitations.

Correspondence and requests for materials should be addressed to A.G. or V.V.

References

1 Rodnina, M.V. and Wintermeyer, W. (2001) Fidelity of aminoacyl-tRNA selection on the ribosome: kinetic and structural mechanisms. *Annual Review of Biochemistry*, **70**, 415–435.

2 Kunkel, T.A. (2004) DNA replication fidelity. *The Journal of Biological Chemistry*, **279**, 16895–16898.

3 Erie, D.A., Hajiseyedjavadi, O., Young, M.C. and von Hippel, P.H. (1993) Multiple RNA polymerase conformations and GreA: control of the fidelity of transcription. *Science*, **262**, 867.

4 Liu, D.R., Magliery, T.J., Pastrnak, M. and Schultz, P.G. (1997) Engineering a tRNA and aminoacyl-tRNA synthetase for the site-specific incorporation of unnatural amino acids into proteins in vivo. *Proceedings of the National Academy of Sciences of the United States of America*, **94**, 10092–10097.

5 Bustamante, C., Smith, S.B., Liphardt, J. and Smith, D. (2000) Single-molecule studies of DNA mechanics. *Current Opinion in Structural Biology*, **10**, 279–285.

6 Davenport, R.J., Wuite, G.J.L., Landick, R. and Bustamante, C. (2000) Single-molecule study of transcriptional pausing and arrest by *E. coli* RNA polymerase. *Science*, **287**, 2497–2500.

7 Greulich, K.O. (2005) Single-Molecule Studies on DNA and RNA. *ChemPhysChem*, **6**, 2459–2471.

8 Wang, M.D. *et al.* (1998) Force and velocity measured for single molecules of RNA polymerase. *Science*, **282**, 902–907.

9 Wuite, G.J., Smith, S.B., Young, M., Keller, D. and Bustamante, C. (2000) Single-molecule studies of the effect of template tension on T7 DNA polymerase activity. *Nature*, **404**, 103–106.

10 Smith, S.B., Cui, Y. and Bustamante, C. (1996) Overstretching B-DNA: The elastic response of individual double-stranded and single-stranded DNA molecules. *Science*, **271**, 795.

11 Smith, S.B., Finzi, L. and Bustamante, C. (1992) Direct mechanical measurements of the elasticity of single DNA molecules by using magnetic beads. *Science*, **258**, 1122.

12 Williams, M.C. and Rouzina, I. (2002) Force spectroscopy of single DNA and RNA molecules. *Current Opinion in Structural Biology*, **12**, 330–336.

13 Bustamante, C., Bryant, Z. and Smith, S.B. (2003) Ten years of tension: single-molecule DNA mechanics. *Nature*, **421**, 423–427.

14 Jeney, S., Stelzer, E.H., Grubmuller, H. and Florin, E.L. (2004) Mechanical properties of single motor molecules studied by three-dimensional thermal force probing in optical tweezers. *ChemPhysChem*, **5**, 1150–1158.

15 Mehta, A.D. (1999) Single-molecule biomechanics with optical methods. *Science*, **283**, 1689–1695.

16 Mogilner, A. and Oster, G. (2003) Polymer motors: pushing out the front and pulling up the back. *Current Biology*, **13**, 721–733.

17 Schnitzer, M.J., Visscher, K. and Block, S.M. (2000) Force production by single

kinesin motors. *Nature Cell Biology*, **2**, 718–723.

18 Strick, T., Allemand, J.F., Croquette, V. and Bensimon, D. (2001) The manipulation of single biomolecules. *Physics Today*, **54**, 46–51.

19 Ha, T. (2001) Single-molecule fluorescence methods for the study of nucleic acids. *Current Opinion in Structural Biology*, **11**, 287–292.

20 Kapanidis, A.N. *et al.* (2006) Initial transcription by RNA polymerase proceeds through a DNA scrunching mechanism. *Science*, **314**, 1144–1147.

21 Keller, R.A. *et al.* (1996) Single-molecule fluorescence analysis in solution. *Applied Spectroscopy*, **50**, 12A–32A.

22 Mullis, K.B. (1993) *The Polymerase Chain Reaction*, Nobel Lecture.

23 van Hest, J.C.M. and Tirrell, D.A. (2001) Protein-based materials, toward a new level of structural control. *Chemical Communications*, **19**, 1897–1904.

24 Fodor, S.P. *et al.* (1993) Multiplexed biochemical assays with biological chips. *Nature*, **364**, 555–556.

25 Merrifield, R.B. (1965) Automated synthesis of peptides. *Science*, **150**, 178–185.

26 Ratner, D.M., Swanson, E.R. and Seeberger, P.H. (2003) Automated synthesis of a protected N-linked glycoprotein core pentasaccharide. *Organic Letters*, **5**, 4717–4720.

27 Ball, P. (2001) It all falls into place. *Nature*, **413**, 667–668.

28 Pavel, I.S. (2005) *Assembly of gold nanoparticles by ribosomal molecular machines*, PhD thesis, Univ. Texas at Austin.

29 Whaley, S.R., English, D.S., Hu, E.L., Barbara, P.F. and Belcher, A.M. (2000) Selection of peptides with semiconductor binding specificity for directed nanocrystal assembly. *Nature*, **405**, 665–668.

30 Chen, H.L. and Goel, A. (2005) in *DNA Computing. Lecture Notes in Computer Science*, Vol. 3384 Springer, Berlin/Heidelberg, pp. 62–75.

31 Hess, H. *et al.* (2005) Molecular self-assembly of 'nanowires'and 'nanospools' using active transport. *Nano Letters*, **5**, 629–633.

32 Winfree, E. and Bekbolatov, R. (2004) in *DNA Computing. Lecture Notes in Computer Science*, Vol 2943 Springer, Berlin/Heidelberg, pp. 126–144.

33 Choi, I.S., Bowden, N. and Whitesides, G.M. (1999) Macroscopic, hierarchical, two-dimensional self-assembly. *Angewandte Chemie (International Edition in English)*, **38**, 3078–3081.

34 Whitesides, G.M. and Boncheva, M. (2002) Beyond molecules: self-assembly of mesoscopic and macroscopic components. *Proceedings of the National Academy of Sciences of the United States of America*, **99**, 4769–4774.

35 Caviston, J.P. and Holzbaur, E.L. (2006) Microtubule motors at the intersection of trafficking and transport. *Trends in Cell Biology*, **16**, 530–537.

36 Howard, J. (2001) *Mechanics of Motor Proteins and the Cytoskeleton*, Sinauer, Sunderland, Massachusetts.

37 Lakadamyali, M., Rust, M. and Zhuang, X. (2006) Ligands for clathrin-mediated endocytosis are differentially sorted into distinct populations of early endosomes. *Cell*, **124**, 997–1009.

38 Lakadamyali, M., Rust, M.J., Babcock, H.P. and Zhuang, X. (2003) Visualizing infection of individual influenza viruses. *Proceedings of the National Academy of Sciences of the United States of America*, **100**, 9280–9285.

39 Månsson, A. and Linke, H. (2007) Controlled Nanoscale Motion. *Proc. Nobel Symp. 131*, Vol 711, Springer, Berlin.

40 Miki, H., Okada, Y. and Hirokawa, N. (2005) Analysis of the kinesin superfamily: insights into structure and function. *Trends in Cell Biology*, **15**, 467–476.

41 Rust, M.J., Lakadamyali, M., Zhang, F. and Zhuang, X. (2004) Assembly of endocytic machinery around individual influenza viruses during viral entry. *Nature Structural & Molecular Biology*, **11**, 567–573.

42 Sotelo-Silveira, J.R., Calliari, A., Kun, A., Koenig, E. and Sotelo, J.R. (2006) RNA trafficking in axons. *Traffic (Copenhagen, Denmark)*, **7**, 508–515.

43 Vale, R.D. (2003) The molecular motor toolbox for intracellular transport. *Cell*, **112**, 467–480.

44 Vallee, R.B. and Sheetz, M.P. (1996) Targeting of motor proteins. *Science*, **271**, 1539–1544.

45 Vale, R.D., Reese, T.S. and Sheetz, M.P. (1985) Identification of a novel force-generating protein, kinesin, involved in microtubule-based motility. *Cell*, **42**, 39–50.

46 Guzik, B.W. and Goldstein, L.S. (2004) Microtubule-dependent transport in neurons: steps towards an understanding of regulation, function and dysfunction. *Current Opinion in Cell Biology*, **16**, 443–450.

47 Gunawardena, S. and Goldstein, L.S. (2001) Disruption of axonal transport and neuronal viability by amyloid precursor protein mutations in Drosophila. *Neuron*, **32**, 389–401.

48 Gunawardena, S. and Goldstein, L.S. (2004) Cargo-carrying motor vehicles on the neuronal highway: transport pathways and neurodegenerative disease. *Journal of Neurobiology*, **58**, 258–271.

49 Mandelkow, E. and Mandelkow, E.-M. (2002) Kinesin motors and disease. *Trends in Cell Biology*, **12**, 585–591.

50 Ström, A.L. *et al.* (23 April 2008) Retrograde axonal transport and motor neuron disease. *Journal of Neurochemistry*, Preprint at ⟨http://www.ncbi.nlm.nih.gov/pubmed/18384644⟩

51 Mukhopadhyay, R. and Hoh, J.H. (2001) AFM force measurements on microtubule-associated proteins: the projection domain exerts a long-range repulsive force. *FEBS Letters*, **505**, 374–378.

52 Mizuno, N. *et al.* (2004) Dynein and kinesin share an overlapping microtubule-binding site. *The EMBO Journal*, **23**, 2459–2467.

53 Vale, R.D. and Toyoshima, Y.Y. (1988) Rotation and translocation of microtubules in vitro induced by dyneins from Tetrahymena cilia. *Cell*, **52**, 459–469.

54 Wang, Z., Khan, S. and Sheetz, M.P. (1995) Single cytoplasmic dynein molecule movements: characterization and comparison with kinesin. *Biophysical Journal*, **69**, 2011–2023.

55 Reck-Peterson, S.L. *et al.* (2006) Single-molecule analysis of dynein processivity and stepping behavior. *Cell*, **126**, 335–348.

56 Seitz, A. and Surrey, T. (2006) Processive movement of single kinesins on crowded microtubules visualized using quantum dots. *The EMBO Journal*, **25**, 267–277.

57 Coppin, C.M., Finer, J.T., Spudich, J.A. and Vale, R.D. (1996) Detection of sub-8-nm movements of kinesin by high-resolution optical-trap microscopy. *Proceedings of the National Academy of Sciences of the United States of America*, **93**, 1913–1917.

58 Svoboda, K., Schmidt, C.F., Schnapp, B.J. and Block, S.M. (1993) Direct observation of kinesin stepping by optical trapping interferometry. *Nature*, **365**, 721–727.

59 Diehl, M.R., Zhang, K., Lee, H.J. and Tirrell, D.A. (2006) Engineering cooperativity in biomotor-protein assemblies. *Science*, **311**, 1468–1471.

60 Hunt, A.J. and Howard, J. (1993) Kinesin swivels to permit microtubule movement in any direction. *Proceedings of the National Academy of Sciences of the United States of America*, **90**, 11653–11657.

61 Clemmens, J. *et al.* (2003) Principles of microtubule guiding on microfabricated kinesin-coated surfaces: chemical and topographic surface patterns. *Langmuir*, **19**, 10967–10974.

62 Reuther, C., Hajdo, L., Tucker, R., Kasprzak, A.A. and Diez, S. (2006) Biotemplated nanopatterning of planar surfaces with molecular motors. *Nano Letters*, **6**, 2177–2183.

63 Clemmens, J., Hess, H., Howard, J. and Vogel, V. (2003) Analysis of microtubule guidance by microfabricated channels coated with kinesin. *Langmuir*, **19**, 1738–1744.

64 Hess, H. *et al.* (2003) Molecular shuttles operating undercover: a new photolithographic approach for the fabrication of structured surfaces supporting directed motility. *Nano Letters*, **3**, 1651–1655.

65 Hiratsuka, Y., Tada, T., Oiwa, K., Kanayama, T. and Uyeda, T.Q. (2001) Controlling the direction of kinesindriven microtubule movements along microlithographic tracks. *Biophysical Journal*, **81**, 1555–1561.

66 Moorjani, S.G., Jia, L., Kackson, T.N. and Hancock, W.O. (2003) Lithographically patterned channels spatially segregate kinesin motor activity and effectively guide microtubule movements. *Nano Letters*, **3**, 633–637.

67 Bunk, R. *et al.* (2003) Actomyosin motility on nanostructured surfaces. *Biochemical and Biophysical Research Communications*, **301**, 783–788.

68 Sundberg, M. *et al.* (2006) Actin filament guidance on a chip: toward high-throughput assays and lab-on-a-chip applications. *Langmuir*, **22**, 7286–7295.

69 Clemmens, J. *et al.* (2004) Motor-protein 'roundabouts': microtubules moving on kinesin-coated tracks through engineered networks. *Lab Chip*, **4**, 83–86.

70 Hess, H., Clemmens, J., Howard, J. and Vogel, V. (2002) Surface imaging by self-propelled nanoscale probes. *Nano Letters*, **2**, 113–116.

71 Stracke, R., Böhm, K.J., Burgold, J., Schacht, H..-J. and Unger, E. (2000) Physical and technical parameters determining the functioning of a kinesin-based cell-free motor system. *Nanotechnology*, **11**, 52–56.

72 Brown, T.B. and Hancock, W.O. (2005) A polarized microtubule array for kinesin-powered nanoscale assembly and force generation. *Nano Letters*, **28**, 571–576.

73 Nitta, T., Tanahashi, A., Hirano, M. and Hess, H. (2006) Simulating molecular shuttle movements: towards computer-aided design of nanoscale transport systems. *Lab Chip*, **6**, 881–885.

74 Vogel, V. and Hess, H. (2007) in *Lecture Notes Proceedings Nobel Symposium*, Vol. 711 Springer, Berlin/Heidelberg, pp. 367–383.

75 Stracke, R., Bohm, K.J., Wollweber, L., Tuszynski, J.A. and Unger, E. (2002) Analysis of the migration behaviour of single microtubules in electric fields. *Biochemical and Biophysical Research Communications*, **293**, 602–609.

76 van den Heuvel, M.G., de Graaff, M.P. and Dekker, C. (2006) Molecular sorting by electrical steering of microtubules in kinesin-coated channels. *Science*, **312**, 910–914.

77 Darnton, N., Turner, L., Breuer, K. and Berg, H.C. (2004) Moving fluid with bacterial carpets. *Biophysical Journal*, **86**, 1863–1870.

78 Hiratsuka, Y., Miyata, M., Tada, T. and Uyeda, T.Q. (2006) A microrotary motor powered by bacteria. *Proceedings of the National Academy of Sciences of the United States of America*, **103**, 13618–13623.

79 Bachand, G.D., Rivera, S.B., Carroll-Portillo, A., Hess, H. and Bachand, M. (2006) Active capture and transport of virus particles using a biomolecular motor-driven, nanoscale antibody sandwich assay. *Small*, **2**, 381–385.

80 Hirabayashi, M. *et al.* (2006) Malachite green-conjugated microtubules as mobile bioprobes selective for malachite green aptamers with capturing/releasing ability. *Biotechnology and Bioengineering*, **94**, 473–480.

81 Martin, B.D. *et al.* (2006) An engineered virus as a bright fluorescent tag and scaffold for cargo proteins: capture and transport by gliding microtubules. *Journal of Nanoscience and Nanotechnology*, **6**, 2451–2460.

82 Muthukrishnan, G., Hutchins, B.M., Williams, M.E. and Hancock, W.O. (2006) Transport of semiconductor nanocrystals by kinesin molecular motors. *Small*, **2**, 626–630.

83 Taira, S. *et al.* (2006) Selective detection and transport of fully matched DNA by DNA-loaded microtubule and kinesin motor protein. *Biotechnology and Bioengineering*, **95**, 533–538.

84 Hess, H., Clemmens, J., Qin, D., Howard, J. and Vogel, V. (2001) Light-controlled molecular shuttles made from motor proteins carrying cargo on engineered surfaces. *Nano Letters*, **1**, 235–239.

85 Hess, H., Howard, J. and Vogel, V. (2002) A piconewton forcemeter assembled from microtubules and kinesins. *Nano Letters*, **2**, 1113–1115.

86 Boal, A.K., Bachand, G.D., Rivera, S.B. and Bunker, B.C. (2006) Interactions between cargo-carrying biomolecular shuttles. *Nanotechnology*, **17**, 349–354.

87 Diez, S. *et al.* (2003) Stretching and transporting DNA molecules using motor proteins. *Nano Letters*, **3**, 1251–1254.

88 Brunner, C., Wahnes, C. and Vogel, V. (2007) Cargo pick-up from engineered loading stations by kinesin driven molecular shuttles. *Lab on a Chip*, **7**, 1263–1271.

89 Ramachandran, S., Ernst, K.H., Bachand, G.D., Vogel, V. and Hess, H. (2006) Selective loading of kinesin-powered molecular shuttles with protein cargo and its application to biosensing. *Small*, **2**, 330.

90 Dinu, C.Z. *et al.* (2006) Parallel manipulation of bifunctional DNA molecules on structured surfaces using kinesin-driven microtubules. *Small*, **2**, 1090–1098.

91 Soldati, T. and Schliwa, M. (2006) Powering membrane traffic in endocytosis and recycling. *Nature Reviews. Molecular Cell Biology*, **7**, 897–908.

92 Bachand, M., Trent, A.M., Bunker, B.C. and Bachand, G.D. (2005) Physical factors affecting kinesin-based transport of synthetic nanoparticle cargo. *Journal of Nanoscience and Nanotechnology*, **5**, 718–722.

93 Du, Y.Z. *et al.* (2005) Motor protein nano-biomachine powered by self-supplying ATP. *Chemical Communications*, 2080–2082.

94 Kufer, S.K., Puchner, E.M., Gumpp, H., Liedl, T. and Gaub, H.E. (2008) Single-molecule cut-and-paste surface assembly. *Science*, **319**, 594–596.

95 Chakravarty, A., Howard, L. and Compton, D.A. (2004) A mechanistic model for the organization of microtubule asters by motor and non-motor proteins in a mammalian mitotic extract. *Molecular Biology of the Cell*, **15**, 2116–2132.

96 Surrey, T., Nedelec, F., Leibler, S. and Karsenti, E. (2001) Physical properties determining self-organization of motors and microtubules. *Science*, **292**, 1167–1171.

97 Doot, R.K., Hess, H. and Vogel, V. (2007) Engineered networks of oriented microtubule filaments for directed cargo transport. *Soft Matter*, **3**, 349–356.

98 Beeg, J. *et al.* (2008) Transport of beads by several kinesin motors. *Biophysical Journal*, **94**, 532.

99 Henry, T., Gorvel, J.P. and Meresse, S. (2006) Molecular motors hijacking by intracellular pathogens. *Cellular Microbiology*, **8**, 23–32.

100 Soo, F.S. and Theriot, J.A. (2005) Large-scale quantitative analysis of sources of variation in the actin polymerization-based movement of *Listeria monocytogenes*. *Biophysical Journal*, **89**, 703–723.

101 Rietdorf, J. *et al.* (2001) Kinesin-dependent movement on microtubules precedes actin-based motility of vaccinia virus. *Nature Cell Biology*, **3**, 992–1000.

102 Smith, G.A., Gross, S.P. and Enquist, L.W. (2001) Herpes viruses use bidirectional fast-axonal transport to spread in sensory neurons. *Proceedings of the National*

Academy of Sciences of the United States of America, **98**, 3466–3470.

103 Döhner, K., Nagel, C.-H. and Sodeik, B. (2005) Viral stop-and-go along microtubules: taking a ride with dynein and kinesins. *Trends in Microbiology*, **13**, 320–327.

104 Sodeik, B. (2002) Unchain my heart, baby let me go: the entry and intracellular transport of HIV. *Cell Biol*, **159**, 393–395.

105 Kulkarni, R.P., Wu, D.D., Davis, M.E. and Fraser, S.E. (2005) Quantitating intracellular transport of polyplexes by spatio-temporal image correlation spectroscopy. *Proceedings of the National Academy of Sciences of the United States of America*, **102**, 7523–7528.

106 Suh, J., Wirtz, D. and Hanes, J. (2003) Efficient active transport of gene nanocarriers to the cell nucleus. *Proceedings of the National Academy of Sciences of the United States of America*, **100**, 3878–3882.

107 Goel, A., Astumian, R.D. and Herschbach, D. (2003) Tuning and switching a DNA polymerase motor with mechanical tension. *Proceedings of the National Academy of Sciences of the United States of America*, **100**, 9699–9704.

108 Donlin, M.J., Patel, S.S. and Johnson, K.A. (1991) Kinetic partitioning between the exonuclease and polymerase sites in DNA error correction. *Biochemistry*, **30**, 538–546.

109 Fersht, A.R., Knill-Jones, J.W. and Tsui, W.C. (1982) Kinetic basis of spontaneous mutation. Misinsertion frequencies, proofreading specificities and cost, of proofreading by DNA polymerases of *Escherichia coli*. *Journal of Molecular Biology*, **156**, 37–51.

110 Hopfield, J.J. (1974) Kinetic proofreading: a new mechanism for reducing errors in biosynthetic processes requiring high specificity. *Proceedings of the National Academy of Sciences of the United States of America*, **71**, 4135–4139.

111 Hopfield, J.J. (1980) The energy relay: a proofreading scheme based on dynamic cooperativity and lacking all characteristic symptoms of kinetic proofreading in DNA replication and protein synthesis. *Proceedings of the National Academy of Sciences of the United States of America*, **77**, 5248.

112 Rodnina, M.V. and Wintermeyer, W. (2001) Ribosome fidelity: tRNA discrimination, proofreading and induced fit. *Trends in Biochemical Sciences*, **26**, 124–130.

113 Wang, D. and Hawley, D.K. (1993) Identification of a $3' \rightarrow 5'$exonuclease activity associated with human RNA polymerase II. *Proceedings of the National Academy of Sciences of the United States of America*, **90**, 843–847.

114 Cavalli, V., Kujala, P., Klumperman, J. and Goldstein, L.S. (2005) Sunday Driver links axonal transport to damage signaling. *The Journal of Cell Biology*, **168**, 775–787.

115 Mandelkow, E.M., Stamer, K., Vogel, R., Thies, E. and Mandelkow, E. (2003) Clogging of axons by tau, inhibition of axonal traffic and starvation of synapses. *Neurobiology of Aging*, **24**, 1079–1085.

116 Goel, A., Ellenberger, T., Frank-Kamenetskii, M.D. and Herschbach, D. (2002) Unifying themes in DNA replication: reconciling single molecule kinetic studies with structural data on DNA polymerases. *Journal of Biomolecular Structure & Dynamics*, **19**, 571–584.

117 Guydosh, N.R. and Block, S.M. (2006) Backsteps induced by nucleotide analogs suggest the front head of kinesin is gated by strain. *Proceedings of the National Academy of Sciences of the United States of America*, **103**, 8054–8059.

118 Spudich, J. (2006) Molecular motors take tension in stride. *Cell*, **126**, 242–244.

119 Vale, R.D. and Milligan, R.A. (2000) The way things move: looking under the hood of molecular motor proteins. *Science*, **288**, 88–95.

120 Veigel, C., Schmitz, S., Wang, F. and Sellers, J.R. (2005) Load-dependent kinetics of myosin-V can explain its high processivity. *Nature Cell Biology*, **7**, 861–869.

121 Maier, B., Bensimon, D. and Croquette, V. (2000) Replication by a single DNA polymerase of a stretched single-stranded DNA. *Proceedings of the National Academy of Sciences of the United States of America*, **97**, 12002–12007.

122 Abbondanzieri, E.A., Greenleaf, W.J., Shaevitz, J.W., Landick, R. and Block, S.M. (2005) Direct observation of base-pair stepping by RNA polymerase. *Nature*, **438**, 460–465.

123 Shaevitz, J.W., Abbondanzieri, E.A., Landick, R. and Block, S. (2003) Backtracking by single RNA polymerase molecules observed at near-base pair resolution. *Nature*, **426**, 684–687.

124 Sinha, D.K., Bhalla, U.S. and Shivashankar, G.V. (2004) Kinetic measurement of ribosome motor stalling force. *Applied Physics Letters*, **85**, 4789–4791.

125 Astumian, R.D. (1997) Thermodynamics and kinetics of a Brownian motor. *Science*, **276**, 917–922.

126 Bustamante, C., Keller, D. and Oster, G. (2001) The physics of molecular motors. *Accounts of Chemical Research*, **34**, 412–420.

127 Goel, A., Frank-Kamenetskii, M.D., Ellenberger, T. and Herschbach, D. (2001) Tuning DNA 'strings': modulating the rate of DNA replication with mechanical tension. *Proceedings of the National Academy of Sciences of the United States of America*, **98**, 8485–8489.

128 Goel, A. and Herschbach, D.R. (2003) Controlling the speed and direction of molecular motors that replicate DNA. *Proc SPIE*, **5110**, 63–68.

129 Mallik, R., Carter, B.C., Lex, S.A., King, S.J. and Gross, S.P. (2004) Cytoplasmic dynein functions as a gear in response to load. *Nature*, **427**, 649–652.

130 Rosenfeld, S.S., Fordyce, P.M., Jefferson, G.M., King, P.H. and Block, S.M. (2003) Stepping and stretching. How kinesin uses internal strain to walk processively. *The Journal of Biological Chemistry*, **278**, 18550–18556.

131 Andricioaei, I., Goel, A., Herschbach, D. and Karplus, M. (2004) Dependence of DNA polymerase replication rate on external forces: a model based on molecular dynamics simulations. *Biophysical Journal*, **87**, 1478–1497.

132 Liu, H. *et al.* (2002) Control of a biomolecular motor-powered nanodevice with an engineered chemical switch. *Nature Mater*, **1**, 173–177.

133 Konishi, K., Uyeda, T.Q. and Kubo, T. (2006) Genetic engineering of a Ca(2 +) dependent chemical switch into the linear biomotor kinesin. *FEBS Letters*, **580**, 3589–3594.

134 Nomura, A., Uyeda, T.Q., Yumoto, N. and Tatsu, Y. (2006) Photo-control of kinesin-microtubule motility using caged peptides derived from the kinesin C-terminus domain. *Chemical Communications*, **1**, 3588–3590.

135 Ionov, L., Stamm, M. and Diez, S. (2006) Reversible switching of microtubule motility using thermoresponsive polymer surfaces. *Nano Letters*, **6**, 1982–1987.

136 van den Heuvel, M.G.L. and Dekker, C. (2007) Motor proteins at work for nanotechnology. *Science*, **317**, 333–336.

137 Browne, W.R. and Feringa, B.L. (2006) Making molecular machines work. *Nature Nanotechnology*, **1**, 25–35.

138 Hess, H. *et al.* (2002) Ratchet patterns sort molecular shuttles. *Applied Physics A*, **75**, 309–313.

Part Four:
Innovative Disease Treatments and Regenerative Medicine

Nanotechnology, Volume 5: Nanomedicine. Edited by Viola Vogel

ISBN: 978-3-527-31736-3

9
Mechanical Forces Matter in Health and Disease: From Cancer to Tissue Engineering

Viola Vogel and Michael P. Sheetz

9.1
Introduction: Mechanical Forces and Medical Indications

One of our earliest experiences showing that mechanical forces matter goes back to when we got our first blisters. Excessive friction causes a tear between the upper layer of the skin – the epidermis – and the layers beneath. When these skin layers – which in healthy skin are held together by cell–cell adhesion complexes – begin to separate, the resultant pocket fills with serum or blood. In some people, who have inherited skin diseases, the blistering occurs much more easily, and studies of point mutations that cause easy blistering have provided considerable insights into the underlying molecular mechanisms. Molecular defects can exist in different intracellular and extracellular proteins that are responsible for weakening the mechanical strength of cell–cell adhesions. The proteins implicated by genetic analysis include keratins, laminins, collagens and integrins [1–3]. Unfortunately, exactly how mutations in these proteins regulate the mechanical stability of the linkages that cells form with their environment remains unknown.

Mechanical forces acting on cells also affect our lives in many other, often unexpected, ways. Regular exercise, for example, not only strengthens our body tone but also offers protection against mortality by delaying the onset of various diseases. It is thought that physical training reduces the chance of chronic heart diseases, atherosclerosis and also type 2 diabetes [4]. But how can exercise have such a profound impact on so many diseases? Chronic low-grade systemic inflammation is a feature of these and many other chronic diseases that have been correlated with elevated levels of several cytokines [5–7]. By yet unknown mechanisms, it is suggested that regular exercise induces anti-inflammatory processes, thus suppressing the production of pro-inflammatory signaling proteins [5, 8].

Many more severe diseases for which we do not have cures also have a mechanical origin, or show abnormalities in cellular mechanoresponses. These range from cancer to cardiovascular disorders, from osteoporosis to other aging-related diseases. In the case of many cancers, the cells grow inappropriately and with the wrong mechanoresponse, which in turn destroys normal tissue mechanics and often also

Nanotechnology, Volume 5: Nanomedicine. Edited by Viola Vogel

ISBN: 978-3-527-31736-3

tissue function [9–11]. While cardiovascular diseases have many forms, cardiac hypertrophy, plaque formation and heart repair are obvious cases where mechanosensory functions are important [12–14]. Abnormal mechanical forces can trigger an aberrant proliferation of endothelial and smooth muscle cells, as observed in the progression of vascular diseases such as atherosclerosis [15]. There is furthermore emerging evidence that immune synapse formation is a mechanically driven process [16]. Finally, damaged tissue is often repaired by new cells that differentiate from pluripotent cells to finally replace and regenerate the damaged regions. Successful healing includes re-establishing the proper mechanical tissue characteristics; even bioscaffolds that are used in reconstructive surgery heal best if they are mechanically exercised [14, 17]. Thus, from molecules to tissue, although the mechanical aspects are recognized as being critical, relatively little has yet been done to correlate mechanical effects with biochemical signal changes, and how these impact clinical outcomes.

There is, therefore, overwhelming evidence that physical and not just biochemical stimuli matter in tissue growth and repair in health and disease. But how do cells sense mechanical forces? A complete answer cannot yet be given as too few techniques have been available in the past to explore this question. However, the broad availability of nanoanalytical and nanomanipulation tools is beginning to have impact. This tool chest provides novel opportunities to decipher how physical and biochemical factors, in combination, can orchestrate the hierarchical control of cell and tissue functions (we will illustrate this point with some concrete examples later in the chapter). The diversity of biological forms in different organisms most likely belies a wide range of mechanosensing mechanisms that are specifically engineered to provide the desired morphology.

To summarize, based on the progress that has been made recently in the field of cell biomechanics, it is now clear that individual cells are dramatically affected in their functions, from growth to differentiation, by the mechanical properties of their environments and by externally applied forces (for reviews, see Refs [18–21]). But the question remains: How do cells sense mechanical forces, and how are mechanical stimuli translated at the nanoscale into biochemical signal changes that ultimately regulate cell function? A few examples are illustrated here of how physical junctions are formed between cells and their environment, how mechanical forces acting on molecules associated with junctions regulate their functional states, and what the downstream implications might be on cell signaling events. Considering the complexity of the puzzle, this chapter cannot provide a comprehensive review; rather, we will focus on describing a few selected molecular players involved in mechanochemical signal conversion, followed by a discussion of the associated signaling pathways and subsequent cellular responses, and then concluding on the role of physical stimuli in various diseases. Once the molecular pathways are identified, and the mechanisms deciphered by which force regulates diverse cell functions, the development of new drugs and therapies will surely emerge. In particular, it is expected that in the future, a number of diseases associated with altered mechanoresponses will be resolved more efficiently by treating the source of the problem, rather than the symptoms.

9.2
Force-Bearing Protein Networks Hold the Tissue Together

The search for proteins that are structurally altered if mechanically stretched, and which could thus serve as force sensors for cells, should start in junctions between cells and their environments. The focus should first be on the junctions that experience the highest tensile forces. The force-bearing elements in tissues are typically the cytoskeleton and extracellular matrix (ECM) fibers, and all the proteins that physically link the cell interior to the exterior. For different tissues, the major force-bearing elements can differ.

9.2.1
Cell–Cell Junctions

Some tissues such as epithelial and endothelial cell layers have barrier functions (Figure 9.1), where the majority of the force is born by the tight cell–cell junctions. These junctions couple the cell–cell adhesion molecules (cadherins) that hold the cells together to the cytoplasmic proteins that ultimately link cadherins to the actin, myosin and intermediate filaments in the cytoskeleton [22].

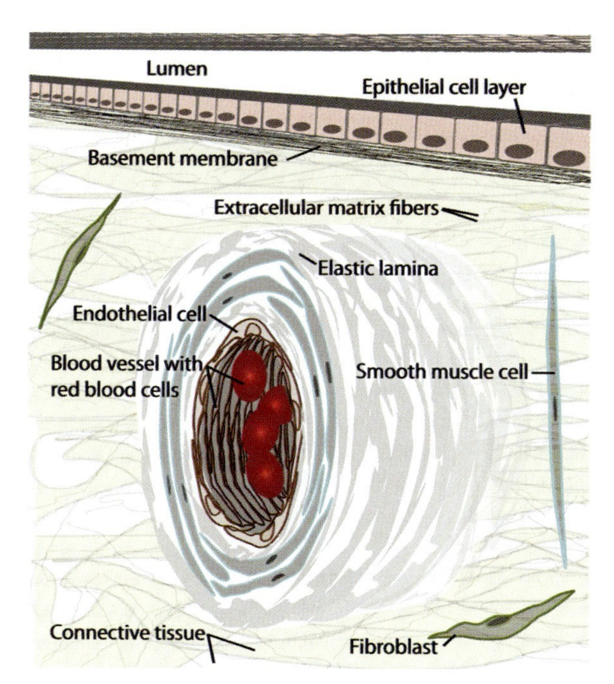

Figure 9.1 Schematic section through tissue, showing how layers of cells form a barrier that separates the connective tissue from the lumen of the lung or of the intestines (epithelial cells), or the blood vessels (endothelial cells). Within endothelial and epithelial cell layers, the cells are tightly connected to each other via cadherin junctions, while integrin junctions anchor cells to the basement membrane as well as to the extracellular matrix (ECM) of connective tissues.

Epithelial tissue lines both, the outside of the body (skin) and the cavities that are connected to the outside, such as the lungs and the gastrointestinal tract. Epithelial cells assume packing geometries in junctional networks that are characterized by different cell shapes, number of neighbor cells and contact areas. The development of specific packing geometries is tightly controlled [23].

Endothelial cells line the tight barrier between the circulating blood and the surrounding vessel wall. A synchronized migration of endothelial cells is required in order to grow blood vessels (the process of *angiogenesis*). When a new blood vessel is forming, for example in response to a lack of oxygen, the endothelial cells must maintain their cell–cell contacts, remain anchored to the basement membrane, and form curved continuous surfaces [24]; otherwise, the walls of the growing vessels would become leaky. Blood vessel formation is thus a tightly regulated process.

9.2.2
Cell–Matrix Junctions

In contrast to the tight cell–cell junctions, cells can also form junctions with surrounding extracellular fibers. The ECM, which is abundant in connective tissue, includes the interstitial matrix and basement membranes. The ECM provides structural support to the cells (Figure 9.1), in addition to performing many other important functions that regulate cell behavior. Cell–matrix contacts are formed by integrins; these molecules can link various ECM proteins, including fibronectin, vitronectin, laminins and collagens, via cytoplasmic adopter proteins to the cytoskeleton. During the formation or regeneration of tissue, major cell movements occur on or through the ECM, such that the cell–matrix junctions enable and facilitate integrin-mediated tissue growth, remodeling and repair processes. Integrins are also required for the assembly of the ECM (for reviews, see Refs [2]). The fibronectin matrix, the assembly of which is upregulated in embryogenesis and wound-healing processes, often serves as an early provisional matrix that is reinforced at later stages, for example, by collagen deposition [27]. Integrins thus mediate the regulatory functions of the ECM on cell migration, growth and differentiation. During wound healing, angiogenesis and tumor invasion, cells often change their expression profiles of fibronectin-binding integrins [28, 29]. Integrin–matrix interactions thus play central roles in regulating cell migration, invasion and extra- and intra-vasation (i.e. moving from the vasculature to the tissue, or *vice versa*), as well as in platelet interaction and wound healing [24, 29–36]. The functional roles of these interactions in health and disease will be discussed in much more detail below.

The forces acting on cell–cell or cell–matrix junctions can either be applied externally, or generated by the contractile cytoskeleton. Shear stresses due to the flow of blood, urine and of other body fluids impart forces on either the endothelial blood vessel linings, the linings of the epithelial urinary tract, as well as on bone cells, respectively, and are known to actively influence cell morphology, function and tissue remodeling (for reviews, see Refs [12, 37–41]). Lung expansion and contraction imposes great strain on lung tissue, and mechanical forces exerted on the lung epithelium are a major regulator of fetal lung development, as well as of the overall pulmonary physiology [42]. Mechanical exercising of the lung also triggers the release of surfactants onto the epithelial surface [43]. Consequently, the levels of force generated and transmitted through cell–cell and cell–matrix junctions can change drastically with time, and between different organ tissues.

The forces that cells apply to their neighbors and matrices are furthermore dependent on the rigidity of their environment. Cell-generated tractile forces are lowest for soft tissues and increase with the rigidity of the organ. The brain is one of the softest tissues, whereas bone cells find themselves in one of the stiffest microenvironments of the body [18]. Yet, in all of these tissues the cells generate forces that provide the basis of active mechanosensing and mechanochemical signal conversion processes. The formation of force-bearing protein networks that connect the contractile cytoskeleton of cells with their surroundings is essential to prevent cell apoptosis of most normal cells.

What is missing is a mechanistic understanding of how the forces that are applied to cells are locally sensed and finally regulate a collective response of many cells to produce the proper tissue morphology and morphological transformations (Figure 9.2). Although it may be general for all tissues, in endothelia there is a

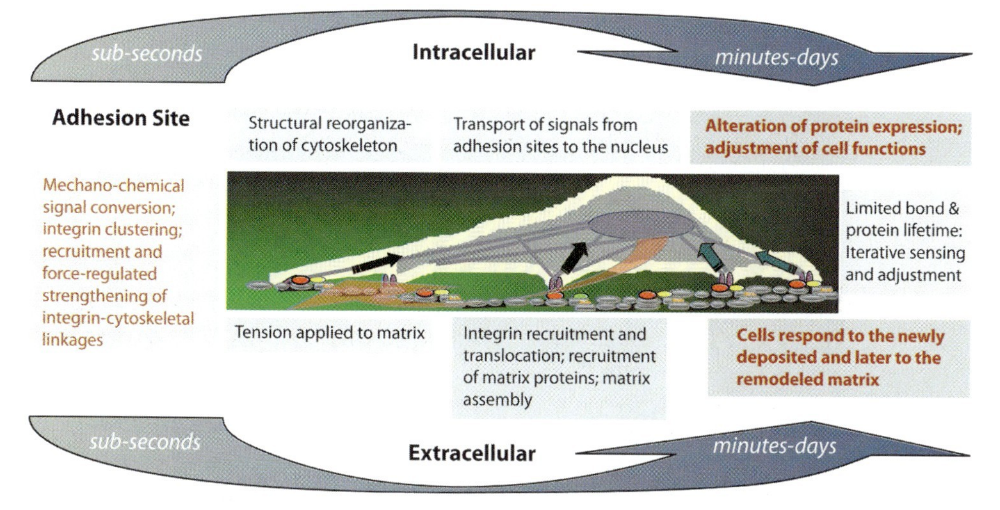

Figure 9.2 Sequential cellular processes of adhesion, mechanosensing and responses with the associated time scales. Initiated by cell adhesion, a cell responds to its environment by subsequent events that involve mechanosensing, reorganization of the cytoskeleton, adjustment of protein expression patterns and, in a secondary feed-back loop, remodeling of the extracellular matrix. Initially, cells will sense the mechanical features of their environment, which will cause rapid motility and signaling responses. As the cell pulls on the environment, it will modify the extracellular matrix and create new signals, such as those originating from fibronectin stretching and unfolding. Intracellular signals will alter the expression pattern of the cell and, over time, the cellular forces and cell-generated matrices will change the cell shape. At any stage, extracellular signals, such as hormones or external mechanical stimuli, can cause acute changes that will set off a further round of cell and matrix modifications.

polarity and bending that must be controlled over many cell lengths. Studies of the development of fly wings and convergent extension in frogs have provided some important clues about mechanisms that can establish an axis in a tissue that would then result in axial contractions. In the fly wing, there are gradients of proteins that affect wing organization by influencing the physical properties [23, 44, 45], and some of those proteins are asymmetrically distributed in the hexagonal wing cells. In many tissues, however, the cells can move and change partners while they change tissue morphology in a stereotypical way, indicating that the multicellular coordination does not solely rely upon stationary protein complexes but rather is sensitive to intercellular forces or curvature.

To better understand the cellular nanomachinery by which cells sense mechanical stimuli, and how forces might synchronize cellular responses, it should be noted that the cellular nanomachinery is subjected both to exogenously applied forces and to cell-generated forces that the cells apply locally to the ECM and neighboring cells. As compressive forces on cells are primarily counterbalanced by the hydrostatic pressure of the cell volume that is contained by the plasma membrane, we will focus here entirely on the impact of tensile forces on proteins and protein networks and the subsequent changes in cell signaling.

When cells stretch their proteins, the protein structural changes may represent one important motif by which mechanical factors can be translated into biochemical signal changes in a variety of tissues and cell types (Figure 9.3). Many proteins are involved in force-bearing networks that connect the cell interior with the exterior, and they all are potential candidate proteins for mechanosensors (for reviews, see Refs [21, 26, 46–54]).

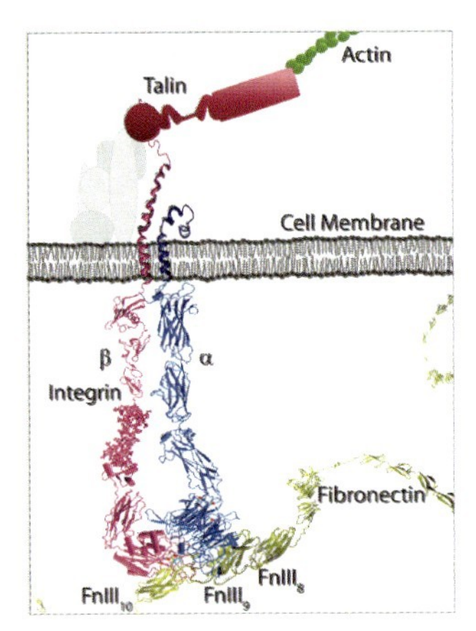

Figure 9.3 The integrin junction connecting the contractile actomyosin cytoskeleton with the extracellular matrix. In the schematic structure shown here, the integrin $\alpha_v\beta_3$ forms a complex with the extracellular matrix protein fibronectin via its cell binding peptide, RGD. In the cell interior, talin couples the cytoplasmic integrin tails to an actin filament. The stretching of talin leads to a reinforcement of the talin–actin linkage through the recruitment of further proteins that are subsequently involved in downstream cell signaling events. Particularly, the recruitment and stretching of p130cas regulates cell signaling events due to its phosphorylation, which is upregulated when stretched.

As cells actively bind, stretch and remodel their surroundings, they use a variety of specialized adhesion structures [25, 56], and their molecular composition will be discussed below (see Section 9.4). Once formed, the first contacts either mature rapidly or break (see Sections 9.5 and 9.7). These structures mechanically link the cell cytoskeleton and force-generating machinery within the cell to the ECM. Intracellular traction can thus generate large forces on the adhesive junctions – forces which are easily visualized as strain applied by cells to stretchable substrates [57–59], as discussed in Section 9.8. In addition, focal contacts are not passively resistant to force, but force actively induces focal contact strengthening through the recruitment of additional focal adhesion proteins, and finally initiates intracellular signaling events [60–64] (Section 9.6). Cell generated forces allow for rigidity sensing (Section 9.9), and causes matrix assembly and remodeling (Section 9.10). The matrix in turn regulates cell motility (Section 9.11). Ultimately, the structure and composition of the adhesions play regulatory roles in tissue formation and remodeling, and also control whether cells derail and evolve into cancer cells or cause other disease conditions (Section 9.12).

9.3 Nanotechnology has Opened a new Era in Protein Research

The advent of nanotech tools, particularly atomic force microscopy (AFM) and optical tweezers [65–67], followed by atomistic simulations of the force-induced unfolding pathways [68], were a major milestone in recognizing the unique mechanical properties of proteins and other biopolymers. The first force measurements on single multimodular proteins were performed on titin, and revealed that the modules cannot be deformed continuously but rather that they ruptured sequentially. But do cells take advantage of switching protein function mechanically? The first functional significance of unfolding proteins upon rapid tissue extension, for example when overstretching a muscle, was seen in them serving as mechanical shock absorbers.

Beyond muscle tissue, protein unfolding might be a much more common theme by which cells sense and transduce a broad range of mechanical forces into distinct sets of biochemical signals that ultimately regulate cellular processes, including adhesion, migration, proliferation, differentiation and apoptosis. The results of recent studies have shown that force-induced protein unfolding does indeed occur in cells and in their surrounding matrices [51–55, 69–71].

9.3.1 Mechanochemical Signal Conversion and Mechanotransduction

How, then, is force translated at the molecular level into biochemical signal changes (mechanochemical signal conversion) that have the potential to alter cellular behavior (mechanotransduction)? Despite all the experimental indications, only limited information is available on how mechanical forces alter the structure–function relationship of proteins and thus coregulate cell-signaling events. After a decade of new insights into single molecule mechanics, a new field is beginning to emerge:

How can the force-induced mechanical unfolding of proteins and other biomolecules switch their functions?

Through careful investigations of the conformational changes of isolated proteins that are mechanically stretched *in vitro*, and through computational simulations that have provided high-resolution structural information of the unfolding pathways of proteins, key design principles are beginning to emerge that describe how intracellular, extracellular and transmembrane proteins might sense mechanical forces and convert them into biochemical signal changes as discussed below (for reviews, see Refs [21, 26, 72, 73]). Stretch has been shown experimentally to expose cryptic phosphorylation sites, resulting in the onset of a major signaling cascade [51], to increase the reactivity of cysteines [52], and also to induce fibronectin fibrillogenesis (for a review, see Ref. [26]). Yeast two-hybrid measurements, crystallographic analyses and high-performance steered molecular dynamics (SMD) calculations all indicate that the exposure of amphipathic helices (e.g. talin, α-actinin) will cause binding to unstrained proteins (vinculin) or to the membrane, as detailed below. Thus, it seems that not a single mechanism can account for all the mechanical activities sensed by cells. Consequently, there is a need to develop a detailed understanding of the mechanical steps in each function of interest, in order to elucidate which of these mechanisms is responsible, or whether a new one must be formulated.

Design principles are also emerging by which such mechanosensory elements are integrated into structural motifs of various proteins, the conformations of which can be switched mechanically (for reviews, see Refs [26, 47, 74–79]). Multidomain proteins that are large and have many interaction sites constitute a major class of potentially force-transducing proteins [26, 80]. Both, matrix and cytoskeletal proteins fall into this class; for example, the cytoskeletal (titin, alpha actinin, filamin, etc.) and membrane skeletal molecules (spectrin, dystrophin, ankyrin) have series of between four and 100 repeat domains that can be stretched over a range of forces. An important feature here is that the repeats are often structurally homologous, but differ in their mechanical stability. Indeed, the differences in the mechanical stability of individual domains determines the time-dependent order in which their structure is altered by force, and consequently the sequence in which the molecular recognition sites are switched by force. Multimodularity thus provides for a mechanism not only for sensing but also for transducing a broad range of strains into a graded alteration of biochemical functionalities. Matrix molecules also have multiple domains and presumably exhibit similar characteristics. In both cases, the stretching of molecules can either reveal sites which can bind to and activate other proteins that could start a signaling cascade, or they can destroy recognition sites that are exposed only under equilibrium conditions [26].

9.3.2
Mechanical Forces and Structure–Function Relationships

As tensile force can stabilize proteins in otherwise short-lived structural intermediates, deciphering how the structure–function relationship of proteins is altered by mechanical forces may well open totally new avenues in biotechnology, systems biology, pharmaceutics and medicine. In order to summarize our current understanding and future opportunities, we will first identify the critical molecules that are involved in linking the cell outside to the inside, and then discuss current knowledge on the effect of force on protein structure and associated force-regulated changes of protein function, and the downstream consequences. Cellular mechanotransduction systems can then transduce these primary physical signals into biochemical responses. More complex physical factors, such as matrix rigidity and the microscale and nanoscale textures of their environments, can be measured by cells through integrated force- and geometry-dependent transduction processes. Thus, it is important to differentiate between the primary sensory processes, the transduction processes and the downstream mechanoresponsive pathways that integrate multiple biochemical signals from sensing and transduction events over space and time, as shown schematically in Figure 9.2. It has also been postulated that cytoskeletal filaments can directly transmit stresses to distant cytoskeletal transduction sites [81, 82], which would involve additional distant mechanosensory and transductional components. Even in those cases, the forces would be focused on sites where primary transduction would occur.

Beyond the unfolding of stretched proteins, there are also other mechanisms in place by which force can alter many biochemical activities (see Box 9.1). The specific force-induced changes in motor protein velocity can lead to stalling their movement or buckling of their respective filaments [74, 83, 84]. Stretch-sensitive ion channels exist where the membrane pressure can regulate the ion current [20, 85–87]. Finally, even the lifetime of the strongest noncovalent bonds that last days under equilibrium, break down within seconds under the tensile force generated by a single kinesin [88, 89]. Not surprisingly, some adhesive bonds have evolved that are not weakened but are strengthened by force; these are also referred to as 'catch bonds' (as reviewed in Refs [90–93]). However, most of these force-regulated processes do not have an evident link to changes in cellular-level functions, or the links are currently not understood. For example, motor protein velocity is not generally linked to mechanically induced changes in cellular function, and neither are the ion currents that accompany the stretch activation of ion channels. Thus, it is unclear whether observed mechanochemical responses are products of the primary transduction of mechanical

Box 9.1
Activities altered by force-induced structural alterations:

- Motor protein velocity
- Stretch-sensitive ion channels (bacteria, hearing, touch)
- Catch bonds (bacterial and lymphocyte rolling and firm adhesion)
- Outside-in cell signaling through stretch-induced alterations of ECM binding sites
- Cytoskeletal protein stretching – phosphorylation by Tyr kinases of cryptic tyrosine repeat domains.

stimulation, or are just part of secondary downstream signaling cascades. Electrophysiological measurements reveal that distinct types of ion channels are mechanically activated [85, 87, 94, 95]; however, the biochemical consequences of channel opening are currently unclear – as are the relationships to downstream mechanoresponsive signaling pathways. Motor proteins will change the rate of movement and ATP hydrolysis in response to load (for a review, see Ref. [96]); however, the molecular pathway linking myosin mutations and cardiac hypertrophy is very unclear [13, 97]. Although catch bonds have a very clear role in enabling cells to adhere to surfaces under flow conditions, the link to subsequent infection or extravasation has not been determined. In any of the systems that employ specific mechanosensors, further investigations are required to determine whether – and/or how – these are functionally linked with specific steps in the cellular functions that are altered by mechanical force.

In the following section, which relates to mechanosensitive processes, we will discuss a few selected proteins that are part of the physical network through which force is transmitted bidirectionally from the cell exterior to the interior, and vice versa. Attention is focused here on the primary changes that have been shown to produce biochemical changes that lead in turn to general signals, although many more possible mechanisms clearly exist.

9.4
Making the Very First Contacts

9.4.1
Molecular Players of Cell–Extracellular Matrix Junctions

Cell motility is regulated by the polymerization of actin which drives the protrusion of the leading edge of the cell. Cells use lamellipodia and filopodia to ‘feel’ their environment and to identify locations to which they can adhere. Lamellipodia are flat, thin extensions of the cell edge that are supported by branched actin networks, while filopodia are finger-like extensions of the cell surface supported by parallel bundles of actin filaments [98]. Both are involved in sensing the environment through cycles of extension and retraction, in the attachment of particles for phagocytosis, in the anchorage of cells on a substratum, and in the response to chemoattractants or other guidance cues [99, 100]. When cells encounter a ligand bound to an extracellular surface, the ligand might bind to a transmembrane protein and ultimately induce coupling of the transmembrane protein to the cytoskeleton. *Integrins* are the key transmembrane proteins that mediate cell matrix interactions. Some integrins can recognize the tripeptide RGD, which is found for example in fibronectin, vitronectin and other matrix molecules, while other integrins bind specifically to collagens and laminins. Once a first bond (or set of bonds) is formed, a competition sets in between the time taken for a bond to break again and the cellular processes that can stabilize an early adhesion. The bond lifetime, however, is significantly decreased if a high tensile force is applied to it [101]. For example, without force, fibronectin can bind to $\alpha_5\beta_1$

integrin for minutes before releasing, whereas a force of approximately 40 pN will cause release in milliseconds [102, 103].

To illustrate some of the general concepts, rather than providing a detailed literature review, we will now briefly describe one of the force-bearing junctions that connects an ECM protein, via integrins, to the cytoskeleton (Figure 9.3).

9.4.1.1 Fibronectin

Fibronectin is a dimeric protein of more than 440 kDa (Figure 9.4), which is a pervasive component of the ECM during development and within healing wounds [24–26, 104, 105]. Fibronectin is composed of three types of repeating

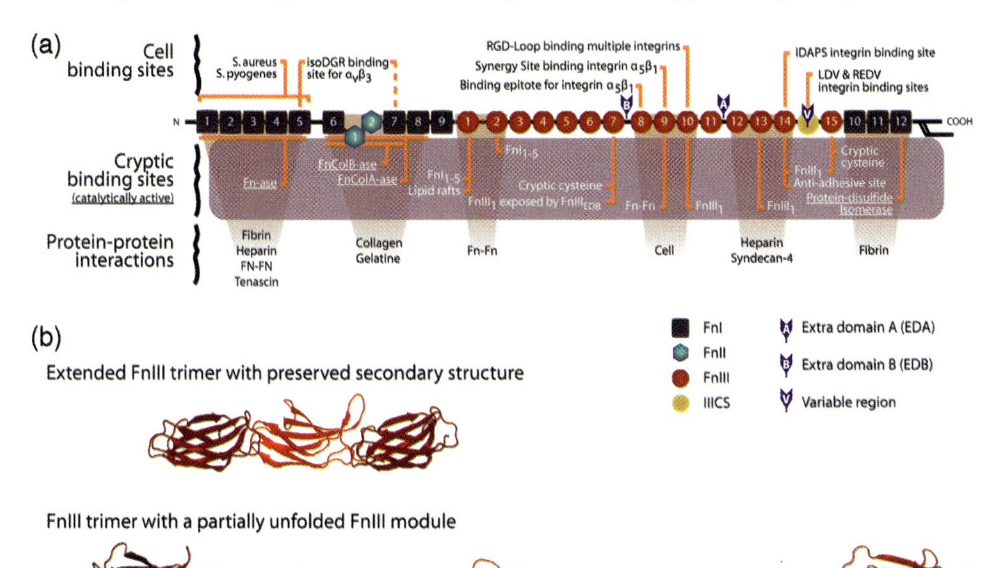

Figure 9.4 Fibronectin's major binding sites and an example of module unfolding under tensile stress [269]. Fibronectins are dimeric molecules composed of over 50 repeats of three different β-sheet modules (FnI, FnII and FnIII). (a) One monomer of as fibronectin found in blood plasma. Fibronectin produced by cells may contain additional alternatively spliced modules, as indicated. Fibronectins contain a large number of molecular recognition and cryptic sites, including the cell-binding site RGD, which is recognized by multiple integrins; the synergy site PHSRN, which is recognized by $\alpha_5\beta_1$ and $\alpha_{IIb}\beta_3$ integrins; the sequence IDAPS at the FnIII13–14 junction in the heparin II binding region of fibronectin, which also supports $\alpha_4\beta_1$-dependent cell adhesion; and the NGR motif in FnI5, which is nonenzymatically converted to isoDGR and can then bind the $\alpha_v\beta_3$ integrin [249]. A similar, highly conserved NGR motif occurs in FnI7, but has not been extensively studied. The cryptic sites include various Fn self-assembly sites, the exposure of which is needed to induce fibronectin fibrillogenesis. Finally, there are two cryptic, nondisulfide-bonded cysteines on each monomer, in modules FnIII7 and FnIII15 which are utilized for site-specific labeling studies by fluorescence resonance energy transfer; (b) Tensile stress applied to Fn fibers causes changes in the quaternary, tertiary and secondary structure of Fn molecules. The figure shows three FnIII modules with intact secondary structure (upper) and with the partial unfolding of one module due to increased tensile stress (lower). (Reproduced with permission from Ref. [269].)

module, each of which has different structural folds, including 12 Fn type I domains, two Fn type II domains, and 15–17 Fn type III domains per Fn monomer. Both, FnI and FnII domains contain two intrachain disulfide bonds, while FnIII domains are not stabilized by disulfides and are hence more susceptible to force-dependent unfolding. Fibronectin displays a number of surface-exposed molecular recognition sites for cells, including integrin binding sites such as the RGD loop, PHSRN synergy site and LDV sequence, as well as binding sites for other ECM components, including collagen, heparin and fibrin. A number of cryptic binding sites and surface-exposed binding sites have been proposed to be exposed or deactivated, respectively, as a result of force-dependent conformational change (as reviewed in Ref. [26]). Interestingly, it is not only fibronectin that contains these modules; in fact, approximately 1% of all mammalian proteins contain FnIII domains that adopt a similar structural fold to the FnIII domains in fibronectin [80].

9.4.1.2 **Integrins**

Integrins – the major cell matrix adhesins – are transmembrane dimers composed of noncovalently bound α and β subunits which associate to form the extracellular, ligand-binding head, followed by two multi-domain 'legs', two single-pass transmembrane helices and two short cytoplasmic tails (Figure 9.5). Although integrins are not constitutively active, their activation is required to form a firm connection with RGD–ligands. Conformational alterations at the ligand binding site of the extracellular integrin head domains propagate all the way to the cytoplasmic integrin

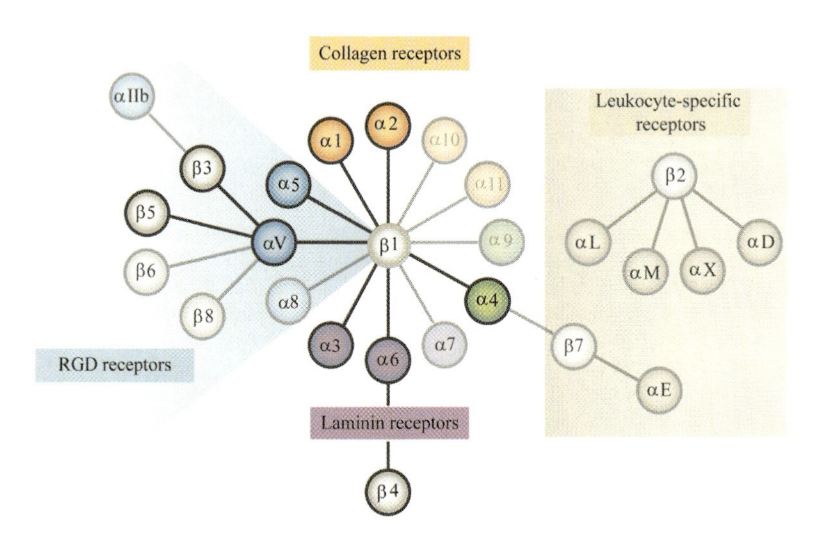

Figure 9.5 The integrin receptor family. Integrins are αβ heterodimers whereby the eight α subunits can assort with 18 β subunits to form 24 distinct integrins. Some integrins recognize the RGD-ligand (blue), while others bind to collagens (orange) or laminins (green), as further discussed in Refs [24, 31, 32]. (Adopted from Refs [31, 32].)

tails, and vice versa, by not-yet understood mechanisms (for reviews, see Refs [31, 32, 35, 106–108]). When a ligand binds to the integrin head, it becomes activated. The activation involves a conformational change that propagates through the extracellular integrin domains, finally forcing the crossed transmembrane helices of the integrin α- and β-subunits to separate, thereby opening up binding sites on their cytoplasmic tails. In contrast, if intracellular events force the crossed integrin tails to separate, then a conformational change will propagate to the extracellular headpiece, thereby priming the integrin head into the high-binding state, even in the absence of an RGD–ligand. This bidirectional conformational coupling between the outside and inside is remarkable, as the integrin molecule is approximately 28 nm long [35, 106–108]. Integrins, however, can also be constitutively activated, for example in the presence of Mn^{2+} ions, by point mutations and via activating monoclonal antibodies [48, 49, 109–111]. Integrin-mediated adhesion often occurs under tensile forces such as fluid flow or myosin-mediated contractions that cells exert to sample the rigidity of their surroundings. In fact, a dynamic mechanism has recently been proposed as to how mechanical forces can accelerate the activation of the RGD–integrin complex [112].

9.4.1.3 **Talin**

Talin is a cytoplasmic protein that can not only activate integrins [113], but also physically links integrins to the contractile cytoskeleton [114, 115], as depicted in Figure 9.6. The talin head has binding sites for integrin β-tails [116], PIP kinase

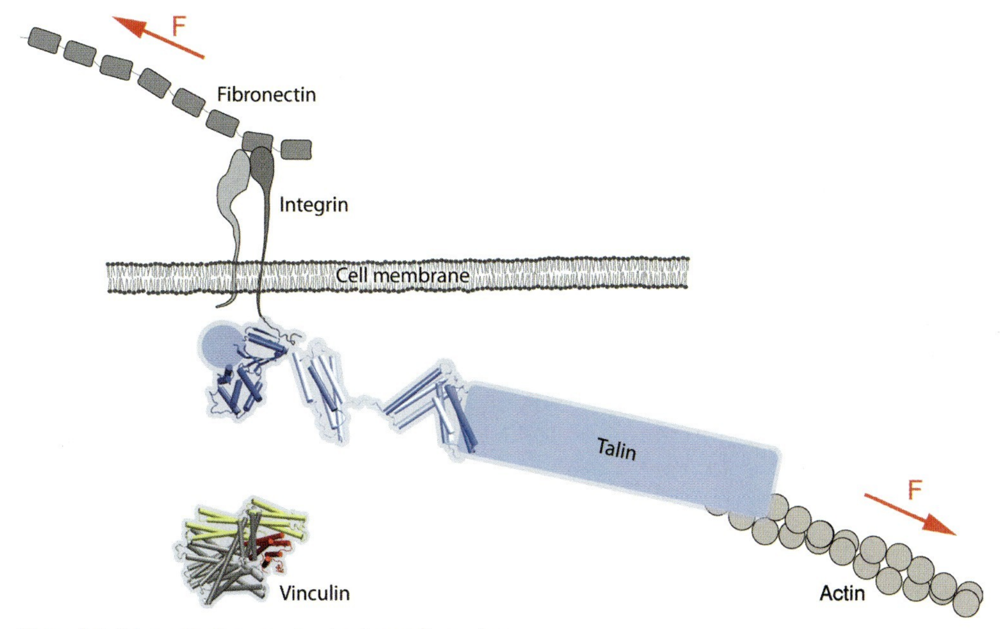

Figure 9.6 Schematic diagram showing how talin anchors integrins to an actin filament. The stretching and partially unfolding of talin (blue) exposes the vinculin binding helices.

γ [117], focal adhesion kinase (FAK) [118], layilin [118] and actin [119] (see also Figure 9.8 below). The 60 nm long talin rod is composed of bundles of amphipathic α-helices [120, 121]. The talin rod contains up to 11 vinculin binding sites (VBSs) [122], including five located within the helices H1–H12, residues 486–889. All of these five binding sites are buried inside helix bundles (native talin shows a considerably lower affinity for vinculin compared to peptide fragments isolated from talin). In addition to the VBSs, the talin rod has binding sites for actin [123] and for integrins [124].

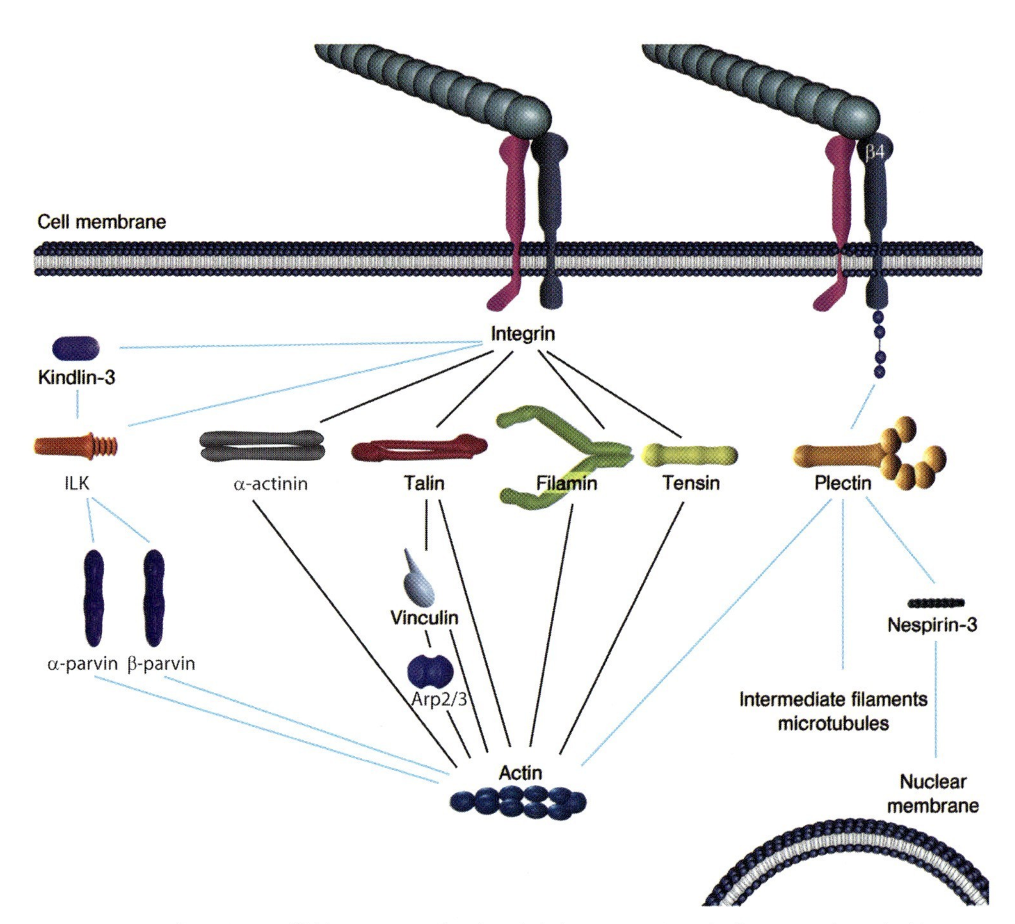

Figure 9.7 Scaffolding proteins that directly link various integrins to actin. Talin, tensin, plectin, filamin and α-actinin were reported to form a single bridge between the various integrins and actin [125–127, 334, 335]. Kindlin-3 was recently added to this list [128], while ILK binds via the formation of a ternary complex with PINCH and parvin [201]. The β_4 integrin has a highly unique intracellular tail which contains four FnIII modules. It can bind via plectin not only to actin, but also intermediate filaments and microtubules, as well as to the nuclear membrane via nesprin-3 [131, 319, 336].

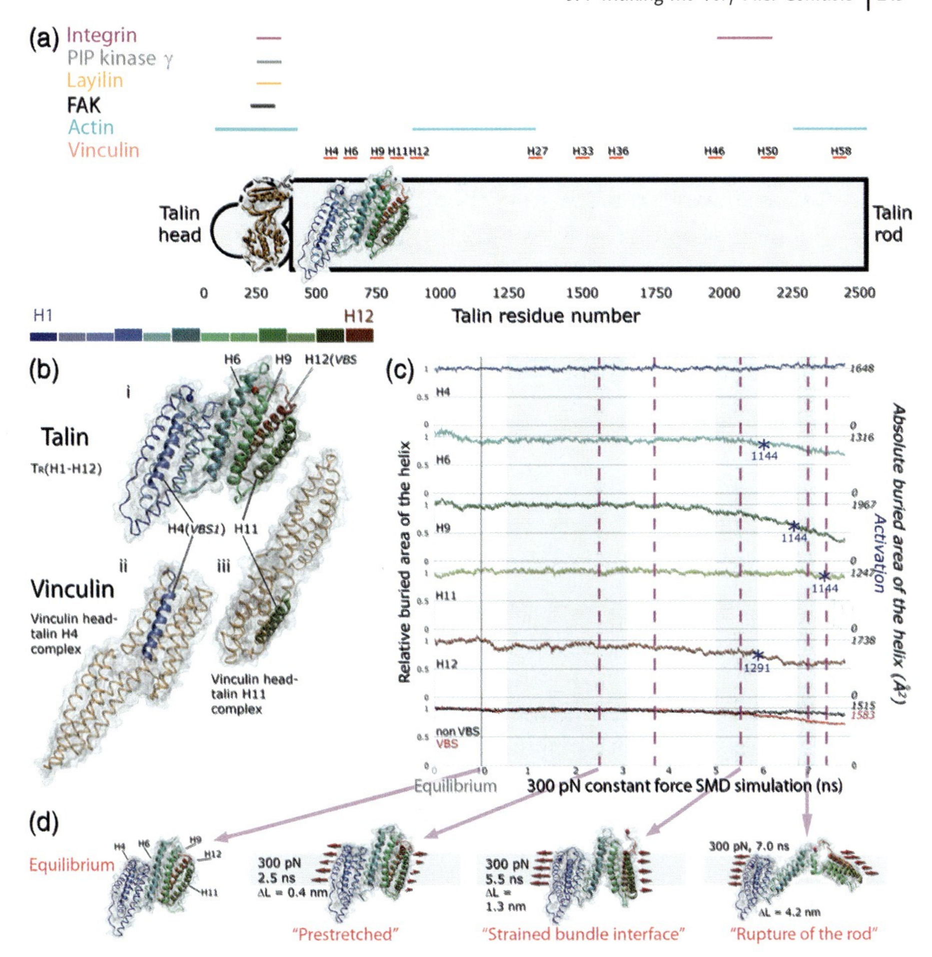

Figure 9.8 Structural mechanism showing how force alters the structure of the N-terminal talin rod comprising helices H1 to H12. (a) Structure of the α-helix bundle of talin, which includes five of the vinculin-binding helices (bold ribbons, namely H4, H6, H9, H11 and H12; (b) The vinculin-binding helices H4 (also referred to as (VBS1)) and H11 in complex with the vinculin head [149] (PDB 1SYQ). The molecular surface is presented in gray; (c) Steered molecular dynamic simulation of the force-induced exposure of the viculin-binding helices to water and the concomitant structural changes shown in (d). Change in the buried surface area of the viculin-binding helices during equilibration and when extended under 300 pN force. The buried areas are shown normalized to the average buried area obtained during equilibration. The respective points of 'activation' – that is, when the buried areas of helices H6, H9, H11 and H12 – in talin equal the experimentally found buried areas of isolated talin helices in complex with the vinculin head, are given as blue asterisks in (c). For H6 and H9, the buried area determined for the H11–vinculin complex is used as a reference because there is no available structure of those helices in complex with vinculin. The buried area of H4 was higher than the buried area of the VH–H4 complex for the whole simulation period. For more detailed information, see Ref. [150].

9.4.1.4 Other Scaffolding Proteins that Provide a Linkage Between Integrins and F-Actin

The physical linkage between integrins and actin can be formed independently by five cytosolic proteins (Figure 9.7). Talin, tensin, plectin, filamin and α-actinin were reported to form a single bridge between the various integrins and actin [125–127], and kindlin-3 was recently added to this list [128]. Talin binds the integrins beta 1, 2, 3 and 5, and weakly to 7 [113]. Tensin and filamin bind to integrins via the same NPxY motif that is recognized by talin [129, 130]. Plectin binds to the laminin-binding integrin β_4 [131], and α-actinin binds to β_1 and β_3 [132, 133]. In addition to integrins, there are ten other membrane-bound adhesion-receptor proteins which bind to either integrins and/or to other adhesion-plaque proteins. Recent data have suggested that certain receptors, for example syndecan [134], can synergize with integrins in adhesion formation [126, 127].

9.4.1.5 Cell Cytoskeleton

Cell–substrate and cell–cell forces are balanced by their interaction partners, except in the case of endothelial cells that experience high fluid flow rates. Thus, the cell cytoskeleton must transmit force across the cell to other sites. This has been observed in the studies of magnetic beads as the propagation of forces to distant substrate sites [135]. There are many ramifications of force propagation in that the cytoskeleton is constantly under tension. Although some of the contractile tension of the cytoskeleton is counterbalanced by the pressure in the cytoplasm, in most cases the intracellular pressure is relatively small (ca. 20 N per m^2) [136]. The majority of the tension is exerted on the actin cytoskeleton, however, we do not yet understand how the spatial distribution of force-bearing adhesions is determined.

9.5 Force-Upregulated Maturation of Early Cell–Matrix Adhesions

9.5.1 Protein Stretching Plays a Central Role

When the integrins latch on to their binding sites in the ECM, the cells apply force to these newly formed adhesion sites, ultimately promoting a rapid bond reinforcement through molecular recruitment. Such recruitment must occur within the lifetime of the initially labile adhesion bond. Key to the reinforcement is *integrin clustering*, followed or paralleled by protein recruitment [125, 137–139]. At least three integrins are needed to form an adhesion [140], and cells show a delayed spreading if the integrins are not sufficiently close [141]. The maturation of adhesion sites seems to involve the stretching and unfolding of proteins, since proteins that are part of such force-bearing linkages might change their structure and, therefore, also their function. One protein which is stretched early in the adhesion process is talin, which links integrins to the cytoskeleton. One of the many proteins that are recruited to newly formed adhesions is vinculin.

9.5.1.1 Vinculin is Recruited to Stretched Talin in a Force-dependent Manner

Upon cell adhesion, talin rapidly accumulates in focal contacts prior to vinculin recruitment [142]. In cases where integrin activation occurs without the application of force, and is thus not part of a force-bearing protein network [139], other adhesion proteins are not recruited. Indeed, the recruitment of vinculin to cell adhesion sites has been shown to be force-dependent [62–64] and to correlate with adhesion strengthening [143] and reduced focal adhesion turnover [144]. Even if not directly shown, this suggested that vinculin recruitment to focal adhesions is upregulated by force [62–64, 145, 146].

Since talin's vinculin-binding helices are buried in its native structure (Figure 9.8b), how might tensile mechanical forces activate them? Some key experimental observations [147–149], together with computational simulations [150] that provided high-resolution structural insights into the force-induced unfolding process of the N-terminal helix bundle of the talin rod which contains five of the vinculin binding sites, suggest the following model of activation.

As the vinculin head consisting of helix bundle is thermodynamically stabilized if it can recruit one additional helix, the vinculin head forms an auto-inhibited complex with its tail domain under equilibrium conditions [151] [PDB: 1TR2]. Instead of binding to itself, the head domain of vinculin can also be stabilized by recruiting other amphipathic helices. For example, isolated vinculin-binding helices of talin can activate vinculin by binding to the vinculin head if added to solution [147, 148, 152, 153]. The release of auto-inhibition is also needed to increase its affinity for actin [154].

Important for the force-activated mechanism is the fact that a larger hydrophobic surface area of talin's vinculin-binding helices can be shielded if they bury themselves in the talin rod rather than in complex with the vinculin head (Figure 9.8c). When mechanically strained, the tightly packed helix bundle of the talin rod breaks into fragments (Figure 9.8d), thereby gradually exposing the buried surface area of the vinculin-binding helices [150]. Once the buried surface area of the vinculin-binding helices in strained talin falls below that shielded if in complex with the vinculin head, the vinculin-binding helices can spontaneously switch their association, breaking off from the strained talin and associating with vinculin; this process is referred to as the *helix swap mechanism* [150]. It was suggested that a vinculin-binding helix would become 'activated' if the buried surface area in mechanically strained talin were to fall below the buried surface area if in complex with vinculin (Figure 9.8c). Vinculin recruitment to talin thus initially increases if talin is incorporated into a force-bearing network formed when a cell adheres to a surface or matrix fibrils [150]. However, as each of the vinculin binding helices is exposed to water at a different time point in the unfolding pathway of the talin (Figure 9.8c), talin can recruit vinculin in a graded response that is upregulated by force. As vinculin can bridge talin and actin, it may reinforce the talin–actin linkage that has been shown previously to be a rather weak bond, breaking at a force of 2 pN [115].

The mechanism described here might not be unique to the talin–vinculin bond, but may be more widespread among other intracellular proteins composed of α-helical bundles. First, when a force breaks away an amphipathic helix from a larger bundle, it might be stabilized by insertion into either the hydrophilic pockets of other proteins or even into the lipid bilayer [148]. Alternatively, other proteins that form helix bundles might also bind vinculin in a force-regulated manner. For example, α-actinin also has a vinculin-binding helix that can form a similar structural complex with vinculin [152, 153, 155, 156]. Similarly to talin, the VBS in α-actinin is buried in the native structure. Identifying the repertoire of mechanisms by which forces can upregulate adhesive interactions has led to the recent discovery of catch bonds, where a receptor–ligand interaction is enhanced when tensile mechanical force is applied between a receptor and its ligand (for reviews, see Refs [90, 93]). In contrast, the force-activated helix-swapping mechanism proposed here requires that the force is applied to just one of the binding partners, thereby activating bond formation with a free ligand. Also in contrast to catch bonds, the ligand need not necessarily form part of the force-bearing protein network at the time the switch is initiated. Thus, while force-induced helix swapping primarily upregulates the bond-formation rate, the catch bond mechanism primarily extends the lifetime of an already existing complex under tension.

9.6 Cell Signaling by Force-Upregulated Phosphorylation of Stretched Proteins

9.6.1 Phosphorylation is Central to Regulating Cell Phenotypes

While bond reinforcement is crucial for the cell to develop a stable adhesion site, the subsequent transformation of mechanical stimuli into biochemical signals is needed to alter cell behavior. But, which molecules act as the major mechanochemical signal converters? Although any experimental demonstration of the stretch-dependence of binding to the cytoskeleton had long been missing, there has always been some concern that the opening of stretch-activated ion channels was the cause of mechanosensation. By using matrix-attached, detergent-extracted cell cytoskeletons, it could

Figure 9.9 Stretch of cytoskeletons activates adhesion protein binding and tyrosine phosphorylation. (a) Diagram of protocol for stretch-dependent binding of cytoplasmic proteins to Triton X-100-insoluble cytoskeletons. L-929 cells were cultured on a collagen-coated silicone substrate, and cytoskeletons prepared by treating with 0.25% Triton X-100/ISO buffer for 2 min. Triton X-100-insoluble cytoskeletons were either left unstretched or stretched (or relaxed from prestretch) with ISO after washing three times. The ISO buffer was replaced with the cytoplasmic lysate solution, incubated for 2 min at room temperature, and washed four times with ISO (b) Tyrosine phosphorylation of many proteins increases upon cytoskeleton stretch. Detergent-extracted cell cytoskeletons showed dramatic increases in phosphotyrosine levels in many different proteins upon stretch. Because soluble kinases have been extracted, it is believed that much of the increased phosphorylation is due to stretch of substrate proteins, such as p130Cas. Thus, it appears that there are many additional proteins that could be involved in sensing stretch of cytoskeletally attached components. (Reproduced from Ref. [160]); (c) Focal contact proteins bind preferentially to stretched cytoskeletons. Western blots of focal contact proteins bound

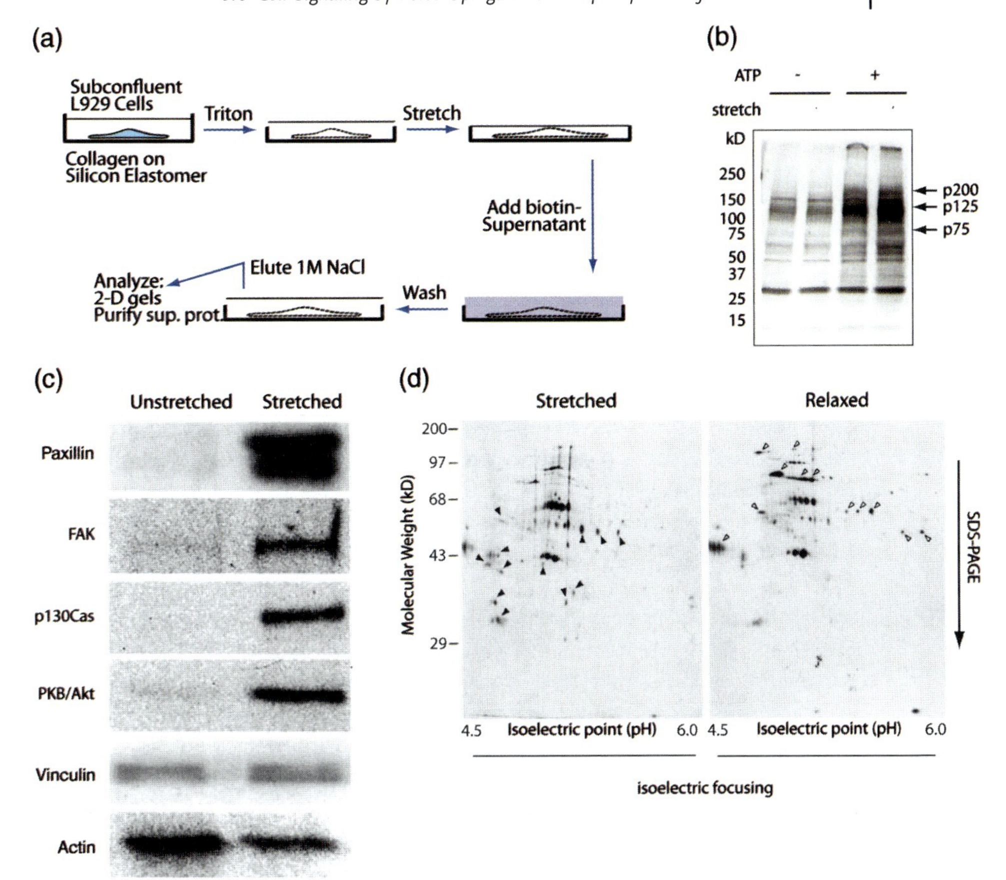

to unstretched and stretched cytoskeletons. L-929 cytoplasmic proteins tagged with a photocleavable biotin (NHS-PC-LC-biotin) were added to Triton X-100-insoluble cytoskeletons of L-929 cells on a stretchable silicone dish [158], and cytoskeletons were stretched or left unstretched (see Figure 9.1). After washing, the bound cytoplasmic proteins were eluted with 1 ml 1 *M* NaCl in HYPO buffer, precipitated with avidin beads (immobilized neutravidin; Pierce Chemical Co.) after sevenfold dilution with HYPO buffer, and released from the bead complex by irradiation with 302 nm UV light (10 min). After photocleavage, proteins were eluted with 120 μl HYPO buffer, and 40 μl of the sample was subjected to 10% SDS–PAGE followed by immunoblotting with antibodies to paxillin, FAK, p130Cas, PKB/Akt (Transduction Laboratories), vinculin (Upstate Biotechnology) or actin (Santa Cruz Biotechnology). Scale bar = 10 μm; (d) 2-D gels of biotinylated proteins that were bound to stretched or relaxed cytoskeletons. The complex of the cytoskeleton with the biotinylated cytoplasmic proteins was solubilized with 1 ml of rehydration buffer (8 *M* urea, 2% CHAPS, 20 m*M* DTT, 0.5% IPG buffer) for isoelectric focusing (the first dimension of 2-D gel electrophoresis). Immobiline dry strip (pH 4–7; Amersham Pharmacia Biotech) was rehydrated with 350 μl of each sample and subjected to isoelectric focusing followed by SDS–PAGE. Biotinylated cytoplasmic proteins in 2-D gels were visualized with affinity blotting using horseradish peroxidase-conjugated streptavidin. Arrowheads mark the spots that were found specifically in Stretched or Relaxed samples. (Reproduced from Ref. [71].)

be shown that different sets of cytoplasmic proteins would bind to cytoskeletons, depending on the extension status (relaxed or stretched) of the cytoskeletons (Figure 9.9), and that binding of the focal adhesion proteins, paxillin, FAK and p130Cas to the cytoskeletons was increased by cytoskeleton stretching [71]. Any increased binding of the cytoplasmic proteins to stretched cytoskeletons would most likely result from the exposure of cryptic binding sites in the cytoskeleton. Whilst it was shown that the binding of another focal adhesion protein – vinculin – remained unchanged in L-929 cells on collagen [71], the force-dependent assembly of vinculin at fibronectin adhesion sites has been reported in other cells [58, 63]. Any specificity derived from the cell type and the substrate to which the cells adhere (including the ECM) appears to account for this discrepancy. In particular, there were no changes in the binding of vinculin to collagen adhesions in intact L-929 cells upon stretching. Subsequent analyses of the range of proteins that were bound to stretched cytoskeletons indicated that both heat-shock proteins and normal focal adhesion proteins would bind to cytoskeletons upon stretching. Both, heat-shock and adhesion stress signals could result from stretch, although the primary signal in the cellular environment is not clear.

9.6.1.1 Stretch-Dependent Binding of Some Cytoplasmic Proteins to Cytoskeletons

Stretch-specific binding studies (Figure 9.9d) indicate that some cytoplasmic proteins will be released from cytoskeletons upon stretching. For example, the binding of actin in a cytoplasmic extract to cytoskeletons was decreased upon cytoskeleton stretching [71]. It is likely that the increase in actin binding to triton (Triton-X-treated) cytoskeletons upon relaxation from a prestressed state is the result of an increase in actin filament assembly (Figure 9.9c), since in intact cells there is an increase in assembly upon relaxation – in that the cell edge becomes very active when a prestretched substrate is relaxed [157]. This indicates that some of the cellular enzyme pathways can be mechanically activated by relaxation, and that some of the binding to the cytoskeletons could result from the activation of enzyme pathways. In addition, cell relaxation-dependent signal activation was observed for Ras [158]. Alternatively, relaxation could enable the refolding of cytoskeletal proteins so that new binding sites would be formed. The binding of cytoplasmic proteins to cytoskeletons could then occur through the relaxation and refolding of cytoskeletal elements. In any case, the cyclic stretching and relaxation of cytoskeletons could play a significant role in controlling the local binding and release of cytoplasmic proteins to the cytoskeletons (Figure 9.9d). The cytoskeleton could thus act as a reservoir such that the mechanical strain would regulate the relative local concentrations of free proteins. This should have an additional impact on biochemical mechanotransduction processes.

Another class of proteins are the *scaffolding proteins* (p130Cas and other candidates that increase in phosphorylation upon cell stretching). These are able to associate with multiple cytoskeletal or signaling complexes through their N- and C-terminal ends, and have multiple phosphorylation sites in a central region (Figure 9.10). The scaffolding proteins appear to have more complex signaling roles, since both of their binding partners and the degree of stretching can change in response to hormone or other signaling pathways. The phosphorylation of Cas requires both an active Src or Abl family kinase, as well as mechanical unfolding of Cas. The phosphotyrosine sites recruit other signaling molecules such as Crk that initiate signaling cascades. The primary transduction event is complicated, however, because the kinase activation step may occur through a force-activated pathway such as a receptor-like protein tyrosine phosphatase or through a hormone receptor. Consequently, primary and secondary force-sensing distinctions can become blurred and involve extrapolation from only a couple of incomplete examples (these are mentioned here only to stimulate further thought on these important mechanotransduction pathways).

9.6.1.2 Tyrosine Phosphorylation as a General Mechanism of Force Sensing

The reversible phosphorylation of intracellular proteins catalyzed by a multitude of protein kinases and phosphatases is central to cell signaling. The recently described phenomenon of substrate priming or stretch-activation of a tyrosine kinase substrate appears to be a major mechanism of force transduction [51]. Anti-phosphotyrosine immunostaining of individual fibroblasts has revealed that tyrosine-phosphorylated proteins are predominantly located at focal adhesions [159–161], where cell-generated forces are concentrated. Furthermore, an adhesion- or stretch-dependent enhancement of tyrosine phosphorylation was observed in many tyrosine phosphorylation sites in T cells [162], fibroblasts (Figure 9.9b) [160, 161] and epithelial cells (Y. Sawada, unpublished observations). In addition, receptor tyrosine kinases have been reported to be tyrosine phosphorylated (i.e. activated) by mechanical stimulation in a ligand-independent manner [163, 164]. These findings indicate that tyrosine phosphorylation plays a general role in adhesion and force-sensing [126, 127]. Due to their hydrophobic character, phosphorylatable tyrosines are typically 'buried' by intramolecular interactions under equilibrium conditions. When such proteins are subjected to stress, however, the buried tyrosines may be exposed, thus enabling them to become phosphorylated. Tyrosine-phosphorylatable proteins also very often carry more than one tyrosine that can be phosphorylated, as does Cas. Multiple repeats of structurally homologous domains are characteristic of many proteins with mechanical functions [26]. Progressive stretching of the molecule can then affect one domain after the other, thus gradually upregulating the response [51]. These observations indicate that substrate priming is a common mechanism for the regulation of tyrosine phosphorylation. As tyrosine phosphorylation appears to be generally involved in force-response (as mentioned above), substrate priming is most likely a universal mechanism of force sensing.

With regards to the force available for stretching molecules in the adhesion complex, the force exerted on one integrin molecule in the adhesion site is estimated to be on the order of 1 pN [58]; this is lower than the forces that allow refolding of many proteins in atomic force microscopy (AFM) experiments [165]. Consistently, the stretching of CasSD (p130Cas substrate domain, the central portion that contains 15 YXXP motifs and is phosphorylated upon stretch) by using AFM gave the appearance of stretching a random coil without any distinct barriers to unfolding (Y. Sawada, J.M. Fernandez and M.P. Sheetz, unpublished observations), suggesting that the Cas substrate domain could be extended by forces below

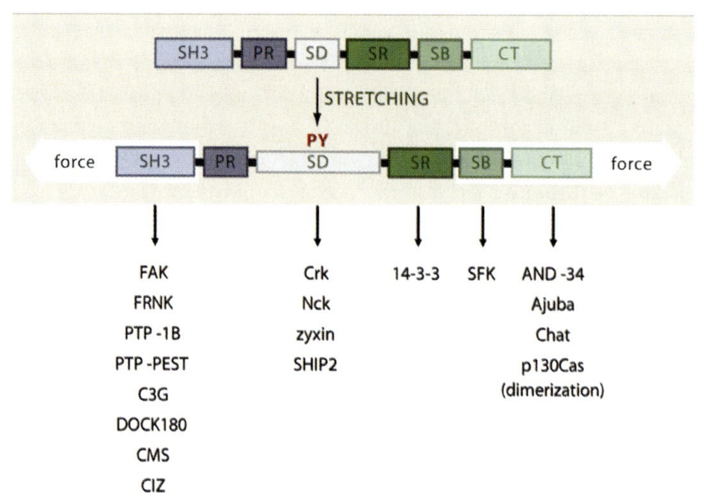

Figure 9.10 Mechanotransduction at focal adhesions through the stretch-dependent phosphorylation of p130cas. The domain structure of the p130CAS molecule is shown (top), before and after stretching. The domains include (from left to right): Src homology 3 (SH3) domain; the proline-rich region (PR); the substrate domain (SD); the serine-rich region (SR), the Src-binding domain (SB); and the C-terminal region. The extension of the substrate domain following stretching and subsequent tyrosine phosphorylation (PY) are indicated. p130Cas and its molecular binding partners are depicted (bottom), including adaptor ('scaffolding') molecules, tyrosine kinases, a serine/threonine kinase, GEFs and tyrosine phosphatases. (Reproduced with permission from Ref. [337].)

the detection limit of AFM (~10 pN). Further, it was observed that CasSD was significantly phosphorylated by Src-kinase with longer incubation times *in vitro*, and that the stretch-dependent enhancement of *in vitro* CasSD-phosphorylation (i.e. fold phosphorylation of stretched/unstretched) is attenuated in longer incubations with Src-kinase (Y. Sawada and M.P. Sheetz, unpublished observations). This indicated that thermal fluctuations of the Cas substrate domain were sufficient to expose tyrosine phosphorylation sites buried in the domain, and raises the possibility that proteins that bind to the native substrate domain like zyxin could stabilize it and inhibit phosphorylation. Thus, the unfolding of p130Cas and its phosphorylation appear to occur at very low applied forces, although the phosphatases that bind to p130Cas could rapidly remove the phosphates.

Finally, tyrosine phosphorylation of the adaptor protein, paxillin, functions as a major switch, regulating the adhesive phenotype of cells [126, 127]. Paxillin, which has binding sites for vinculin and p130cas, can be phosphorylated by tyrosine kinases (including FAK and ABL) and dephosphorylated by the phosphatase Shp2 [126, 127]. Whilst phosphorylated paxillin enhanced lamellipodial protrusions, nonphosphorylated paxillin is essential for fibrillar adhesion formation and for fibronectin fibrillogenesis. The modulation of tyrosine phosphorylation of paxillin thus regulates both the assembly and turnover of adhesion sites. Whilst the method by which force regulates paxillin recruitment and phosphorylation remains unknown, enzymatic activities appeared to be unnecessary for the reversible, stretch-dependent binding of paxillin as the removal of ATP and inhibition of phosphatases did not block paxillin (Figure 9.9c) binding and release from stretched and relaxed cytoskeletons, respectively [71].

Other intracellular proteins have also been shown to be structurally altered by tensile forces. Some proteins (including nonmuscle myosin IIA, vimentin and spectrin), when mechanically stretched within a living cell [52], expose free cysteines, the functional significance of which is not yet known. Yet, key to all of these mechanisms – from force-induced exposure of otherwise buried residues to helix swapping – is that force induces a structural alteration that allows another protein to bind only if the binding partner is mechanically strained.

9.7 Dynamic Interplay between the Assembly and Disassembly of Adhesion Sites

9.7.1 Molecular Players of the Adhesome

It is essential that cells are able constantly to sense their environment and respond to alterations in mechanical parameters. Thus, while one set of cues and molecular players is required that will promote and drive the assembly of an adhesion complex, another set is responsible for regulating their disassembly. Cell adhesion to the ECM triggers the formation of integrin-mediated contacts and ultimately the reorganization of the actin cytoskeleton. The formation of matrix adhesions is a hierarchical process, consisting of several sequential molecular events during maturation (for reviews, see Refs [24, 125, 166–170]). The very first contacts are formed by matrix-specific integrins, and this leads to the immediate recruitment of talin and of phosphorylated paxilin [171]. This event of building the first integrin connection with actin filaments is followed by FAK activation and the force-sensitive recruitment of vinculin [58, 172–174] and the recruitment of α-actinin (α-actinin crosslinks actin filaments). Vinculin has binding sites for vasodilator-stimulated phosphoprotein (VASP) [175] and FAK [176], both of which coregulate actin assembly (via recruitment of profilin/G-actin complex to talin as well as Arp2/3, respectively). pp125FAK also functions as a key regulator of fibronectin receptor-stimulated cell migration events through the recruitment of both SH2 and SH3 domain-containing signaling proteins to sites of integrin receptor clustering [177, 178]. While the adhesions mature further, zyxin and tensin are recruited [167], and zyxin upregulates actin polymerization [179, 180]. The transition from paxillin-rich focal complexes to definitive, zyxin-containing focal adhesions, takes place only after the leading edge stops advancing or retracts [181]. A decrease in cellular traction forces on focal adhesions then leads to an increased off-rate for zyxin [182].

Tensin plays a central role in fibronectin fibrillogenesis which is upregulated by enhanced cell contractility [183]. As with talin, tensin binds to the NPxY

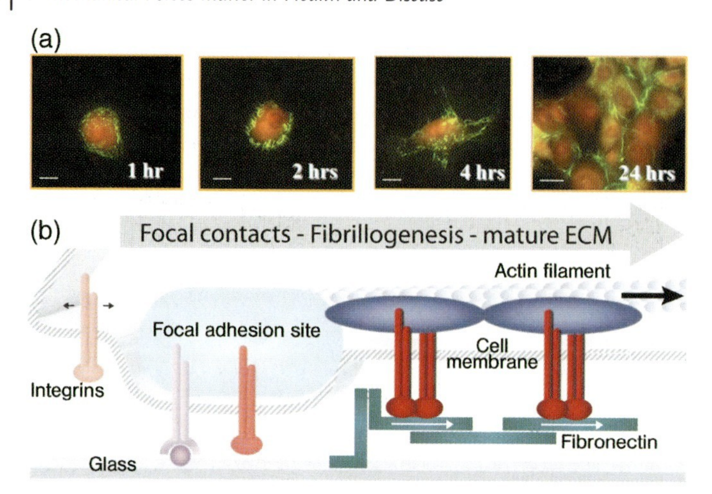

Figure 9.11 Sequential steps in the formation of fibrillar adhesions and fibronectin fibrillogenesis. (a) Time sequence of fibronectin fibril assembly for fibroblasts seeded on fibronectin-coated glass surfaces. (From Ref. [69].) The cells harvested photolabeled plasma fibronectin from solution and incorporated it into newly formed fibers (green). The autofluorescence of the cell bodies is shown in red; (b) Proposed mechanism by which fibronectin-bound integrins translocate out of the focal adhesion sites to form elongated fibrillar adhesions. This process is thought to initiate fibronectin fibrillogenesis. (Adopted from Ref. [167]).

motif of the cytoplasmic β-integrin tails. Fibronectin fibrillogenesis sets in when fibronectin-bound α5β1 integrins coupled via tensin to actin filaments are pulled out of the focal adhesions to form fibrillar adhesions [184] (Figure 9.11). $\alpha_5\beta_1$ integrins translocate along actin fibers, while the other integrins stay back in the adhesion contacts. Fibrillar adhesions translocate centripetally at a mean rate of 18 μm per hour in an actomyosin-dependent manner [56], and evidence is mounting that the stretching of fibronectin induces its polymerization into fibers [69, 70, 185–187]. While phosphorylation of the NPxY tyrosine disrupts talin binding, it has a negligible effect on tensin binding. This suggests that the tyrosine phosphorylation of integrins, which occurs during the maturation of focal adhesions, could act as a switch to promote the formation of fibrillar adhesions [130]. Tyrosine phosphorylation of paxillin regulates both the assembly and turnover of adhesion sites. Moreover, phosphorylated paxillin enhanced lamellipodial protrusions, whereas nonphosphorylated paxillin was essential for fibrillar adhesion formation and for fibronectin fibrillogenesis [126, 127].

The question is, therefore, how are all these processes linked to cell contractility? Rho family GTPases, the major substrates of which are myosin light chain and myosin phosphatase, upregulate myosin activity [15, 188–191], and thus play central roles in integrin signaling [192]. RhoA in particular has been linked with upregulated fibronectin fibrillogenesis [193, 194]. In this context, it is important to note that it is not the intracellular but the extracellular domain of integrin β_1 that controls RhoA activity [194, 195], potentially via its colocalization with syndecan-4 [134].

The formation of such adhesions depends on actomyosin contractility, matrix rigidity [58, 62, 196–198], and the spacing of the integrin ligands on the surface to which the cell adheres [140, 141]. Rigidity sensing is mediated through a mechanism which is further discussed below, where the receptor-like tyrosine phosphatase alpha (RPTPalpha) colocalizes with α_v-integrins at the leading edge of the cell and regulates the activation of Src family kinases [173, 174]. Src family kinases, particularly Fyn, phosphorylate p130Cas in a force-dependent manner [197]; thus, actomyosin contractility enables both fibronectin fibrillogenesis and rigidity sensing. Fibronectin fibrillogenesis, however, occurs via $\alpha_5\beta_1$ integrins, while rigidity sensing is mediated by α_v-integrins.

Finally, microfilament and microtubule networks are significantly reorganized by cyclic stretching, and the cytoskeletal reorganization plays an important role in stretch-induced gene transfer and expression [199]. Integrin-linked kinase (ILK) activity thereby plays an important role in Rac- and Cdc42-mediated actin cytoskeleton reorganization and gene transcription [200, 201]. ILK is a component in focal adhesions that interacts with the cytoplasmic domains of integrins, recruits adaptor proteins that link integrins to the actin cytoskeleton, and phosphorylates the serine/threonine kinases PKB/Akt and GSK-3beta [202]. Finally, the disassembly of adhesion sites is critical, especially at the rear of the cell to enable its forward locomotion [166, 203–205].

The fine-tuning of cell adhesion and detachment, however, requires far more proteins than the few discussed so far. The 'integrin adhesome' consists of a complex network of 156 so-far identified components that are linked in modular complexes and are modified by 690 interactions that have been identified to date [126, 127]. Three major protein families comprise the physical framework of the adhesome, the membrane-anchored adhesion receptors, adaptor or scaffolding proteins, and actin regulators (Figure 9.12). The remaining families consist of mostly enzymes that have roles in regulating the assembly and turnover of the adhesion sites, as well as signaling from the adhesion site into the cell [126, 127]. There are two proteolytic systems associated with the adhesome (ubiquitin E3 ligase protein (Cbl) and calpain), each acting on multiple substrates. Cbl is regulated by tyrosine kinases, and calpain by serine/threonine kinases [126, 127]. Calpain also degrades two tyrosine phosphatases – one of these, shp1, is a regulator of Cbl. Cbl is activated by tyrosine phosphorylation, whilst through its E3-ligase activity it downregulates tyrosine kinase signaling and promotes the proteasomal degradation of integrins [126, 127]. This regulation of binding interactions is very important. Anchoring of adhesion components through multiple links might suggest a robust scaffold structure. In contrast, it must be a highly dynamic, regulated structure that needs to respond to external stimuli and to support morphogenesis and cell migration [126, 127].

Several functional 'subnets' are involved in switching on or off many of the molecular interactions within the network, consequently affecting cell adhesion, migration and cytoskeletal organization. An examination of the adhesome network motifs reveals a relatively small number of key motifs, dominated by three-component complexes in which a scaffolding molecule recruits both a signaling molecule and its downstream target [126, 127]. The authors estimate that more than half of

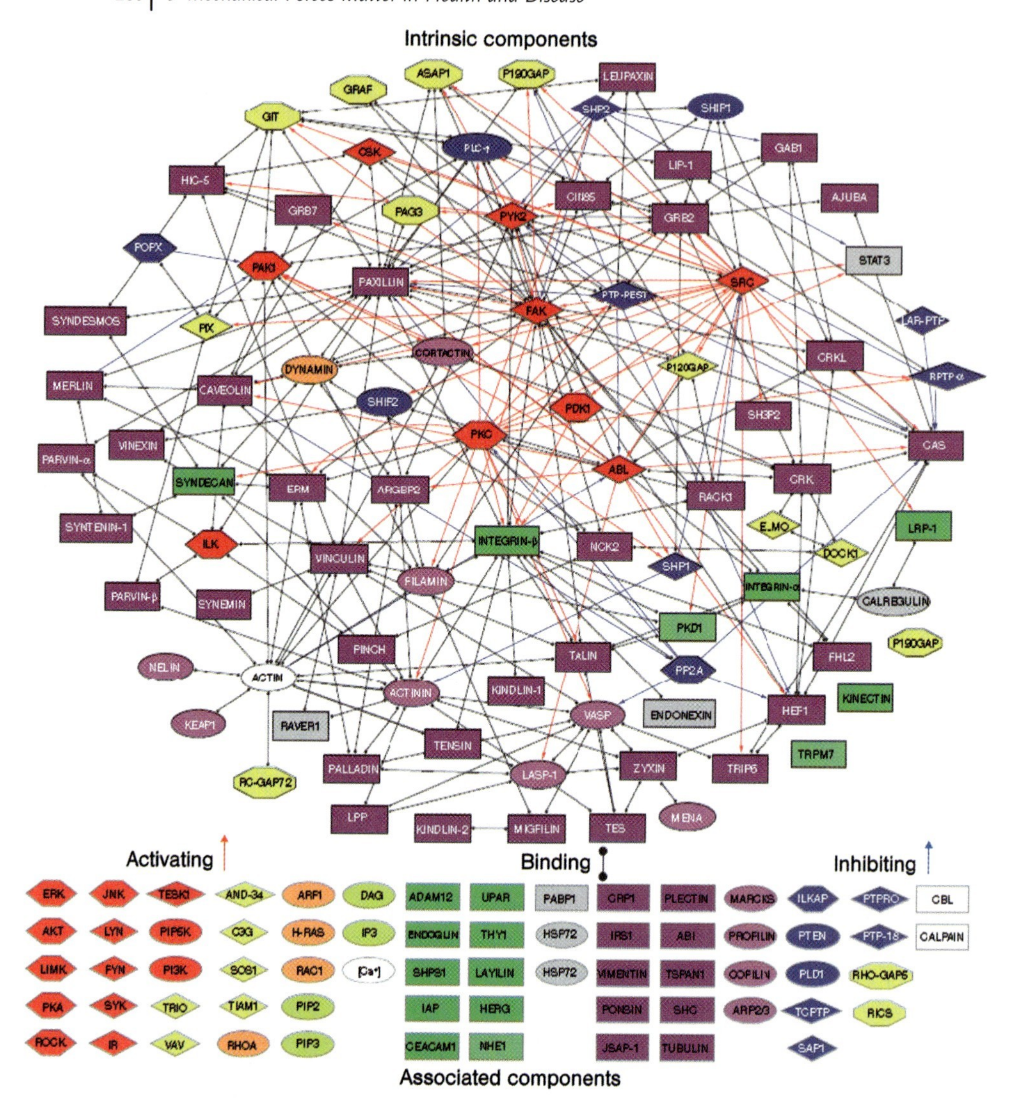

Figure 9.12 Interactions between all intrinsic components of the adhesome and a grouped list of the associated components [126, 127]. Black lines with full circles at their ends denote nondirectional binding interactions; blue arrows represent directional inhibition (e.g. dephosphorylation, G-protein inactivation, proteolysis); red arrows represent directional activation (e.g. phosphorylation, G-protein activation) interactions. The nodes are shape- and color-coded according to the function of the proteins. Intrinsic components are surrounded by a black frame and associated components by a gray frame. (Reproduced with permission from Refs [126, 127].)

the links interconnecting different adhesome components can be switched on or off by signaling elements. There are several types of regulated interaction switches, including conformational switches, GTPase switches, lipid switches, proteolytic switches and PY–SH2 switches [126, 127].

9.8 Forces that Cells Apply to Mature Cell–Matrix and Cell–Cell Junctions

9.8.1 Insights Obtained from Micro- and Nanofabricated Tools

Major experimental tools were developed to probe, with high spatial and temporal resolution, the forces that cells apply to two-dimensional (2-D) surfaces [58, 59, 206–212]. For example, the deflection of microfabricated pillars makes it possible to observe the complete spatial pattern of actin–myosin-driven traction forces applied to the substrate, as shown in Figure 9.13 [59]. This and other tools enable research groups to determine how the linkage between the ECM and the cytoskeleton is stabilized by mechanical force [62, 115, 173, 174, 213–221].

If the force that cells apply to a newly formed junction is too high, the junction will break instantaneously. Thus, the generation and sensing of force is critical for the correct formation of the organism and functions of its tissues. First, however, we should define what is meant by the basic physical parameters of stress and strain in the cellular context, since the anchoring of a cell to an environment is critical for its survival. If firmly anchored to the ECM, the integrins couple the matrix to the contractile machinery of the cell; in this way the major cellular forces are generated by myosin II filaments pulling on actin in early spreading cells [222]. While the level of force orthogonal to the membrane plane that can be supported by the fluid lipid

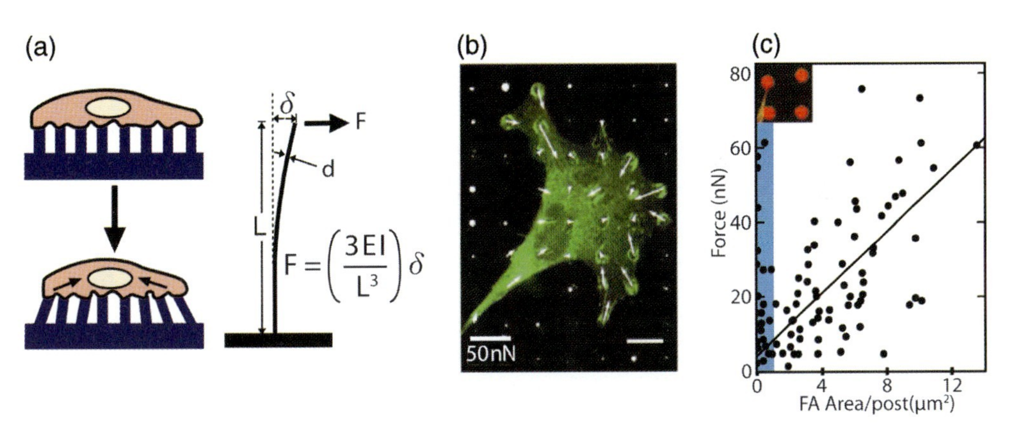

Figure 9.13 Measurement of contractile forces that cells apply to substrates [59]. (a) Cell culture on arrays of polydimethylsiloxane (PDMS) posts covered with fibronectin (as produced by microcontact printing); (b) Confocal images of immunofluorescence staining of a smooth muscle cell on posts. Cells deflected posts maximally during the 1–2 h period after plating, and were fully spread after 2 h. Scale bars indicate 10 μm. The positions of the tips of the posts were used to calculate the force exerted by cells (white arrows). The lengths of arrows indicate the magnitude of the calculated force (top right arrow indicates 50 nN); (c) Plot of the force generated on each post as a function of total area of focal adhesion staining per post. Each point represents the force and area of vinculin staining associated with each post; focal adhesions from five cells were analyzed. The shaded region (blue) indicates the adhesions smaller than 1 μm^2 (inset). (Reproduced with permission from Ref. [59].)

membrane alone is quite small (typically 10–20 pN for a circular area of 100 nm diameter; i.e. a membrane tether) for mammalian cells [223], the plasma membrane is supported by internal and external filamentous proteins that have links to the cytoskeleton. As discussed above, mechanical reinforcement of the very first contacts that a cell forms with the ECM is thus essential. When the adhesions have matured, they can typically withstand forces of a few nN per square micrometer (see Figure 9.13). Tensile forces are then transmitted to the cytoskeleton network in the cell, which can disseminate the force to many or a few sites, even on the opposite side of the cell [224]. Major cellular forces are generated by myosin II filaments pulling on actin throughout the cytoskeleton in early spreading cells, and in the periphery of epithelial cells, particularly around damaged areas in the tissue [160]. In mature epithelial cells, networks of intermediate filaments are generated that bear much of the force when the tissues are stretched.

Far more is currently known about the mechanical characteristics of cell–ECM contacts than about the mechanical properties of cell–cell contacts [21, 126, 127, 225]. The formation of tight cell–cell junctions, however, is critical for many developmental processes such as formation of the gut, kidney, breast and many other epithelial tissues, and is mediated by homologous cadherin–cadherin bonds [22]. These bonds must be dynamic because there are movements of cells relative to each other in epithelial monolayers [23, 226]. Further, when a cell dies, its neighbors rapidly close the gap by first forming an actomyosin collar around the hole; the collar then contracts to cover the hole [160]. Similarly, in the process of convergent extension during embryogenesis, cells converge along one axis while being able to move relative to one another. This is the major morphological movement responsible for organizing the spinal cord axis [227]. Many of the important morphological changes in development involve the contraction of epithelial cells, from the early formation of the gut to the later formation of kidney tubules. In all cases, there is evidence that although the individual cells can move independently, the whole tissue still acts as a unit to undergo a coordinated morphological change. The molecular basis of the mechanical coordination in epithelial or endothelial cell monolayers is not known, but the precision of movement implies that a rapid feedback mechanism is present. It is thus very interesting to note that when force-sensing micropillars were coated with cadherins rather than with fibronectin, the mechanical stresses transduced through cadherin-adhesions were of the same order of magnitude as those previously characterized for focal adhesions on fibronectin [225]. So, the question is, what is the relative importance of cell–cell and cell–matrix contacts in different tissues on transducing mechanical stimuli into altering the downstream behavior? In both tissue types, cell–cell interactions predominate. Endothelial cells require cell–cell contacts, while vascular endothelial cells utilize cadherin engagement to transduce stretch into proliferative signals [15]. Hence, stretch stimulated Rac1 activity in endothelial cells, whereas RhoA was activated by stretch in smooth muscle cells.

Finally, tissue remodeling often reflects alterations in local mechanical conditions that result in an integrated response among the different cell types that share – and thus cooperatively manage – the surrounding ECM and its remodeling. The question therefore is whether mechanical stresses can be communicated between different cell types to synergize a matrix remodeling response. When normal stresses were imposed on bronchial epithelial cells in the presence of unstimulated lung fibroblasts, it could be shown that mechanical stresses can be communicated from stressed to unstressed cells to elicit a remodeling response. Thus, the integrated response of two cocultured cell types to mechanical stress mimics the key features of airways remodeling as seen in asthma: namely, an increase in production of fibronectin, collagen types III and V and matrix metalloproteinase type 9 (MMP-9) [228].

9.9 Sensing Matrix Rigidity

9.9.1 Reciprocity of the Physical Aspects of the Extracellular Matrix and Intracellular Events

As long as the cell–matrix and cell–cell linkages hold tight, intracellular motile activity will interrogate the matrix, and the subsequent cellular activity will depend on the physical properties of the extracellular fibrillar network, and vice versa. The interrogation involves a mechanical testing of the rigidity as well as the geometry of the environment through the normal cell motility processes (Box 9.2). When the rigidity is determined, the cell will respond appropriately. For example, the extracellular network structure is remodeled if it is of the same or a softer compliance than the intracellular network [18]. Rigidity is an important part of the environment of a cell, and different tissues have different rigidities as well as different levels of activity (this point will be discussed later). There is considerable interest in learning how this is achieved at a molecular level.

A number of reports have indicated that matrix rigidity is a critical factor regulating fundamental cell processes, including differentiation and growth. Discher's group recently showed that the differentiation of mesenchymal stem cells is heavily dependent on the rigidity of the matrix to which cells adhere [19]. The group reported that mesenchymal stem cells preferentially become neurogenic on soft substrates, while they preferentially commit to myogenic and osteogenic differentiation on intermediate and rigid substrates, respectively. The cellular response to rigidity has been seen at the time scale of seconds for submicron latex beads [213] and during cell spreading [229]. Fibroblast migration toward rigid substrates indicates further that the process has important ramifications for *in vivo* motile activities [230]. From the signal transduction point of view, these observations indicate that the sensing of different rigidities can be very rapid and may have profound effects on cell function at a variety of levels. Rigidity is a rather complicated parameter for the cell to sense because measurements of both force and displacement must be combined. In order for a cell to sense matrix rigidity, the cell must actively pull on the matrix; thus, the cell must actively test the rigidity of its environment. As a corollary, the cell in an inactive state will not be able to develop a rigidity response. Although these statements apply for single cells *in vitro*, the situation is more complicated in a tissue environment where neighboring cells and

Box 9.2
Cell Forces

- *Tensile Forces:* As the cytoskeleton of the cell is generally contractile, the transmembrane forces on cells are primarily tensile. Large forces in the movement of organisms or tissues are typically generated by linkages from the cytoskeleton to the ECM or neighboring cells through integrins or cadherins, respectively. In contrast, the lipid bilayer of the plasma membrane is fluid and can be distorted with relative ease. Forces are exerted typically on noncovalent protein–protein bonds (one exception is the transglutaminase linkage of lysines to glutamines on collagen or fibrin). Typical protein–protein bonds can sustain about 1 pN per bond (e.g. 1 pN per integrin–fibronectin bond in living tissue, where the bond holds for a matter of seconds). Although those forces seem low, they need to be maintained for long periods, and even the very high-affinity avidin–biotin bond only has a lifetime of about 5 s under a force of 5 pN. Typical forces that fibroblasts can exert on fibronectin-coated surfaces are on the order of 1–5 nNper μm cell edge.

- *Compressive Forces:* Compressive forces are primarily resisted by the hydrostatic pressure of the cytoplasm of the cell (particularly true in plant systems). The other type of compressive force is that generated as cells extend lamellipodia, filopodia or pseudopodia, and those processes push on neighboring cells or the environment.

Tissue Rigidity

- *Resting Rigidity:* A tissue at rest has a rigidity that is defined by the Young's modulus, E, which can be determined from the slope of the tensile stress versus the tensile strain curve:

$$E = \text{tensile stress}/\text{tensile strain} = \delta\sigma/\delta\varepsilon = \delta(F/A_0)/\delta(L/L_0)$$

where: E is the Young's modulus (modulus of elasticity) measured in Pascals; F is the force applied to the object; A_0 is the original cross-sectional area through which the force is applied; δL is the amount by which the length of the object changes; and L_0 is the original length of the object.

- *Rigidity Sensing:* To measure rigidity, cells develop force on matrix over time. This means that the mechanism of sensing rigidity must compare force and displacement during a given time period. If a cell pulls on an object, the total displacement is the sum of the matrix and of the intracellular displacements. Since the cell directly probes the relative molecular displacements in the adhesion site, it should not matter whether the force is applied from the outside or the inside. This statement, however, is true for soft surfaces only if the external force is applied rapidly before the cell has displaced the substrate. The time window is important since the cell does not pull with a constant force but gradually increases the force after making a first contact. The physiologically relevant time window in this context typically lasts for just 1–2 seconds. The real question is whether the cell ramps up the force until a max force applied to the adhesion site is reached, or alternatively a maximal stress, or until a certain relative displacement of the intracellular rigidity sensors within adhesion sites is reached which might be in the order of 130 nm. If the latter is the case, the cells pull with larger forces on rigid objects attempting to reach the critical displacement of the intracellular rigidity sensors.

tissue modulation can produce stresses and strains on cellular components in a sustained manner. The complexity of the rigidity measurement can potentially be an important part of tissue assembly.

In terms of the mechanism by which cells can sense rigidity, they must measure the force needed for a given displacement (rigidity) or the displacement for a given force (compliance). The rigidity of tissues can be based upon the rigidity of either the matrix or of the cells themselves. Most of the focus of *in vitro* studies has been on the effect of matrix rigidity rather than cell rigidity, because that is easier to manipulate. By using elastic pillars of different rigidity as substrates, Saez *et al.* showed that the forces exerted by the cells increased linearly with rigidity of the pillars. Thus, the cells deform the pillars by the same amount (on the order of 130 nm) over an almost 100-fold change in rigidity (Figure 9.14) [231]. This observation suggests that cells can sense the displacement of the cell–substrate anchor sites and continue to develop higher forces on rigid substrates until the proper displacement is reached. Because displacement must be measured from the cell edge or the initial site of pulling, the sensor molecules must be anchored both at the edge (initial site of pulling) and at the integrin that is being pulled. In terms of a possible physical mechanism (see Figure 9.14), the movement of a component relative to a stationary component could produce a signal to stop further recruitment or further movement. As many molecules can easily span 130 nm, the movement of the actin (anchored to the integrin) past myosin or some other relative molecular displacement could be linked to a signaling process. If the displacements were less than 130 nm, then a signal for myosin filament contraction and/or assembly could be generated. If the displacements were greater than 130 nm, then further myosin activation would be blocked because the sensor would be physically separated from the enzyme that modifies it or the sensor could be fully stretched. Because the distance is molecular and yet micrometer-sized contacts were produced on rigid pillars, the measurement of rigidity must continually be made over time by activating new regions to contract and allowing old regions to relax. This can be compatible with a previously described model of rigidity sensing through p130Cas by assuming that force must increase until p130Cas is displaced from the kinase in (Figure 9.14) [21, 197]. However, other models, which are more closely linked to the movement of actin relative to myosin, may be used for longer-term rigidity sensing. Multiple mechanisms probably exist to enable different types of cells to properly sense rigidity in different environments. In many cases, there is a strong need for a local feedback between the level of actin and myosin recruited and the rigidity in the epithelial cells. Hence, additional experiments are required to identify possible molecules that could be the rulers in such a system. A consequence of the displacement sensing is that higher

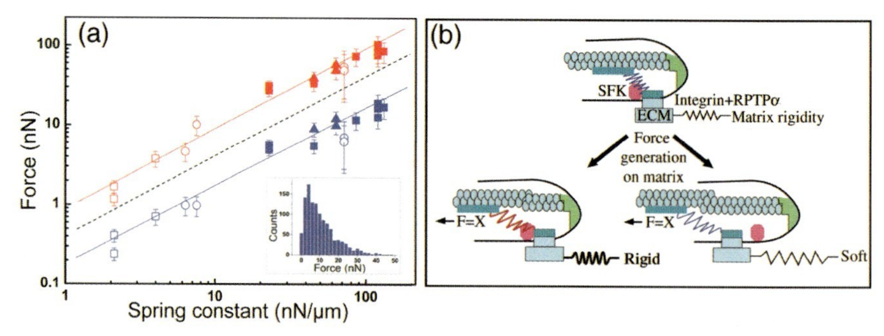

Figure 9.14 Rigidity sensing; evidence for a displacement sensor and one plausible molecular mechanism. (a) Linear relationship between rigidity and average force generated indicates that cells sense displacement [231]. Log-log plot of the force as a function of substrate rigidity. F (blue) and F_{max} (red) within an island of cells are represented for different surface densities: (ratio of the post surface over the total surface) 10% (circles); 22% (squares); 40% (triangles). Open and solid symbols correspond to pillars of 1 and 2 μm in diameter, respectively. The slope of the dashed line is 1. Inset: Typical histogram of force distribution (spring constant 64 nN μm^{-1}); (b) Relative displacement model for rigidity sensing by substrate priming. In this scheme, the displacement of the cytoskeleton–integrin–matrix complex from the active Src-family kinase (Fyn) is the signal to stop further phosphorylation. An initial force signal activates RPTPα at the leading edge that then recruits Fyn to a stationary lipid domain through its palmitoylation. Contraction of the cytoskeleton stretches Fyn substrates such as p130Cas, priming it for phosphorylation by Fyn. In rigid substrates, additional force-generating links are recruited by continued phosphorylation to develop the higher forces needed to displace the p130Cas from the kinase. (Adapted from Ref. [197], where the force needed to stretch p130Cas was emphasized; however, the results of Saez *et al.* [232] and other studies indicated that displacement was measured.)

rigidities will cause higher forces that will in turn enhance the intensity at upstream signals in mechanotransduction (e.g. tyrosine phosphorylation) [51] and may result in a greater activation of downstream signals on rigid substrates.

Another, possibly important, factor which affects the rigidity response is the distance between the cytoskeleton and cell–substrate anchor sites. As elasticity is defined by the magnitude of deformation (e.g. change of dimension) per unit force, cell–substrate anchor sites will be displaced by a larger distance on soft substrates for the same force (see Figure 9.14). If that is the case, the distance between cytoskeleton and cell–substrate anchor sites (the length of the matrix plus the integrin and the actin-binding molecules) would be larger on rigid substrates than on soft substrates [233, 234]. However, larger displacement of cell–substrate anchor sites has not yet been demonstrated, and the uniform cellular deformation of elastic substrates of vastly different rigidities [231] implies that the displacement of anchor sites is not necessarily larger on soft substrate.

9.9.1.1 Time Dependence and Rigidity Responses

When cells pull on the substrate to sense rigidity, they use the rearward movement of actin to generate the force. Actin moves rearward at a rate of 30–120 nm s^{-1}, which means that displacements of 130 nm will take more than 1 s. In the models of rigidity sensing that have been discussed, a rapid rise in force that is sustained could elicit a rigid substrate response. *In vivo,* if the cell experiences tensile forces from the neighboring cells or the matrix during the period where it is pulling on the matrix, then the matrix can appear rigid because the force will rise rapidly. In experiments where the force was increased rapidly on fibronectin beads with a soft laser tweezers, the bead appeared to be in a rigid laser tweezers, as was predicted. Similarly, *in vivo* many tissues experience external forces on roughly a second time-scale that could develop a rigidity response. Thus, rigidity-dependent growth could be stimulated *in vivo* by tissue contractions.

The time dependence of the assembly of components in the integrin–cytoskeleton complex might affect the rigidity response, since different components bind and detach during the life cycle of an adhesion site [184, 235]. Primary connections between integrins and the cytoskeleton, and their reinforcement, depend on talin (which is probably one of the first proteins in adhesion sites) [64, 115], on α-actinin (which crosslinks actin filaments) and on zyxin (which enters the adhesion site during its maturation) [181, 236]. The transition from mature focal adhesion to fibrillar adhesion is characterized by the segregation of tensin and specific integrins [56]. Because the ECM–integrin–cytoskeleton connection is a viscoelastic material (i.e. it is not purely elastic) [237], the time required to reach the threshold force for rigidity responses probably differs depending on the stiffness of the ECM. Accordingly, a soft optical trap could mimic the effects of a rigid trap on the stabilization of the integrin–cytoskeleton linkages if externally applied forces rise rapidly [233]. In lamellipodia, the cytoskeletal-dependent radial transport of a contractile signal directs the timing of contraction and, probably, adhesion site initiation to stabilize protrusive events [229] (Figure 9.15). Consequently, the formation of cell contacts with the ECM is not a continuous process but rather involves cycles of contraction and relaxation.

9.9.1.2 **Position and Spacing Dependence of the Rigidity Responses**

The position dependence of rigidity responses is exemplified by the fact that structural and signaling proteins that are necessary for rigidity responses are placed at strategic locations – for example, at the cell edge during protrusive events and at early adhesion sites. Many proteins involved in rigidity responses and/or phosphotyrosine signaling, including talin [64, 115], integrins ($\alpha_v\beta_3$) [115, 238], paxillin [173, 174], α-actinin [173, 174, 229], RPTPα [173, 174], Rap1 [239] and p130Cas [240], are localized at the leading edge of the cell, ready to respond to any contraction generated by the cell or by the ECM. There is a position-dependent binding-and-release cycle of fibronectin–integrin–cytoskeleton interactions, with preferential binding occurring at the active edges of motile fibroblasts and release at 0.5–3 μm back from the edge [241]. Interestingly, reinforcement occurs preferentially at the edge in rigid tweezers [233], whereas weak connections that break easily are favored by nonrigid tweezers and at sites >1 μm back from the leading edge [115, 233]. At the molecular level, the reinforcement of integrin–cytoskeleton interactions are limited to linkages that have experienced force, and not those nearby (<1 μm) [213].

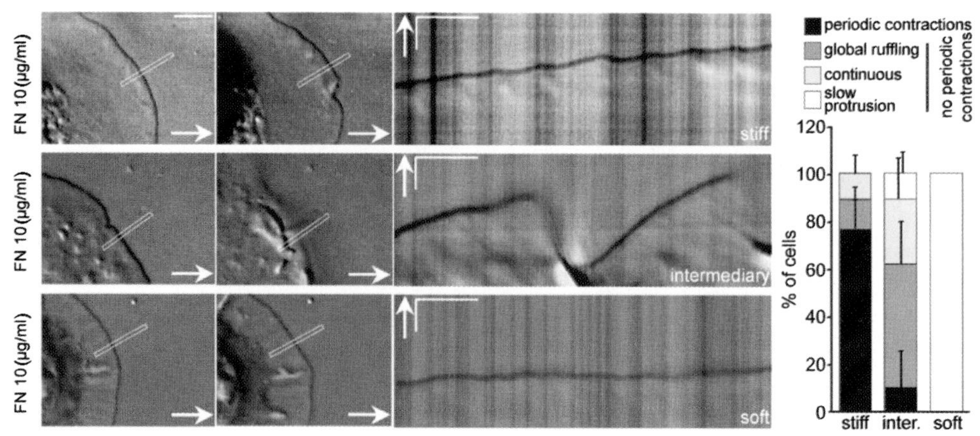

Figure 9.15 Contraction of spreading cells results in periodic contractions and further spreading on rigid surfaces, but no further spreading on soft surfaces. On intermediary stiff surfaces coated with the same concentration of covalently attached fibronectin, cells attempt to spread but lose adhesion. (Reproduced with permission from Ref. [229].)

Finally, many tissues experience periodic stretch *in vivo* during normal activity. When the tissue is inactive, it often experiences atrophy; this is obviously true for muscle, bone and connective tissue. The greatest problem for space travelers is that astronauts typically lose 1–2% of their bone mass for each month in space, even though they may exercise regularly [242]. Similarly, the skin on the feet or hands thickens with use or labor, and thins with disuse. Thus, force from activity or rigidity appears to be a global regulator of tissue function, and an understanding of the mechanisms whereby force is transduced into biochemical signals is an important area of future research.

At the subcellular level, there are many forces that must be regulated to produce normal cell morphology and the proper distribution of organelles. Although the protein composition of many genomes and even individual cell types is known, relatively little knowledge exists of how those proteins are assembled into functional complexes. Individual proteins, typically 5–20 nm in diameter, are assembled into larger functional complexes that can be considered as subcellular machines, controlling and regulating complex cellular functions, from reading and translating the genetic blueprint to the synthesis and transport of proteins, from cell migration to cell division, from cell differentiation to cell death. Those subcellular complexes range in size from ribosomes to the lamellipodial machines that drive ECM assembly and remodeling [243], including collagen fiber rearrangements [244]. It is important to be able to dissect the steps into these subcellular processes to enable greater understanding of the sequence of coherent molecular events [245] (Figure 9.16).

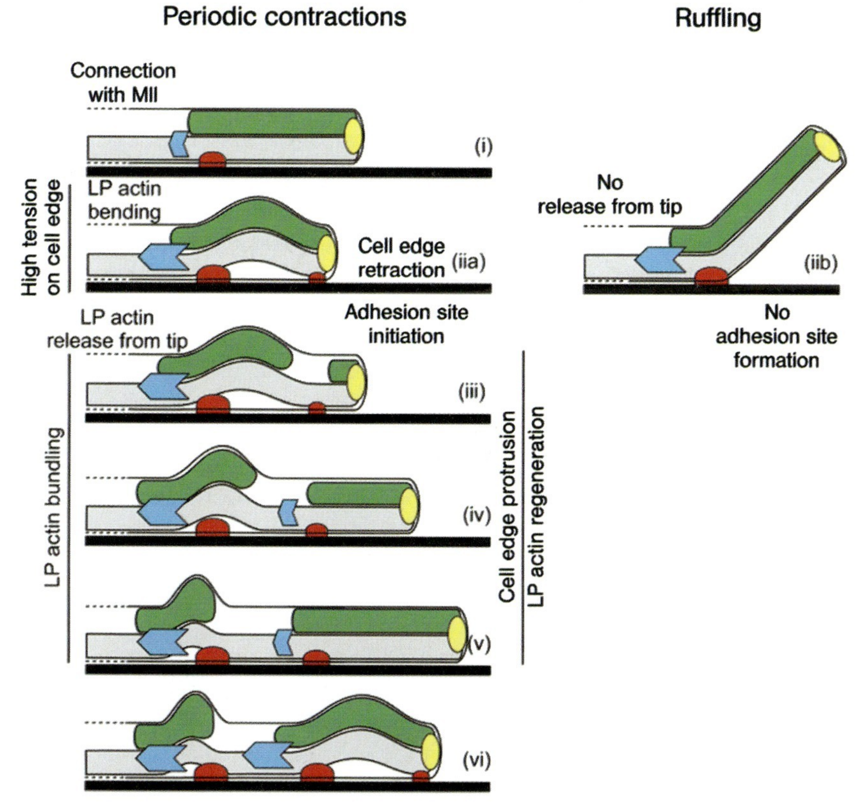

Figure 9.16 Schematic representation of lamellipodial (LP)–actin periodic regeneration. The LP actin (green) is above the LM (gray). Polymerization at the front of the LP actin network causes the back of the network to grow towards the back of the LP contractile module until it reaches an adhesion site (i) where a MII cluster (blue) forms. MII pulls the LP actin, generating high tension on the cell front, causing LP bending, edge retraction and initiation of new adhesion sites (red) on the extracellular matrix (ECM; black rectangle) (iia). The LP actin continues to be pulled until it is released from the tip (iii) and edge protrusion restarts. A new LP actin network immediately resumes growth, which suggests that the actin polymerization machinery (yellow) is still present at the cell tip (iv). The released LP actin, still pulled by MII, further condenses into a bundle at the back of the previous adhesion site (v), while the newly growing LP actin reaches the next adhesion site and the cycle begins anew (vi). LP ruffling (iib) occurred in the case when the total bond energy connecting LP actin to the edge was greater than the bond energy of nascent adhesion sites to the ECM. (Reproduced with permission from Ref. [245]).

9.10 Cellular Response to Initial Matrix Conditions

9.10.1 Assembly, Stretching and Remodeling of the Extracellular Matrix

Once the first contacts have been made by the cell with its surrounding, it will often soon begin to assemble its own matrix. Initially, cells build a provisional matrix which is rich in fibronectin, which plays a particularly important role in early embryogenesis and in wound healing [104]. In order that the assembly process is started shortly after attachment – and even before the genetic machinery upregulates the expression of matrix proteins – the cells harvest fibronectin from body fluids or the cell medium. Only at later stages do the cells start to produce and secrete their own fibronectin, which has some structural and functional differences compared to plasma fibronectin [246]. The complex fibrillogenesis process begins when the cells apply force to fibronectin molecules (see Figure 9.11). Crucial to the assembly process is integrin $\alpha_5\beta_1$, which specifically recognizes only fibronectin among all other matrix proteins, due to fibronectin's unique synergy site that is sitting on the $FIII_9$ module adjacent to the RGD site (see Figure 9.4). Fibronectin fibrils then emerge as the fibronectin-bound $\alpha_5\beta_1$ integrin complexes are translocated along with actin fibers away from the cell periphery [25, 56, 126, 127, 184, 194]. The tensin-mediated $\alpha_5\beta_1$ integrin translocation thus initiates fibronectin fibrillogenesis on the cell surface. However, an integrin-mediated activation step is not the only way by which fibronectin fibrillogenesis can be initiated. While integrins serve as handles by which cells can apply force to fibronectin molecules, fibronectin fibrillogenesis can also be induced in the absence of integrins, as long as fibronectin molecules are stretched, for example by shear [247] or physical entrapment [248]. The RGD-sequence is thus not necessary for fibronectin fibrillogenesis [249]. It was shown recently that the fibronectin conformation in artificially pulled fibronectin fibers is similar to that found in cell-generated matrix fibers [53–55, 250, 251]. Artificially pulled fibronectin fibers can thus serve as physiologically significant model systems. Finally, src-kinase activity not only regulates rigidity sensing but is also essential in fibronectin fibrillogenesis. Src-induced phosphorylation of paxillin at Y118 is required for assembly of the FN matrix by fibroblasts, as well as for maintaining the attachment of FN matrix fibrils to the cell surface [252].

9.10.1.1 Switching the Biochemistry Displayed by the Matrix by Stretching and Unfolding of Matrix Proteins

When the cells have assembled the extracellular matrix fibers, a number of questions remain unanswered. First, do the cells respond only to the rigidity of the matrix fibers, or can the cell-generated tension alter the biochemistry displayed by the matrix proteins? Can molecular recognition sites that confer biochemical specificity to proteins be altered by stretching proteins? Is it possible that cell-generated tension is sufficient not just to strain but to mechanically unfold those proteins that form part of the force-bearing protein ECM networks in living tissue? Conclusive evidence that cell-generated forces are sufficient to unfold ECM proteins, and that the unfolding imparts new functional switches, rely most importantly on visualization techniques to probe protein conformations *ex vivo* and in cell culture.

Fluorescence resonance energy transfer (FRET) between multiple energy donors and acceptors (Figure 9.17) was indeed used to show that fibronectin in cell culture is exposed to cell-induced dynamic levels of stress which lead to partial fibronectin unfolding [53–55, 69, 70]. Fibronectin unfolding is hypothesized to mediate a variety of functions, ranging from altered mechanisms for cell binding

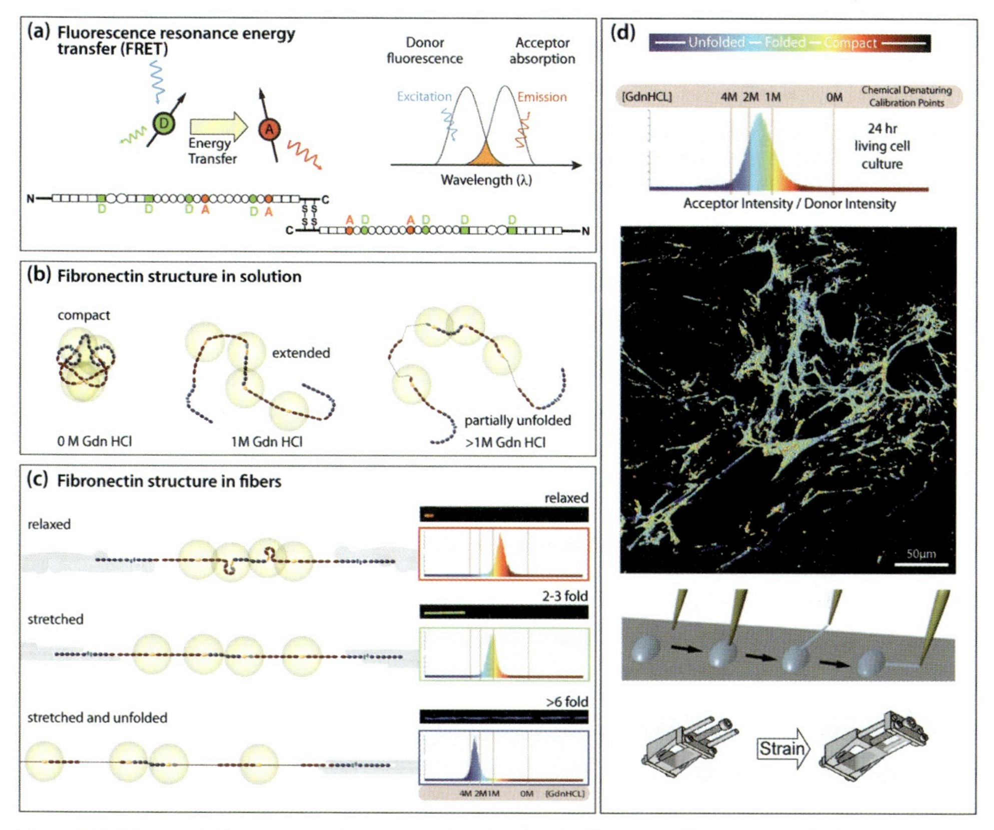

Figure 9.17 The use of fluorescence resonance energy transfer (FRET) to probe matrix stretching and unfolding in cell culture and artificially prepared fibronectin fibers. (a) The two free cysteines per fibronectin monomer, located on modules $FnIII_7$ and $FnIII_{15}$ (see Figure 9.4) are site-specifically labeled with the acceptors, while the donors are randomly distributed along the protein [69, 70]. The distance over which the acceptors can couple with potential energy donors are shown as yellow circles; (b) Schematic drawing of the fibronectin quaternary structure in solution and under denaturing conditions; (c) While the ultrastructure of fibronectin fibers remains unknown, FRET indicates that some quaternary structure is present when the fibers are fully relaxed. When the fibronectin fibers are stretched 200–300% of their equilibrium length, a first loss of secondary structure is observed [53–55, 250]; (d) Image of cell-made fibronectin matrix where trace amounts of FRET-labeled fibronectin were added to the cell culture medium. A broad distribution of different average fibronectin conformations can be seen. False colors have been used to visualize altered FRET ratios. Correlating FRET with mechanical strain was made possible by depositing fibronectin fibers onto stretchable PDMS sheets [250]. The fibers were therefore manually pulled out of a droplet of concentrated fibronectin solution. All of the conformations that can be seen in a broad range of strained artificial fibers coexist in every single field of view of a living cell culture.

to fibronectin to exposure of sites with enzymatic functions [21, 26, 53–55, 69, 70]. These cellular studies are important as they demonstrate that the nonequilibrium conformations of proteins can be stabilized by force and are thus physiologically relevant.

With regards to the translation of mechanical forces into biochemical signal changes, it should be noted that a wide variety of proteins (which are part of force-bearing protein networks linking the intracellular cytoskeleton and ECM) have multimodular structures where each individual module often carries one or more unique binding sites. However, what are the advantages of linking numerous modules equipped with different functionalities into macromolecules? The answer might be found in the following consideration. If domains of multimodular proteins were to possess similar mechanical stabilities, these domains would rupture in a stochastic sequence. But, if the domains have significantly different mechanical stabilities, as observed for fibronectin's FnIII modules [253–257], then sequential domain unfolding would lead to a well-defined graded sequence in which the various molecular recognition sites displayed by those domains would be altered as a function of mechanical strain. The significance of having a hierarchy of mechanical stabilities is thus central to mechanochemical signal conversion: as the mechanical hierarchy defines the sequence in which bonds or modules break, a sequence of stretching of different domains could be translated into a sequence of biochemical signal changes. This is also a likely explanation for why so many proteins contain buried cryptic sites in their hydrophobic interior that are only exposed when the protein unfolds [51, 52, 254, 256, 258–265]. Multidomain proteins are thus ideally suited to serve as mechanochemical signal converters to translate a large range of forces into sequential strain-specific functional changes [26].

Studies of the mechanical characteristics of proteins have thus revealed new insights into protein engineering principles. While fragments or a few domains of these molecules have often been employed in biotechnology or tissue engineering, mimicking only partial aspects of the whole molecule, engineers must ask what functional aspects might be missed if materials and surfaces were to be functionalized with peptides that contained only single molecular recognition motifs, as has been achieved for example with the cell-adhesive tripeptide RGD from the 450 kDa fibronectin molecule, instead of using the full-length protein. Such engineered systems would show a rather different mechanoresponse, and would not make it possible to translate a range of forces into a range of graded biochemical signal changes.

While the biochemistry of the quasi-equilibrium states of proteins is well understood, it is experimentally challenging to investigate how the structure–function relationship of proteins is altered when they are stretched. Deciphering the physiological significance of force-induced unfolding of fibronectin, and whether any of its versatile functions are up- or downregulated in a mechanoresponsive manner, however, has been hampered due to a lack of appropriate assays. In response to the growing need for quantitative biochemical and cellular assays that address whether the ECM acts as a mechanochemical signal converter to coregulate cellular mechanotransduction processes, we have developed a new assay where plasma fibronectin fibers are manually deposited onto elastic sheets, and force-induced changes in protein conformation are monitored using FRET (Figure 9.17). Our aim had been to develop a mechanical strain assays where the conformation of fibronectin can be adjusted externally on demand, while the force-induced protein extension is monitored optically [250, 251]. To probe how the alterations of the structure of stretched Fn impact biochemical interactions and cellular behavior, such a strain assay needs to be amenable to cell culture environments. To tune the conformation of fibronectin, fibers were drawn manually from a concentrated fibronectin solution [266–268]. When adding trace amounts of FRET-labeled fibronectin into the solution, followed by a step where the freshly drawn fibers are deposited onto polymeric sheets that are mounted into stretch devices, fibers can be generated that have a far more narrow conformational distribution, as found in native matrix [250, 251]. Furthermore, the mechanical strain can be externally adjusted, which enabled protein-binding studies to be conducted as a function of the strain of fibrillar fibronectin [250, 251]. An image of conformationally tuned fibronectin fibers can be seen in Figure 9.17. These assays further open the doors to the question of whether – and how – cell phenotypes are regulated by force-induced alterations of fibronectin conformation.

9.10.1.2 **Cell Responses to Initial Biomaterial Properties and Later to Self-Made Extracellular Matrix**

While the differentiation of mesenchymal stem cells has been correlated with the rigidity of the substrate on which they were initially plated [19], does the rigidity response of the cell change as it assembles its own matrix hours or days after the cells have been seeded on a substrate with defined rigidity? For example, the initial rigidity of a substrate has been proposed to determine whether or not mammary epithelial cells upregulate integrin expression and differentiate into a malignant phenotype [145], and also to dictate whether mesenchymal stem cells differentiate into bone, muscle or neuronal tissue [19]. In those experiments, the macroscopic materials properties of the substrate were typically correlated with outcome after four to ten days of cell culturing. While the mechanical properties of a substrate or engineered scaffold have indeed been correlated with various aspects of cell behavior, the underlying mechanisms how substrate rigidity ultimately regulates the long-term responses have not yet been defined.

Once cells have been seeded onto synthetic matrices they rapidly begin to assemble their own matrix, and will ultimately touch, feel and respond to their self-made ECM. A few hours after cells have attached to surfaces and begun to assemble their matrix, the physical characteristics of the newly assembled matrix do indeed depend on the rigidity of the substrate. After 4 h of cell culture, the FRET data showed that the fibronectin matrix was indeed more unfolded on a rigid (33 kPa) compared to a soft substrate (7 kPa) (Figure 9.18a) [269]. Surprisingly, however, after only one day the fibroblasts that were initially seeded onto glass had produced sufficient matrix so as to sit on a much softer biopolymer cushion. The cells then assembled a matrix that was far less unfolded than the matrix they made during the first 4 h on glass, as probed by adding FRET-labeled fibronectin only during the last 23–24 h after seeding. This newly made matrix is comparable to the cells feeling a 7 kPa substrate [269]. Most interestingly, the aging matrix changes its physical properties. When the cells were seeded onto glass and allowed to assemble matrix for three days, the matrix deposited during the first 24 h was highly unfolded, while the younger matrix was far less

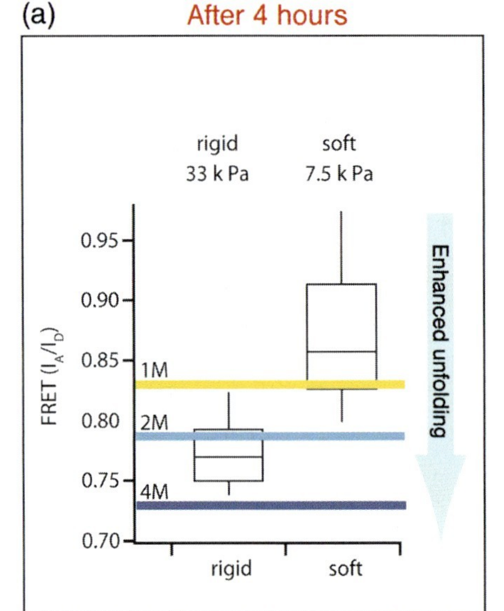

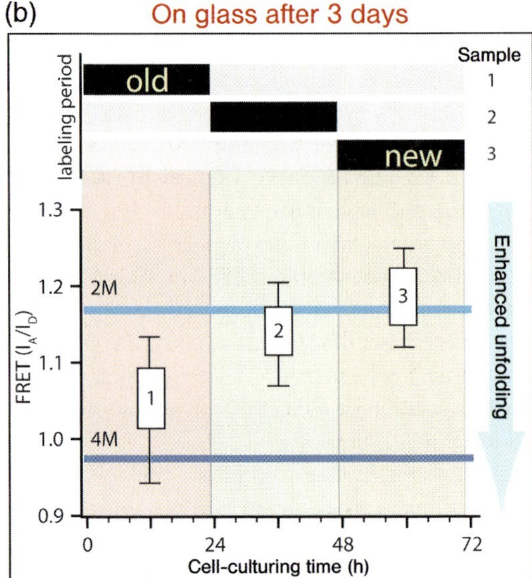

Figure 9.18 The tension of the extracellular matrix (and thus the mechanical strain of fibronectin) is upregulated with the rigidity of the substrate surface and changes as the matrix ages. (a) Fibronectin matrix assembly and unfolding on rigid and soft polyacrylamide surfaces 4 h after seeding the fibroblasts. The probabilities of finding certain FRET ratios are shown as boxes, where the maximum is given as the center line in the box, and the upper and lower ends of the boxes represent the 25th to 75th percentiles, and the 'whiskers' show the 2nd and 98th percentiles. FRET from fibrils on the rigid surface falls far below the FRET signature observed for fibronectin in solution at mild denaturing conditions (1 *M* GnHCl), thereby indicating that Fn is partially unfolded on rigid surfaces [53–55, 70, 250, 251]. FRET values on the soft surface indicate that the secondary structure of fibronectin is intact; (b) Three-day cell culture where FRET-labeled fibronectin was added for only limited time periods, as indicated in the upper bar graph. When the cells are seeded on glass and allowed to assemble matrix for three days, the matrix deposited during the first 24 h was highly unfolded, while the younger matrix was far less unfolded. Thus, the physical properties of matrix change as the matrix ages. (Adopted from Ref. [269].)

unfolded (Figure 9.18b). Thus, the matrix was progressively more unfolded as it aged, while the newly deposited matrix showed little unfolding. These data provided the first evidence that matrix maturation occurs, and that aging is associated with an increased stretching of fibronectin fibrils. Matrix assembly and remodeling involves at least partial unfolding of the secondary structure of fibronectin modules. Consequently, matured and aged matrix may display different physical and biochemical characteristics, and is structurally distinguishable from newly deposited matrix. A comparison of the conformation of Fn in these three-dimensional (3-D) matrices with those constructed by cells on rigid and flexible polyacrylamide surfaces suggests that cells in maturing matrices experience a microenvironment of gradually increased rigidity [269]. A future goal must be to understand the physiological consequences of matrix unfolding on cell function, including cancer and stem cell differentiation.

9.11
Cell Motility in Response to Force Generation and Matrix Properties

The relationship between force generation and motility is not simple. Fibroblasts that develop high forces on substrates do not move rapidly [222], whereas neutrophils that move at the highest rates reported for cells (about 40 $\mu m\,min^{-1}$ or 700 $nm\,s^{-1}$) generate very low forces on substrates [53–55]. There is considerable interest in the extravasation of cancer cells moving out of the bloodstream into tissues in the process of metastasis, although many of the steps in that process involve proteolysis of the matrix and deformation of the cell to pass through small gaps in the endothelium [270–272]. Although force generation is needed for motility, it is not the most important factor – indeed, cell polarization, matrix proteases, directional signals and the deformation of the cytoplasm are often as important.

Cell motility depends on substrate density and rigidity, and therefore also on the processes that respond to rigidity [273]. Many of the proteins involved in rigidity response have been linked to motility disorders, including cancer as well as malformations in development and neuronal connectivity. Src family kinases (SFKs) [274, 275], FAK [173, 174, 276]), the SH2 domain-containing phosphatase SHP-2 [173, 174] and RPTPs) [173, 174] are important components of the force-dependent signal transduction pathways that lead to the assembly of adhesion sites. The force-dependent initiation of adhesion sites and their rapid reinforcement occurs in protruding portions of cells, where adhesion sites can transmit cell propulsive forces [179, 277]. In extending regions of the cell, forces are generated on integrins by actin rearward flow rather than stress fibers. At the trailing end of the cell, mature focal adhesions create passive resistance during cell migration. To overcome this resistance, high forces must be generated by nascent adhesion sites [179]. However, in some static cells, higher forces are correlated with mature focal adhesions [58]. At the cell rear, traction stresses induce the disassembly instead of the reinforcement of focal adhesions and linked stress fibers [189, 278–280]; this is dependent on mechanosensitive ion channels and calcium signaling in keratocytes and astrocytoma cells. SFKs, FAK and PEST domain-enriched tyrosine phosphatase (PTP-PEST) are also crucial factors in adhesion site disassembly [229, 274, 281]. Recent studies have shown that the rigidity of 3-D matrices affects the migration rate differently from 2-D matrices, in that the less-rigid matrices cause an increase in migration rate [271]. At a biochemical level, the actin depolymerizing protein cofilin has been implicated in the motility of fibroblasts *in vitro*, and it is downstream in the biochemical signaling pathway of the integrins that have been activated at the leading edge [282–285]. These varying results indicate that different modalities of force generation and rigidity response at the cell front versus rear of the cell or in 3-D versus 2-D matrices can correlate with position-dependent regulation of phosphotyrosine signaling, and that different mechanisms of rigidity responses based on phosphotyrosine signaling can independently direct cell morphology as well as motility. Many observations point to tight links between morphology, migration, rigidity responses and tyrosine kinase activity.

9.12
Mechanical Forces and Force Sensing in Development and Disease

During the development and regeneration of tissues, forces act on and are propagated throughout most tissues. Such forces provide a local and global mechanism to shape cells and tissues, and to maintain homeostasis. Forces play a critical role by which cells interact with their environment and gain environmental feedback that regulates cell behavior. The signal for wound healing is often the loss of tissue integrity and the concomitant loss of force. Further, use of tissue and the periodic generation of force are often tied to the growth of that tissue, whereas inactivity is tied to the atrophy of the tissue. Bedridden patients suffer from a loss of muscle tissue and other aspects of atrophy. Similarly, with aging, osteoporosis, as well as many other cardiovascular diseases, mechanical changes and inappropriate responses of cells to mechanical changes, are critical and give rise to many symptoms. The size of the organism and its form are also set, at least in part, through a physical feedback between individual cells and their neighbors that is dynamic. As the cells grow and divide in development, they are constantly moving and even changing neighbors on occasion. The force-bearing cytoskeleton is actively remodeling and must clearly be responsive to changes in the level of force, or else the tissue would relax or contract too much. Contractile activity in individual cells can change the turgor of the tissue, and that parameter is under control of the signaling pathways that activate myosin contraction.

Consequently, forces play a critical role in health and disease in controlling the outcome of many biological processes (Figure 9.19). One obvious case is in cancer, where the cells ignore normal environmental cues and grow aberrantly, although there are many other examples (including problems with angiogenesis and tissue repair). Thus, it is speculated that in the future there will be an increasing emphasis

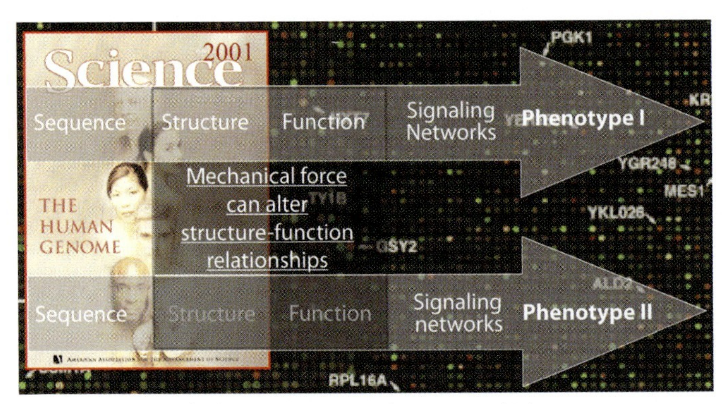

Figure 9.19 From the human genome to quantitative biology. The path is more complicated than originally thought, as mechanical force provides Nature with an additional dimension of regulating protein function. A switch in protein function might then alter cell signaling and thus the cell phenotype. The background shows the cover of *Science* announcing that details of the human genome had been resolved.

on the mechanical treatment of clinical problems and the targeting of therapeutics to mechanical response pathways. For the more effective treatment of diseases, there needs to be a greater recognition of the role that mechanical factors play in the development of the organism, as well as in the onset and progression of diseases. In other words, the genes code for a set of proteins that have developed the proper mechanical responses to shape the organism reproducibly. Similarly, disease-related alterations can result in alterations in mechanoresponsive pathways that are a major part of the disease. A much better understanding of those pathways is needed for proper treatment and therapy.

9.12.1 Cancer and Cell Transformation

Many cancer biologists have realized that cancer is inherently associated with a diseased mechanosensing and mechanoresponsive system. Many cancer cells ignore the normal environmental signals that regulate growth, and many of those cues are mechanical. For example, one of the early hallmarks of cancer cells is that they are often transformed, which was defined as the ability of those cells to grow on soft agar whereas normal cells required a rigid substrate [9–11]. Early observations linked transformation to uncontrolled cellular growth and to profound alterations in cell shape, as well as to the deregulation of tyrosine kinase and phosphatase activity. The first defined oncogene, *v-Src*, encodes an altered form of an important cellular tyrosine kinase, c-Src [286, 287]. In most studies on tumor cells, changes in morphology – but not cytoskeletal dynamics – have been reported.

The progression of cells from normal to a cancerous or even metastatic state is reflected in an increased softening of the cells, as probed by laser traps on suspended single cells [288]. Malignant fibroblasts, for example, have 30–35% less actin than normal cells. Transformed cells in culture often are rounder in morphology than primary cells. Tumor cells are also generally less adhesive than normal cells, and deposit less ECM [289], and the resulting loosened matrix adhesions may contribute to the ability of tumor cells to leave their original position in the tissue. In transformed cells, many aspects of nuclear and cell morphology as well as migration are altered. Focal adhesions can be replaced by podosomes and in addition, stress fibers can be absent [290]. Some transformed cells acquire anchorage independence – that is, they can grow without attachment to a substrate, suggesting a rigidity response deregulation. For example, transformed cells generate weak, poorly coordinated traction forces [291] but increased contractility. Thus, the one generality is that transformed cells can grow inappropriately, ignoring the mechanical cues of the environment that neighboring normal cells will follow to maintain appropriate tissue morphology. Although other factors, such as hormonal signals, form part of many cancers, the inappropriate mechanical responsiveness of cancer cells must also be considered as an important part of the process.

Cancer progression leads to a loss of tissue differentiation due to abnormal cell proliferation rates. Even if isolated malignant cells are associated with an increased softness of their overall cytoskeleton, it is equaly significant that tumors have a stiffer

ECM [145, 292, 293]. Malignant transformations of the breast, for example, are associated with dramatic changes in gland tension that include an increased ECM stiffness of the surrounding stroma [293]. Chronically increased mammary gland tension may influence tumor growth, perturb tissue morphogenesis, facilitate tumor invasion, and alter tumor survival and treatment responsiveness. However, changes in environmental factors (i.e. changes in ECM rigidity) and internal force generation (i.e. inappropriate rigidity responses) might be key factors in determining a transformed cell morphology and malignant phenotype [145]. For example, tumors are stiffer than normal tissue because they have a stiff stroma and elevated Rho-dependent cytoskeletal tension that drives focal adhesions, disrupts adherens junctions, perturbs tissue polarity, enhances growth, and hinders lumen formation [145]. Matrix stiffness thereby perturbs epithelial morphogenesis by clustering integrins to enhance ERK activation and increase ROCK-generated contractility and focal adhesions, thereby promoting malignant behavior [145].

Metastatic cells escape the tumor by invading the surrounding tissue, entering the circulation and finally attaching to previously unaffected tissues in often remote locations. In 1889, Stephen Paget published an article in *The Lancet* that described the propensity of various types of cancer to form metastases in specific organs, and proposed that these patterns were due to the "... dependence of the seed (the cancer cell) on the soil (the secondary organ)" [294]. This has often been linked to the chemical environment of the secondary organ, although recent results have indicated that it could also be a result of the mechanical environment in the secondary organ [295]. It was found that lung metastases from human breast cancer cells would grow better on soft fibronectin substrates than on hard, whereas bone metastases would grow better on hard than on soft (A. Kostic and M.P. Sheetz, unpublished results). Metastasis is an inefficient process, and many cancer cells are shed but few actually grow into a tumor at a new site. One reason for this is that the new site might not have the appropriate mechanical properties. At another level, tumor cells are generally less adhesive than normal cells and deposit less ECM [289]. The resulting loosened matrix adhesions, combined with the softened cytoskeleton, may contribute to the ability of tumor cells to leave their original position in the tissue and squeeze through tiny holes.

Many of the molecules discussed above play key roles in cancer progression, and also metastasis. Integrin-mediated cell adhesion leads to the activation of FAK and c-Src, after which the FAK–Src complex binds to and can phosphorylate various adaptor proteins such as p130Cas and paxillin. The results of recent studies have shown that the FAK–Src complex is activated in many tumor cells, and generates signals leading to tumor growth and metastasis (as reviewed in Ref. [296]). Tyrosine phosphorylation of paxillin regulates both the assembly and turnover of adhesion sites. Phosphorylated paxillin enhanced lamellipodial protrusions and thus promoted cell migration [126, 127]; the migration of tumor cells in 3-D matrices is then governed by matrix stiffness, along with cell–matrix adhesion and proteolysis [271]. As discussed above, the phosphorylation of p130Cas is upregulated when cells are located on a more rigid substrate.

The overall survival of breast cancer patients is inversely correlated with the levels of p130Cas (BCAR1) in the tumors [297] indicating that increased levels of p130Cas in tumor cells contributed to patient death. It was subsequently found that cell migration was activated by p130Cas and the associated GEF (AND-34 or BCAR3), which indicated that metastasis was favored by elevated p130Cas [298]. Both, p130Cas and a p130Cas binding protein, AND34 (BCAR3), will increase the epithelial to mesenchymal transition when overexpressed [298]. p130Cas appears to have a central role in cell growth and motility; in many cases, it is dramatically altered in its phosphotyrosine levels in correlation with transformation [299–302] and metastasis [303]. In the specific case of lung tumors, metastasis was increased following removal of the primary tumor, and required p130Cas expression. Further, the substrate domain YxxP tyrosines were needed for both invasive and metastatic properties of the cells [304]. Even the invasion of Matrigel and the formation of large podosome structures required the YxxP tyrosines. Thus, it is suggested that the inappropriate growth of cancer cells may be partly due to changes in the normal force and rigidity-sensing pathways that can alter the cellular program. This means, in turn, that the protein mechanisms involved in controlling mechanical responses can be good targets for therapies in cancer. In addition, mechanical treatments can possibly alter the course of cancers. Several levels can be identified in the process of mechanosensing, transduction and response where alterations in cancer cells could result in abnormal growth control. For example, c-Src, Fyn and Yes knockout cells are each missing three important Src family kinases, and will grow on soft agar while not sensing any difference between soft and hard agar. However, the restoration of Fyn activity will enable the cells to sense rigidity, and they will no longer grow on soft agar [198]. Thus, an understanding of the mechanisms of force and rigidity sensing can provide an important perspective on cancer.

9.12.2 **Angiogenesis**

The growth of new blood vessels – that is, angiogenesis – is crucial not only in tissue growth and remodeling but also in wound healing and cancer. Vascular development requires correct interactions among endothelial cells, pericytes and surrounding cells [24]. Thus, the formation of new blood vessels might be compromised if any of these interactions – including cell–matrix interactions, both with basement membranes and with surrounding ECMs – are perturbed. Equally important, the injury-mediated degradation of the ECM can lead to changes in matrix–integrin interactions, causing an impaired reactivity of the endothelial cells that will lead to vascular wall remodeling. Consequently, alterations in integrin signaling, growth factor signaling, and even of the architecture and composition of the ECM, might all affect vascular development. As in other motility processes, angiogenesis involves a very stereotypical set of movements of the endothelial cells that result in the formation of capillary tubes.

The role and mechanisms by which mechanical forces promote angiogenesis remain unclear. It is notable, however, that angiogenesis is regulated by integrin signaling [305–308]. Angiogenesis is furthermore promoted by vascular endothelial growth factor (VEGF). As tumor neovascularization plays critical roles for the development, progression and metastasis of cancers, new therapeutic approaches to treat malignancies have been aimed at controlling angiogenesis by monoclonal antibodies targeting VEGF, as well as with several tyrosine kinase inhibitors targeting VEGF-related pathways (for a review, see Ref. [309]).

VEGF binds to its transmembrane receptor by stimulated complex formation between VEGF receptor-2 and β_3 integrin. Prior studies have suggested, for example, that α_v-integrins ($\alpha_v\beta_3$ and $\alpha_v\beta_5$) could act as negative regulators of angiogenesis (as discussed in Refs [31, 32]). Neovascularization is impaired in mutant mice where the β_3 integrins were unable to undergo tyrosine phosphorylation [310]. The lack of integrin phosphorylation suppressed the complex formation with VEGF. Furthermore, the phosphorylation of VEGF receptor-2 was significantly reduced in cells expressing mutant β_3 compared to wild-type, leading to an impaired integrin activation in these cells. With its binding locations at both the N and C termini, VEGF also binds to fibronectin fibers [311, 312] and, when bound, has been shown to increase cell migration, proliferation and differentiation [311, 313, 314]. A reduced extracellular pH is one of the key signals that can induce angiogenesis. By demonstrating that VEGF binding to fibronectin is dependent on pH, and that released VEGF sustained biological activity, Goerges *et al.* [315] suggested that cells may use a lowered pH as a localized mechanism of controlled VEGF release [316]. Goerges and colleagues also suggested that VEGF might be stored in the ECM via interactions with fibronectin and heparan sulfate in tissues that are in need of vascularization, so that it can aid in directing the dynamic process of growth and migration of new blood vessels. If – and how – VEGF signaling is regulated by mechanical force, however, remains unclear.

Tumor blood vessels have an altered integrin expression profile, and both blood and lymphatic vessels have pathological lesions [28]. In contrast to healthy tissue, integrin β_4 signaling in tumor blood vessels promotes the onset of the invasive phase of pathological angiogenesis [317], while loss of the β_4 integrin subunit reduces tumorigenicity [318]. Integrin β_4 binds to laminin (Figure 9.5), which is enriched in basement membranes, but not to the RGD-ligand as exposed in fibronectin. Another difference from the RGD-binding integrins is that integrin β_4 connects to the cytoskeleton via plectin (not talin, as illustrated in Figure 9.7) [319], and little is known about the mechanoresponsivity of that linkage.

Another open question is whether degradation of the ECM is regulated by force. Exploring this question is of particular relevance since, in addition to serving as an anchoring scaffold and storage for growth factors, a group of angiogenesis regulators are derived from fragments of ECM or blood proteins. Endostatin, antithrombin and anastellin are members of this group of substances. Some of these compounds are currently undergoing clinical trials as inhibitors of tumor angiogenesis [320], as well as synthetic peptides modeled after these anti-angiogenic proteins, such as Anginex [321]. RGD-containing breakdown products of the ECM may also cause sustained vasodilation [87].

9.12.3
Tissue Engineering

Deciphering the mechanisms by which ECMs might sense and transduce mechanical stimulation into functional alterations of cell behavior and fate is also a critical issue in advancing tissue engineering and regenerative medicine, as our abilities to interface synthetic materials with living soft tissue to promote angiogenesis and healing are still rather limited. Beyond artificial skin, few advances have been made when using synthetic scaffolds to grow functional soft organs or organ patches that can at least support some crucial physiological organ functions. The conventional thinking was that once surfaces are coated with the correct set of biomolecules, the cells might recognize them as 'biological'. If it will indeed be confirmed that cells have the ability to dynamically regulate the biochemical display of the surrounding matrix on demand by applying force, the currently pursued more 'static' approaches for designing tissue engineering scaffolds do neglect force as major regulatory factor of ECM function [322]. If the cells are in contact with synthetic surfaces, their ability to dynamically regulate the conformations of matrix proteins by stretching them might be compromised. A full appreciation of the engineering principles of adhesion molecules, and of the complexity by which ECM can respond to cell contractility [26], might thus lead to new approaches for how to better engineer the interface between cells and synthetic materials.

The engineering of scaffolds that can regenerate soft organs or support some soft organ functions remains a daunting task. Scaffolds derived from natural tissues or matrix proteins have so far shown significantly better clinical performance than their synthetic counterparts [323, 324]. The creation of a bioartificial heart, for example, requires the engineering of cardiac architecture, as well as of appropriate cellular constituents and pump function. A major breakthrough in bioartificial heart engineering has recently been reported [14]. While many approaches have been attempted in the past, some success was achieved by using decellularized organ-specific matrices as scaffolds. For this, hearts were first decellularized by coronary perfusion with detergents; this preserved the underlying ECM, the vascular architecture, competent acellular valves and an intact chamber geometry (Figure 9.20). After decellularization, collagens I and III, laminin and fibronectin remained within the thinned heart matrix. The fiber composition and orientation of the myocardial ECM were preserved, the ventricular ECM was retained, and the vascular basal membranes remained intact. In order to mimic cardiac cell composition, these decellularized heart matrices were reseeded first with cardiac and later endothelial cells [14], with macroscopic contractions being observed at day 4. By day 8, under physiological load and electrical stimulation, the constructs could generate a pump function (equivalent to about 2% of adult or 25% of 16-week fetal heart function). Notably however, perfusion and physiological stimulation were absolutely needed for tissue formation and to regain tissue function. The authors speculated that such organs, if matured even further, could become transplantable either in part (e.g. as a ventricle for congenital heart disease such as hypoplastic left heart syndrome) or as an entire donor heart in end-stage heart failure. The technique was subsequently applied to a variety of mammalian organs, including lung, liver, kidney and muscle. Ongoing studies are directed towards optimizing reseeding strategies to promote the dispersion of cells throughout the construct, *in vitro* conditions required for organ maturation, and the choice of stem or progenitor cells necessary to generate either autologous or off-the-shelf bioartificial solid organs for transplantation. In contrast,

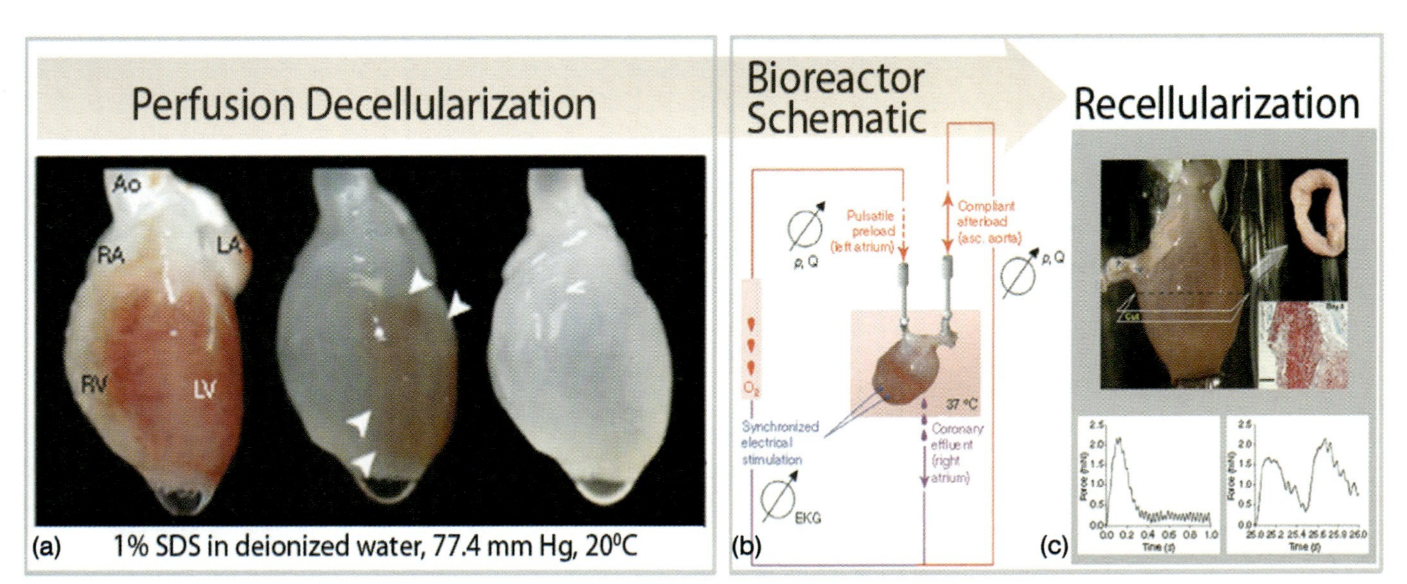

Figure 9.20 Decellularization and recellularization of a working heart-like construct. (a) Perfusion decellularization of whole rat hearts. Photographs of cadaveric rat hearts mounted on a Langendorff apparatus. Ao = aorta; LA = left atrium; LV = left ventricle; RA = right atrium; RV = right ventricle. Retrograde perfusion of cadaveric rat heart using sodium dodecyl sulfate (SDS) over 12 h. The heart becomes more translucent as cellular material is washed out from the right ventricle, followed by the atria and finally the left ventricle; (b) Schematic of working heart bioreactor showing cannulation of the left atrium and ascending (asc.) aorta. The heart is exposed to physiological preload, afterload and intraventricular pressure, and electrically stimulated at 5–20 V. Oxygenated medium containing serum and antibiotics enters through the left atrium and exits through the aortic valve. Pulsatile distention of the left ventricle and a compliance loop attached to the ascending aorta provide physiological coronary perfusion and afterload. Coronary perfusate (effluent) exits through the right atrium; (c) Formation of a working perfused bioartificial heart-like construct by recellularization of decellularized cardiac extracellular matrix. Top, recellularized whole rat heart at day 4 of perfusion culture in a working heart bioreactor. Upper insert: cross-sectional ring harvested for functional analysis (day 8). Lower insert: Masson's trichrome staining of a ring thin section, showing cells throughout the thickness of the wall. Scale bar = 100 mm. Bottom: force generation in left ventricular rings after 1 Hz (left) and 2 Hz (right) electrical stimulation.

stem cells directly injected into scarred tissue, for example into an infarcted heart, are not properly directed by the matrix to form new myocardium [325]. Rather than injecting cells into rigidified scar tissue, tissue regeneration was far more effective when transplanting an entire monolayer sheet of mesenchymal stem cells [326]. The engrafted cell sheet gradually grew to form a thick stratum that included newly formed vessels, undifferentiated cells and few cardiomyocytes which might be promoted by the more favorable microenvironment that the cells would find if transplanted as sheets rather than being injected individually.

Early studies have already shown that matrix crosslinking would compromise the ability of tissue-derived matrices for use in functional reconstruction [327]. One of several reasons for this might be that crosslinking alters the rigidity of the matrix, and an upregulated rigidity response of the reseeded cells might interfere with regaining tissue function. Another – not necessarily exclusive – possibility is that crosslinking would inhibit the protein conformation changes caused by stretching the matrix fibers. It has been found that fibronectin fibers, for example, can be stretched on average more than five times their equilibrium length before they break [250, 251], whereas crosslinked fibers show a markedly decreased extensibility. Crosslinked cell-derived matrices cause an upregulated cellular rigidity response and alter the biophysical properties of the matrix that the newly seeded cells are generating [328]. Thus, force-induced protein unfolding in a newly deposited matrix is, at least in part, upregulated by crosslinking, with all the functional implications as discussed above. Another aspect of native matrix, that might be compromised by crosslinking, is its ability to serve as a scaffold for storing cytokines and growth factors and to release them upon demand. Integrins were also shown recently to play a central role in activating the matrix-bound cytokine transforming growth factor-beta 1 (TGF-β1) by cell-generated tension acting on the matrix [329]. TGF-β1 controls tissue homeostasis in embryonic and normal adult tissues, and also contributes to the development of fibrosis, cancer, autoimmune and vascular diseases. In most of these conditions, active TGF-β1 is generated by dissociation from a large latent protein complex that sequesters latent TGF-β1 in the fibronectin-containing ECM [330]. The studies of Wipff and colleagues might suggest that matrix stiffness could regulate the equilibrium between storage and release of a host of matrix-bound growth factors [331].

Finally, the fact that not only the intact ECM but also its breakdown products have regulatory functions, can be actively exploited in tissue engineering. Low-molecular-weight peptides derived from the ECM, for example, can act as chemo-attractants for primary endothelial cells [138]. ECM extracts were found to have antimicrobial activity [332], and fragments of ECM or blood proteins, including endostatin, antithrombin and anastellin, may serves as inhibitors of angiogenesis [321]. Moreover, these angiostatic peptides use plasma fibronectin to home to the angiogenic vasculature [321]. Finally, uncharacterized digestive products of the ECM seem to act as strong inflammatory mediators [333]. Extensive future investigations are required in order to provide a full comprehension of the multifaceted regulatory roles of the ECM and its constituents, and how forces might coregulate many of these functions.

Consequently, learning how to switch the structure–function relationship of proteins by force has far-reaching potential not only in tissue engineering but also in biotechnology, and for the development of new drugs that might target proteins stretched into nonequilibrium states.

Acknowledgments

We gratefully acknowledge the many discussions with colleagues and our students, and thank in particular Sheila Luna for the graphics. Financial support was provided by the Nanotechnology Center for Mechanics in Regenerative Medicine (an NIH Roadmap Nanomedicine Development Center), the Volkswagen Stiftung, and various grants from NIH and ETH Zurich.

References

1 Korge, B.P. and Krieg, T. (1996) The molecular basis for inherited bullous diseases. *Journal of Molecular Medicine (Berlin, Germany)*, **74** (2), 59–70.

2 McGrath, J.A. (1999) Hereditary diseases of desmosomes. *Journal of Dermatological Science*, **20** (2), 85–91.

3 Jonkman, M.F., Pas, H.H., Nijenhuis, M., Kloosterhuis, G. and Steege, G. (2002) Deletion of a cytoplasmic domain of integrin beta4 causes epidermolysis bullosa simplex. *The Journal of Investigative Dermatology*, **119** (6), 1275–1281.

4 Pedersen, B.K. (2006) The anti-inflammatory effect of exercise: its role in diabetes and cardiovascular disease control. *Essays in Biochemistry*, **42**, 105–117.

5 Petersen, A.M. and Pedersen, B.K. (2005) The anti-inflammatory effect of exercise. *Journal of Applied Physiology (Bethesda, Md: 1985)*, **98** (4), 1154–1162.

6 Fries, R.S., Mahboubi, P., Mahapatra, N.R., Mahata, S.K., Schork, N.J., Schmid-Schoenbein, G.W. and O'Connor, D.T. (2004) Neuroendocrine transcriptome in genetic hypertension: multiple changes in diverse adrenal physiological systems. *Hypertension*, **43** (6), 1301–1311.

7 Harrison, D.G., Widder, J., Grumbach, I., Chen, W., Weber, M. and Searles, C. (2006) Endothelial mechanotransduction, nitric oxide and vascular inflammation. *Journal of Internal Medicine*, **259** (4), 351–363.

8 McGarry, J.D. (2002) Banting lecture 2001: dysregulation of fatty acid metabolism in the etiology of type 2 diabetes. *Diabetes*, **51** (1), 7–18.

9 Quigley, J.P. (1979) Phorbol ester-induced morphological changes in transformed chick fibroblasts: evidence for direct catalytic involvement of plasminogen activator. *Cell*, **17** (1), 131–141.

10 Giguere, L. and Gospodarowicz, D. (1983) Effect of Rous sarcoma virus transformation of rat-1 fibroblasts upon their growth factor and anchorage requirements in serum-free medium. *Cancer Research*, **43** (5), 2121–2130.

11 McClure, D.B. (1983) Anchorage-independent colony formation of SV40 transformed BALB/c-3T3 cells in serum-free medium: role of cell- and serum-derived factors. *Cell*, **32** (3), 999–1006.

12 Riha, G.M., Lin, P.H., Lumsden, A.B., Yao, Q. and Chen, C. (2005) Roles of hemodynamic forces in vascular cell differentiation. *Annals of Biomedical Engineering*, **33** (6), 772–779.

13 Jacques, A.M., Briceno, N., Messer, A.E., Gallon, C.E., Jalilzadeh, S., Garcia, E., Kikonda-Kanda, G., Goddard, J., Harding, S.E., Watkins, H., Esteban, M.T., Tsang,

V.T., McKenna, W.J. and Marston, S.B. (2008) The molecular phenotype of human cardiac myosin associated with hypertrophic obstructive cardiomyopathy. *Cardiovascular Research*, **79**, 481–491.

14 Ott, H.C., Matthiesen, T.S., Goh, S.K., Black, L.D., Kren, S.M., Netoff, T.I. and Taylor, D.A. (2008) Perfusion-decellularized matrix: using nature's platform to engineer a bioartificial heart. *Nature Medicine*, **14** (2), 213–221.

15 Liu, W.F., Nelson, C.M., Tan, J.L. and Chen, C.S. (2007) Cadherins, RhoA, and Rac1 are differentially required for stretch-mediated proliferation in endothelial versus smooth muscle cells. *Circulation Research*, **101** (5), e44–e52.

16 Sims, T.N., Soos, T.J., Xenias, H.S., Dubin-Thaler, B., Hofman, J.M., Waite, J.C., Cameron, T.O., Thomas, V.K., Varma, R., Wiggins, C.H., Sheetz, M.P., Littman, D.R. and Dustin, M.L. (2007) Opposing effects of PKCtheta and WASp on symmetry breaking and relocation of the immunological synapse. *Cell*, **129** (4), 773–785.

17 Badylak, S.F. (2007) The extracellular matrix as a biologic scaffold material. *Biomaterials*, **28** (25), 3587–3593.

18 Discher, D.E., Janmey, P. and Wang, Y.L. (2005) Tissue cells feel and respond to the stiffness of their substrate. *Science*, **310** (5751), 1139–1143.

19 Engler, A.J., Sen, S., Sweeney, H.L. and Discher, D.E. (2006) Matrix elasticity directs stem cell lineage specification. *Cell*, **126** (4), 677–689.

20 Ingber, D.E. (2006) Cellular mechano-transduction: putting all the pieces together again. *The FASEB Journal*, **20** (7), 811–827.

21 Vogel, V. and Sheetz, M. (2006) Local force and geometry sensing regulate cell functions. *Nature Reviews - Molecular Cell Biology*, **7** (4), 265–275.

22 Hartsock, A. and Nelson, W.J. (2008) Adherens and tight junctions: structure, function and connections to the actin cytoskeleton. *Biochimica et Biophysica Acta*, **1778** (3), 660–669.

23 Farhadifar, R., Roper, J.C., Aigouy, B., Eaton, S. and Julicher, F. (2007) The influence of cell mechanics, cell-cell interactions, and proliferation on epithelial packing. *Current Biology*, **17** (24), 2095–2104.

24 Hynes, R.O. (2007) Cell-matrix adhesion in vascular development. *Journal of Thrombosis and Haemostasis*, **5** (Suppl. 1), 32–40.

25 Mao, Y. and Schwarzbauer, J.E. (2005) Fibronectin fibrillogenesis, a cell-mediated matrix assembly process. *Matrix Biology*, **24**, 389–399.

26 Vogel, V. (2006) Mechanotransduction involving multimodular proteins: converting force into biochemical signals. *Annual Review of Biophysics and Biomolecular Structure*, **35**, 459–488.

27 Kadler, K.E., Hill, A. and Canty-Laird, E.G. (2008) Collagen fibrillogenesis: fibronectin, integrins, and minor collagens as organizers and nucleators. *Current Opinion in Cell Biology*, **20**, 495–501.

28 Ruoslahti, E. (2004) Vascular zip codes in angiogenesis and metastasis. *Biochemical Society Transactions*, **32** (Pt 3), 397–402.

29 Zaman, M.H. (2007) Understanding the molecular basis for differential binding of integrins to collagen and gelatine. *Biophysical Journal*, **92** (2), L17–L19.

30 Mizejewski, G.J. (1999) Role of integrins in cancer: survey of expression patterns. *Proceedings of the Society for Experimental Biology and Medicine*, **222** (2), 124–138.

31 Hynes, R.O. (2002) Integrins: bidirectional, allosteric signaling machines. *Cell*, **110** (6), 673–687.

32 Hynes, R.O. (2002) A reevaluation of integrins as regulators of angiogenesis. *Nature Medicine*, **8** (9), 918–921.

33 Ginsberg, M.H., Partridge, A. and Shattil, S.J. (2005) Integrin regulation. *Current Opinion in Cell Biology*, **17** (5), 509–516.

34 Arnaout, M.A., Goodman, S.L. and Xiong, J.P. (2007) Structure and mechanics of integrin-based cell adhesion. *Current Opinion in Cell Biology*, **19** (5), 495–507.

35 Luo, B.H., Carman, C.V. and Springer, T.A. (2007) Structural basis of integrin regulation and signalling. *Annual Review of Immunology*, **25**, 619–647.

36 Petrich, B.G., Marchese, P., Ruggeri, Z.M., Spiess, S., Weichert, R.A., Ye, F., Tiedt, R., Skoda, R.C., Monkley, S.J., Critchley, D.R. and Ginsberg, M.H. (2007) Talin is required for integrin-mediated platelet function in hemostasis and thrombosis. *The Journal of Experimental Medicine*, **204** (13), 3103–3111.

37 Han, Y., Cowin, S.C., Schaffler, M.B. and Weinbaum, S. (2004) Mechanotransduction and strain amplification in osteocyte cell processes. *Proceedings of the National Academy of Sciences of the United States of America*, **101** (47), 16689–16694.

38 Davies, P.F., Spaan, J.A. and Krams, R. (2005) Shear stress biology of the endothelium. *Annals of Biomedical Engineering*, **33** (12), 1714–1718.

39 Chien, S. (2008) Effects of disturbed flow on endothelial cells. *Annals of Biomedical Engineering*, **36** (4), 554–562.

40 Robling, A.G., Castillo, A.B. and Turner, C.H. (2006) Biomechanical and molecular regulation of bone remodelling. *Annual Review of Biomedical Engineering*, **8**, 455–498.

41 Wang, Y., McNamara, L.M., Schaffler, M.B. and Weinbaum, S. (2007) A model for the role of integrins in flow induced mechanotransduction in osteocytes. *Proceedings of the National Academy of Sciences of the United States of America*, **104** (40), 15941–15946.

42 Hooper, S.B. and Wallace, M.J. (2006) Role of the physicochemical environment in lung development. *Clinical and Experimental Pharmacology and Physiology*, **33** (3), 273–279.

43 Vlahakis, N.E. and Hubmayr, R.D. (2003) Response of alveolar cells to mechanical stress. *Current Opinion in Critical Care*, **9** (1), 2–8.

44 Classen, A.K., Anderson, K.I., Marois, E. and Eaton, S. (2005) Hexagonal packing of Drosophila wing epithelial cells by the planar cell polarity pathway. *Developmental Cell*, **9** (6), 805–817.

45 Marois, E. and Eaton, S. (2007) RNAi in the Hedgehog signaling pathway: pFRiPE, a vector for temporally and spatially controlled RNAi in *Drosophila*. *Methods in Molecular Biology (Clifton, NJ)*, **397**, 115–128.

46 Bao, G. and Suresh, S. (2003) Cell and molecular mechanics of biological materials. *Nature Materials*, **2** (11), 715–725.

47 Bustamante, C., Chemla, Y.R., Forde, N.R. and Izhaky, D. (2004) Mechanical processes in biochemistry. *Annual Review of Biochemistry*, **73**, 705–748.

48 Chen, C.S., Tan, J. and Tien, J. (2004) Mechanotransduction at cell-matrix and cell-cell contacts. *Annual Review of Biomedical Engineering*, **6**, 275–302.

49 Chen, J., Takagi, J., Xie, C., Xiao, T., Luo, B.H. and Springer, T.A. (2004) The relative influence of metal ion binding sites in the I-like domain and the interface with the hybrid domain on rolling and firm adhesion by integrin alpha4beta7. *The Journal of Biological Chemistry*, **279** (53), 55556–55561.

50 Orr, A.W., Helmke, B.P., Blackman, B.R. and Schwartz, M.A. (2006) Mechanisms of mechanotransduction. *Developmental Cell*, **10** (1), 11–20.

51 Sawada, Y., Tamada, M., Dubin-Thaler, B.J., Cherniavskaya, O., Sakai, R., Tanaka, S. and Sheetz, M.P. (2006) Force sensing by mechanical extension of the Src family kinase substrate p130Cas. *Cell*, **127** (5), 1015–1026.

52 Johnson, C.P., Tang, H.Y., Carag, C., Speicher, D.W. and Discher, D.E. (2007) Forced unfolding of proteins within cells. *Science*, **317** (5838), 663–666.

53 Smith, L.A., Aranda-Espinoza, H., Haun, J.B., Dembo, M. and Hammer, D.A. (2007) Neutrophil traction stresses are concentrated in the uropod during migration. *Biophysical Journal*, **92** (7), L58–L60.

54 Smith, M.L., Gourdon, D., Little, W.C., Kubow, K.E., Eguiluz, R.A., Luna-Morris, S. and Vogel, V. (2007) Force-induced unfolding of fibronectin in the extracellular matrix of living cells. *PLoS Biology*, **5** (10), e268.

55 Smith, M.L., Gourdon, D., Little, W.C., Kubow, K.E., Eguiluz, R.A., Luna-Morris, S. and Vogel, V. (2007) Force-induced unfolding of fibronectin in the extracellular matrix of living cells. *Public Library of Science Biology*, **5** (10).

56 Vogel, V., Sheetz, M.P. (2009) Cell Fate Regulation by Coupling Mechanical Cycles to Biochemical Signaling Pathways. *Current Opinion Cell Biol.* **21** (1):in press.

57 Harris, A.K., Wild, P. and Stopak, D. (1980) Silicone rubber substrata: a new wrinkle in the study of cell locomotion. *Science*, **208** (4440), 177–179.

58 Balaban, N.Q., Schwarz, U.S., Riveline, D., Goichberg, P., Tzur, G., Sabanay, I., Mahalu, D., Safran, S., Bershadsky, A., Addadi, L. and Geiger, B. (2001) Force and focal adhesion assembly: a close relationship studied using elastic micropatterned substrates. *Nature Cell Biology*, **3** (5), 466–472.

59 Tan, J.L., Tien, J., Pirone, D.M., Gray, D.S., Bhadriraju, K. and Chen, C.S. (2003) Cells lying on a bed of microneedles: an approach to isolate mechanical force. *Proceedings of the National Academy of Sciences of the United States of America*, **100** (4), 1484–1489.

60 Chrzanowska-Wodnicka, M. and Burridge, K. (1996) Rho-stimulated contractility drives the formation of stress fibers and focal adhesions. *The Journal of Cell Biology*, **133** (6), 1403–1415.

61 Helfman, D.M., Levy, E.T., Berthier, C., Shtutman, M., Riveline, D., Grosheva, I., Lachish-Zalait, A., Elbaum, M. and Bershadsky, A.D. (1999) Caldesmon inhibits nonmuscle cell contractility and interferes with the formation of focal adhesions. *Molecular Biology of the Cell*, **10** (10), 3097–3112.

62 Riveline, D., Zamir, E., Balaban, N.Q., Schwarz, U.S., Ishizaki, T., Narumiya, S., Kam, Z., Geiger, B. and Bershadsky, A.D. (2001) Focal contacts as mechanosensors: externally applied local mechanical force induces growth of focal contacts by an mDia1-dependent and ROCK-independent mechanism. *The Journal of Cell Biology*, **153** (6), 1175–1186.

63 Galbraith, C.G., Yamada, K.M. and Sheetz, M.P. (2002) The relationship between force and focal complex development. *The Journal of Cell Biology*, **159** (4), 695–705.

64 Giannone, G., Jiang, G., Sutton, D.H., Critchley, D.R. and Sheetz, M.P. (2003) Talin1 is critical for force-dependent reinforcement of initial integrin-cytoskeleton bonds but not tyrosine kinase activation. *The Journal of Cell Biology*, **163** (2), 409–419.

65 Kellermayer, M.S., Smith, S.B., Granzier, H.L. and Bustamante, C. (1997) Folding-unfolding transitions in single titin molecules characterized with laser tweezers. *Science*, **276** (5315), 1112–1116.

66 Rief, M., Gautel, M., Oesterhelt, F., Fernandez, J.M. and Gaub, H.E. (1997) Reversible unfolding of individual titin immunoglobulin domains by AFM. *Science*, **276** (5315), 1109–1112.

67 Tskhovrebova, L., Trinick, J., Sleep, J.A. and Simmons, R.M. (1997) Elasticity and unfolding of single molecules of the giant muscle protein titin. *Nature*, **387** (6630), 308–312.

68 Lu, H., Isralewitz, B., Krammer, A., Vogel, V. and Schulten, K. (1998) Unfolding of titin immunoglobulin domains by steered molecular dynamics simulation. *Biophysical Journal*, **75**, 662–671.

69 Baneyx, G., Baugh, L. and Vogel, V. (2001) Coexisting conformations of fibronectin in cell culture imaged using fluorescence resonance energy transfer. *Proceedings of the National Academy of Sciences of the United States of America*, **98** (25), 14464–14468.

70 Baneyx, G., Baugh, L. and Vogel, V. (2002) Fibronectin extension and unfolding within cell matrix fibrils controlled by cytoskeletal tension. *Proceedings of the National Academy of Sciences of the United States of America*, **99** (8), 5139–5143.

71 Sawada, Y. and Sheetz, M.P. (2002) Force transduction by Triton cytoskeletons. *The Journal of Cell Biology*, **156** (4), 609–615.

72 Clausen-Schaumann, H., Seitz, M., Krautbauer, R. and Gaub, H.E. (2000) Force spectroscopy with single bio-molecules. *Current Opinion in Chemical Biology*, **4** (5), 524–530.

73 Fredberg, J.J. and Kamm, R.D. (2006) Stress transmission in the lung: Pathways from organ to molecule. *Annual Review of Physiology*, **68**, 507–541.

74 Khan, S. and Sheetz, M.P. (1997) Force effects on biochemical kinetics. *Annual Review of Biochemistry*, **66**, 785–805.

75 Bershadsky, A.D., Balaban, N.Q. and Geiger, B. (2003) Adhesion-dependent cell mechanosensitivity. *Annual Review of Cell and Developmental Biology*, **19**, 677–695.

76 Silver, F.H. and Siperko, L.M. (2003) Mechanosensing and mechanochemical transduction: how is mechanical energy sensed and converted into chemical energy in an extracellular matrix? *Critical Reviews in Biomedical Engineering*, **31** (4), 255–331.

77 Martinac, B. (2004) Mechanosensitive ion channels: molecules of mechanotransduction. *Journal of Cell Science*, **117** (Pt 12), 2449–2460.

78 Kung, C. (2005) A possible unifying principle for mechanosensation. *Nature*, **436** (7051), 647–654.

79 Shemesh, T., Geiger, B., Bershadsky, A.D. and Kozlov, M.M. (2005) Focal adhesions as mechanosensors: A physical mechanism. *Proceedings of the National Academy of Sciences of the United States of America*, **102** (35), 12383–12388.

80 Hytönen, V.P., Smith, M.L. and Vogel, V. (2009) Translating mechanical force into discrete biochemical signal changes: multimodularity imposes unique properties to mechanotransductive proteins. *Mechanotransduction* (eds R. Kamm and M.R.K. Mofrad), Cambridge University Press, in press.

81 Ingber, D.E. (2005) Mechanical control of tissue growth: function follows form. *Proceedings of the National Academy of Sciences of the United States of America*, **102** (33), 11571–11572.

82 Wang, N. and Suo, Z. (2005) Long-distance propagation of forces in a cell. *Biochemical and Biophysical Research Communications*, **328** (4), 1133–1138.

83 Gittes, F., Meyhofer, E., Baek, S. and Howard, J. (1996) Directional loading of the kinesin motor molecule as it buckles a microtubule. *Biophysical Journal*, **70** (1), 418–429.

84 Block, S.M., Asbury, C.L., Shaevitz, J.W. and Lang, M.J. (2003) Probing the kinesin reaction cycle with a 2D optical force clamp. *Proceedings of the National Academy of Sciences of the United States of America*, **100** (5), 2351–2356.

85 Sukharev, S.I., Blount, P., Martinac, B., Blattner, F.R. and Kung, C. (1994) A large-conductance mechanosensitive channel in *E. coli* encoded by mscL alone. *Nature*, **368** (6468), 265–268.

86 Gullingsrud, J. and Schulten, K. (2004) Lipid bilayer pressure profiles and mechanosensitive channel gating. *Biophysical Journal*, **86** (6), 3496–3509.

87 Wu, X., Yang, Y., Gui, P., Sohma, Y., Meininger, G.A., Davis, G.E., Braun, A.P. and Davis, M.J. (2008) Potentiation of large conductance, Ca^{2+}-activated K^{+} (BK) channels by alpha5beta1 integrin activation in arteriolar smooth muscle. *The Journal of Physiology*, **586** (6), 1699–1713.

88 Merkel, R., Nassoy, P., Leung, A., Ritchie, K. and Evans, E. (1999) Energy landscapes of receptor-ligand bonds explored with dynamic force spectroscopy. *Nature*, **397** (6714), 50–53.

89 Hess, H., Howard, J. and Vogel, V. (2002) A piconewton forcemeter assembled from microtubules and kinesins. *Nano Letters*, **2** (10), 1113–1115.

90 Zhu, C. and McEver, R.P. (2005) Catch bonds: physical models and biological functions. *Molecular and Cellular Biomechanics*, **2** (3), 91–104.

91 Evans, E.A. and Calderwood, D.A. (2007) Forces and bond dynamics in cell adhesion. *Science*, **316** (5828), 1148–1153.

92 Sokurenko, E., Vogel, V. and Thomas, W.E. (2008) Catch bond mechanism of force-enhanced adhesion: counter-intuitive, elusive but . . . widespread? *Cell Host and Microbe*, **4**, 314–323.

93 Thomas, W.E., Vogel, V. and Sokurenko, E. (2008) Biophysics of catch bonds. *Annual Review of Biophysics*, **37**, 399–416.

94 Bianchi, L. (2007) Mechanotransduction: touch and feel at the molecular level as modeled in *Caenorhabditis elegans*. *Molecular Neurobiology*, **36**, 254–271.

95 Christensen, A.P. and Corey, D.P. (2007) TRP channels in mechanosensation: direct or indirect activation? *Nature Reviews in Neuroscience*, **8**, 510–521.

96 Kolomeisky, A.B. and Fisher, M.E. (2007) Molecular motors: a theorist's perspective. *Annual Review of Physical Chemistry*, **58**, 675–695.

97 Bartman, T., Walsh, E.C., Wen, K.K., McKane, M., Ren, J., Alexander, J., Rubenstein, P.A. and Stainier, D.Y. (2004) Early myocardial function affects endocardial cushion development in zebrafish. *PLoS Biology*, **2**, E129.

98 Vignjevic, D. and Montagnac, G. (2008) Reorganisation of the dendritic actin network during cancer cell migration and invasion. *Seminars in Cancer Biology*, **18** (1), 12–22.

99 Gupton, S.L. and Gertler, F.B. (2007) Filopodia: the fingers that do the walking. *Science's STKE: Signal Transduction Knowledge Environment*, **2007** (400), re5.

100 Medalia, O., Beck, M., Ecke, M., Weber, I., Neujahr, R., Baumeister, W. and Gerisch, G. (2007) Organization of actin networks in intact filopodia. *Current Biology*, **17** (1), 79–84.

101 Evans, E. (2001) Probing the relation between force--lifetime--and chemistry in single molecular bonds. *Annual Review of Biophysics and Biomolecular Structure*, **30**, 105–128.

102 Lehenkari, P.P. and Horton, M.A. (1999) Single integrin molecule adhesion forces in intact cells measured by atomic force microscopy. *Biochemical and Biophysical Research Communications*, **259** (3), 645–650.

103 Sun, Z., Martinez-Lemus, L.A., Trache, A., Trzeciakowski, J.P., Davis, G.E., Pohl, U. and Meininger, G.A. (2005) Mechanical properties of the interaction between fibronectin and {alpha}5{beta}1 integrins on vascular smooth muscle cells studied using atomic force microscopy. *American Journal of Physiology. Heart and Circulatory Physiology*, **289**, 2526–2535.

104 Hynes, R.O. (1990) *Fibronectins*, Springer-Verlag, New York, NY.

105 Pankov, R. and Yamada, K.M. (2002) Fibronectin at a glance. *Journal of Cell Science*, **115** (Pt 20), 3861–3863.

106 Mould, A.P. and Humphries, M.J. (2004) Regulation of integrin function through conformational complexity: not simply a knee-jerk reaction? *Current Opinion in Cell Biology*, **16** (5), 544–551.

107 Arnaout, M.A., Mahalingam, B. and Xiong, J.P. (2005) Integrin structure, allostery, and bidirectional signaling. *Annual Review of Cell and Developmental Biology*, **21**, 381–410.

108 Wegener, K.L., Partridge, A.W., Han, J., Pickford, A.R., Liddington, R.C., Ginsberg, M.H. and Campbell, I.D. (2007) Structural basis of integrin activation by talin. *Cell*, **128** (1), 171–182.

109 Mould, A.P., Barton, S.J., Askari, J.A., Craig, S.E. and Humphries, M.J. (2003) Role of ADMIDAS cation-binding site in ligand recognition by integrin alpha 5 beta 1. *The Journal of Biological Chemistry*, **278** (51), 51622–51629.

110 Adair, B.D., Xiong, J.P., Maddock, C., Goodman, S.L., Arnaout, M.A. and Yeager, M. (2005) Three-dimensional EM structure of the ectodomain of integrin {alpha}V{beta}3 in a complex with fibronectin. *The Journal of Cell Biology*, **168** (7), 1109–1118.

111 Mould, A.P., Travis, M.A., Barton, S.J., Hamilton, J.A., Askari, J.A., Craig, S.E., Macdonald, P.R., Kammerer, R.A., Buckley, P.A. and Humphries, M.J. (2005) Evidence that monoclonal antibodies directed against the integrin beta subunit plexin/semaphorin/integrin domain stimulate function by inducing receptor extension. *The Journal of Biological Chemistry*, **280** (6), 4238–4246.

112 Puklin-Faucher, E., Gao, M., Schulten, K. and Vogel, V. (2006) How the headpiece hinge angle is opened: New insights into the dynamics of integrin activation. *The Journal of Cell Biology*, **175** (2), 349–360.

113 Calderwood, D.A. (2004) Talin controls integrin activation. *Biochemical Society Transactions*, **32** (Pt 3), 434–437.

114 Calderwood, D.A. and Ginsberg, M.H. (2003) Talin forges the links between integrins and actin. *Nature Cell Biology*, **5** (8), 694–697.

115 Jiang, G., Giannone, G., Critchley, D.R., Fukumoto, E. and Sheetz, M.P. (2003) Two-piconewton slip bond between fibronectin and the cytoskeleton depends on talin. *Nature*, **424** (6946), 334–337.

116 Calderwood, D.A., Zent, R., Grant, R., Rees, D.J., Hynes, R.O. and Ginsberg, M.H. (1999) The talin head domain binds to integrin beta subunit cytoplasmic tails and regulates integrin activation. *The Journal of Biological Chemistry*, **274** (40), 28071–28074.

117 Barsukov, I.L., Prescot, A., Bate, N., Patel, B., Floyd, D.N., Bhanji, N., Bagshaw, C.R., Letinic, K., Di Paolo, G., De Camilli, P., Roberts, G.C. and Critchley, D.R. (2003) Phosphatidylinositol phosphate kinase type 1gamma and beta1-integrin cytoplasmic domain bind to the same region in the talin FERM domain. *The Journal of Biological Chemistry*, **278** (33), 31202–31209.

118 Borowsky, M.L. and Hynes, R.O. (1998) Layilin, a novel talin-binding transmembrane protein homologous with C-type lectins, is localized in membrane ruffles. *The Journal of Cell Biology*, **143** (2), 429–442.

119 Lee, H.S., Bellin, R.M., Walker, D.L., Patel, B., Powers, P., Liu, H., Garcia-Alvarez, B., de Pereda, J.M., Liddington, R.C., Volkmann, N., Hanein, D., Critchley, D.R. and Robson, R.M. (2004) Characterization of an actin-binding site within the talin FERM domain. *Journal of Molecular Biology*, **343** (3), 771–784.

120 Molony, L., McCaslin, D., Abernethy, J., Paschal, B. and Burridge, K. (1987) Properties of talin from chicken gizzard smooth muscle. *The Journal of Biological Chemistry*, **262** (16), 7790–7795.

121 McLachlan, A.D., Stewart, M., Hynes, R.O. and Rees, D.J. (1994) Analysis of repeated motifs in the talin rod. *Journal of Molecular Biology*, **235** (4), 1278–1290.

122 Gingras, A.R., Ziegler, W.H., Frank, R., Barsukov, I.L., Roberts, G.C., Critchley, D.R. and Emsley, J. (2005) Mapping and consensus sequence identification for multiple vinculin binding sites within the talin rod. *The Journal of Biological Chemistry*, **280** (44), 37217–37224.

123 Hemmings, L., Rees, D.J., Ohanian, V., Bolton, S.J., Gilmore, A.P., Patel, B., Priddle, H., Trevithick, J.E., Hynes, R.O. and Critchley, D.R. (1996) Talin contains three actin-binding sites each of which is adjacent to a vinculin-binding site. *Journal of Cell Science*, **109**, 2715–2726.

124 Xing, B., Jedsadayanmata, A. and Lam, S.C. (2001) Localization of an integrin binding site to the C terminus of talin. *The Journal of Biological Chemistry*, **276** (48), 44373–44378.

125 Wiesner, S., Lange, A. and Fassler, R. (2006) Local call: from integrins to actin assembly. *Trends in Cell Biology*, **16** (7), 327–329.

126 Zaidel-Bar, R., Itzkovitz, S., Ma'ayan, A., Iyengar, R. and Geiger, B. (2007)

Functional atlas of the integrin adhesome. *Nature Cell Biology*, **9** (8), 858–867.

127 Zaidel-Bar, R., Milo, R., Kam, Z. and Geiger, B. (2007) A paxillin tyrosine phosphorylation switch regulates the assembly and form of cell-matrix adhesions. *Journal of Cell Science*, **120** (Pt 1), 137–148.

128 Moser, M., Nieswandt, B., Ussar, S., Pozgajova, M. and Fassler, R. (2008) Kindlin-3 is essential for integrin activation and platelet aggregation. *Nature Medicine*, **14** (3), 325–330.

129 Kiema, T., Lad, Y., Jiang, P., Oxley, C.L., Baldassarre, M., Wegener, K.L., Campbell, I.D., Ylanne, J. and Calderwood, D.A. (2006) The molecular basis of filamin binding to integrins and competition with talin. *Molecular Cell*, **21** (3), 337–347.

130 McCleverty, C.J., Lin, D.C. and Liddington, R.C. (2007) Structure of the PTB domain of tensin1 and a model for its recruitment to fibrillar adhesions. *Protein Science: A Publication of the Protein Society*, **16** (6), 1223–1229.

131 Litjens, S.H. and de Pereda, J.M. and Sonnenberg, A. (2006) Current insights into the formation and breakdown of hemidesmosomes. *Trends in Cell Biology*, **16** (7), 376–383.

132 Otey, C.A., Vasquez, G.B., Burridge, K. and Erickson, B.W. (1993) Mapping of the alpha-actinin binding site within the beta 1 integrin cytoplasmic domain. *The Journal of Biological Chemistry*, **268** (28), 21193–21197.

133 Greenwood, J.A., Theibert, A.B., Prestwich, G.D. and Murphy-Ullrich, J.E. (2000) Restructuring of focal adhesion plaques by PI 3-kinase. Regulation by PtdIns (3,4,5)-p(3) binding to alpha-actinin. *The Journal of Cell Biology*, **150** (3), 627–642.

134 Mostafavi-Pour, Z., Askari, J.A., Parkinson, S.J., Parker, P.J., Ng, T.T. and Humphries, M.J. (2003) Integrin-specific signaling pathways controlling focal adhesion formation and cell migration. *The Journal of Cell Biology*, **161** (1), 155–167.

135 Hu, S. and Wang, N. (2006) Control of stress propagation in the cytoplasm by prestress and loading frequency. *Molecular & Cellular Biomechanics*, **3** (2), 49–60.

136 Dai, J. and Sheetz, M.P. (1999) Membrane tether formation from blebbing cells. *Biophysical Journal*, **77** (6), 3363–3370.

137 Katz, B.Z., Miyamoto, S., Teramoto, H., Zohar, M., Krylov, D., Vinson, C., Gutkind, J.S. and Yamada, K.M. (2002) Direct transmembrane clustering and cytoplasmic dimerization of focal adhesion kinase initiates its tyrosine phosphorylation. *Biochimica et Biophysica Acta*, **1592** (2), 141–152.

138 Li, R., Bennett, J.S. and Degrado, W.F. (2004) Structural basis for integrin alphaII beta3 clustering. *Biochemical Society Transactions*, **32** (Pt 3), 412–415.

139 Cluzel, C., Saltel, F., Lussi, J., Paulhe, F., Imhof, B.A. and Wehrle-Haller, B. (2005) The mechanisms and dynamics of (alpha) v(beta)3 integrin clustering in living cells. *The Journal of Cell Biology*, **171** (2), 383–392.

140 Coussen, F., Choquet, D., Sheetz, M.P. and Erickson, H.P. (2002) Trimers of the fibronectin cell adhesion domain localize to actin filament bundles and undergo rearward translocation. *Journal of Cell Science*, **115** (Pt 12), 2581–2590.

141 Cavalcanti-Adam, E.A., Volberg, T., Micoulet, A., Kessler, H., Geiger, B. and Spatz, J.P. (2007) Cell spreading and focal adhesion dynamics are regulated by spacing of integrin ligands. *Biophysical Journal*, **92** (8), 2964–2974.

142 De Pasquale, J.A. and Izzard, C.S. (1991) Accumulation of talin in nodes at the edge of the lamellipodium and separate incorporation into adhesion plaques at focal contacts in fibroblasts. *The Journal of Cell Biology*, **113** (6), 1351–1359.

143 Gallant, N.D., Michael, K.E. and Garcia, A.J. (2005) Cell adhesion strengthening: contributions of adhesive area, integrin binding, and focal adhesion assembly. *Molecular Biology of the Cell*, **16** (9), 4329–4340.

144 Chandrasekar, I., Stradal, T.E., Holt, M.R., Entschladen, F., Jockusch, B.M. and Ziegler, W.H. (2005) Vinculin acts as a sensor in lipid regulation of adhesion-site turnover. *Journal of Cell Science*, **118** (Pt 7), 1461–1472.

145 Paszek, M.J., Zahir, N., Johnson, K.R., Lakins, J.N., Rozenberg, G.I., Gefen, A., Reinhart-King, C.A., Margulies, S.S., Dembo, M., Boettiger, D., Hammer, D.A. and Weaver, V.M. (2005) Tensional homeostasis and the malignant phenotype. *Cancer Cell*, **8** (3), 241–254.

146 Ziegler, W.H., Liddington, R.C. and Critchley, D.R. (2006) The structure and regulation of vinculin. *Trends in Cell Biology*, **16** (9), 453–460.

147 Izard, T., Evans, G., Borgon, R.A., Rush, C.L., Bricogne, G. and Bois, P.R. (2004) Vinculin activation by talin through helical bundle conversion. *Nature*, **427** (6970), 171–175.

148 Papagrigoriou, E., Gingras, A.R., Barsukov, I.L., Bate, N., Fillingham, I.J., Patel, B., Frank, R., Ziegler, W.H., Roberts, G.C., Critchley, D.R. and Emsley, J. (2004) Activation of a vinculin-binding site in the talin rod involves rearrangement of a five-helix bundle. *The EMBO Journal*, **23** (15), 2942–2951.

149 Fillingham, I., Gingras, A.R., Papagrigoriou, E., Patel, B., Emsley, J., Critchley, D.R., Roberts, G.C. and Barsukov, I.L. (2005) A vinculin binding domain from the talin rod unfolds to form a complex with the vinculin head. *Structure (Camb.)*, **13** (1), 65–74.

150 Hytonen, V.P. and Vogel, V. (2008) How force might activate talin's vinculin binding sites: SMD reveals a structural mechanism. *PLoS Computational Biology*, **4** (2), e24.

151 Chen, H., Cohen, D.M., Choudhury, D.M., Kioka, N. and Craig, S.W. (2005) Spatial distribution and functional significance of activated vinculin in living cells. *The Journal of Cell Biology*, **169** (3), 459–470.

152 Bois, P.R., O'Hara, B.P., Nietlispach, D., Kirkpatrick, J. and Izard, T. (2006) The vinculin binding sites of talin and alpha-actinin are sufficient to activate vinculin. *The Journal of Biological Chemistry*, **281**, 7228–7236.

153 Bois, P.R., O'Hara, B.P., Nietlispach, D., Kirkpatrick, J. and Izard, T. (2006) The vinculin binding sites of talin and alpha-actinin are sufficient to activate vinculin. *The Journal of Biological Chemistry*, **281** (11), 7228–7236.

154 Johnson, R.P. and Craig, S.W. (1995) F-actin binding site masked by the intramolecular association of vinculin head and tail domains. *Nature*, **373** (6511), 261–264.

155 Chen, H., Choudhury, D.M. and Craig, S.W. (2006) Coincidence of actin filaments and talin is required to activate vinculin. *The Journal of Biological Chemistry*, **281** (52), 40389–40398.

156 Kelly, D.F., Taylor, D.W., Bakolitsa, C., Bobkov, A.A., Bankston, L., Liddington, R.C. and Taylor, K.A. (2006) Structure of the alpha-actinin-vinculin head domain complex determined by cryo-electron microscopy. *Journal of Molecular Biology*, **357** (2), 562–573.

157 Raucher, D. and Sheetz, M.P. (2000) Cell spreading and lamellipodial extension rate is regulated by membrane tension. *The Journal of Cell Biology*, **148** (1), 127–136.

158 Sawada, Y., Nakamura, K., Doi, K., Takeda, K., Tobiume, K., Saitoh, M., Morita, K., Komuro, I., De Vos, K., Sheetz, M. and Ichijo, H. (2001) Rap1 is involved in cell stretching modulation of p38 but not ERK or JNK MAP kinase. *Journal of Cell Science*, **114** (Pt 6), 1221–1227.

159 Kirchner, J., Kam, Z., Tzur, G., Bershadsky, A.D. and Geiger, B. (2003) Live-cell monitoring of tyrosine phosphorylation in focal adhesions following microtubule disruption. *Journal of Cell Science*, **116** (Pt 6), 975–986.

160 Tamada, M., Sheetz, M.P. and Sawada, Y. (2004) Activation of a signaling cascade by cytoskeleton stretch. *Developmental Cell*, **7** (5), 709–718.

161 Ballestrem, C., Erez, N., Kirchner, J., Kam, Z., Bershadsky, A. and Geiger, B. (2006) Molecular mapping of tyrosine-phosphorylated proteins in focal adhesions using fluorescence resonance energy transfer. *Journal of Cell Science*, **119** (Pt 5), 866–875.

162 Bunnell, S.C., Hong, D.I., Kardon, J.R., Yamazaki, T., McGlade, C.J., Barr, V.A. and Samelson, L.E. (2002) T cell receptor ligation induces the formation of dynamically regulated signaling assemblies. *The Journal of Cell Biology*, **158** (7), 1263–1275.

163 Jin, Z.G., Ueba, H., Tanimoto, T., Lungu, A.O., Frame, M.D. and Berk, B.C. (2003) Ligand-independent activation of vascular endothelial growth factor receptor 2 by fluid shear stress regulates activation of endothelial nitric oxide synthase. *Circulation Research*, **93** (4), 354–363.

164 Shimizu, N., Yamamoto, K., Obi, S., Kumagaya, S., Masumura, T., Shimano, Y., Naruse, K., Yamashita, J.K., Igarashi, T. and Ando, J. (2008) Cyclic strain induces mouse embryonic stem cell differentiation into vascular smooth muscle cells by activating PDGF receptor beta. *Journal of Applied Physiology (Bethesda, Md: 1985)*, **104** (3), 766–772.

165 Fisher, T.E., Marszalek, P.E., Oberhauser, A.F., Carrion-Vazquez, M. and Fernandez, J.M. (1999) The micro-mechanics of single molecules studied with atomic force microscopy. *The Journal of Physiology*, **520** (Pt 1), 5–14.

166 Ridley, A.J., Schwartz, M.A., Burridge, K., Firtel, R.A., Ginsberg, M.H., Borisy, G., Parsons, J.T. and Horwitz, A.R. (2003) Cell migration: integrating signals from front to back. *Science*, **302** (5651), 1704–1709.

167 Zaidel-Bar, R., Cohen, M., Addadi, L. and Geiger, B. (2004) Hierarchical assembly of cell-matrix adhesion complexes. *Biochemical Society Transactions*, **32** (Pt 3), 416–420.

168 Dobereiner, H.G., Dubin-Thaler, B.J., Giannone, G. and Sheetz, M.P. (2005) Force sensing and generation in cell phases: analyses of complex functions. *Journal of Applied Physiology (Bethesda, Md: 1985)*, **98** (4), 1542–1546.

169 Bershadsky, A.D., Ballestrem, C., Carramusa, L., Zilberman, Y., Gilquin, B., Khochbin, S., Alexandrova, A.Y., Verkhovsky, A.B., Shemesh, T. and Kozlov, M.M. (2006) Assembly and mechanosensory function of focal adhesions: experiments and models. *European Journal of Cell Biology*, **85** (3–4), 165–173.

170 Lele, T.P., Thodeti, C.K. and Ingber, D.E. (2006) Force meets chemistry: analysis of mechanochemical conversion in focal adhesions using fluorescence recovery after photobleaching. *Journal of Cellular Biochemistry*, **97** (6), 1175–1183.

171 Miyamoto, S., Akiyama, S.K. and Yamada, K.M. (1995) Synergistic roles for receptor occupancy and aggregation in integrin transmembrane function. *Science*, **267** (5199), 883–885.

172 Humphries, J.D., Wang, P., Streuli, C., Geiger, B., Humphries, M.J. and Ballestrem, C. (2007) Vinculin controls focal adhesion formation by direct interactions with talin and actin. *The Journal of Cell Biology*, **179** (5), 1043–1057.

173 von Wichert, G., Haimovich, B., Feng, G.S. and Sheetz, M.P. (2003) Force-dependent integrin-cytoskeleton linkage formation requires downregulation of focal complex dynamics by Shp2. *The EMBO Journal*, **22** (19), 5023–5035.

174 von Wichert, G., Jiang, G., Kostic, A., De Vos, K., Sap, J. and Sheetz, M.P. (2003) RPTP-alpha acts as a transducer of mechanical force on alphav/beta3-integrin-cytoskeleton linkages. *The Journal of Cell Biology*, **161** (1), 143–153.

175 Critchley, D.R., Holt, M.R., Barry, S.T., Priddle, H., Hemmings, L. and Norman, J. (1999) Integrin-mediated cell adhesion: the cytoskeletal connection. *Biochemical Society Symposium*, **65**, 79–99.

176 Gu, J., Tamura, M., Pankov, R., Danen, E.H., Takino, T., Matsumoto, K. and Yamada, K.M. (1999) Shc and FAK differentially regulate cell motility and directionality modulated by PTEN. *The Journal of Cell Biology*, **146** (2), 389–403.

177 Sieg, D.J., Hauck, C.R. and Schlaepfer, D.D. (1999) Required role of focal adhesion kinase (FAK) for integrin-stimulated cell migration. *Journal of Cell Science*, **112** (Pt 16), 2677–2691.

178 Selhuber-Unkel, C., Lopez-Garcia, M., Kessler, H. and Spatz, J.P. (2008) Cooperativity in adhesion cluster formation during initial cell adhesion. *Biophysical Journal*.

179 Beningo, K.A., Dembo, M., Kaverina, I., Small, J.V. and Wang, Y.L. (2001) Nascent focal adhesions are responsible for the generation of strong propulsive forces in migrating fibroblasts. *The Journal of Cell Biology*, **153** (4), 881–888.

180 Hirata, H., Tatsumi, H. and Sokabe, M. (2008) Mechanical forces facilitate actin polymerization at focal adhesions in a zyxin-dependent manner. *Journal of Cell Science*.

181 Zaidel-Bar, R., Ballestrem, C., Kam, Z. and Geiger, B. (2003) Early molecular events in the assembly of matrix adhesions at the leading edge of migrating cells. *Journal of Cell Science*, **116** (Pt 22), 4605–4613.

182 Lele, T.P., Pendse, J., Kumar, S., Salanga, M., Karavitis, J. and Ingber, D.E. (2005) Mechanical forces alter zyxin unbinding kinetics within focal adhesions of living cells. *Journal of Cellular Physiology*.

183 Zamir, E., Katz, B.Z., Aota, S., Yamada, K.M., Geiger, B. and Kam, Z. (1999) Molecular diversity of cell-matrix adhesions. *Journal of Cell Science*, **112** (Pt 11), 1655–1669.

184 Katz, B.Z., Zamir, E., Bershadsky, A., Kam, Z., Yamada, K.M. and Geiger, B. (2000) Physical state of the extracellular matrix regulates the structure and molecular composition of cell-matrix adhesions. *Molecular Biology of the Cell*, **11** (3), 1047–1060.

185 Wu, C., Keivens, V.M., O'Toole, T.E., McDonald, J.A. and Ginsberg, M.H. (1995) Integrin activation and cytoskeletal interaction are essential for the assembly of a fibronectin matrix. *Cell*, **83** (5), 715–724.

186 Cali, G., Mazzarella, C., Chiacchio, M., Negri, R., Retta, S.F., Zannini, M., Gentile, F., Tarone, G., Nitsch, L. and Garbi, C. (1999) RhoA activity is required for fibronectin assembly and counteracts beta1B integrin inhibitory effect in FRT epithelial cells. *Journal of Cell Science*, **112** (Pt 6), 957–965.

187 Pankov, R., Cukierman, E., Katz, B.Z., Matsumoto, K., Lin, D.C., Lin, S., Hahn, C. and Yamada, K.M. (2000) Integrin dynamics and matrix assembly: tensin-dependent translocation of alpha(5)beta(1) integrins promotes early fibronectin fibrillogenesis. *The Journal of Cell Biology*, **148** (5), 1075–1090.

188 Chen, B.H., Tzen, J.T., Bresnick, A.R. and Chen, H.C. (2002) Roles of Rho-associated kinase and myosin light chain kinase in morphological and migratory defects of focal adhesion kinase-null cells. *The Journal of Biological Chemistry*, **277** (37), 33857–33863.

189 Burridge, K. and Wennerberg, K. (2004) Rho and Rac take center stage. *Cell*, **116** (2), 167–179.

190 McBeath, R., Pirone, D.M., Nelson, C.M., Bhadriraju, K. and Chen, C.S. (2004) Cell shape, cytoskeletal tension, and RhoA regulate stem cell lineage commitment. *Developmental Cell*, **6** (4), 483–495.

191 Yoneda, A., Multhaupt, H.A. and Couchman, J.R. (2005) The Rho kinases I and II regulate different aspects of myosin II activity. *The Journal of Cell Biology*, **170** (3), 443–453.

192 Barry, S.T., Flinn, H.M., Humphries, M.J., Critchley, D.R. and Ridley, A.J. (1997) Requirement for Rho in integrin signalling. *Cell Adhesion and Communication*, **4** (6), 387–398.

193 Zhong, C., Chrzanowska-Wodnicka, M., Brown, J., Shaub, A., Belkin, A.M. and Burridge, K. (1998) Rho-mediated contractility exposes a cryptic site in fibronectin and induces fibronectin matrix assembly. *The Journal of Cell Biology*, **141** (2), 539–551.

194 Danen, E.H., Sonneveld, P., Brakebusch, C., Fassler, R. and Sonnenberg, A. (2002) The fibronectin-binding integrins alpha5beta1 and alphavbeta3 differentially modulate RhoA-GTP loading, organization of cell matrix adhesions, and fibronectin fibrillogenesis. *The Journal of Cell Biology*, **159** (6), 1071–1086.

195 Miao, H., Li, S., Hu, Y.L., Yuan, S., Zhao, Y., Chen, B.P., Puzon-McLaughlin, W., Tarui, T., Shyy, J.Y., Takada, Y., Usami, S. and Chien, S. (2002) Differential regulation of Rho GTPases by beta1 and beta3 integrins: the role of an extracellular domain of integrin in intracellular signaling. *Journal of Cell Science*, **115** (Pt 10), 2199–2206.

196 Giannone, G. and Sheetz, M.P. (2006) Substrate rigidity and force define form through tyrosine phosphatase and kinase pathways. *Trends in Cell Biology*, **16** (4), 213–223.

197 Kostic, A. and Sheetz, M.P. (2006) Fibronectin rigidity response through Fyn and p130Cas recruitment to the leading edge. *Molecular Biology of the Cell*, **17** (6), 2684–2695.

198 Kostic, A., Sap, J. and Sheetz, M.P. (2007) RPTPalpha is required for rigidity-dependent inhibition of extension and differentiation of hippocampal neurons. *Journal of Cell Science*, **120** (Pt 21), 3895–3904.

199 Geiger, R.C., Taylor, W., Glucksberg, M.R. and Dean, D.A. (2006) Cyclic stretch-induced reorganization of the cytoskeleton and its role in enhanced gene transfer. *Gene Therapy*, **13** (8), 725–731.

200 Filipenko, N.R., Attwell, S., Roskelley, C. and Dedhar, S. (2005) Integrin-linked kinase activity regulates Rac- and Cdc42-mediated actin cytoskeleton reorganization via alpha-PIX. *Oncogene*, **24**, 5837–5849.

201 Legate, K.R., Montanez, E., Kudlacek, O. and Fassler, R. (2006) ILK, PINCH and parvin: the tIPP of integrin signalling. *Nature Reviews - Molecular Cell Biology*, **7** (1), 20–31.

202 Sakai, T., Li, S., Docheva, D., Grashoff, C., Sakai, K., Kostka, G., Braun, A., Pfeifer, A., Yurchenco, P.D. and Fassler, R. (2003) Integrin-linked kinase (ILK) is required for polarizing the epiblast, cell adhesion, and controlling actin accumulation. *Genes and Development*, **17** (7), 926–940.

203 Sheetz, M.P., Felsenfeld, D., Galbraith, C.G. and Choquet, D. (1999) Cell migration as a five-step cycle. *Biochemical Society Symposium*, **65**, 233–243.

204 Wehrle-Haller, B. and Imhof, B. (2002) The inner lives of focal adhesions. *Trends in Cell Biology*, **12** (8), 382–389.

205 Kirfel, G., Rigort, A., Borm, B. and Herzog, V. (2004) Cell migration: mechanisms of rear detachment and the formation of migration tracks. *European Journal of Cell Biology*, **83** (11–12), 717–724.

206 Galbraith, C.G. and Sheetz, M.P. (1997) A micromachined device provides a new bend on fibroblast traction forces. *Proceedings of the National Academy of Sciences of the United States of America*, **94** (17), 9114–9118.

207 Pelham, R.J. Jr and Wang, Y. (1997) Cell locomotion and focal adhesions are regulated by substrate flexibility. *Proceedings of the National Academy of Sciences of the United States of America*, **94** (25), 13661–13665.

208 Sterba, R.E. and Sheetz, M.P. (1998) Basic laser tweezers. *Methods in Cell Biology*, **55**, 29–41.

209 Beningo, K.A. and Wang, Y.L. (2002) Flexible substrata for the detection of cellular traction forces. *Trends in Cell Biology*, **12** (2), 79–84.

210 LeDuc, P., Ostuni, E., Whitesides, G. and Ingber, D. (2002) Use of micropatterned adhesive surfaces for control of cell behaviour. *Methods in Cell Biology*, **69**, 385–401.

211 Prechtel, K., Bausch, A.R., Marchi-Artzner, V., Kantlehner, M., Kessler, H. and Merkel, R. (2002) Dynamic force spectroscopy to probe adhesion strength of living cells. *Physical Review Letters*, **89** (2), 028101.

212 du Roure, O., Saez, A., Buguin, A., Austin, R.H., Chavrier, P., Siberzan, P. and Ladoux, B. (2005) Force mapping in epithelial cell migration. *Proceedings of the National Academy of Sciences of the United States of America*, **102** (7), 2390–2395.

213 Choquet, D., Felsenfeld, D.P. and Sheetz, M.P. (1997) Extracellular matrix rigidity causes strengthening of integrin-cytoskeleton linkages. *Cell*, **88** (1), 39–48.

214 Geiger, P.C., Cody, M.J., Macken, R.L., Bayrd, M.E., Fang, Y.H. and Sieck, G.C. (2001) Mechanisms underlying increased force generation by rat diaphragm muscle fibers during development. *Journal of Applied Physiology (Bethesda, Md: 1985)*, **90** (1), 380–388.

215 Tseng, Y., Kole, T.P. and Wirtz, D. (2002) Micromechanical mapping of live cells by multiple-particle-tracking microrheology. *Biophysical Journal*, **83** (6), 3162–3176.

216 Mack, P.J., Kaazempur-Mofrad, M.R., Karcher, H., Lee, R.T. and Kamm, R.D. (2004) Force-induced focal adhesion translocation: effects of force amplitude and frequency. *American Journal of Physiology. Cell Physiology*, **287** (4), C954–C962.

217 Ballestrem, C. and Geiger, B. (2005) Application of microscope-based FRET to study molecular interactions in focal adhesions of live cells. *Methods in Molecular Biology (Clifton, NJ)*, **294**, 321–334.

218 Katsumi, A., Naoe, T., Matsushita, T., Kaibuchi, K. and Schwartz, M.A. (2005) Integrin activation and matrix binding mediate cellular responses to mechanical stretch. *The Journal of Biological Chemistry*, **280** (17), 16546–16549.

219 Vallotton, P., Danuser, G., Bohnet, S., Meister, J.J. and Verkhovsky, A.B. (2005) Tracking retrograde flow in keratocytes: news from the front. *Molecular Biology of the Cell*, **16** (3), 1223–1231.

220 Sniadecki, N.J., Desai, R.A., Ruiz, S.A. and Chen, C.S. (2006) Nanotechnology for cell-substrate interactions. *Annals of Biomedical Engineering*, **34** (1), 59–74.

221 Sniadecki, N.J., Anguelouch, A., Yang, M.T., Lamb, C.M., Liu, Z., Kirschner, S.B., Liu, Y., Reich, D.H. and Chen, C.S. (2007) Magnetic microposts as an approach to apply forces to living cells. *Proceedings of the National Academy of Sciences of the United States of America*, **104** (37), 14553–14558.

222 Cai, Y., Biais, N., Giannone, G., Tanase, M., Jiang, G., Hofman, J.M., Wiggins, C.H., Silberzan, P., Buguin, A., Ladoux, B. and Sheetz, M.P. (2006) Nonmuscle myosin IIA-dependent force inhibits cell spreading and drives F-actin flow. *Biophysical Journal*, **91** (10), 3907–3920.

223 Sheetz, M.P. (2001) Cell control by membrane-cytoskeleton adhesion. *Nature Reviews - Molecular Cell Biology*, **2** (5), 392–396.

224 Wang, N., Butler, J.P. and Ingber, D.E. (1993) Mechanotransduction across the cell surface and through the cytoskeleton. *Science*, **260** (5111), 1124–1127.

225 Ganz, A., Lambert, M., Saez, A., Silberzan, P., Buguin, A., Mege, R.M. and Ladoux, B. (2006) Traction forces exerted through N-cadherin contacts. *Biology of the Cell/Under the Auspices of the European Cell Biology Organization*, **98** (12), 721–730.

226 Ehrlich, J.S., Hansen, M.D. and Nelson, W.J. (2002) Spatio-temporal regulation of Rac1 localization and lamellipodia dynamics during epithelial cell-cell adhesion. *Developmental Cell*, **3** (2), 259–270.

227 Davidson, L.A., Marsden, M., Keller, R. and Desimone, D.W. (2006) Integrin alpha5beta1 and fibronectin regulate polarized cell protrusions required for *Xenopus* convergence and extension. *Current Biology*, **16** (9), 833–844.

228 Swartz, M.A., Tschumperlin, D.J., Kamm, R.D. and Drazen, J.M. (2001) Mechanical stress is communicated between different cell types to elicit matrix remodelling. *Proceedings of the National Academy of Sciences of the United States of America*, **98** (11), 6180–6185.

229 Giannone, G., Dubin-Thaler, B.J., Dobereiner, H.G., Kieffer, N., Bresnick, A.R. and Sheetz, M.P. (2004) Periodic lamellipodial contractions correlate with rearward actin waves. *Cell*, **116** (3), 431–443.

230 Lo, C.M., Wang, H.B., Dembo, M. and Wang, Y.L. (2000) Cell movement is guided by the rigidity of the substrate. *Biophysical Journal*, **79** (1), 144–152.

231 Saez, A., Buguin, A., Silberzan, P. and Ladoux, B. (2005) Is the mechanical activity of epithelial cells controlled by deformations or forces? *Biophysical Journal*, **89** (6), L52–L54.

232 Saez, A., Ghibaudo, M., Buguin, A., Silberzan, P. and Ladoux, B. (2007) Rigidity-driven growth and migration of epithelial cells on microstructured anisotropic substrates. *Proceedings of the National Academy of Sciences of the United States of America*, **104**, 8281–8286.

233 Jiang, G., Huang, A.H., Cai, Y., Tanase, M. and Sheetz, M.P. (2006) Rigidity sensing at the leading edge through alphavbeta3 integrins and RPTPalpha. *Biophysical Journal*, **90** (5), 1804–1809.

234 Guo, Y., Hsu, S., Sawhney, H.S., Kumar, R. and Shan, Y. (2007) Robust object matching for persistent tracking with heterogeneous features. *IEEE Transactions on Pattern Analysis and Machine Intelligence*, **29** (5), 824–839.

235 Webb, D.J., Donais, K., Whitmore, L.A., Thomas, S.M., Turner, C.E., Parsons, J.T. and Horwitz, A.F. (2004) FAK-Src signalling through paxillin, ERK and MLCK regulates adhesion disassembly. *Nature Cell Biology*, **6** (2), 154–161.

236 Laukaitis, C.M., Webb, D.J., Donais, K. and Horwitz, A.F. (2001) Differential dynamics of alpha 5 integrin, paxillin, and alpha-actinin during formation and disassembly of adhesions in migrating cells. *The Journal of Cell Biology*, **153** (7), 1427–1440.

237 Bausch, A.R., Ziemann, F., Boulbitch, A.A., Jacobson, K. and Sackmann, E. (1998) Local measurements of viscoelastic parameters of adherent cell surfaces by magnetic bead microrheometry. *Biophysical Journal*, **75** (4), 2038–2049.

238 Kiosses, W.B., Shattil, S.J., Pampori, N. and Schwartz, M.A. (2001) Rac recruits high-affinity integrin alphavbeta3 to lamellipodia in endothelial cell migration. *Nature Cell Biology*, **3** (3), 316–320.

239 Arthur, W.T., Quilliam, L.A. and Cooper, J.A. (2004) Rap1 promotes cell spreading by localizing Rac guanine nucleotide exchange factors. *The Journal of Cell Biology*, **167** (1), 111–122.

240 Di Stefano, P., Cabodi, S., Boeri Erba, E., Margaria, V., Bergatto, E., Giuffrida, M.G., Silengo, L., Tarone, G., Turco, E. and Defilippi, P. (2004) P130Cas-associated protein (p140Cap) as a new tyrosine-phosphorylated protein involved in cell spreading. *Molecular Biology of the Cell*, **15** (2), 787–800.

241 Nishizaka, T., Shi, Q. and Sheetz, M.P. (2000) Position-dependent linkages of fibronectin-integrin-cytoskeleton. *Proceedings of the National Academy of Sciences of the United States of America*, **97** (2), 692–697.

242 Oganov, V.S. (2004) Modern analysis of bone loss mechanisms in microgravity. *The Journal of Gravitational Physiology*, **11** (2), P143–P150.

243 Dallas, S.L., Chen, Q. and Sivakumar, P. (2006) Dynamics of assembly and reorganization of extracellular matrix proteins. *Current Topics in Developmental Biology*, **75**, 1–24.

244 Meshel, A.S., Wei, Q., Adelstein, R.S. and Sheetz, M.P. (2005) Basic mechanism of three-dimensional collagen fibre transport by fibroblasts. *Nature Cell Biology*, **7** (2), 157–164.

245 Giannone, G., Dubin-Thaler, B.J., Rossier, O. *et al.* (2007) Lamellipodial actin mechanically links myosin activity with adhesion-site formation. *Cell*, **128**, 561–575.

246 Astrof, S., Crowley, D., George, E.L., Fukuda, T., Sekiguchi, K., Hanahan, D. and Hynes, R.O. (2004) Direct test of potential roles of EIIIA and EIIIB alternatively spliced segments of fibronectin in physiological and tumor angiogenesis. *Molecular and Cellular Biology*, **24** (19), 8662–8670.

247 Brown, R.A., Blunn, G.W. and Ejim, O.S. (1994) Preparation of orientated fibrous mats from fibronectin: composition and stability. *Biomaterials*, **15** (6), 457–464.

248 Baneyx, G. and Vogel, V. (1999) Self-assembly of fibronectin into fibrillar networks underneath dipalmitoyl phosphatidylcholine monolayers: role of lipid matrix and tensile forces. *Proceedings of the National Academy of Sciences of the United States of America*, **96** (22), 12518–12523.

249 Takahashi, S., Leiss, M., Moser, M., Ohashi, T., Kitao, T., Heckmann, D., Pfeifer, A., Kessler, H., Takagi, J., Erickson, H.P. and Fassler, R. (2007) The RGD motif in fibronectin is essential for development but dispensable for fibril assembly. *The Journal of Cell Biology*, **178** (1), 167–178.

250 Little, W.C., Smith, M.L., Ebneter, U. and Vogel, V. (2008) Assay to mechanically tune and optically probe fibrillar fibronectin conformations from fully relaxed to breakage. *Matrix Biology*, **27**, 451–465.

251 Klotzsch, E., Smith, M.L., Kubow, K.E., Muntwyler, S., Little, W.C., Beyeler, F., Gourdon, D., Nelson, B.J. and Vogel, V. (2009) Fibronectin self-assembles when stretched into the most elastic biological fibers diplaying force-regulated molecular recognition switches. submitted.

252 Wierzbicka-Patynowski, I., Mao, Y. and Schwarzbauer, J.E. (2007) Continuous requirement for pp60-Src and phospho-paxillin during fibronectin matrix assembly by transformed cells. *Journal of Cellular Physiology*, **210** (3), 750–756.

253 Paci, E. and Karplus, M. (1999) Forced unfolding of fibronectin type 3 modules: an analysis by biased molecular dynamics simulations. *Journal of Molecular Biology*, **288** (3), 441–459.

254 Craig, D., Krammer, A., Schulten, K. and Vogel, V. (2001) Comparison of the early stages of forced unfolding for fibronectin type III modules. *Proceedings of the National Academy of Sciences of the United States of America*, **98** (10), 5590–5595.

255 Oberhauser, A.F., Badilla-Fernandez, C., Carrion-Vazquez, M. and Fernandez, J.M. (2002) The mechanical hierarchies of fibronectin observed with single-molecule AFM. *Journal of Molecular Biology*, **319** (2), 433–447.

256 Craig, D., Gao, M., Schulten, K. and Vogel, V. (2004) Tuning the mechanical stability of fibronectin type III modules through sequence variations. *Structure (Camb.)*, **12** (1), 21–30.

257 Ng, S.P., Rounsevell, R.W., Steward, A., Geierhaas, C.D., Williams, P.M., Paci, E. and Clarke, J. (2005) Mechanical unfolding of TNfn3: the unfolding pathway of a fnIII domain probed by protein engineering, AFM and MD simulation. *Journal of Molecular Biology*, **350** (4), 776–789.

258 Krammer, A., Lu, H., Isralewitz, B., Schulten, K. and Vogel, V. (1999) Forced unfolding of the fibronectin type III module reveals a tensile molecular recognition switch. *Proceedings of the National Academy of Sciences of the United States of America*, **96** (4), 1351–1356.

259 Marszalek, P.E., Lu, H., Li, H., Carrion-Vazquez, M., Oberhauser, A.F., Schulten,

K. and Fernandez, J.M. (1999) Mechanical unfolding intermediates in titin modules. *Nature*, **402** (6757), 100–103.

260 Gao, M., Craig, D., Vogel, V. and Schulten, K. (2002) Identifying unfolding intermediates of FN-III(10) by steered molecular dynamics. *Journal of Molecular Biology*, **323** (5), 939–950.

261 Krammer, A., Craig, D., Thomas, W.E., Schulten, K. and Vogel, V. (2002) A structural model for force regulated integrin binding to fibronectin's RGD-synergy site. *Matrix Biology*, **21** (2), 139–147.

262 Gao, M., Craig, D., Lequin, O., Campbell, I.D., Vogel, V. and Schulten, K. (2003) Structure and functional significance of mechanically unfolded fibronectin type III1 intermediates. *Proceedings of the National Academy of Sciences of the United States of America*, **100** (25), 14784–14789.

263 Andresen, M., Wahl, M.C., Stiel, A.C., Grater, F., Schafer, L.V., Trowitzsch, S., Weber, G., Eggeling, C., Grubmuller, H., Hell, S.W. and Jakobs, S. (2005) Structure and mechanism of the reversible photoswitch of a fluorescent protein. *Proceedings of the National Academy of Sciences of the United States of America*, **102** (37), 13070–13074.

264 Grater, F., Shen, J., Jiang, H., Gautel, M. and Grubmuller, H. (2005) Mechanically induced titin kinase activation studied by force-probe molecular dynamics simulations. *Biophysical Journal*, **88** (2), 790–804.

265 Lee, E.H., Hsin, J., Mayans, O. and Schulten, K. (2007) Secondary and tertiary structure elasticity of titin Z1Z2 and a titin chain model. *Biophysical Journal*, **93** (5), 1719–1735.

266 Ejim, O.S., Blunn, G.W. and Brown, R.A. (1993) Production of artificial-orientated mats and strands from plasma fibronectin: a morphological study. *Biomaterials*, **14** (10), 743–748.

267 Wojciak-Stothard, B., Denyer, M., Mishra, M. and Brown, R.A. (1997) Adhesion, orientation, and movement of cells cultured on ultrathin fibronectin fibers. *In Vitro Cellular & Developmental Biology - Animal*, **33** (2), 110–117.

268 Ahmed, Z. and Brown, R.A. (1999) Adhesion, alignment, and migration of cultured Schwann cells on ultrathin fibronectin fibres. *Cell Motility and the Cytoskeleton*, **42** (4), 331–343.

269 Antia, M., Baneyx, G., Kubow, K.E. and Vogel, V. (2008) Fibronectin in aging extracellular matrix fibrils is progressively unfolded by cells and elicits an enhanced rigidity response. *Faraday Discsussions*, **139**, 229–249.

270 Wolf, K. and Friedl, P. (2005) Functional imaging of pericellular proteolysis in cancer cell invasion. *Biochimie*, **87** (3–4), 315–320.

271 Zaman, M.H., Trapani, L.M., Siemeski, A., Mackellar, D., Gong, H., Kamm, R.D., Wells, A., Lauffenburger, D.A. and Matsudaira, P. (2006) Migration of tumor cells in 3D matrices is governed by matrix stiffness along with cell-matrix adhesion and proteolysis. *Proceedings of the National Academy of Sciences of the United States of America*, **103** (29), 10889–10894.

272 Wolf, K., Wu, Y.I., Liu, Y., Geiger, J., Tam, E., Overall, C., Stack, M.S. and Friedl, P. (2007) Multi-step pericellular proteolysis controls the transition from individual to collective cancer cell invasion. *Nature Cell Biology*, **9** (8), 893–904.

273 Janmey, P.A. and Weitz, D.A. (2004) Dealing with mechanics: mechanisms of force transduction in cells. *Trends in Biochemical Sciences*, **29** (7), 364–370.

274 Felsenfeld, D.P., Schwartzberg, P.L., Venegas, A., Tse, R. and Sheetz, M.P. (1999) Selective regulation of integrin--cytoskeleton interactions by the tyrosine kinase Src. *Nature Cell Biology*, **1** (4), 200–206.

275 Volberg, T., Romer, L., Zamir, E. and Geiger, B. (2001) pp60(c-src) and related tyrosine kinases: a role in the assembly and reorganization of matrix adhesions.

Journal of Cell Science, **114** (Pt 12), 2279–2289.

276 Wang, H.B., Dembo, M., Hanks, S.K. and Wang, Y. (2001) Focal adhesion kinase is involved in mechanosensing during fibroblast migration. *Proceedings of the National Academy of Sciences of the United States of America*, **98** (20), 11295–11300.

277 Munevar, S., Wang, Y.L. and Dembo, M. (2001) Distinct roles of frontal and rear cell-substrate adhesions in fibroblast migration. *Molecular Biology of the Cell*, **12** (12), 3947–3954.

278 Perrin, B.J. and Huttenlocher, A. (2002) Calpain. *The International Journal of Biochemistry and Cell Biology*, **34** (7), 722–725.

279 Kole, T.P., Tseng, Y., Jiang, I., Katz, J.L. and Wirtz, D. (2005) Intracellular mechanics of migrating fibroblasts. *Molecular Biology of the Cell*, **16** (1), 328–338.

280 Pankov, R., Endo, Y., Even-Ram, S., Araki, M., Clark, K., Cukierman, E., Matsumoto, K. and Yamada, K.M. (2005) A Rac switch regulates random versus directionally persistent cell migration. *The Journal of Cell Biology*, **170** (5), 793–802.

281 Frame, M.C. and Brunton, V.G. (2002) Advances in Rho-dependent actin regulation and oncogenic transformation. *Current Opinion in Genetics and Development*, **12** (1), 36–43.

282 Ghosh, M., Song, X., Mouneimne, G., Sidani, M., Lawrence, D.S. and Condeelis, J.S. (2004) Cofilin promotes actin polymerization and defines the direction of cell motility. *Science*, **304** (5671), 743–746.

283 Danen, E.H., van Rheenen, J., Franken, W., Huveneers, S., Sonneveld, P., Jalink, K. and Sonnenberg, A. (2005) Integrins control motile strategy through a Rho-cofilin pathway. *The Journal of Cell Biology*, **169** (3), 515–526.

284 Mouneimne, G., DesMarais, V., Sidani, M., Scemes, E., Wang, W., Song, X., Eddy, R. and Condeelis, J. (2006) Spatial and temporal control of cofilin activity is required for directional sensing during chemotaxis. *Current Biology*, **16** (22), 2193–2205.

285 Sidani, M., Wessels, D., Mouneimne, G., Ghosh, M., Goswami, S., Sarmiento, C., Wang, W., Kuhl, S., El-Sibai, M., Backer, J.M., Eddy, R., Soll, D. and Condeelis, J. (2007) Cofilin determines the migration behavior and turning frequency of metastatic cancer cells. *The Journal of Cell Biology*, **179** (4), 777–791.

286 Levinson, A.D., Oppermann, H., Levintow, L., Varmus, H.E. and Bishop, J.M. (1978) Evidence that the transforming gene of avian sarcoma virus encodes a protein kinase associated with a phosphoprotein. *Cell*, **15** (2), 561–572.

287 Parker, R.C., Varmus, H.E. and Bishop, J.M. (1984) Expression of v-src and chicken c-src in rat cells demonstrates qualitative differences between pp60v-src and pp60c-src. *Cell*, **37** (1), 131–139.

288 Guck, J., Schinkinger, S., Lincoln, B., Wottawah, F., Ebert, S., Romeyke, M., Lenz, D., Erickson, H.M., Ananthakrishnan, R., Mitchell, D., Kas, J., Ulvick, S. and Bilby, C. (2005) Optical deformability as an inherent cell marker for testing malignant transformation and metastatic competence. *Biophysical Journal*, **88** (5), 3689–3698.

289 Ruoslahti, E. (1999) Fibronectin and its integrin receptors in cancer. *Advances in Cancer Research*, **76**, 1–20.

290 Gimona, M. (2008) The microfilament system in the formation of invasive adhesions. *Seminars in Cancer Biology*, **18** (1), 23–34.

291 Mierke, C.T., Rosel, D., Fabry, B. and Brabek, J. (2008) Contractile forces in tumor cell migration. *European Journal of Cell Biology*, **87**, 669–676.

292 Huang, S. and Ingber, D.E. (2005) Cell tension, matrix mechanics, and cancer development. *Cancer Cell*, **8** (3), 175–176.

293 Paszek, M.J. and Weaver, V.M. (2004) The tension mounts: mechanics meets morphogenesis and malignancy. *Journal* *of Mammary Gland Biology and Neoplasia*, **9** (4), 325–342.

294 Paget, S. (1889) The distribution of secondary growths in cancer of the breast. *Cancer Metastasis Reviews*, **8** (2), 98–102.

295 Beacham, D.A. and Cukierman, E. (2005) Stromagenesis: the changing face of fibroblastic microenvironments during tumor progression. *Seminars in Cancer Biology*, **15** (5), 329–341.

296 Mitra, S.K. and Schlaepfer, D.D. (2006) Integrin-regulated FAK-Src signaling in normal and cancer cells. *Current Opinion in Cell Biology*, **18** (5), 516–523.

297 Dorssers, L.C., Grebenchtchikov, N., Brinkman, A. *et al.* (2004) The prognostic value of BCAR1 in patients with primary breast cancer. *Clinical Cancer Research*, **10**, 6194–6202.

298 Riggins, R.B., De Berry, R.M., Toosarvandani, M.D. and Bouton, A.H. (2003) Src-dependent association of Cas and p85 phosphatidylinositol 3′-kinase in v-crk-transformed cells. *Molecular Cancer Research*, **1**, 428–437.

299 Nakamoto, T., Sakai, R., Honda, H., Ogawa, S., Ueno, H., Suzuki, T., Aizawa, S., Yazaki, Y. and Hirai, H. (1997) Requirements for localization of p130cas to focal adhesions. *Molecular and Cellular Biology*, **17** (7), 3884–3897.

300 Nievers, M.G., Birge, R.B., Greulich, H., Verkleij, A.J., Hanafusa, H., van Bergen en Henegouwen, P. M. (1997) v-Crk-induced cell transformation: changes in focal adhesion composition and signalling. *Journal of Cell Science*, **110** (Pt 3), 389–399.

301 Sakai, R., Nakamoto, T., Ozawa, K., Aizawa, S. and Hirai, H. (1997) Characterization of the kinase activity essential for tyrosine phosphorylation of p130Cas in fibroblasts. *Oncogene*, **14** (12), 1419–1426.

302 Kirsch, K., Kensinger, M., Hanafusa, H. and August, A. (2002) A p130Cas tyrosine phosphorylated substrate domain decoy disrupts v-crk signalling. *BMC Cell Biology*, **3**, 18.

303 Gotoh, T., Cai, D., Tian, X., Feig, L.A. and Lerner, A. (2000) p130Cas regulates the activity of AND-34, a novel Ral, Rap1, and R-Ras guanine nucleotide exchange factor. *Journal of Biological Chemistry*, **275**, 30118.

304 Brabek, J., Constancio, S.S., Siesser, P.F., Shin, N.Y., Pozzi, A. and Hanks, S.K. (2005) Crk-associated substrate tyrosine phosphorylation sites are critical for invasion and metastasis of SRC-transformed cells. *Molecular Cancer Research*, **3**, 307–315.

305 Ingber, D.E. and Folkman, J. (1989) Mechanochemical switching between growth and differentiation during fibroblast growth factor-stimulated angiogenesis in vitro: role of extracellular matrix. *The Journal of Cell Biology*, **109** (1), 317–330.

306 Larsen, M., Wei, C. and Yamada, K.M. (2006) Cell and fibronectin dynamics during branching morphogenesis. *Journal of Cell Science*, **119** (Pt 16), 3376–3384.

307 Heil, M. and Schaper, W. (2007) Insights into pathways of arteriogenesis. *Current Pharmaceutical Biotechnology*, **8** (1), 35–42.

308 Mammoto, A., Mammoto, T. and Ingber, D.E. (2008) Rho signaling and mechanical control of vascular development. *Current Opinion in Hematology*, **15** (3), 228–234.

309 Furuya, M. and Yonemitsu, Y. (2008) Cancer neovascularization and proinflammatory microenvironments. *Current Cancer Drug Targets*, **8** (4), 253–265.

310 Mahabeleshwar, G.H., Feng, W., Phillips, D.R. and Byzova, T.V. (2006) Integrin signaling is critical for pathological angiogenesis. *The Journal of Experimental Medicine*, **203** (11), 2495–2507.

311 Wijelath, E.S., Murray, J., Rahman, S., Patel, Y., Ishida, A., Strand, K., Aziz, S., Cardona, C., Hammond, W.P., Savidge, G.F., Rafii, S. and Sobel, M. (2002) Novel vascular endothelial growth factor binding domains of fibronectin enhance vascular endothelial growth factor biological activity. *Circulation Research*, **91** (1), 25–31.

312 Wijelath, E.S., Rahman, S., Namekata, M., Murray, J., Nishimura, T., Mostafavi-Pour, Z., Patel, Y., Suda, Y., Humphries, M.J. and Sobel, M. (2006) Heparin-II domain of fibronectin is a vascular endothelial growth factor-binding domain: enhancement of VEGF biological activity by a singular growth factor/matrix protein synergism. *Circulation Research*, **99** (8), 853–860.

313 Miralem, T., Steinberg, R., Price, D. and Avraham, H. (2001) VEGF(165) requires extracellular matrix components to induce mitogenic effects and migratory response in breast cancer cells. *Oncogene*, **20** (39), 5511–5524.

314 Wijelath, E.S., Rahman, S., Murray, J., Patel, Y., Savidge, G. and Sobel, M. (2004) Fibronectin promotes VEGF-induced CD34 cell differentiation into endothelial cells. *Journal of Vascular Surgery*, **39** (3), 655–660.

315 Georges, P.C. and Janmey, P.A. (2005) Cell type-specific response to growth on soft materials. *Journal of Applied Physiology*, **98**, 1547–1553.

316 Goerges, A.L. and Nugent, M.A. (2004) pH regulates vascular endothelial growth factor binding to fibronectin: a mechanism for control of extracellular matrix storage and release. *The Journal of Biological Chemistry*, **279** (3), 2307–2315.

317 Nikolopoulos, S.N., Blaikie, P., Yoshioka, T., Guo, W. and Giancotti, F.G. (2004) Integrin beta4 signaling promotes tumor angiogenesis. *Cancer Cell*, **6** (5), 471–483.

318 Bon, G., Folgiero, V., Bossi, G., Felicioni, L., Marchetti, A., Sacchi, A. and Falcioni, R. (2006) Loss of beta4 integrin subunit reduces the tumorigenicity of MCF7 mammary cells and causes apoptosis upon hormone deprivation. *Clinical Cancer Research*, **12** (11 Pt 1), 3280–3287.

319 Steinbock, F.A. and Wiche, G. (1999) Plectin: a cytolinker by design. *Biological Chemistry*, **380** (2), 151–158.

320 Yi, M., Sakai, T., Fassler, R. and Ruoslahti, E. (2003) Antiangiogenic proteins require plasma fibronectin or vitronectin for in vivo activity. *Proceedings of the National Academy of Sciences of the United States of America*, **100** (20), 11435–11438.

321 Akerman, M.E., Pilch, J., Peters, D. and Ruoslahti, E. (2005) Angiostatic peptides use plasma fibronectin to home to angiogenic vasculature. *Proceedings of the National Academy of Sciences of the United States of America*, **102**, 2040–2045.

322 Vogel, V. and Baneyx, G. (2003) The tissue engineering puzzle: a molecular perspective. *Annual Review of Biomedical Engineering*, **5**, 441–463.

323 Badylak, S.F. (2002) The extracellular matrix as a scaffold for tissue reconstruction. *Seminars in Cell and Developmental Biology*, **13**, 377–383.

324 Urech, L., Bittermann, A.G., Hubbell, J.A. and Hall, H. (2005) Mechanical properties, proteolytic degradability and biological modifications affect angiogenic process extension into native and modified fibrin matrices in vitro. *Biomaterials*, **26**, 1369–1379.

325 Leor, J., Gerecht, S., Cohen, S., Miller, L., Holbova, R., Ziskind, A., Shachar, M., Feinberg, M.S., Guetta, E. and Itskovitz-Eldor, J. (2007) Human embryonic stem cell transplantation to repair the infarcted myocardium. *Heart*, **93**, 1278–1284.

326 Miyahara, Y., Nagaya, N., Kataoka, M., Yanagawa, B., Tanaka, K., Hao, H., Ishino, K., Ishida, H., Shimizu, T., Kangawa, K., Sano, S., Okano, T., Kitamura, S. and Mori, H. (2006) Monolayered mesenchymal stem cells repair scarred myocardium after myocardial infarction. *Nature Medicine*, **12** (4), 459–465.

327 Gilbert, T.W., Stewart-Akers, A.M. and Badylak, S.F. (2007) A quantitative method for evaluating the degradation of biologic scaffold materials. *Biomaterials*, **28** (2), 147–150.

328 Kubow, K.E., Klotzsch, E., Smith, M.L., Gourdon, D., Little, W., Vogel, V. Rigidity and not fibronectin conformation controls extracellular matrix assembly by fibroblasts reseeded into de-cellularized ECM scaffolds, submitted.

329 Wipff, P.J. and Hinz, B. (2008) Integrins and the activation of latent transforming growth factor beta1: An intimate relationship. *European Journal of Cell Biology*, **87** (8–9), 601–615.

330 Dallas, S.L., Sivakumar, P., Jones, C.J., Chen, Q., Peters, D.M., Mosher, D.F., Humphries, M.J. and Kielty, C.M. (2005) Fibronectin regulates latent transforming growth factor-beta (TGF beta) by controlling matrix assembly of latent TGF beta-binding protein-1. *Journal of Biological Chemistry*, **280**, 18871–18880.

331 Wells, R.G. and Discher, D.E. (2008) Matrix elasticity, cytoskeletal tension, and TGF-beta: the insoluble and soluble meet. *Science Signaling*, **1** (10), pe13.

332 Sarikaya, A., Record, R., Wu, C.C., Tullius, B., Badylak, S. and Ladisch, M. (2002) Antimicrobial activity associated with extracellular matrices. *Tissue Engineering*, **8** (1), 63–71.

333 Schmid-Schonbein, G.W. and Hugli, T.E. (2005) A new hypothesis for microvascular inflammation in shock and multiorgan failure: self-digestion by pancreatic enzymes. *Microcirculation (New York, NY: 1994)* **12** (1), 71–82.

334 Liu, S., Calderwood, D.A. and Ginsberg, M.H. (2000) Integrin cytoplasmic domain-binding proteins. *Journal of Cell Science*, **113** (Pt 20), 3563–3571.

335 Brakebusch, C. and Fassler, R. (2003) The integrin-actin connection, an eternal love affair. *The EMBO Journal*, **22** (10), 2324–2333.

336 Wilhelmsen, K., Litjens, S.H., Kuikman, I., Tshimbalanga, N., Janssen, H., van den Bout, I., Raymond, K. and Sonnenberg, A. (2005) Nesprin-3, a novel outer nuclear membrane protein, associates with the cytoskeletal linker protein plectin. *The Journal of Cell Biology*, **171** (5), 799–810.

337 Geiger, B. (2006) A role for p130Cas in mechanotransduction. *Cell*, **127** (5), 879–881.

10
Stem Cells and Nanomedicine: Nanomechanics of the Microenvironment

Florian Rehfeldt, Adam J. Engler, and Dennis E. Discher

10.1
Introduction

Tissue cells in our body adhere to other cells and matrix and have evolved to require such attachment. While it has been known for some time that adhesion is needed for viability and normal function of most tissue cells, it has only recently been appreciated that adhesive substrates *in vivo* – namely other cells and the extracellular matrix (ECM) – are generally compliant. The only rigid substrate in most mammals is calcified bone. While the biochemical milieu for a given cell generally contains a wide range of important and distinctive soluble factors (e.g. neuronal growth factor, epidermal growth factor, fibroblast growth factor, erythropoietin), the physical environment may also possess very different *elasticity* from one tissue to another. It is well accepted that cells can 'smell' or 'taste' the soluble factors and respond via specific receptor pathways; however, it is also increasingly clear that cells 'feel' the mechanical properties of their surroundings. Regardless of the adhesion mechanism – that is, cadherins binding to adjacent cells or integrins binding to the ECM – cells engage their contractile actin/myosin cytoskeleton to exert forces on the environment, and this drives a *feedback* with responses that range from structural remodeling to differentiation. In this chapter, we aim to provide a brief overview of the diversity of *in vivo* micro/nano-environments in the human body, and also seek to describe some *mechanosensitive phenomena*, particularly with regards to adult stem cells cultured in *in vitro* systems and intended to mimic the elastic properties of native tissues.

10.2
Stem Cells in Microenvironment

10.2.1
Adult Stem Cells

Adult stem cells are distinct from embryonic stem cells (ESCs). Two properties are required for a stem cell: *self-renewal* and *pluripotency*. Stem cells must be able to

Nanotechnology, Volume 5: Nanomedicine. Edited by Viola Vogel

ISBN: 978-3-527-31736-3

divide 'indefinitely' (or at least many times compared to other cells) and also maintain their undifferentiated state. They must be potent to differentiate into various lineages. The fertilized egg cell is *totipotent* because all possible cell types in the body are derived from it. Adult stem cells or somatic stem cells are *multipotent* and are not derived directly from eggs, sperm or early embryos, as are ESCs. Among the adult stem cells found in fully developed organisms, two classes are of paramount importance for both basic scientific inquiry and possible medical application: mesenchymal stem cells (MSCs) and hematopoietic stem cells (HSCs), both of which can be obtained from adult bone marrow (Figure 10.1).

For more than 50 years, it has been known that HSCs are present in the bone marrow and can differentiate into all of the different blood cell types. Becker and coworkers identified a certain type of cell in mouse bone marrow that, when transplanted into mice which had been heavily irradiated to kill any endogenous cells, would reconstitute the various HSCs: red blood cells, white blood cells, platelets, and so on [1]. Since then, the transplantation of HSCs has become a routine medical treatment for many blood diseases, such as leukemia. However, controversies persist regarding the differentiation and de-differentiation of HSCs, particularly whether or not they can become pluripotent, being able to differentiate into neurons, muscle or bone [2–5].

MSCs also reside in the bone marrow, and can certainly differentiate into various types of solid tissue cells such as muscle, bone, cartilage and fat. From human bone marrow, Pittenger and coworkers successfully isolated truly multipotent MSCs and also demonstrated *in vitro* differentiation into various lineages [6]. Using media cocktails – often composed of glucocorticoids such as dexamethasone – many other groups have since standardized their differentiation into different tissue lineages [7–13].

Unlike the adult stem cells, ESCs are derived from a newly fertilized egg that has divided sufficiently to form a blastocyst. Once isolated from the inner mass of the blastocyst, the ESCs are then cultured and expanded *in vitro* to produce a sufficient number of cells for study. These ESCs are pluripotent in that *all* organismal cell types in the developing embryo emerge from them. This cell type is therefore considered to be the most promising for cell therapy and regenerative medicine. On the other hand, major problems such as immune rejection are significant obstacles as ESCs will generally not be from the same organism.

Figure 10.1 *In vivo* microenvironments of adult stem cells: the physiological range of stiffness. (a) Range of physiological elasticity of native cells and tissue: mesenchymal stem cells (MSCs) reside in the bone marrow (see panel (b)), but can egress from their niche into the bloodstream and travel to different tissues and organs, facing new environments with a wide range of stiffness. Nerve cells and brain tissue (around and below 1 kPa) are softest, adipocytes are assumed to be slightly stiffer, while muscle has an intermediate stiffness (~10 kPa). The elasticity of chondrocytes is speculated to fall between myocytes and osteoids (precalcified bone precursors) that are very stiff ($E \sim 30$–50 kPa) prior to mineralization into bone (~MPa to GPa); (b) Hematopoietic stem cells (HSCs), as distinct from the MSC differentiation pathway, that also reside in the bone marrow niche are the precursor cells of all blood lineage-type cells. Within the bone marrow and between the marrow and the blood there are various gradients, for example, oxygen concentration, biochemical factors and, of course, viscoelasticity.

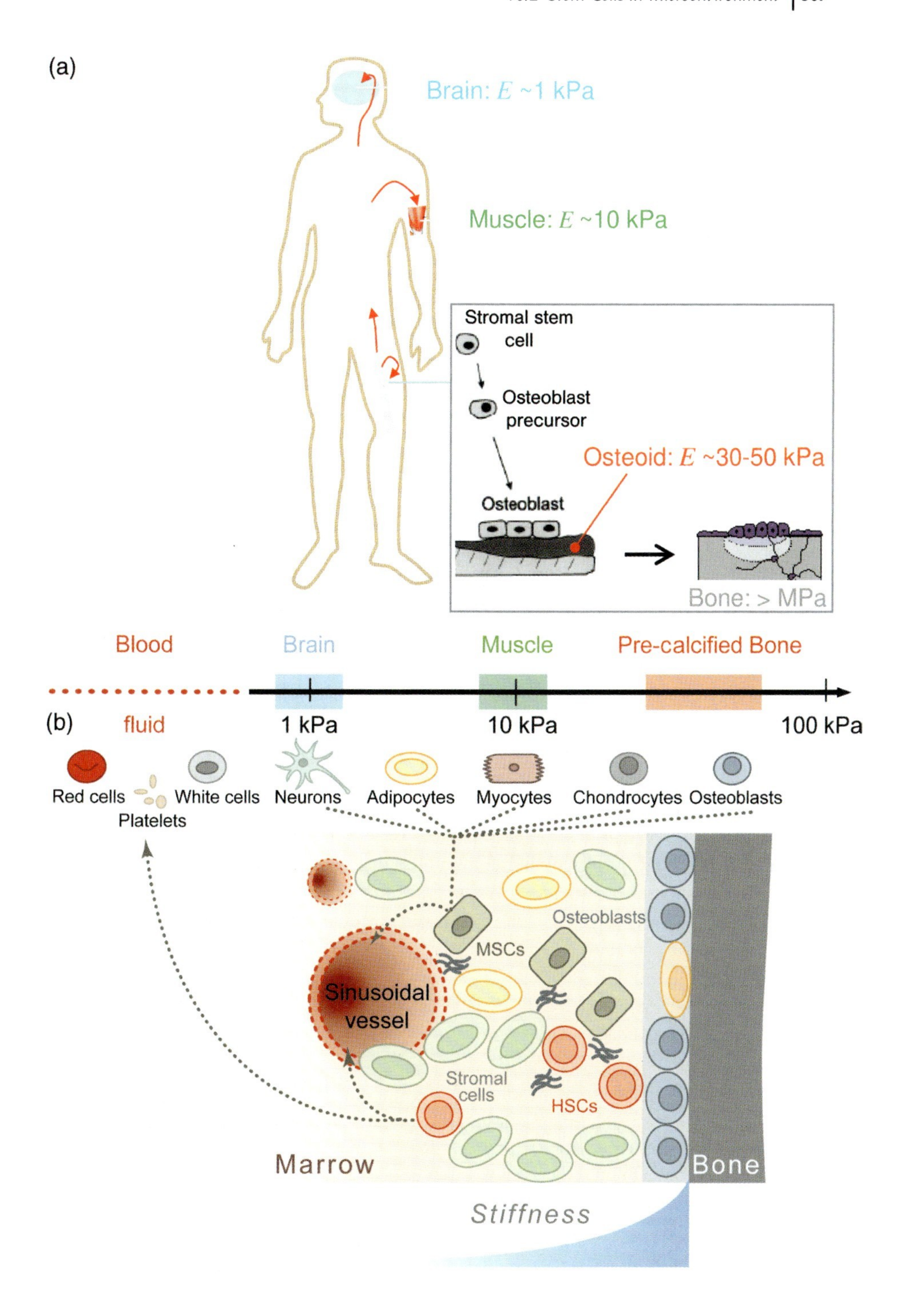

(a)
Brain: E ~1 kPa
Muscle: E ~10 kPa
Stromal stem cell
Osteoblast precursor
Osteoid: E ~30-50 kPa
Osteoblast
Bone: > MPa
Blood
fluid
Brain
Muscle
Pre-calcified Bone
1 kPa
10 kPa
100 kPa
(b)
Red cells
Platelets
White cells
Neurons
Adipocytes
Myocytes
Chondrocytes
Osteoblasts
Osteoblasts
MSCs
Sinusoidal vessel
Stromal cells
HSCs
Marrow
Bone
Stiffness

Regardless of the type of stem cell, it is essential to understand how the stem cell differentiates and in which ways it interacts with its environment, both within the niche and potential target tissues. Biochemical factors are key but not singular factors: mechanical cues in the microenvironment have only recently been recognized as contributing to the fate of MSCs.

In this chapter we focus on the nanomechanics of adult stem cells, as these cells interact with a substrate of well-defined stiffness. Stiffness – or, more formally, Young's elastic modulus, E (measured in Pascal (Pa)) – is a potential issue because of its physiological variation. That is, the only tissue in the body that is rigid is bone – all other solid tissues are soft, with elastic moduli in the range of 0.1 to 100 kPa (Figure 10.1a). Over the past decade, it has become particularly evident that the ability of an adherent cell to exert forces and build up tension reflects this elastic resistance of the surrounding microenvironment. This mechanosensitivity is based on the tension generated by ubiquitously expressed non-muscle myosins II (NMM II) which are the molecular motors that drive transduction. As with other class II myosins, such as skeletal muscle myosin (that moves all of our limbs against everyday), NMM II assembles into filaments. These NMM II minifilaments of ~300 nm length are bipolar with heads on either side that bind and actively traverse actin filaments [14]. In many situations involving moderately stiff substrates, actin–myosin assemblies are visible in cells as stress fibers, which is a prototypical contractile unit seen at least within cells grown on glass coverslips and other rigid substrates.

Of the three isoforms of NMM II (a–c), only non-muscle myosin IIa (NMM·IIa) appears prominent near the leading edge of crawling cells where cells probe their microenvironment, and is responsible for the bulk of force generation in non-muscle cells [15]. NMM IIa is also the only isoform that is essential in the developing embryo: [16] NMM IIa knockout mice fail to exhibit any functional differentiation beyond germ layers, in that they do not undergo gastrulation, and the null embryos die by day 7 *in utero*. Embryoid bodies grown in suspension culture also appear flat and flaccid, rather than spherical, which suggests weak and unstable cell–cell contacts, even though cadherins are clearly expressed. Such results highlight the fact that cell motors – as an active part of cell mechanics – are important even in the earliest stages of differentiation and development.

10.2.2
Probing the Nanoelasticity of Cell Microenvironments

Cells probe the elasticity of their microenvironment on the micro- and nanometer scale, whether the surroundings are tissue, ECM or an artificial culture substrate. Since this micro/nano-length scale is the range that cells can feel their surrounding, an appropriate experimental tool is needed to measure the elasticity on the same length scale. Perhaps the most suitable and pervasive technique for this is atomic force microscopy (AFM), which has been most widely used for imaging but can also make accurate force and elasticity measurements. The atomic force microscope exploits a microfabricated cantilever with a probe tip of well-defined geometry. This tip is pressed into the sample and the indentation depth as well as the required

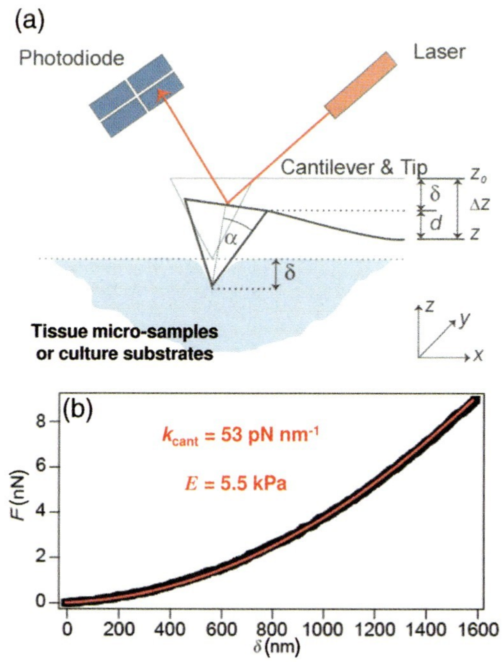

Figure 10.2 Probing the microelasticity by AFM. (a) Schematic of the measurement principle of Young's elastic modulus (*E*) of a hydrogel with AFM. A probe of well-defined geometry located at the tip of a cantilever with known spring constant *k* is pressed into the sample, and the deflection of a reflected laser beam is recorded with a four-quadrant photodiode. The required force *F* can be calculated by multiplying the deflection by the spring constant; (b) Force–indentation curve of a polyacrylamide gel with an elasticity of 5.5 kPa. The black points depict measured data; the solid red line represents the best fit of a modified Hertz model. (Reprinted with permission from Rehfeldt, F. *et al.* (2007) Cell responses to the mechano-chemical microenvironment – Implications for regenerative medicine and drug delivery. *Advanced Drug Delivery Reviews*, **59**, 1329–1339; © 2007, Elsevier.)

force is measured. The Young's elastic modulus *E* of the material surface can then be calculated from classical expressions and compared to bulk measurements, such as results from classic tensile tests.

AFM was developed during the 1980s by Binnig and coworkers [17], and was originally designed to investigate surfaces at the atomic scale with nanometer resolution. The instrument's ability to operate in a fluid environment has made it increasingly important for biological samples, and today it is used frequently to measure the nanomechanical properties of fre sh tissue samples [18], hydrogels for cell culture [18], and even living cells [19] as well as single proteins [20]. Many commercial AFM instruments allow for a precise measurement of forces, and have the ability to raster the sample at nanometer resolution, which permits mapping of a sample's elasticity.

The basic principle of AFM for determining the elasticity of a sample is sketched in Figure 10.2a. The cantilever's tip is pressed into the surface and the deflection of the bent cantilever is monitored by a laser beam reflected onto a four-quadrant photodiode. The exerted force F can be calculated by multiplying the calibrated spring constant k of the cantilever with the measured deflection d.

$$F = k \cdot d$$

The fundamental problem of the deformation of two elastic solids was first described by Hertz in 1881 [21], and the classic Hertz model was subsequently refined and modified by Sneddon to take special geometries into account [22]. A pyramidal tip that is commonly used for imaging but also works for elasticity measurements is most simply approximated as a conical probe with an opening angle α. For this geometry, E can be calculated from

$$E = \frac{\pi(1-\nu^2)F}{\delta^2 \, 2\tan\alpha}.$$

Here, F is the applied force of the tip, δ is the indentation depth of the probe into the sample, and ν is the Poisson ratio of the sample that is separately measured or can be estimated typically as 0.3–0.5.

Figure 10.2b shows a representative force indentation curve for a polyacrylamide hydrogel with a modulus of $E = 5.5$ kPa, an elasticity typical of soft tissues. The measured data points (black thick line) span a range of surface indentation (0–2000 nm) and surface forces (0–10 nN) that are typical of matrix displacements and cell-generated forces at focal adhesions [23]. AFM experiments involve measurements of the same surface THAT a cell would engage, and the results fit very well to a modified Hertz model (thin red line). The elastic modulus determined by this type of experiment only reflects the low frequency (~0.1–10 Hz) or quasi-static elasticity of the material, which is relevant to studies such as cardiac myocyte beating [23]. Additional techniques can address frequency-dependent viscoelasticity, which can be important for the interactions of cells and their surroundings. Several studies have shown a differential behavior of cells subject to an external static or oscillating force field, and a simple theoretical model seems to describe the cell response [24]. Though it seems unlikely that frequencies in the MHz range or higher will have an effect, timescales from milliseconds to hours are likely relevant when compared to processes of assembly and disassembly of actin filaments, microtubules, focal contacts and focal adhesions [25].

Frequency-dependent rheology measurements that not only encompass the static elasticity as a low-frequency limit but also measure dynamic viscous properties, are commonly performed with bulk samples using rheometers. Here, the hydrogel sample is placed between two parallel plates or a plate and a cone with very small angle (around 1°) and a well-defined stress is applied while the strain within the sample is measured. Using such a rheometer, Storm *et al.* investigated the complex rheology of several biopolymers (collagen, fibrin, neurofilaments, actin, vimentin) and polyacrylamide (PA) hydrogels, and found a nonlinear increase of the complex shear modulus with higher strain – so-called *strain-hardening behavior* [26]. In contrast to the biological gels, PA hydrogels exhibit a constant shear modulus over a large strain range. The nonlinear behavior found for biopolymers has clear, albeit unproven, implications for cell–matrix interactions. With these instruments the bulk measures of the storage and loss modulus of the material can be determined, but they do not access the rheology on the cellular scale. AFM can be similarly used if, after the probe is indented into the sample (cells, gels, etc.), a sinusoidal signal is superimposed to measure the frequency-dependent viscoelasticty. Mahaffy *et al.* used this technique to determine the viscoelastic parameters for cells and PA gels, and found good agreement with bulk measurements, at least for PA gels [27]. Additional particle-based techniques include magnetic tweezers or magnetic twisting rheology, optical tweezers and two-particle passive rheology. Although beyond the scope of this chapter, Hoffman *et al.* have shown that the use of all these different microrheology tools converges towards a "...consensus mechanics of cultured mammalian cells" [28].

10.2.3 Physical Properties of *Ex-Vivo* Microenvironments

The microelasticity of freshly isolated tissue samples can also be determined using AFM, revealing tissue inhomogeneities vis-à-vis a lateral mapping of mechanical properties that macroscopic measurements cannot address. Thus, AFM serves as a mechanical analogue to histology as it could thus reveal microelasticity differences across diseased tissues, such as fibrotic regions.

At the whole-body scale, the elasticities of normal, soft tissue vary considerably (see Figure 10.1 and Table 10.1). Aside from bodily fluids such as blood, that obviously have zero elasticity, perhaps the softest solid tissue is the brain, with an elastic modulus of just $E \sim 0.1$–1 kPa [29–31]. Native mammary gland tissue has a similar elasticity ($E \sim 0.2$ kPa) [32]. The lateral elasticity of muscle, in its relaxed state, is significantly stiffer, with AFM probing yielding $E \sim 10$ kPa. Even at the subcellular level, myofibrils isolated from rabbit skeletal muscle have $E = 11.5 \pm 3.5$ kPa in the relaxed state [33]; this is consistent with the transverse stiffness of rat skeletal muscle measured under *in vivo* conditions ($E = 15.6 \pm 5.4$ kPa) [34], as well as the elasticity of both *ex-vivo* mouse *extensor digitorum longus* (EDL) muscle ($E \approx 12 \pm 4$ kPa) and one-week cultures of C2C12 myotubes ($E \approx 12$–15 kPa) [35]. Bone is of course the stiffest material in the body, with a modulus in the region of GPa after calcification; however, bone is a composite of protein plus mineral, and precalcified bone or 'osteoid' is

Table 10.1 Elastic modulus E of normal and diseased* tissues.

Type of tissue	Elastic modulus E [kPa]	Reference(s)
Brain (macroscopic)	0.1–1	[29–31]
Mammary gland tissue	~0.2	[32]
Muscle (passive, lateral)	~10	[33,34]
Osteoid (secreted film in culture)	30–50	[36–38]
*Mammary gland tumor tissue	~4	[32]
*Dystrophic muscle	~18	[39]
*Infarcted myocardium (surface)	~55	[42]
*Granulation tissue	~50	[41]

a heavily crosslinked network of collagen-I plus other matrix proteins such as osteocalcin [36–38] with an elastic modulus of 30–50 kPa. All of the measurements above indicate that the physiological range of the elasticity of soft solid tissues ranges from 0.1 kPa up to 100 kPa.

Abnormal or diseased tissue can exhibit an elasticity which is significantly different from normal tissue. 'Sclerosis' is Greek for 'hardening of tissue', and is a descriptor of many diseases, such as atherosclerosis, that refers to a hardening of the arteries. In the context of muscle, one AFM study has shown that the muscle of mice with dystrophy has an elevated stiffness of 18 ± 6 kPa in comparison to $E \approx 12 \pm 4$ kPa for healthy tissue [39], a difference which previously had only been described in bulk [40]. Commonly found granulation tissue after wound healing exhibits an elastic modulus of $E = 50 \pm 30$ kPa, which is three- to fivefold higher than that of normal tissue [41]. Following a myocardial infarction, the myocardium is also remodeled with fibrosis that stiffens the tissue significantly: AFM measurements have shown healthy rats with a normal myocardial $E = 10$–20 kPa that increases threefold to $E = 55 \pm 15$ kPa within the infarcted region [42]. Another significant increase in tissue stiffness from the native to the diseased state is observed in tumorigenesis. The average mammary gland tumor stiffens 20-fold to $E \approx 4 \pm 1$ kPa compared to 0.2 kPa for the normal tissue [32]. In addition to fibrosis, cell tension generally stiffens the cells and tissues; this is apparent even with isolated myofibrils that exhibit an order of magnitude increase in stiffness (relaxed: $E = 11.5 \pm 3.5$ kPa to rigor: $E = 84.0 \pm 18.1$ kPa) [33]. Whether hardening is part of the key initial causes, or simply a late after-effect of a disease, is presently unclear. Nonetheless, it seems reasonable to hypothesize that hardening contributes to the cycle of disease in many cases, with a likely basis in how cells feel normal soft tissue versus abnormally hard (sclerotic) tissue. Addressing tissue hardening in regenerative medicine would seem an important aspect of therapy.

10.3 *In Vitro* Microenvironments

Different tissue cells reside in a range of very different microenvironments *in vivo*. In order to assess biological questions in rigor, however, it is necessary to culture cells, and the first reports of culturing tissue cells *in vitro* date back to the start of the twentieth century. With time, techniques and protocols have become increasingly sophisticated, allowing the growth and maintenance of a wide variety of primary cells and cell lines. The introduction of serum-free media with well-defined soluble supplements during the 1970s opened up the possibility to investigate the effects of growth factors and other factors on cultured cells. Polystyrene (PS) cell culture flasks facilitated large-scale sterile cultures, while glass coverslips have allowed very high-resolution microscopy of live cells in culture. However, both types of material substrates – as ubiquitous as they are – are very rigid with an elastic modulus in the range of MPa or GPa, which is many orders of magnitude higher than the mechanical properties that cells encounter *in vivo* (see Figure 10.1 and Table 10.1). The difference contributes to the distinct differences between cells cultured *in vitro* compared to those cultured *in vivo*. For example, cells grown on tissue culture plastic or glass very often exhibit 'stress fibers' that are not found *in vivo* and seem to reflect the mechanical stresses applied isometrically to rigid substrates. In the same way that more sophisticated media cocktails have been formulated and continue to be generated in order to dissect the different biochemical stimuli that affect tissue cells, there is growing effort with different substrates to better mimic the various physical and mechanical properties that cells encounter in soft tissues.

10.3.1 Cells Probe and Feel their Mechanical Microenvironment

In 1980, Harris and coworkers demonstrated that most cell types actively exert forces on their substrates [43]. The culture of 3T3 fibroblasts on thin silicone rubber films showed that these cells actively deform these films, generating a wrinkling pattern (Figure 10.3a). Opas demonstrated a decade later that chick retinal pigmented epithelial (RPE) cells exhibit a differential response to substrates that are rigid or viscoelastic, despite a similar surface composition [44]. On a thick compliant Matrigel substrate, the cells did not spread and remained heavily pigmented. In contrast, on the rigid glass substrate that was covalently coated with soluble basement membrane (Matrigel), the cells spread, developed stress fibers, vinculin- and talin-rich focal contacts, and expressed the dedifferentiated phenotype. These results were perhaps the first to suggest the mechanosensitivity of cells to substrate flexibility, although the study was far from conclusive: rather, it was limited to only two conditions with

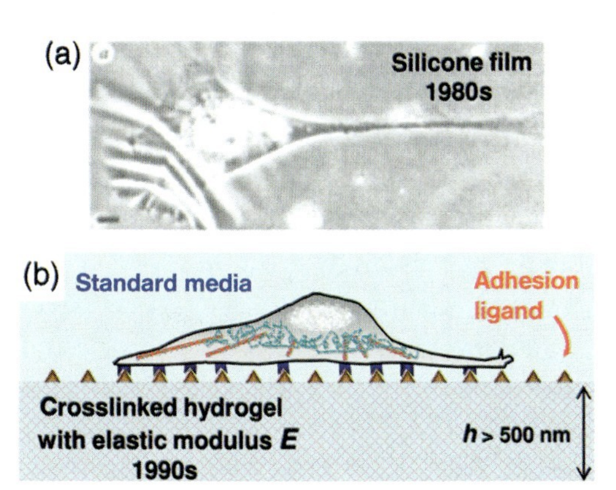

Figure 10.3 Cells pull and feel their mechanical environment. (a) 3T3 Fibroblasts cultured on a thin silicone rubber film exert forces that result in substrate wrinkling. Scale bar = 10 μm. (Reprinted with permission from Harris, A.K., Stopak, D. and Wild, P. (1981) Fibroblast traction as a mechanism for collagen morphogenesis. *Nature*, **290** (5803), 249–251; © 1981, Macmillan Publishers Ltd); (b) *In vitro* model system for cell culture on a flexible substrate. The elastic substrate with its Young's modulus E is coated with a ligand (e.g. collagen-I) for cell attachment. The cell attaches to this ligand and senses the elasticity via tension of actin–myosin complexes and/or stress fibers that are coupled to the substrate via integrins and other cell-surface receptors.

obvious compositional differences, namely rigid functionalized glass and soft Matrigel, and there were no quantitative measurements of substrate elasticity.

The study of cell mechanics on flexible substrates was significantly advanced in 1997 by Pelham and Wang, with their seminal studies on epithelial and fibroblast cell lines cultured on a range of collagen-I coated, elastic PA hydrogels [45]. An adhesive ligand in such studies is always needed because PA gels are not adhesive to cells (i.e. they do not engage integrins); collagen-I is a logical first choice for an ECM ligand because it is one of the most abundant proteins in mammals – which means of course that cells are very likely to encounter this protein in an organism. By using different crosslinker concentrations, a set of gels ranging in elastic modulus with an otherwise identical surface was generated. With this *in vitro* culture system, distinct differences were exhibited by cells on soft and stiff matrices: cells were seen to spread more on stiffer substrates, and also exhibited more typical focal adhesions on stiffer substrates. It was also clear that non-muscle myosin is a key player in generating force in the mechanosensitivity. When exposed to the common myosin inhibitors of the time (2,3-butanedione monoxime or KT5926), the cells could no longer distinguish between soft and stiff substrates. The effects on cell motility also have become clear: fibroblast cells that approached the transition from the soft side could easily migrate to the stiffer side, with a simultaneous increase in their spread area and traction forces [46]. In comparison, cells on the stiffer side of the gel often turned around or retracted as they reached the border. This dependence of cell movement on purely physical properties of the substrate has been termed *durotaxis,* and clearly shows that cells probe and feel the mechanics of their microenvironment.

Although these initial studies very elegantly demonstrated the differential responses of cells to substrate elasticity, the precise connections to *in vivo* microenvironments, as well as the role that diseases play in matrix stiffness, remained unclear and required further exploration. Measurements of elasticity were also only approximate: they were made by estimating indentations with steel balls of known weight. Different tissues often exhibit characteristic elasticities and can have significant alterations in disease (as discussed in Section 10.1). Engler and coworkers cultured myoblasts on the same collagen-I-coated PA gel systems with a wide range of AFM-determined elasticities, and showed that fusion into myotubes was not significantly affected, whereas myosin–actin striation was most prominent within cells grown on substrates with $E = 12 \pm 5$ kPa, which corresponds to the native muscle elasticity [39]. Gel substrates that were too stiff, as well as rigid glass, inhibited the formation of striated actin–myosin fibers. Striation was weak in myotubes on 18 kPa gels emulating dystrophic muscle, suggesting a significant counterinfluence against differentiation in disease.

In the context of wound healing, Goffin *et al.* examined fibroblast adhesion and cytoskeletal organization in cells on surfaces of different rigidity [41]. These cells play an essential role in wound healing and tissue remodeling as they migrate to wounded tissue and can develop stress fibers and tension to facilitate wound closure and healing. Using substrates that simulate normal soft tissue and stiffened wounded tissue, more contractile and differentiated myofibroblasts were only seen on the stiffer substrates. In addition, 'supermature' focal adhesions (suFAs) were found to develop only on rigid substrates and to exert a fourfold higher stress on the matrix than was exerted under more typical focal adhesions formed on 11 kPa gels. It was proposed that this is a means by which the matrix influences the tension that the cells apply and therefore helps to steer the wound-healing process.

Similar findings were recently reported for liver-derived portal fibroblasts and their differentiation to myofibroblasts *in vitro* on PA substrates in the presence or absence of transforming growth factor-β (TGF-β) [47]. When these fibroblasts differentiate towards myofibroblasts – as occurs in response to an acute liver injury – they start to express α-smooth muscle actin (α-SMA) and form stress fibers on rigid surfaces (>3 kPa), but not on very soft (400 Pa) gels that resemble the elasticity of native rat or human liver tissue. When treated with 100 pM TGF-β the portal fibroblasts began to express α-SMA even on the soft matrix, although they did not develop organized stress fibers; stiffer matrices were required for cell spreading and stress fiber organization. Cells treated with 5 μ*M* TGF-β receptor kinase inhibitor did not differentiate on any of the substrates, which suggests that TGF-β functions as an essential contractile inducer in these cells (opposite to myosin inhibitors), leading to higher α-SMA expression and stress fiber organization as stiffer substrates. Both, biochemical and biophysical stimuli are thus part of the complex interplay of mechanosensing.

10.3.2
Cells React to External Forces

The responses of cells to physical cues in their microenvironment – namely elasticity and geometry – are not the only physical factors of importance. Cells also react to external forces, and forces are found throughout an organism. Muscles contract and relax, joints are compressed during standing, walking and running, and the average human heart beats 72 times each minute to keep our blood circulating, which leads to shear stresses on the surfaces of endothelial cells. The effects of external forces on cells have been widely studied, although it is often not clear to what extent the force-generation capabilities of cells are again part of the response.

Several studies have documented the influence of both static and dynamic strains on cells. For example, C2C12 mouse myoblasts cultured for several days on a substrate that is subject to a continuous, unidirectional stretching leads to alignment and elongation of the cells [35]. This applied static strain mimics the *in vivo* conditions of long bone growth and muscle development. Another investigation of fibroblast morphological changes in collagen matrices under a mechanical load [48] revealed cell alignment with the direction of the external force to minimize their exposure to the strain. In contrast to the parallel alignment in the case of static strain, several studies have also reported a more perpendicular orientation for rapidly oscillating external forces [49–52]. Experimental results such as these provide insight into potential mechanisms in development and repair of connective tissue.

Experiments are many, but theories are few and would benefit in understanding and predicting. Safran and coworkers have modeled the cell as a contractile dipole in an external force field [24]. This coarse-grain model of a cell aligns parallel to a static or quasistatic external oscillating force field, but it orients perpendicular to the field if the frequency is too high to follow. This is analogous to an electric dipole in an oscillating electromagnetic field. The model not only agrees with experimental evidence but also demonstrates the applicability of basic physical concepts to cell mechanics.

10.3.3 Adult Stem Cell Differentiation

The impact of substrate elasticity on cell behavior is now evident in many studies. One last – but central – example for this chapter perhaps highlights the potent influence of elastic matrix effects, namely the differentiation of adult stem cells (MSCs) [53]. The usual method for inducing the differentiation of MSCs towards any particular lineage (e.g. adipocytes, chondrocytes, myocytes, osteocytes) is to use media cocktails based on steroids and/or growth factors [6–13]. Our approach has been to use a single, 10% serum-containing media and to vary only the stiffness of the culture substrate in sparsely plated cultures. These cells are exposed to serum *in vivo*, but during natural processes of emigration from the marrow to repair and maintain tissue, they also encounter different micromechanical environments. It is this latter aspect of environment that we sought to mimic.

MSCs were plated on collagen-I PA hydrogels of different elasticity E (Figure 10.3b) and found to exhibit after just 4 h a significantly different morphology that becomes even more pronounced over the next 24 to 96 h (Figure 10.4a). The cells spread more with increasing substrate stiffness, as found with other cells [45], but they also take on different morphologies. As the cells are multipotent, it was of further interest to assess whether substrate mechanics could also influence gene transcription, and therefore differentiation. Immunostaining for early lineage specific proteins indeed revealed that the neurogenic marker, β3 tubulin was only present on the softest 1 kPa gels, the myogenic transcription factor MyoD was most prominent on the 11 kPa gels, and an early osteogenic transcription factor, CBFα1, was detectable only on the stiffest 34 kPa substrates. Remarkably, these stiffnesses that induced differentiation correspond to the elasticities that the various lineages would experience in their respective native tissues: quantitative analyses of differentiation markers emphasizes the finding that adult stem cells adjust their lineage towards the stiffness of their surrounding tissues (Figure 10.4b).

Figure 10.4 (***Continued***) increase of any of the three proteins. Dashed green and orange curves depict the substrate-dependent upregulation of the markers for already committed cells [C2C12 (muscle) and hFOB (bone)] exhibiting the same qualitative behavior at a higher intensity due to their committed nature; (c) Transcription profiling array shows selective upregulation of several lineage-specific genes due to matrix elasticity. Values for MSCs cultured on PA gels for one week were normalized by β-actin and then further normalized with data obtained from naïve, undifferentiated MSCs before plating. Red depicts relative upregulation; green shows downregulation. Gene transcripts of the different lineages are upregulated only on the substrates with the appropriate stiffness; blebbistatin treatment inhibits this upregulation and thus differentiation. (Reprinted with permission from Engler, A.J. *et al.* (2006) Matrix elasticity directs stem cell lineage specification. *Cell*, **126**, 677–689; © 2006, Elsevier).

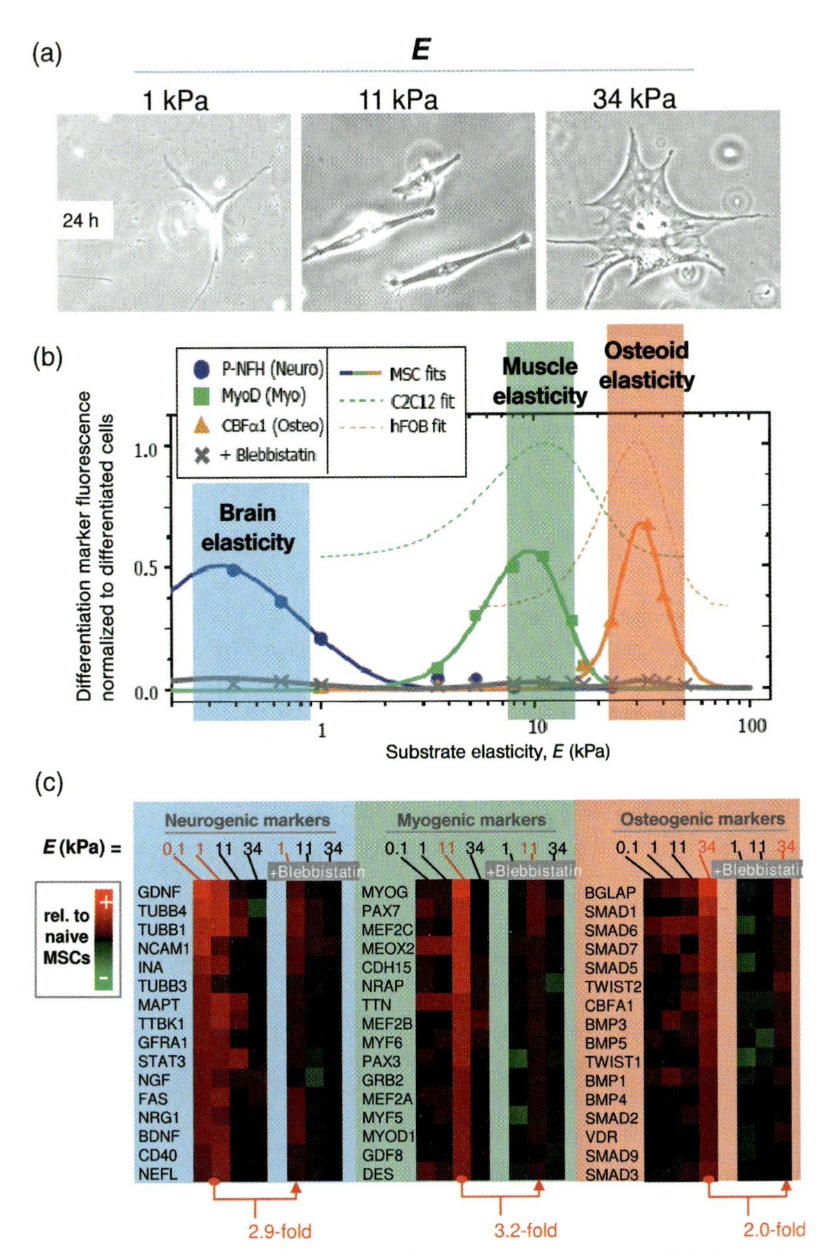

Figure 10.4 Differentiation of adult stem cells guided by matrix elasticity. (a) Mesenchymal stem cells (MSCs) on collagen-I coated PA gels with different elasticities ($E = 1$ kPa; 11 kPa; 34 kPa) show distinct morphology at 24 h after plating; (b) Quantitative immunofluorescence of the lineage markers (blue, P-NFH (neuro); green, MyoD (muscle); orange, CBFα1 (osteo)) reveals the stiffness-dependent differentiation of the MSCs. The multipotent stem cells upregulate differentiation markers only on substrates yielding a stiffness in the range of the native tissue, respectively. The gray curve of blebbistatin-treated cells shows no selective

Treatment of the MSCs with blebbistatin, a potent, recently synthesized NMM II inhibitor, largely blocked the expression of any of the differentiation markers, and again highlighted the key role of this motor in sensing the substrate in mechanoguided differentiation. Repeating the same experiment with two cell lines that were already committed (C2C12 mouse myoblasts and hFOB, human osteoblasts) showed a similar upregulation of the differentiation marker according to the tissue-level elasticity, but there was also a higher, baseline level of expression that reflected the fact that these cells were already committed. This led to a new hypothesis of differentiation mechanisms suggesting that both biochemical and biophysical stimuli influenced the differentiation of these *multipotent* adult stem cells.

Transcript profiling of some of the most commonly accepted lineage markers was used to more broadly assess lineage specification by matrix. The top-16 genes profiled for neuro-, myo- and osteo-genesis show selective upregulation of several relative to the naïve undifferentiated MSCs before plating (Figure 10.4c). Consistent with protein markers, it can be shown that β3 tubulin is the sixth-ranked gene on the softest 0.1–1 kPa gels, MyoD is the 14th-ranked gene on the 11 kPa gels, and CBFα1 is the seventh-ranked gene on the stiffest 34 kPa matrices. Also consistent with the downregulation of protein with blebbistatin treatment, the transcripts also exhibited a downregulation of about two to threefold.

Further examination of differentially expressed genes is revealing. Neural growth factors such as glial-derived neurotrophic factor (GDNF) and nerve growth factor (NGF) are upregulated on softer matrices. GDNF is interesting because its most prominent feature is its ability to support the survival of dopaminergic and motor neurons. The latter neuronal populations die during the course of amyotrophic lateral sclerosis (ALS). Myostatin (GDF8) is upregulated on the 11 kPa myogenic matrix and secreted by skeletal muscle cells; it is understood to circulate and act as a negative regulator of muscle mass, slowing down the myogenesis of muscle stem cells. Several bone growth factors (e.g. bone morphogenetic proteins: BMP 3, 4, 5) are upregulated on the stiffest osteogenic matrices. These proteins are interesting as potent osteoinductive growth factors belonging to the TGF-β superfamily, which was described in Section 10.2.1 as promoting stress fibers in fibroblasts on stiff matrices. This is very consistent with stress fiber assembly seen also in the MSCs [53]. Follow-up studies are certainly required to assess the secretion of these factors as well as autocrine–paracrine loops, although matrix elasticity is clearly the initiating factor throughout. Additionally, the many transcription factor genes listed (e.g. STATs, MYFs, MEFs and SMADs), as well as the many cytoskeletal and adhesion transcripts (e.g. NCAM, TTN and BGLAP (or osteocalcin)) make for a compelling story of how these MSCs physically interact with their microenvironment and reprogram their gene expression accordingly.

10.3.4
Implications for Regenerative Medicine

MSCs are believed to have considerable potential for cell therapies and regenerative medicine. Taking into account the impact of the microenvironment (as described above), it perhaps becomes clear how important it is to carefully assess potential applications of these cells.

One application which currently is undergoing major exploration is the injection of purified and enriched MSCs into a stiffened infarct of the heart – a technique known as *cellular cardiomyoplasty*. The hope is that these adult stem cells will differentiate to cardiomyocytes and improve contractile function, although recent animal models and even clinical trials have yielded mixed results at best [42,54–56]. For example, in one rat infarct model, the injection of human-MSCs was found to marginally improve myocardial compliance as determined using AFM [42], but the MSCs did not regenerate the infarcted heart muscle tissue. Working strictly with a mouse model, Fleischmann and coworkers also injected MSCs into an infarcted myocardium [55] and, two to four weeks after injection, identified encapsulated calcifications and ossifications in the infarcted zone. These compartments were clearly restricted to the scarred region of the infarct where the elastic modulus E is much higher than that of native cardiac muscle. Interestingly, when MSCs were injected into intact non-infarcted hearts, calcifications and ossifications occurred only on the scar tissue along the injection channel. These findings were in excellent agreement with the *in vitro* studies of Engler *et al.* [53], where osteogenesis of MSCs was found on matrices having an elasticity in the range of 30–50 kPa (Figure 10.4) – that is, the stiffness of postinfarct scar tissue.

For future experiments and clinical trials, it will be of paramount importance to clearly dissect all of the possible cell stimuli in order to at least avoid negative implications for the patient such as calcifications. Our cells live in a 'world' of biophysics and biochemistry, and it seems necessary to understand and control parameters of both sides.

10.4 Future Perspectives

This chapter could only highlight some of many studies on the implications of the *mechano*-chemical environment of cells, even this small selection underscores the importance of a better understanding of the interactions between cells and environment to improve the design of therapeutic applications. Adult stem cells are probably one of the most promising candidates for successful tissue regeneration, given their multipotency, availability and limited ethical considerations, although their interactions with the microenvironment must be taken into account. Further studies must elucidate the complex interplay of biochemistry and biophysics, and should identify ways to influence either side with stimuli from the other. As a prime example, approaches to repair the infarcted heart reveal how new strategies are needed to overcome the physical limitations of a fibrotic tissue. Perhaps it is possible to alter the cell's perception of the surrounding stiffness so that adult stem cells could develop towards a suitable myogenic lineage (instead of osteogenic)? This is clearly a large playground for future studies of what are ultimately diseases that couple to cell mechanics.

Acknowledgments

F.R. gratefully acknowledges the Feodor Lynen fellowship from the Alexander von Humboldt foundation, and thanks André E.X. Brown for critical reading of the manuscript and Andrea Rehfeldt for help with the illustrations. A.J.E. and D.E.D. acknowledge the NIH and NSF for support via NRSA and R01 funding, respectively.

References

1 Becker, A.J., Till, J.E. and McCulloch, E.A. (1963) *Nature*, **197**, 452.

2 Corbel, S.Y., Lee, A., Yi, L., Duenas, J., Brazelton, T.R., Blau, H.M. and Rossi, F.M.V. (2003) *Nature Medicine*, **9**, 1528.

3 Hess, D.C., Abe, T., Hill, W.D., Studdard, A.M., Carothers, J., Masuya, M., Fleming, P.A., Drake, C.J. and Ogawa, M. (2004) *Experimental Neurology*, **186**, 134.

4 Roybon, L., Ma, Z., Asztely, F., Fosum, A., Jacobsen, S.E.W., Brundin, P. and Li, J.Y. (2006) *Stem Cells (Dayton, Ohio)*, **24**, 1594.

5 Deten, A., Volz, H.C., Clamors, S., Leiblein, S., Briest, W., Marx, G. and Zimmer, H.G. (2005) *Cardiovascular Research*, **65**, 52.

6 Pittenger, M.F., Mackay, A.M., Beck, S.C., Jaiswal, R.K., Douglas, R., Mosca, J.D., Moorman, M.A., Simonetti, D.W., Craig, S. and Marshak, D.R. (1999) *Science*, **284**, 143.

7 Caplan, A.I. (1991) *Journal of Orthopaedic Research*, **9**, 641.

8 Hofstetter, C.P., Schwarz, E.J., Hess, D., Widenfalk, J., El Manira, A., Prockop, D.J. and Olson, L. (2002) *Proceedings of the National Academy of Sciences of the United States of America*, **99**, 2199.

9 Kondo, T., Johnson, S.A., Yoder, M.C., Romand, R. and Hashino, E. (2005) *Proceedings of the National Academy of Sciences of the United States of America*, **102**, 4789.

10 McBeath, R., Pirone, D.M., Nelson, C.M., Bhadriraju, K. and Chen, C.S. (2004) *Developmental Cell*, **6**, 483.

11 Kuznetsov, S.A., Krebsbach, P.H., Satomura, K., Kerr, J., Riminucci, M., Benayahu, D. and Robey, P.G. (1997) *Journal of Bone and Mineral Research*, **12**, 1335.

12 Prockop, D.J. (1997) *Science*, **276**, 71.

13 Yoo, J.U., Barthel, T.S., Nishimura, K., Solchaga, L., Caplan, A.I., Goldberg, V.M. and Johnstone, B. (1998) *Journal of Bone and Joint Surgery - American Volume*, **80**, 1745.

14 Verkhovsky, A.B., Svitkina, T.M. and Borisy, G.G. (1995) *Journal of Cell Biology*, **131**, 989.

15 Cai, Y.F., Biais, N., Giannone, G., Tanase, M., Jiang, G.Y., Hofman, J.M., Wiggins, C.H., Silberzan, P., Buguin, A., Ladoux, B. and Sheetz, M.P. (2006) *Biophysical Journal*, **91**, 3907.

16 Conti, M.A., Even-Ram, S., Liu, C.Y., Yamada, K.M. and Adelstein, R.S. (2004) *Journal of Biological Chemistry*, **279**, 41263.

17 Binnig, G., Quate, C.F. and Gerber, C. (1986) *Physical Review Letters*, **56**, 930.

18 Engler, A.J., Richert, L., Wong, J.Y., Picart, C. and Discher, D.E. (2004) *Surface Science*, **570**, 142.

19 Radmacher, M. (2002) Measuring the elastic properties of living cells by the atomic force microscope, in *Methods in Cell Biology* (eds B.P. Jena and H.J.K. Horber), Vol. 68, Academic Press, San Diego, pp. 67–90.

20 Ludwig, M., Rief, M., Schmidt, L., Li, H., Oesterhelt, F., Gautel, M. and Gaub, H.E. (1999) *Applied Physics A - Materials Science and Processing*, **68**, 173.

21 Hertz, H. (1881) *Journal für Die Reine und Angewandte Mathematik*, **92**, 156.

22 Sneddon, I.N. (1965) *International Journal of Engineering Science*, **3**, 47.

23 Balaban, N.Q., Schwarz, U.S., Riveline, D., Goichberg, P., Tzur, G., Sabanay, I., Mahalu, D., Safran, S., Bershadsky, A., Addadi, L. and Geiger, B. (2001) *Nature Cell Biology*, **3**, 466.

24 De, R. Zemel, A. and Safran, S.A. (2007) *Nature Physics*, **3**, 655.

25 von Wichert, G., Haimovich, B., Feng, G.S. and Sheetz, M.P. (2003) *EMBO Journal*, **22**, 5023.

26 Storm, C., Pastore, J.J., MacKintosh, F.C., Lubensky, T.C. and Janmey, P.A. (2005) *Nature*, **435**, 191.

27 Mahaffy, R.E., Shih, C.K., MacKintosh, F.C. and Kas, J. (2000) *Physical Review Letters*, **85**, 880.

28 Hoffman, B.D., Massiera, G., Van Citters, K.M. and Crocker, J.C. (2006) *Proceedings of the National Academy of Sciences of the United States of America*, **103**, 10259.

29 Gefen, A. and Margulies, S.S. (2004) *Journal of Biomechanics*, **37**, 1339.

30 Lu, Y.B., Franze, K., Seifert, G., Steinhauser, C., Kirchhoff, F., Wolburg, H., Guck, J., Janmey, P., Wei, E.Q., Kas, J. and Reichenbach, A. (2006) *Proceedings of the National Academy of Sciences of the United States of America* **103**, 17759.

31 Georges, P.C., Miller, W.J., Meaney, D.F., Sawyer, E.S. and Janmey, P.A. (2006) *Biophysical Journal*, **90**, 3012.

32 Paszek, M.J., Zahir, N., Johnson, K.R., Lakins, J.N., Rozenberg, G.I., Gefen, A., Reinhart-King, C.A., Margulies, S.S., Dembo, M., Boettiger, D., Hammer, D.A. and Weaver, V.M. (2005) *Cancer Cell*, **8**, 241.

33 Yoshikawa, Y., Yasuike, T., Yagi, A. and Yamada, T. (1999) *Biochemical and Biophysical Research Communications* **256**, 13.

34 Bosboom, E.M.H., Hesselink, M.K.C., Oomens, C.W.J., Bouten, C.V.C., Drost, M.R. and Baaijens, F.P.T. (2001) *Journal of Biomechanics*, **34**, 1365.

35 Collinsworth, A.M., Torgan, C.E., Nagda, S.N., Rajalingam, R.J., Kraus, W.E. and Truskey, G.A. (2000) *Cell and Tissue Research*, **302**, 243.

36 Morinobu, M., Ishijima, M., Rittling, S.R., Tsuji, K., Yamamoto, H., Nifuji, A., Denhardt, D.T. and Noda, M. (2003) *Journal of Bone and Mineral Research*, **18**, 1706.

37 Andrades, J.A., Santamaria, J.A., Nimni, M.E. and Becerra, J. (2001) *International Journal of Developmental Biology*, **45**, 689.

38 Holmbeck, K., Bianco, P., Caterina, J., Yamada, S., Kromer, M., Kuznetsov, S.A., Mankani, M., Robey, P.G., Poole, A.R., Pidoux, I., Ward, J.M. and Birkedal-Hansen, H. (1999) *Cell*, **99**, 81.

39 Engler, A.J., Griffin, M.A., Sen, S., Bönnemann, C.G., Sweeney, H.L. and Discher, D.E. (2004) *Journal of Cell Biology*, **166**, 877.

40 Stedman, H.H., Sweeney, H.L., Shrager, J.B., Maguire, H.C., Panettieri, R.A., Petrof, B., Narusawa, M., Leferovich, J.M., Sladky, J.T. and Kelly, A.M. (1991) *Nature*, **352**, 536.

41 Goffin, J.M., Pittet, P., Csucs, G., Lussi, J.W., Meister, J.J. and Hinz, B. (2006) *The Journal of Cell Biology*, **172**, 259.

42 Berry, M.F., Engler, A.J., Woo, Y.J., Pirolli, T.J., Bish, L.T., Jayasankar, V., Morine, K.J., Gardner, T.J., Discher, D.E. and Sweeney, H.L. (2006) *American Journal of Physiology - Heart and Circulatory Physiology*, **290**, H2196.

43 Harris, A.K. Wild, P. and Stopak, D. (1980) *Science*, **208**, 177.

44 Opas, M. (1989) *Developmental Biology*, **131**, 281.

45 Pelham, R.J. and Wang, Y.L. (1997) *Proceedings of the National Academy of Sciences of the United States of America*, **94**, 13661.

46 Lo, C.M., Wang, H.B., Dembo, M. and Wang, Y.L. (2000) *Biophysical Journal*, **79**, 144.

47 Li, Z.D., Dranoff, J.A., Chan, E.P., Uemura, M., Sevigny, J. and Wells, R.G. (2007) *Hepatology (Baltimore, Md)*, **46**, 1246.

48 Eastwood, M., Mudera, V.C., McGrouther, D.A. and Brown, R.A. (1998) *Cell Motility and the Cytoskeleton*, **40**, 13.

49 Shirinsky, V.P., Antonov, A.S., Birukov, K.G., Sobolevsky, A.V., Romanov, Y.A., Kabaeva, N.V., Antonova, G.N. and Smirnov, V.N. (1989) *Journal of Cell Biology*, **109**, 331.

50 Hayakawa, K. Sato, N. and Obinata, T. (2001) *Experimental Cell Research*, **268**, 104.

51 Kurpinski, K., Chu, J., Hashi, C. and Li, S. (2006) *Proceedings of the National Academy of Sciences of the United States of America*, **103**, 16095.

52 Cha, J.M., Park, T.N., Noh, T.H. and Suh, T. (2006) *Artificial Organs*, **30**, 250.

53 Engler, A.J., Sen, S., Sweeney, H.L. and Discher, D.E. (2006) *Cell*, **126**, 677.

54 Lee, M.S. Lill, M. and Makkar, R.R. (2004) *Reviews in Cardiovascular Medicine*, **5**, 82.

55 Breitbach, M., Bostani, T., Roell, W., Xia, Y., Dewald, O., Nygren, J.M., Fries, J.W.U., Tiemann, K., Bohlen, H., Hescheler, J., Welz, A., Bloch, W., Jacobsen, S.E.W. and Fleischmann, B.K. (2007) *Blood*, **110**, 1362.

56 Murry, C.E., Soonpaa, M.H., Reinecke, H., Nakajima, H., Nakajima, H.O., Rubart, M., Pasumarthi, K.B.S., Virag, J.I., Bartelmez, S.H., Poppa, V., Bradford, G., Dowell, J.D., Williams, D.A. and Field, L.J. (2004) *Nature*, **428**, 664.

11
The Micro- and Nanoscale Architecture of the Immunological Synapse

Iain E. Dunlop, Michael L. Dustin, and Joachim P. Spatz

11.1
Introduction

In vivo, biological cells come into direct physical contact with other cells, and with extracellular matrices in a wide variety of contexts. These contact events are in turn used to pass an enormous variety of cell signals, often by bringing ligand–receptor pairs on adjacent cells into contact with each other. Whereas, some traditional outlooks on cell signaling arguably focused strongly on these individual ligation events as triggers for signaling cascades, it is now becoming clear that this is insufficient. Rather, in some cases where signal-activating ligands are found on cell or matrix surfaces *in vivo*, the properties of each surface as a whole need to be considered if the events leading to signaling are to be fully understood. That is, the strength of signaling – and whether signaling occurs at all – can depend on factors such as the spatial distribution of signaling-inducing ligands that are presented on a surface, the mobility of these ligands, the stiffness of the substrate, and the force and contact time between the surface and the cell being stimulated [1]. The effects of such surface properties on the activation of cell signaling pathways can often be studied by bringing the cells into contact with artificial surfaces, the properties of which can be controlled and systematically varied, so that the effects of such properties on signaling pathway activation can be observed. These studies have been successfully conducted in the context of signaling pathways associated with cell behaviors such as fibroblast adhesion to the extracellular matrix (ECM) [2, 3] and rolling adhesion of leukocytes [4, 5]. One important system in which cell–cell communication has been studied is the immunological synapse formed between T lymphocytes and tissue cells at multiple stages of the immune response.

We first introduce the role of T cells in the immune response and the concept of an immunological synapse (for an introduction to immunological concepts, see Ref. [6]). T cells are an important component of the mammalian adaptive immune system, and each of the billions of T cells in a mammal expresses a unique receptor generated by the recombination of variable genomic segments. This can then serve as a substrate

Nanotechnology, Volume 5: Nanomedicine. Edited by Viola Vogel

ISBN: 978-3-527-31736-3

for the selection of pathogen-specific T cells suitable for combating only infections by identical or similar pathogens, the proteomes of which share a particular short peptide sequence, known as the T cell's *agonist peptide*. There are three main subclasses of T cell, classified according to their effector functions: helper-, killer- and regulatory T cells. Broadly speaking, helper T cells act to stimulate and maintain the immune response in the vicinity of an infection, whereas killer T cells are responsible for detecting and destroying virus-infected cells. Regulatory T cells play a role in the prevention of autoimmune disease. In this chapter we will concern ourselves almost entirely with the activation of helper T cells. As the number of possible pathogens is enormous, the body does not maintain large stocks of T cells of a wide variety of specificities, but rather maintains small numbers of inactive T cells of each possible specificity in locations such as the lymph nodes and the spleen. When a pathogen is detected in the body, specialist antigen-presenting cells (APCs) travel to these locations and locate the correct T cells to combat the infection. This causes the T cells to become *activated*, whereupon they proliferate, producing a large number of T cells that travel to the infected tissues to carry out their antipathogen roles. Activation of the T cell occurs during direct physical contact between the T cell and the APC, and proceeds via the formation of a stable contact region between the T cell and the APC, known as the immunological synapse. (The term 'synapse' was applied due to a number of shared features with neurological synapses, such as stability, a role for adhesion molecules and directed transfer of information between cells [7].) It is known that one of the central requirements for activation is the ligation of T-cell receptors (TCRs) on the T-cell surface by peptide-major histocompatibility complex protein (p-MHC) complexes on the APC surface. The MHC may be thought of as a molecule in which the APC mounts short peptides made by breaking down all proteins in the cytoplasm (MHC class I) or in the endocytic pathway (MHC class II), including both pathogenic and 'self' proteins. As MHC class II molecules are relevant for helper T cells, we will focus on these from here on. Each TCR molecule strongly binds MHC molecules that mount the agonist peptide, and weakly binds MHC molecules that mount a subset of self-peptides. These strong and weak interactions synergize to generate sustained signals in the T cell. Thus, the APC activates only the correct T cells to combat the particular infection that is under way due to the necessary role of the agonist peptide in T-cell activation, but does so with great sensitivity (early in infection) due to the contribution of the self-peptides [8].

In addition to the initial activation process, helper T cells may encounter other agonist p-MHC-bearing APCs later in the infection process with which they can also form immunological synapses. This is particularly important in the infected tissues, where helper T cells coordinate responses by many immune cell types. The signaling from these synapses effectively informs the T cells that the infection is still in progress, encouraging them to continue countering it locally. Although there are differences between the initial activation process and these subsequent restimulations, similar signaling methods may underlie them both, and we will usually not be concerned with such distinction in this chapter. Although most of the agonist peptide-specific T cells will die when the infectious agent has been eliminated from the body, a small subset will live on as 'memory' T cells and can facilitate the mounting of a response to reinfection with the same or closely related agents at a later time [9–11]. In fact, memory T cells are the basis of vaccination, and the process by which they are reactivated is likely to be similar in its requirements for immunological synapse formation.

Although TCR–p-MHC ligation is necessary for T-cell activation, there is evidence that the structure of the T cell–APC contact zone on a wide variety of length scales from tens of micrometers down to one to a few nanometers plays a role in determining the strength of activation signaling. Artificial surfaces functionalized with p-MHC and other immune cell proteins have been used to study structures that arise in the contact zone, and their effect on the activation process.

In this chapter, we will review the emerging body of work in which surface functionalization and lithography techniques have been used to produce artificial surfaces that have shed light on the nature and dynamics of the immunological synapse. We will first consider the structure of the immunological synapse on the micrometer scale, including the so-called supramolecular activation clusters (SMACs). These are essentially segregated areas in which different ligated receptor species predominate. Although SMACs have been widely studied, it now seems unlikely that they are the critical structures in providing activation signaling, with smaller-scale microclusters consisting of around 5–20 TCR–p-MHC pairs bound closely together being of greater significance [12]; these microclusters will also be discussed. We will then describe experiments that demonstrate the importance of the spatial distribution of molecules on the nanometer scale – that is, one to a few protein molecules – using materials such as soluble p-MHC oligomers to stimulate T cells. By illustrating the importance of the nanoscale, these results should motivate future studies in which T cells are brought into contact with surfaces that are patterned on a nanometer scale with p-MHC and other immunological synapse molecules. We will then discuss an emerging nanolithography technique that could plausibly be used to perform such studies, namely block copolymer micellar nanolithography. Finally, we will consider the possibility of making direct therapeutic use of micro- and nanopatterned T-cell-activating surfaces, and conclude that the most likely application is in *adoptive cell therapy*. In this method, T cells are removed from a patient, expanded *in vitro*, and returned to the patient to combat a disease – most commonly a cancerous tumor. It has been suggested that the success of adoptive cell therapy can depend heavily on the detailed phenotype of the returned T cells; the use of micro- and nanopatterned surfaces for *in vitro* T-cell activation could help to control their phenotype.

11.2 The Immunological Synapse

11.2.1 Large-Scale Structure and Supramolecular Activation Clusters (SMACs)

The immunogical synapse is a complex structure, which features a number of important ligand–receptor interactions in addition to the crucial TCR–p-MHC interactions. The artificial substrates discussed here are based on a simplified model of the synapse that includes two of the most significant ligand–receptor pairs: TCR with p-MHC, and lymphocyte function-associated antigen 1 (LFA-1) with intracellular adhesion molecule 1 (ICAM-1). LFA-1 is an integrin-family protein, the function of which is to control T cell to APC adhesion. LFA-1 is expressed on T cell surfaces and binds ICAM-1 on the APC surface.

A major contribution to the understanding of the immunological synapse has been derived from studies in which T cells are allowed to settle on glass substrates on which lipid bilayers have been deposited. These bilayers contain some lipid molecules that are bound to protein constructs containing the extracellular portions of p-MHC and ICAM-1, respectively. Due to the fluidity of the lipid bilayer, the p-MHC and ICAM-1 are mobile, creating a simplified model of the APC surface (see for example, Refs [12–15]). Although a simplified system, this model reproduces features of immunological synapses observed *in vivo* with some types of APC, for example the so-called *B cells*; differences between these synapses and those observed between T cells and so-called *dendritic cells* (another type of APC) may be due to the dendritic cell cytoskeleton's restricting and controlling of p-MHC and/or ICAM-1 motion [16].

On the largest length scales, the evolution of the synapse can be seen to proceed in three stages, as illustrated in Figure 11.1 [13]. In the first stage, the T cell is migrating over the model bilayer surface; this corresponds to an *in vivo* T cell forming transient contacts with passing APCs. A central core of adhesive LFA-1–ICAM-1 contacts forms, around which the cytoskeleton deforms to produce an area of very close contact between the cell and the substrate, in which TCR with agonist p-MHC pairs can readily form. This cytoskeletal deformation is important, as TCR and p-MHC molecules are both rather short and consequently easily prevented from coming into contact by abundant larger glycosolated membrane proteins [13]. Signaling arising from the formation of TCR–p-MHC pairs causes LFA-1 molecules to change their

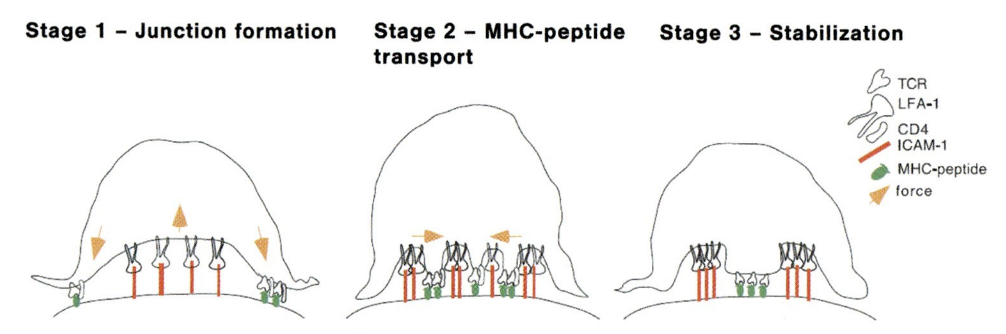

Figure 11.1 Schematic showing the three stages in the formation of the immunological synapse. A detailed description is provided in the text. Briefly, in stage 1, a central area of ICAM-1 ligating LFA-1 forms, around which the cytoskeleton rearranges to give a narrow zone in which p-MHCs can readily ligate TCRs. In stage 2, p-MHC-ligated TCRs move to the center of the contact zone, leading to stage 3, where a central area rich in TCRs and p-MHCs (the cSMAC) is surrounded by an annular area rich in ICAM-1-ligated LFA-1 (the pSMAC). (In the present chapter, to simplify the discussion, the role of CD4 is not described.) (Reproduced from *Science* **1999**, *285*, 221–227 [13]. Reprinted with permission from AAAS.)

shape so that they bind ICAM-1 strongly, which in turn causes the cell to stop migrating and thus stabilizes the synapse. *In vivo*, this mechanism enables the APCs to adhere strongly to T cells of the correct specificity, for the periods of up to several hours that may be necessary for full activation to take place. However, it simultaneously prevents the APCs from forming time-wasting, long-lasting contacts with other T cells.

In the second stage of immunological synapse evolution, p-MHC-ligated TCRs migrate to the middle of the contact zone, possibly due to actin-based transport, leading to the third stage, where a stable central region of bound TCR–p-MHC pairs, the central supramolecular activation cluster (cSMAC), is surrounded by a ring of ICAM-1 bound to LFA-1, the peripheral supramolecular activation cluster (pSMAC), which in turn is surrounded by an area of very close contact between the cell and the surface, suitable for the formation of new TCR–p-MHC pairs, the distal supramolecular activation cluster (dSMAC). As the primary purpose of the LFA-1–ICAM-1 bond is to bind the T cell to the APC, most of the lines of adhesive force between the cells pass through the dSMAC where these molecules are highly present, as shown by the arrows in Figure 11.1. Close examination of the structure and dynamics of cytoskeletal actin in the cell, as well as the distribution of LFA-1–ICAM-1 pairs, has shown that the dSMAC and pSMAC may be thought of as respectively analogous to *lamellapodium* and *lammella* structures that are exhibited by many motile cells [17], such as fibroblasts moving across the ECM. In the case of the fibroblast, $\alpha_v\beta_3$ integrin molecules (analogous to LFA-1) bind to the ECM surface in the lamellapodium, which is pushed out in the direction of desired motion. A characteristic feature is that the actin in the lamellapodium/dSMAC is organized into two stacked layers, whereas that in the lamella/pSMAC is organized into one layer only. By a combination of actin polymerization at the periphery and depolymerization at the cell center, the cell center effectively pulls on the anchored integrin molecules and thus moves towards them. In the case of the immunological synapse, the same actin polymerization and depolymerization occur, but because the dSMAC extends out in all directions, and because the ICAM-1 molecules are mobile in the APC lipid membrane (in contrast to integrin-binding elements of the ECM), the center of the T cell remains stationary, although there is a constant motion of actin towards the center of the cell [18]. Some important implications of this effect will be described in Section 11.2.2.

In a recent study conducted by Doh and Irvine, photolithographic methods were used to produce a substrate that encouraged T cells to form a cSMAC/pSMAC-like structure, but without using mobile ligands [19]. Rather, a surface was patterned with shaped patches of anti-CD3, a type of antibody that binds TCR and can thus simulate the effect of p-MHC, against a background of ICAM-1. This study employed a novel method for patterning surfaces with two proteins using photolithography, by using a photoresist that can be processed using biological buffers [20]. The photoresist used in this method is a random terpolymer with a methacrylate backbone, and methyl, *o*-nitrobenzyl and poly(ethylene glycol) (PEG) ($M_n \approx 600$ Da) side groups randomly distributed along the chain, where some of the PEG side chains are terminated with biotin. The PEG chains make the polymer somewhat hydrophilic, while the *o*-nitrobenzyl group can be cleaved to a carboxylic acid-bearing group by ultraviolet

(UV) light. The photoresist is spincoated onto a catonic substrate (in this case aminosilane-functionalized glass) so that, if the resist is exposed to UV light, and then rinsed with pH 7.4 buffer so that it contains negatively charged carboxylic acid groups, the negative charge of these groups causes the majority of the photoresist to be soluble and thus to be washed away. This will leave behind a thin layer of resist molecules, the negatively charged groups of which are ionically bound to positively charged amine groups on the glass surface. The sequence of events in preparing the patterned surface used in the T-cell studies [19] is shown in Figure 11.2. The photoresist layer was first exposed to UV through a photomask and then washed with pH 7.4 buffer to remove all but the thin residual polymer layer from the areas to be patterned with anti-CD3. After further UV irradiation of the whole surface, streptavidin followed by biotinylated anti-CD3 was deposited at pH 6.0 (at which the resist is stable) with the streptavidin binding the anti-CD3 to the biotin sites on the resist polymer surfaces. A second wash at pH 7.4 removed most of the resist from the non-anti-CD3 functionalized area, leaving a thin biotinylated polymer layer to which streptavidin followed by biotinylated ICAM-1 could be attached.

The T cells that were allowed to settle on surfaces bearing widely spaced circles of anti-CD3 6 μm in diameter against a background of ICAM-1 (prepared using this method) migrated until they encountered an anti-CD3 circle, and then formed a central cSMAC-like area of TCRs bound to anti-CD3, surrounded by a pSMAC-like ring of LFA-1 bound to ICAM-1 [19]. Molecules known to be associated respectively

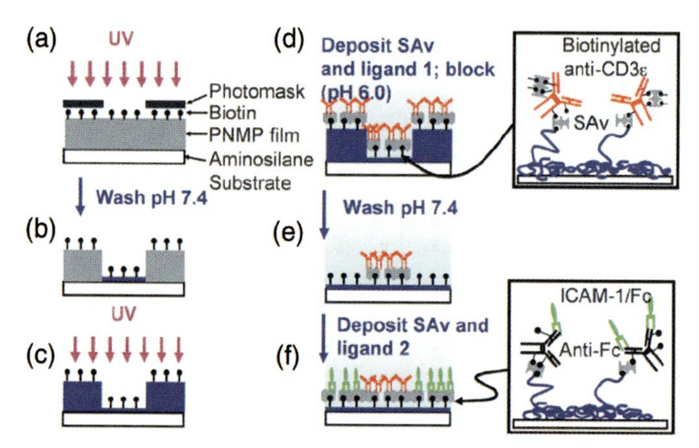

Figure 11.2 Schematic of the photolithographic production of surfaces patterned with two proteins on a micrometer scale, using a novel photoresist that can be processed in biological buffers [19]. A detailed description is provided in the text. Briefly: (a) A resist layer created by spincoating onto a cationic substrate was UV-irradiated through a photomask; (b) A wash at pH 7.4 removed all but a molecularly thin layer of resist from the irradiated areas; (c) The sample was uniformly UV-irradiated; (d) Biotinylated anti-CD3 was bound to the biotin functional groups in the resist layer via streptavidin (SAv); (e) Washing at pH 7.4 removed all but a molecularly thin layer of resist from the originally unirradiated areas; (f) ICAM-1-Fc was bound to the biotin functional groups in the resist layer via biotinylated anti-Fc and streptavidin. (Reproduced from *Proc. Nat. Acad. Sci. U.S.A.* **2006**, *103*, 5700 [19]. Copyright 2006 National Academy of Sciences, U.S.A.)

with LFA-1 (talin) and TCR (protein kinase C, θ isoform (PKC-θ)) signaling localized in the 'pSMAC'- and 'cSMAC'-like regions respectively, in accordance with some previous observations of T cell–APC contact zones [21]. T cells that formed these model 'cSMAC–pSMAC' structures showed elevated levels of intracellular free calcium (an early sign of activation), and eventually proliferated and showed increased cytokine production with respect to cells on control surfaces, thus confirming that full activation had taken place.

Although the cSMAC–pSMAC–dSMAC model corresponds well to *in vivo* images of some T cell–APC contacts (notably where a so-called B cell is used as the APC [21]), when the important dendritic cell APC type is used, a different structure is seen, which may be conceptualized as a multifocal pattern with several smaller cSMAC-like zones. This type of pattern, which could conceivably arise from the dendritic cell cytoskeleton's imposing constraints on the mobility of TCR [16], was reproduced by Doh and Irvine [19], by using their photolithographic technology to produce groups of four small anti-CD3 spots (spots 2 μm diameter, spot centers placed at corners of a ~5 μm cube). The T cells that encountered such groups indeed formed multifocal contacts.

Similar multifocal contacts have also been produced by Mossmann *et al.*, who used electron-beam lithography (EBL) to produce chromium 'walls' (about 100 nm wide and 5 nm high) on a silicon dioxide substrate on which a lipid bilayer containing ICAM-1 and p-MHC was then deposited. In this way, the ICAM-1 and p-MHC were able to move freely laterally up to, but not through, the walls. The p-MHC molecules could thus be confined to several large regions, resulting in the formation of a multifocal pattern with several miniature cSMAC-like regions [15, 16], as shown in Figure 11.3.

The T-cell structures produced on the anti-CD3-patterned substrates of Doh and Irvine [19] might be thought of as a good model of an activating T cell, with the cSMAC as the principal source of activation signaling from ligated TCR. However, as will be seen below, evidence has emerged which suggests that the cSMAC is not an important source of activation signaling, which rather comes primarily from TCR–p-MHC microclusters, and the physiological relevance of the model substrate of Doh and Irvine [19] may be questioned in this respect. It is possible that the interfacial line between anti-CD3- and ICAM-1-coated areas in the studies of Doh and Irvine [19] served the same function as the dSMAC in immunological synapses formed on B cells on supported planar bilayers. The important generalization is that T cells may be highly adaptable as part of their evolution to navigate a wide variety of anatomic sites and interact with essentially any cell in the body to combat continually evolving pathogens. Hence, one important role of nanotechnology may be to test the limits of adaptability and understand the fundamental recognition elements and how they may be manipulated.

11.2.2
TCR–p-MHC Microclusters as Important Signaling Centers

In the model shown in Figure 11.1, an early stage of immunological synapse formation was the ligation of TCR by p-MHC in the peripheral area around the

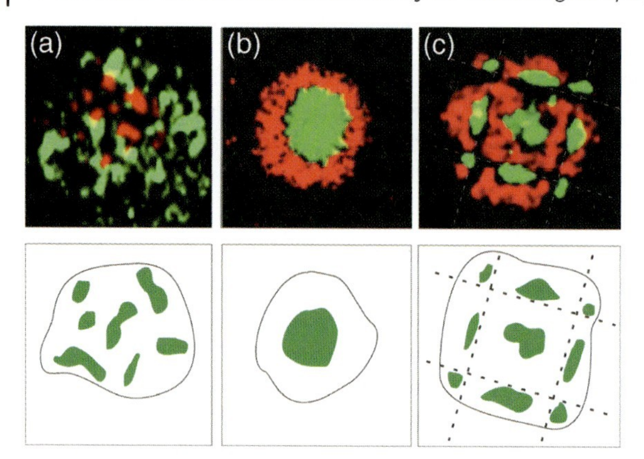

Figure 11.3 Fluorescence micrographs of immunological synapses, with TCRs labeled green (top row) and schematics (bottom row, green show TCR locations, solid back lines show cell outline). (a) A fixed synapse between a T cell and a dendritic cell (red shows PKC-θ, which is not important for our purposes here). TCRs gather at several separate focal points; (b) A synapse formed between a T cell and a supported lipid bilayer, where the lateral motion of p-MHC and ICAM-1 in the bilayer is unconstrained; red shows ICAM-1. TCRs gather in one large cluster, the cSMAC; (c) A synapse formed between a T cell and a supported lipid bilayer where the lateral mobility of p-MHC and ICAM-1 in the bilayer is constrained by chromium 'walls', indicated by dashed lines (white in micrograph, black in schematic) . (ICAM-1 is labeled red.) TCRs gather at several focal points. Note the presence of multiple TCR foci in (a) and (c), which suggests that a reduced lateral mobility of p-MHC on the dendritic cell surface might be responsible for the multifocal nature of the T cell with dendritic cell synapse (a). The central square formed by the chromium 'walls' in (c) has dimensions $2 \times 2\,\mu$m. (Micrographs reproduced from *Curr. Opin. Immunol.*, *18*, 512–516, M.L. Dustin, S.Y. Tseng, R. Varma, G. Campi, T cell-dendritic cell immunological synapses. Copyright (2006), with permission from Elsevier [16]; panels (b) and (c) are originally from **2005**, *310*, 1191–1193. Reprinted with permission from AAAS [15].)

central region of initial ICAM-1 adhesion, with the resulting TCR–p-MHC pairs then migrating to the center of the contact zone to finally form the cSMAC. A closer examination of this system in fact shows the TCR–p-MHC pairs combining to form microclusters throughout the contact zone, which then combine to form the cSMAC [22]. By using highly sensitive total internal reflection fluorescence microscopy (TIRFM), it has also proved possible to image a subsequent continuous 'rain' of microclusters, each consisting of between approximately five and 20 p-MHC–TCR pairs, that form in the peripheral dSMAC region, and then move radially inwards eventually joining the cSMAC [12, 23]. This motion is likely to occur because the TCR are indirectly connected to actin filaments, which are moving continuously inwards in the dSMAC and pSMAC (as discussed in Section 11.2.1). Experiments in which an antibody that disrupts TCR binding to p-MHC was added after the T cells had formed a stable cSMAC on a lipid bilayer surface (such that at early times the formation of new microclusters was prevented but the cSMAC was not yet disrupted) suggest strongly that activation signaling arises from the microclusters rather than from the cSMAC, as signaling almost completely ceased at a time when the cSMAC was still intact [12]. It therefore seems likely that, rather than functioning as a signaling device, the cSMAC in fact plays other roles. In particular, it has been observed that significant numbers of

TCR are endocytosed in the cSMAC. Some of these may be recycled through the cell for reincorporation into the dSMAC, ensuring a continuous supply of TCR and thus enabling signaling to continue for a long time, while others may be degraded [14, 24].

If TCR–p-MHC microclusters arising in the peripheral dSMAC give rise to activation signaling, which switches off as they join the cSMAC, there are two possible hypotheses: the initial signaling may decrease either with time, or with proximity to the center of the contact zone. It proved possible to distinguish between these hypotheses by using the chromium 'walls' of Mossmann *et al.* to divide the contact zone into many small areas, thus preventing microclusters from moving large distances. Using this approach, it was shown that, while the signaling from each microcluster has a finite lifetime, such lifetime decreased strongly with proximity to the center of the contact zone. This showed that spatial factors do play a role, and help to confirm the picture of the cSMAC as a non-signaling region [15].

In contrast to the studies just mentioned, in which TCR–p-MHC microclusters formed spontaneously by the coming together or pulling together of mobile p-MHC molecules in a lipid bilayer [12, 15], Anikeeva *et al.* effectively created artificial TCR–p-MHC microclusters by exposing T cells to a solution of *quantum dots*. These are fluorescent semiconductor nanocrystals, to which p-MHC molecules have been bound, with the binding mechanism being the ligation of zinc ions on the semiconductor surfaces by carboxylic acid groups belonging to six histidine residues inserted at the base of the p-MHC molecule [25]. Approximately 12 p-MHC molecules were found per quantum dot, as determined by the measurement of nonradiative energy transfer between the quantum dot and fluorophores bound to the p-MHC molecules. This suggested that the number of ligated TCRs in the artificial microclusters might have been about six, comparable to the size of the smallest signaling microclusters observed in one of the previously mentioned bilayer studies [12]. The stimulation of T cells with appropriate p-MHC-functionalized quantum dots caused activation signaling to occur. Although this study does not relate directly to our theme of T-cell activation by artificial substrates, we mention it here because it is indicative of the potential value of studies performed using p-MHC molecules bound to nanospheres. In fact, it will be seen below that surfaces functionalized with nanospheres may play an important role in future studies.

In addition to TCR–p-MHC microclusters, LFA-1–ICAM-1 microclusters have also been observed; the latter seem to form in the dSMAC and to move inwards before eventually joining thread-like LFA-1–ICAM-1 structures in the pSMAC [18]. Figure 11.4 summarizes schematically the localization of TCRs, p-MHCs, LFA-1 and ICAM-1 in the three SMACs. The structure of the cytoskeletal actin in these regions, as discussed in Section 11.2.1, is also shown.

11.3 The Smallest Activating Units? p-MHC Oligomers

As discussed in Section 11.2, signaling from p-MHC-ligated TCRs seems to depend on the ligated TCRs coming together to form microclusters, rather than ligation alone being enough for signaling. It transpires that activation signaling can indeed not be initiated by a single ligated TCR, but that the coming together of at least two ligated TCRs is necessary for signaling. This was demonstrated by Boniface *et al.*, who combined biotin-functionalized p-MHCs with naturally tetravalent streptavidin molecules to produce p-MHC monomers, dimers, trimers and tetramers. T cells exposed to the p-MHC monomer showed no activation signaling, whereas significant signaling was already present in the case of the dimer, and the strength increased

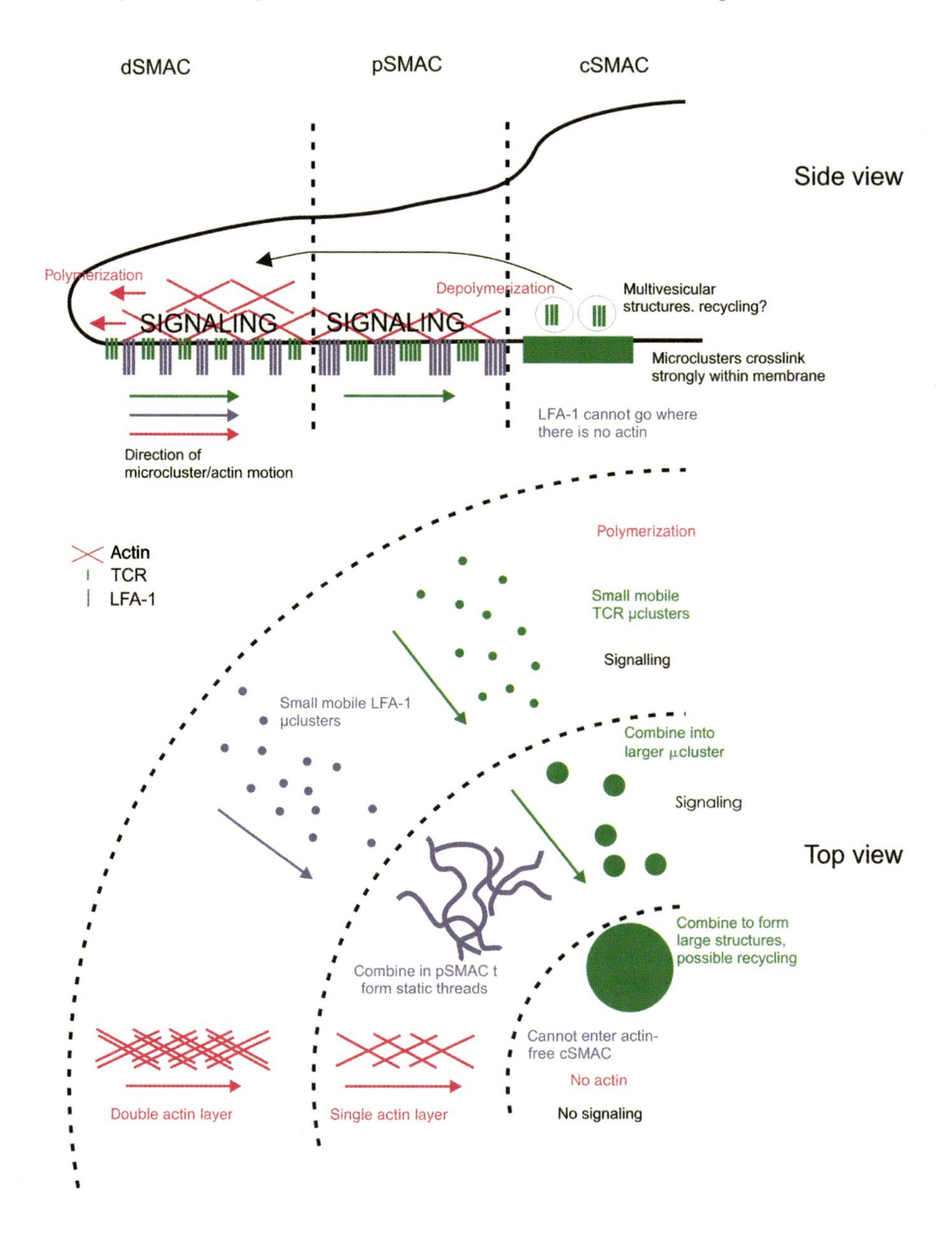

when trimer or tetramer was used [26]. This suggests that some degree of TCR 'clustering' is necessary for T-cell activation signaling. This could conceivably indicate that part of the signaling mechanism requires the close proximity of the cytoplasmic parts of neighboring TCRs.

Interestingly, doubt was cast on the finding that TCR clustering is required for the activation signal when, in an experiment using APCs in which all of the agonist p-MHC molecules had been fluorescently labeled, activation signaling was observed to be initiated by a T cell where the contact zone with the APC surface contained only one agonist p-MHC molecule [27]. This apparent contradiction may have been resolved by Krogsgaard *et al.*, who obtained a T-cell activation signal by stimulating cells with a synthetic heterodimer consisting of one MHC molecule with agonist peptide and one with self-peptide (i.e. peptide found within the proteome of the T cell-producing mammal, in this case a mouse) [28]. Krogsgaard *et al.* argued that such heterodimers may play a role in *in vivo* activation. This controversy and its resolution underlines the roles that molecules other than agonist p-MHC may play in *in vivo* T-cell activation; one of the principal advantages of experiments performed on artificial substrates is that 'clean' experiments can be performed, without the possible intrusion of unknown ligands. The ability of T cells to respond to mixed stimulations by agonist and self-peptide-loaded MHC molecules may be important for the functioning of the immune system, as it could increase the likelihood of T-cell activation by APC surfaces that present only small amounts of agonist peptide [8].

If TCR clustering is indeed required for T-cell activation, then it is interesting to ask how close together the TCRs need to be drawn in order for signaling to occur. A significant contribution towards answering this question was made by Cochran *et al.*, who used p-MHC molecules genetically engineered to contain free cystine residues to produce p-MHC dimers; the dimers were created by reacting the cystine residues

Figure 11.4 Summary of our current understanding of the structure of the immunological synapse. Schematic top and side views of the T cell only are provided here: it is assumed that all or most TCRs and LFA-1 molecules shown are ligated by p-MHCs or ICAM-1 on the opposing APC surface (not shown). In the top view, TCRs, LFA-1 and actin are shown in separate locations purely for visual clarity. In the dSMAC, which is analogous to a lamellapodium in a migrating cell, and thus contains two stacked layers of cytoskeletal actin, microclusters of TCR–p-MHC and LFA-1–ICAM-1 form and are transported towards the cell center by the inward motion of actin filaments, as indicated by arrows below the cell. The actin filament motion is caused by depolymerization at the edge of the cSMAC, and polymerization at the edge of the cell. The direction of actin filament growth is indicated by arrows within the cell. In the pSMAC, TCR–p-MHC microclusters merge to form somewhat larger microclusters, and continue migrating inwards, whereas LFA-1–ICAM-1 microclusters merge into a thread-like structure of mutually associated LFA–ICAM-1 pairs, and thus cease moving. The pSMAC is analogous to a lamella in a migrating cell, and thus contains only a single layer of cytoskeletal actin. In the cSMAC, TCR–pMHC pairs merge into a large mass of mutually associated TCR–pMHC pairs. Significant quantities of TCRs are endocytosed by the T cell, and some of these may be recycled through the cell back to the dSMAC, where the process can begin again, enabling signaling to be maintained over a long period of time. LFA-1–ICAM-1 pairs do not enter the actin-free cSMAC in significant quantities. TCR–pMHC microclusters in the dSMAC and pSMAC participate in T-cell activation signaling; there is no signaling due to TCR–p-MHC in the cSMAC.

with maleidomide groups on polypeptide crosslinkers of various lengths. The activation response from T cells decreased as the length of the spacer between the bases of the p-MHC molecules was increased from <1 to 9 nm [29].

11.4 Molecular-Scale Nanolithography

It can be seen from the above discussions that the clustering of p-MHC-ligated TCRs is critical to the initiation of T cell activation signaling. In this section, we describe possible future experiments aimed at further examining these effects, using the technology of *block copolymer micellar nanolithography*. This has recently become available, and enables surfaces to be patterned on the nanometer scale with single-protein molecules such as p-MHCs. Here, we will describe the technique in detail and review its previous uses in cell signaling studies. We will also discuss how the technique can be used, in combination with chemistry and protein engineering methods, to perform experiments to further our understanding of the immunological synapse.

11.4.1 Block Copolymer Micellar Nanolithography

The concept of block copolymer nanolithography is illustrated in Figure 11.5 [30–32]. Here, poly(styrene block 2-vinyl pyridine) forms a micellar solution in toluene with the hydrophilic 2-vinyl pyridine (2VP) block making up the micelle core. Hydrogen tetrachloroaurate (III) ($HAuCl_4$) is added, and complexes the 2VP, producing a gold-rich micelle core. When a flat substrate with a chemically suitable flat surface such as silicon oxide is immersed in the micellar solution and then withdrawn, the approximately spherical micelles form a two-dimensional (2-D) close-packed array on the substrate surface, with the capillary force due to the retreating toluene interface possibly playing a role in forcing them into this configuration. The micelle-coated surface is then exposed to a hydrogen plasma; this removes the polymeric material and reduces the gold ions, producing metallic gold particles at the former sites of the micelle cores. The result is a hexagonal array of gold particles on a (usually) silicon dioxide background. The size of the gold particles can be controlled between approximately 3 nm and 10 nm by varying the amount of added ($HAuCl_4$), while the spacing between adjacent particles can be controlled by varying the length of the styrene block of the original diblock copolymer, and to some extent also by varying the speed of withdrawal of the substrate from the micellar solution. Present investigations have produced interparticle spacings in the range of approximately 15–250 nm, although it seems likely that spacings below 15 nm should also be achievable.

For experiments to study the stimulation of biological cells, the gold nanoparticles can be functionalized with biological ligands using thiol chemistry, while the silicon dioxide surface in between the spheres can be differently functionalized using silane

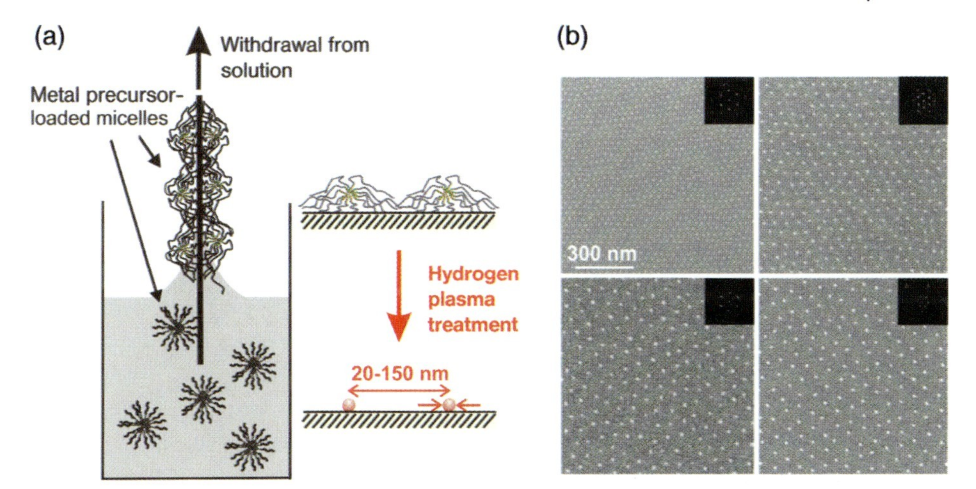

Figure 11.5 Block copolymer micellar nanolithography. Full details are provided in the text. Briefly: (a) Schematic: Micelles of which the cores are loaded with gold ions form a 2-D close-packed layer on a suitable substrate surface that is withdrawn at a suitable speed from the micellar solution. Treatment with a hydrogen plasma removes organic material, resulting in a hexagonally ordered array of gold nanoparticles, with the interparticle spacing being determined by the original polymer block molecular weights and the speed of withdrawal from the solution; (b) Scanning electron microscopy images of surfaces patterned with gold nanoparticles produced using diblock copolymers, where the two blocks have various different molecular weights: the variation in the lattice parameter can be seen. The ordered nature of the patterns is demonstrated by the sharp peaks in the numerically calculated 2-D Fourier transforms of the images (insets at top right of images). ((a) is from R. Glass, M. Arnold, J. Blummel, A. Kuller, M. Moller, J.P. Spatz: Micro-Nanostructured Interfaces Fabricated by the Use of Inorganic Block Copolymer Micellar Monolayers as Negative Resist for Electron-Beam Lithography. *Adv. Funct. Mat.* **2003**, *13*, 569–575 [39]. (b) is from M. Arnold, E.A. Calvalcanti-Adam, R. Glass, J. Blummel, W. Eck, M. Kantlehner, H. Kessler, J.P. Spatz: Activation of Integrin Function by Nanopatterned Adhesive Interfaces. *ChemPhysChem* [2] **2005**, *5*, 383–388. (a) (b) copyright Wiley-VCH Verlag GmbH & Co. KGaA. Reproduced with permission.)

chemistry. The power of this technique is well illustrated by the studies of Arnold *et al.*, who functionalized the gold nanoparticles with cyclic arginine-glycine-aspartate (RGD) peptide molecules that were bound to the gold via a thiol-functionalized linker [2]. These RGD peptides bind strongly to $\alpha_v\beta_3$ integrins, which are membrane-bound receptors that play an important role in the initiation of adhesion by fibroblasts to the ECM. The large size of $\alpha_v\beta_3$ integrins ensured that only one integrin could bind to each gold particle, so that the interparticle spacing could be used as a measure of the minimum separation between adjacent ligated integrin molecules. Experiments showed that fibroblasts adhered readily to substrates with an interparticle spacing of 58 ± 7 nm or less, but did not adhere to substrates with an interparticle spacing of 73 ± 8 nm or more. This suggested that some clustering of ligated integrins is necessary for the initiation of adhesion signaling in fibroblasts, and that the critical spacing below which integrins may be considered to be 'clustered' lies between 58 nm and 73 nm. Additionally, actin-rich protein clusters known as *focal adhesions* that form at sites of $\alpha_v\beta_3$ integrin-mediated adhesion, and which may be considered as local indicators of adhesion signaling, were observed to form only when the interparticle spacing was higher than this critical value.

It is important to note that, while other lithographic techniques such as EBL [3] and dip-pen nanolithography [33–35], have been used to immobilize biological ligands on surfaces, to the best of our knowledge, block copolymer micellar lithography is the only lithographic method that has thus far been used to spatially isolate individual ligand receptor interactions. This is most likely due to its ability to reliably produce particle sizes as small as 3 nm. This limit compares favorably with, for example, the lower size limit for reliable structure production using EBL with conventional poly (methyl methacrylate) (PMMA) resists, which is about 10 nm [36], and the smallest protein feature that has been created to date using dip-pen nanolithography, which is about 25–40 nm [33].

In view of the apparent requirement for the clustering of ligated TCRs if T-cell activation signaling were to occur, it would clearly be very interesting to perform an analogous experiment to that just described [2], but to study TCR rather than $\alpha_v\beta_3$ integrin clustering. In order to perform such an experiment, each gold nanoparticle would need to be functionalized with a single molecule of p-MHC, and the effect of the interparticle spacing on the activation signaling behavior of T cells brought into contact with such surfaces determined. If TCR clustering were indeed necessary for T-cell activation signaling, then one would expect to observe no signaling when the interparticle spacing was high, with signaling perhaps onsetting as the interparticle spacing was reduced below a critical value. Indeed, the above-mentioned studies of Cochran *et al.* suggested that this spacing should range between 1 and 15 nm [29].

The binding of p-MHCs to the gold dots could be achieved by creating a recombinant MHC construct containing an appropriately located free cystine residue that could react directly with the gold. Alternatively, protein constructs containing multiple consecutive histidine residues have been successfully bound to gold nanospheres on block copolymer micellar nanolithography-patterned surfaces by binding thiol-functionalized nitrilotriacetic acid (NTA) molecules, and allowing the carboxylic acid groups of the NTA and histidine to simultaneously coordinate the same nickel cation. Functionalization of the silicon dioxide surface between the gold nanospheres should also be considered. In the integrin-clustering experiments of Arnold *et al.*, the area between the gold nanoparticles was functionalized with protein-repellent PEG molecules that were end-functionalized with trimethoxysilane groups (this enabled them to bind covalently to the silicon dioxide surface). This PEG functionalization ensured that the cells under study interacted with the surface only via receptor interactions with ligand-functionalized gold particles, and not via nonspecific attractions, as well as resisting the deposition of cell-secreted proteins onto the silicon dioxide surface [2]. In the context of experiments to study T-cell activation, it might be advantageous to functionalize the area between the gold nanoparticles with a combination of PEG molecules to reduce the effect of nonspecific cell-surface attractions, and ICAM-1, to bring about the LFA-1-mediated cell adhesion that is a critical feature of the immunological synapse. This could be achieved by incorporating functional groups into the PEG layer that could be bound specifically to ICAM-1; an example would be to incorporate biotin groups into the PEG layer that bind via streptavidin to biotinylated ICAM-1 molecules. The incorporation of biotin into surface-grafted PEG layers has been achieved [37, 38]; I.E. Dunlop *et al.*, unpublished results], and its incorporation into the PEG layer between the gold nanoparticles should be readily achievable.

11.4.2
Micronanopatterning by Combining Block Copolymer Micellar Nanolithography and Electron-Beam Lithography

Surfaces that are structured on both the micrometer and nanometer scales can be produced using a method that combines block copolymer micellar nanolithography with EBL.

The principle of the method is shown schematically in Figure 11.6 [39]. After having deposited a close-packed monolayer of $HAuCl_4$-loaded block copolymer micelles onto the substrate, part of the layer is exposed to an electron beam, which causes the polymer molecules to become highly crosslinked. The substrate is then rinsed with acetone to remove all noncrosslinked polymer, leaving micelles behind only in the area that was exposed to the electron beam. These micelles are then exposed to a hydrogen plasma, which leads to hexagonally arranged gold nanoparticles in the normal manner. It is thus possible, by using a steerable electron beam (such as that in the scanning electron microscope) to pattern only parts of a surface using block copolymer micellar nanolithography, and thus to produce patches of patterning containing controlled numbers of gold nanoparticles (see Figure 11.6).

Surfaces prepared using this method could enable experiments that address questions relating to the number of p-MHC molecules or clusters required for T-cell activation signaling to be addressed. For example, if p-MHC-ligated TCR dimers are sufficient to cause T-cell activation signaling, then it is interesting to ask whether one dimer would be sufficient to produce a detectable activation signal, as suggested by the cell–cell contact experiments of Irvine *et al.* [27], and how the signaling strength would depend on the number of dimers with which a cell interacts. Additionally, the suggestion of Varma *et al.*, that signaling might arise primarily from microclusters of between roughly five and 20 p-MHC-ligated TCRs, suggests that it would be interesting to examine the effect on signaling intensity of the microcluster size, and the number of microclusters per cell. Simulated 'microclusters' containing precisely controlled numbers of molecules could be produced using patches of gold nanoparticles, similar to that shown in Figure 11.6; here, each gold nanoparticle could bear one p-MHC molecule and the interparticle spacing could be chosen sufficiently small that a T cell could 'see' the resulting ligated TCRs as being clustered. Alternatively, microclusters could be simulated by allowing several p-MHCs to bind to one larger nanosphere, as in the experiments of Anikeeva *et al.* mentioned above, in which p-MHCs were bound to soluble quantum dots [25].

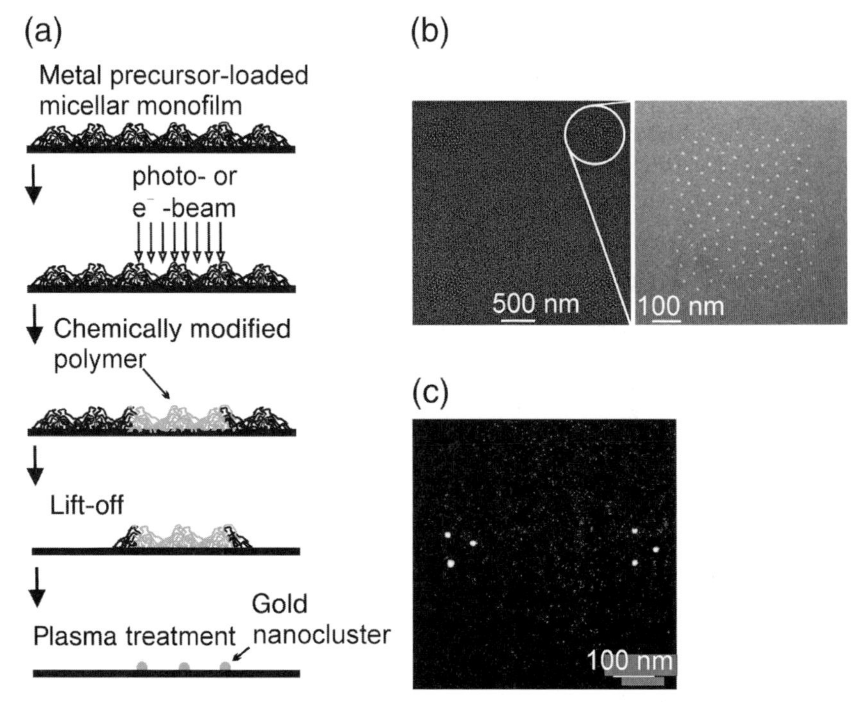

Figure 11.6 Block copolymer micellar nanolithography combined with electron beam lithography to produce patches of substrate nanopatterned with controlled numbers of nanoparticles. (a) Schematic: parts of a close-packed film of gold-loaded micelles are irradiated with a steerable electron beam. Washing in acetone 'lifts off' unirradiated polymer before plasma treatment produces hexagonal arrays of gold particles in the treated areas; (b) (c) Scanning electron microscopy images of different surfaces produced using this method, showing the possibility of producing widely spaced patches of patterned surface, where each patch contains a similar number of gold nanoparticles. ((a) Adapted from R. Glass, M. Arnold, J. Blummel, A. Kuller, M. Moller, J.P. Spatz: Micro-Nanostructured Interfaces Fabricated by the Use of Inorganic Block Copolymer Micellar Monolayers as Negative Resist for Electron-Beam Lithography. *Adv. Funct. Mat.* **2003**, *13*, 569–575. Copyright Wiley-VCH Verlag GmbH & Co. KGaA. Reproduced with permission [39]. (b) (c) Adapted from *Methods Cell Biol.*, *83*, 89–111, J.P. Spatz and B. Geiger: Molecular engineering of cellular environments: Cell adhesion to nano-digital surfaces. Copyright (2007), with permission from Elsevier [53].)

11.5
Therapeutic Possibilities of Immune Synapse Micro- and Nanolithography

So far, we have mostly considered the use of micro- and nanolithographically patterned T-cell-activating surfaces as tools to investigate the functioning of the immune synapse. It is also interesting to consider the potential of these technologies for direct use in clinical therapies. Although it is possible to imagine incorporating T-cell-activating surfaces into medical implants, in order to encourage a specific and local immune response (e.g. against a tumor), it is likely that the first therapeutic use of such surfaces will be in the context of *adoptive cell transfer*.

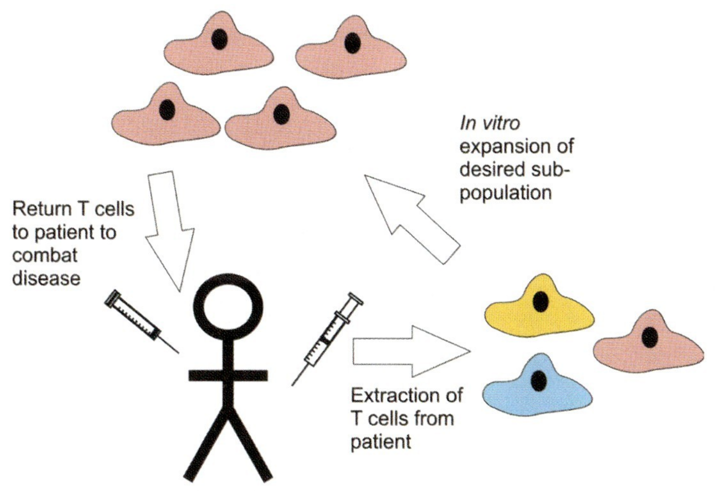

Figure 11.7 The principle of adoptive T-cell therapy.

The principle of the techniques is shown schematically in Figure 11.7. Here, T cells are removed from a patient and a selected subpopulation is deliberately activated, causing it to expand *in vitro*, before being returned to the patient's body. The returned T cells should then produce a strong immune response to the disease being treated. Although adoptive T cell transfer may prove useful in combating certain viral infections, much research has focused on the treatment of cancerous tumors, where the subpopulation of cells to be expanded should clearly be selected to be responsive to tumor-related antigens (for reviews, see Refs [40–42]). One approach is to use extracted tumor cells directly to selectively activate T cells of an appropriate specificity; this leads to a population of T cells that are specific for a variety of epitopes contained in the tumor [43]. Alternatively, epitopes that are known to be tumor-associated can be chosen, and T cells that are specific to those epitopes activated using artificial MHC–peptide constructs [40]. Here, we will focus on the second approach, as micro- and nanopatterned biomimetic surfaces functionalized with p-MHCs and costimulatory molecules might be of value in this context.

The identification of tumor-specific antigens is key if the adoptive cell therapy is to target a tumor, without damaging the healthy tissue: this approach is of particular value in tumors that are virus-induced, where antigens derived from viral proteins can be used [44]. Equally, many tumors express significantly mutated proteins that could be targeted, although the individual genetic analysis of a patient's tumor could prove expensive [41, 45]. Alternatively, antigens can be chosen from proteins that are known to be overexpressed in particular types of tumor, or even from healthy but tissue-specific proteins in tissues that are not necessary for survival, such as the prostate gland [41, 46].

In order to activate T cells using synthetic p-MHCs it is not necessary to use sophisticated spatially patterned substrates; the p-MHCs could simply be bound to a surface with no control of its spatial distribution. However, the results of recent studies of adoptive cell therapy have emphasized that T-cell activation is not a simple on-off event; depending on the details of the activation method, as well as the prior history of the T cell, a huge variety of subtly different phenotypes can be obtained. Moreover, the differences between these phenotypes can determine the outcome of treatment [40]. An important factor here is the strength of T-cell stimulation with p-MHCs. T cells that are fairly strongly stimulated tend to differentiate to an *effector* or *effector memory* T-cell phenotype which will combat infection but not give rise to a long-lived population of T cells *in vivo*. In contrast, less-strongly stimulated cells tend to differentiate to longer-lived *central memory* T cells, which are more likely to act as progenitors of a large, long-lived T-cell population [40, 47]. It has been suggested that adoptive cell therapy can be more effective if central memory, rather than effector or effector memory, T cells are used [48]. P-MHC micro- and nanopatterned surfaces could clearly be used to control the activation 'dose' delivered to each T cell by, for example, producing spatially separated activating 'patches', each of which contains a given number of p-MHCs, along with appropriate adhesion molecules and cofactors. As discussed above, the spatial distribution of p-MHCs on an activating surface can play a role in determining immunological synapse structure and also the degree of T-cell activation; spatially structured p-MHC-functionalized surfaces may therefore be of use in controlling the phenotype of T cells used for adoptive cell therapy.

A number of other factors, in addition to the nature of the p-MHC stimulus, have been identified as important in the preparation of T cells for adoptive cell therapy [40]. For example, it may be necessary to selectively activate either helper or killer T cells, and it is certainly important not to activate regulatory T cells which act to suppress the immune response to the target epitope [42, 49]. Also, certain effects of *in vitro* culture may cause T cells to senesce in ways that resemble the weakening of the immune system on aging, thus reducing their therapeutic effectiveness [40, 50, 51]. Both of these issues have been addressed by activating T cells using costimulatory molecules simultaneously with p-MHCs. Given the spatial structuring of the immunological synapse, using lithographic methods to determine the relative positions of p-MHCs and costimulatory molecules (as described above for p-MHCs and ICAM-1) might well lead to a better control of the final T-cell phenotype. Recently, when using microcontact printing to generate patterns of costimulatory anti-CD28 and TCR-activating anti-CD3, it was shown that multiple peripheral anti-CD28 foci were better than one large central spot with the same amount of anti-CD28 for the stimulation of T-cell interleukin-2 production [52].

To summarize, adoptive cell therapy based on the *ex vivo* activation of T cells shows promise as an anti-cancer therapy, but better control of the detailed phenotype of the activated T cells is desired. Lithographic patterning of activating surfaces with p-MHCs and costimulatory molecules may contribute to attaining this control.

11.6 Conclusions

Studies performed by bringing T cells into contact with artificial surfaces that mimic aspects of the APC surfaces have contributed greatly to our understanding of the immunological synapse, and such surfaces may be of therapeutic value in the future. Among the most informative experiments have been those performed using substrates bearing lipid bilayers that contain mobile ICAM-1 and p-MHCs. Photolithographic methods have been used to control the mobility of molecules within such bilayers, stimulating possible effects of the APC cytoskeleton and enabling the effects of reduced mobility on signaling by p-MHC-ligated TCR microclusters to be investigated. In separate studies, photolithographic methods that enable the patterning of surfaces with multiple proteins have been used to bring about artificial SMAC-like structures. The importance of studying p-MHC-ligated TCR clustering effects at the nanometer scale is attested to by evidence from several studies in which T cells were stimulated with soluble p-MHC oligomers, and substrates that are patterned with single p-MHC molecules on the nanometer scale will accordingly be required for the next generation of such studies. Block copolymer micellar nanolithography represents a suitable technique for generating such substrates and, when combined with EBL. will enable the production of surfaces patterned on both nanometer and micrometer length scales. Tcell activation experiments performed on such substrates are likely to play a role in extending our understanding of the immunological synapse. Both, micro- and nanopatterned substrates may also be used for *ex vivo* T cell activation in the context of T cell adoptive immunotherapy, where T cells removed from a patient are activated and expanded *ex vivo* before being returned to combat disease, notably cancer. The use of these substrates may also help to gain close control of the phenotypes of *ex vivo*-activated T cells, leading to more effective treatments.

Acknowledgments

The authors thank Thomas O. Cameron and Rajat Varma for useful discussions. This chapter was partially supported by the National Institutes of Health through the NIH Roadmap for Medical Research (PN2 EY016586) (I.E.D., M.L.D., J.P.S.) and by the Max Planck Society (I.E.D., J.P.S.). I.E.D. acknowledges a Humboldt Research Fellowship.

References

1 Vogel, V. and Sheetz, M. (2006) *Nature Reviews Molecular Cell Biology*, **7**, 265.

2 Arnold, M., Cavalcanti-Adam, E.A., Glass, R., Blummel, J., Eck, W., Kantlehner, M., Kessler, H. and Spatz, J.P. (2004) *Chemphyschem*, **5**, 383.

3 Cherniavskaya, O., Chen, C.J., Heller, E., Sun, E., Provezano, J., Kam, L., Hone, J., Sheetz, M.P. and Wind, S.J. (2005) *Journal of Vacuum Science & Technology B*, **23**. 2972.

4 Taite, L.J., Rowland, M.L., Ruffino, K.A., Smith, B.R.E., Lawrence, M.B. and West, J.L. (2006) *Annals of Biomedical Engineering*, **34**, 1705.

5 Chen, S.Q., Alon, R., Fuhlbrigge, R.C. and Springer, T.A. (1997) *Proceedings of the National Academy of Sciences of the United States of America*, **94**, 3172.

6 Janeway, C.A.J., Travers, P., Walport, M. and Schlomchik, M.J. (2005) *Immunobiology: The Immune System in*

Health and Disease, 6th edn, Garland Science Publishing, New York, N.Y., USA.
7 Dustin, M.L. and Colman, D.R. (2002) *Science*, **298**, 785.
8 Davis, M.M., Krogsgaard, M., Huse, M., Huppa, J.B., Lillemeier, B.F. and Li, Q.J. (2007) *Annual Review of Immunology*, **25**, 681.
9 Pulendran, B. and Ahmed, R. (2006) *Cell*, **124**, 849.
10 Gourley, T.S., Wherry, E.J., Masopust, D. and Ahmed, R. (2004) *Seminars in Immunology*, **16**, 323.
11 Crotty, S. and Ahmed, R. (2004) *Seminars in Immunology*, **16**, 197.
12 Varma, R., Campi, G., Yokosuka, T., Saito, T. and Dustin, M.L. (2006) *Immunity*, **25**, 117.
13 Grakoui, A., Bromley, S.K., Sumen, C., Davis, M.M., Shaw, A.S., Allen, P.M. and Dustin, M.L. (1999) *Science*, **285**, 221.
14 Lee, K.H., Dinner, A.R., Tu, C., Campi, G., Raychaudhuri, S., Varma, R., Sims, T.N., Burack, W.R., Wu, H., Kanagawa, O., Markiewicz, M., Allen, P.M., Dustin, M.L., Chakraborty, A.K. and Shaw, A.S. (2003) *Science*, **302**, 1218.
15 Mossman, K.D., Campi, G., Groves, J.T. and Dustin, M.L. (2005) *Science*, **310**, 1191.
16 Dustin, M.L., Tseng, S.Y., Varma, R. and Campi, G. (2006) *Current Opinion in Immunology*, **18**, 512.
17 Sims, T.N., Soos, T.J., Xenias, H.S., Dubin-Thaler, B., Hofman, J.M., Waite, J.C., Cameron, T.O., Thomas, V.K., Varma, R., Wiggins, C.H., Sheetz, M.P., Littman, D.R. and Dustin, M.L. (2007) *Cell*, **129**, 773.
18 Kaizuka, Y., Douglass, A.D., Varma, R., Dustin, M.L. and Vale, R.D. (2007) *Proceedings of the National Academy of Sciences of the United States of America*, **104**, 20296.
19 Doh, J. and Irvine, D.J. (2006) *Proceedings of the National Academy of Sciences of the United States of America*, **103**, 5700.
20 Doh, J. and Irvine, D.J. (2004) *Journal of the American Chemical Society*, **126**, 9170.
21 Monks, C.R.F., Freiberg, B.A., Kupfer, H., Sciaky, N. and Kupfer, A. (1998) *Nature*, **395**, 82.
22 Freiberg, B.A., Kupfer, H., Maslanik, W., Delli, J., Kappler, J., Zaller, D.M. and Kupfer, A. (2002) *Nature Immunology*, **3**, 911.
23 Campi, G. Varma, R. and Dustin, M.L. (2005) *Journal of Experimental Medicine*, **202**, 1031.
24 Lee, K.H., Holdorf, A.D., Dustin, M.L., Chan, A.C., Allen, P.M. and Shaw, A.S. (2002) *Science*, **295**, 1539.
25 Anikeeva, N., Lebedeva, T., Clapp, A.R., Goldman, E.R., Dustin, M.L., Mattoussi, H. and Sykulev, Y. (2006) *Proceedings of the National Academy of Sciences of the United States of America*, **103**, 16846.
26 Boniface, J.J., Rabinowitz, J.D., Wulfing, C., Hampl, J., Reich, Z., Altman, J.D., Kantor, R.M., Beeson, C., McConnell, H.M. and Davis, M.M. (1998) *Immunity*, **9**, U7.
27 Irvine, D.J., Purbhoo, M.A., Krogsgaard, M. and Davis, M.M. (2002) *Nature*, **419**, 845.
28 Krogsgaard, M., Li, Q.J., Sumen, C., Huppa, J.B., Huse, M. and Davis, M.M. (2005) *Nature*, **434**, 238.
29 Cochran, J.R., Cameron, T.O., Stone, J.D., Lubetsky, J.B. and Stern, L.J. (2001) *Journal of Biological Chemistry*, **276**, 28068.
30 Spatz, J.P. Sheiko, S. and Moller, M. (1996) *Macromolecules*, **29**, 3220.
31 Spatz, J.P. Roescher, A. and Moller, M. (1996) *Advanced Materials*, **8**, 337.
32 Spatz, J.P., Mossmer, S., Hartmann, C., Moller, M., Herzog, T., Krieger, M., Boyen, H.G., Ziemann, P. and Kabius, B. (2000) *Langmuir*, **16**, 407.
33 Li, B., Zhang, Y., Hu, J. and Li, M.Q. (2005) *Ultramicroscopy*, **105**, 312.
34 Lee, K.B. Lim, J.H. and Mirkin, C.A. (2003) *Journal of the American Chemical Society*, **125**, 5588.
35 Lee, K.B., Kim, E.Y., Mirkin, C.A. and Wolinsky, S.M. (2004) *Nano Letters*, **4**, 1869.

36 Vieu, C., Carcenac, F., Pepin, A., Chen, Y., Mejias, M., Lebib, A., Manin-Ferlazzo, L., Couraud, L. and Launois, H. (2000) *Applied Surface Science*, **164**, 111.

37 Morgenthaler, S., Zink, C., Stadler, B., Voros, J., Lee, S., Spencer, N.D. and Tosatti, S.G.P. (2006) *Biointerphases*, **1**, 156.

38 You, Y..-Z. and Oupicky, D. (2007) *Biomacromolecules*, **8**, 98.

39 Glass, R., Arnold, M., Blummel, J., Kuller, A., Moller, M. and Spatz, J.P. (2003) *Advanced Functional Materials*, **13**, 569.

40 June, C.H. (2007) *Journal of Clinical Investigation*, **117**, 1204.

41 June, C.H. (2007) *Journal of Clinical Investigation*, **117**, 1466.

42 Gattinoni, L., Powell, D.J., Rosenberg, S.A. and Restifo, N.P. (2006) *Nature Reviews Immunology*, **6**, 383.

43 Milone, M.C. and June, C.H. (2005) *Clinical Immunology*, **117**, 101.

44 Straathof, K.C.M., Bollard, C.M., Popat, U., Huls, M.H., Lopez, T., Morriss, M.C., Gresik, M.V., Gee, A.P., Russell, H.V., Brenner, M.K., Rooney, C.M. and Heslop, H.E. (2005) *Blood*, **105**, 1898.

45 Sjoblom, T., Jones, S., Wood, L.D., Parsons, D.W., Lin, J., Barber, T.D., Mandelker, D., Leary, R.J., Ptak, J., Silliman, N., Szabo, S., Buckhaults, P., Farrell, C., Meeh, P., Markowitz, S.D., Willis, J., Dawson, D., Willson, J.K.V., Gazdar, A.F., Hartigan, J., Wu, L., Liu, C.S., Parmigiani, G., Park, B.H., Bachman, K.E., Papadopoulos, N., Vogelstein, B., Kinzler, K.W. and Velculescu, V.E. (2006) *Science*, **314**, 268.

46 Pardoll, D.M. (1999) *Proceedings of the National Academy of Sciences of the United States of America*, **96**, 5340.

47 Sallusto, F. Geginat, J. and Lanzavecchia, A. (2004) *Annual Review of Immunology*, **22**, 745.

48 Klebanoff, C.A., Gattinoni, L., Torabi-Parizi, P., Kerstann, K., Cardones, A.R., Finkelstein, S.E., Palmer, D.C., Antony, P.A., Hwang, S.T., Rosenberg, S.A., Waldmann, T.A. and Restifo, N.P. (2005) *Proceedings of the National Academy of Sciences of the United States of America*, **102**, 9571.

49 Curiel, T.J., Coukos, G., Zou, L.H., Alvarez, X., Cheng, P., Mottram, P., Evdemon-Hogan, M., Conejo-Garcia, J.R., Zhang, L., Burow, M., Zhu, Y., Wei, S., Kryczek, I., Daniel, B., Gordon, A., Myers, L., Lackner, A., Disis, M.L., Knutson, K.L., Chen, L.P. and Zou, W.P. (2004) *Nature Medicine*, **10**, 942.

50 Zhou, J.H., Shen, X.L., Huang, J.P., Hodes, R.J., Rosenberg, S.A. and Robbins, P.F. (2005) *Journal of Immunology*, **175**, 7046.

51 Monteiro, J., Batliwalla, F., Ostrer, H. and Gregersen, P.K. (1996) *Journal of Immunology*, **156**, 3587.

52 Shen, K., Thomas, V.K., Dustin, M.L. and Kam, L.C. (2008) *Proceedings of the National Academy of Sciences of the United States of America*, **105**, 7791.

53 Spatz, J.P. and Geiger, B. (2007) *Methods in Cell Biology*, **83**, 89.

12
Bone Nanostructure and its Relevance for Mechanical Performance, Disease and Treatment

Peter Fratzl, Himadri S. Gupta, Paul Roschger, and Klaus Klaushofer

12.1
Introduction

The human skeleton not only serves as an ion reservoir for calcium homeostasis but also has an obvious mechanical function in supporting and protecting the body. These functions place serious requirements on the mechanical properties of bone, which should be stiff enough to support the body's weight and tough enough to prevent easy fracturing. Such outstanding mechanical properties are achieved by a very complex hierarchical structure of bone tissue, which has been described in a number of reviews [1–3]. Starting from the macroscopic structural level, bones can have quite diverse shapes, depending on their respective function. Long bones, such as the femur or the tibia, are found in the body's extremities and provide stability against bending and buckling. In other cases, for example in the vertebra or the head of the femur, the applied load is mainly compressive, and in such cases the bone shell is filled with highly porous cancellous bone (see Figure 12.1). Several levels of hierarchy are visible in this figure, with trabeculae or osteons in the hundred-micron range (Figure 12.1b and c), a lamellar structure in the micron range (Figure 12.1d), collagen fibrils of 50–200 nm diameter (Figure 12.1e), and collagen molecules as well as bone mineral particles with just a few nanometers thickness.

This hierarchical structure is largely responsible for the outstanding mechanical properties of bone. At the nanoscale, both collagen and mineral – and also their structural arrangement – play a crucial role. In this chapter we review the structure of bone at the nanoscale, and describe some recent findings concerning the influence of bone on deformation and fracture. We also outline some approaches to studying biopsy specimens in diseases and in treatments that are known to influence bone at the nanoscale.

Nanotechnology, Volume 5: Nanomedicine. Edited by Viola Vogel

ISBN: 978-3-527-31736-3

12.2 Nanoscale Structure of Bone

At the nanometer scale, bone is a composite of a collagen-rich organic matrix and mineral nanoparticles made from carbonated hydroxyapatite. The structure and properties of bone have recently been reviewed [2]. The basic building block of the bone material is a mineralized collagen fibril of between 50 and 200 nm diameter (Figure 12.1e and f). Collagen type I is the organic constituent of these fibrils in bone and in many biological tissues, including tendon, ligaments skin and cornea. The collagen molecules are triple helices with a length of about 300 nm, and are assembled within the cell. After secretion, the globular ends of the molecules are cleaved off enzymatically and the (apart from short telopeptide ends) triple helical molecules [4, 5] undergo a self-assembly process that leads to a staggered arrangement of parallel molecules. This in turn creates a characteristic pattern of low-density gap zones that are 35 nm long and high-density overlap zones 32 nm long within the fibril [6]; hence, the effective periodicity (D) will be 67 nm (Figure 12.1f). The collagen fibrils are filled and coated by mineral crystallites [7, 8]; the latter are mainly flat plates [9] that are mostly arranged parallel to each other in a fibril, and parallel to the long axis of the collagen fibrils [10]; however, they may not always be parallel between different fibrils [7]. The crystallites have a periodicity in axial packing density along the fibrils of the same 67 nm dimension [11] by which adjacent collagen molecules are staggered (Figure 12.1f). Crystal formation is triggered by collagen or (more likely) other noncollagenous proteins which act as nucleation centers [12]. After nucleation, the plate-like crystals become elongated but extremely thin [7, 9, 13, 14],

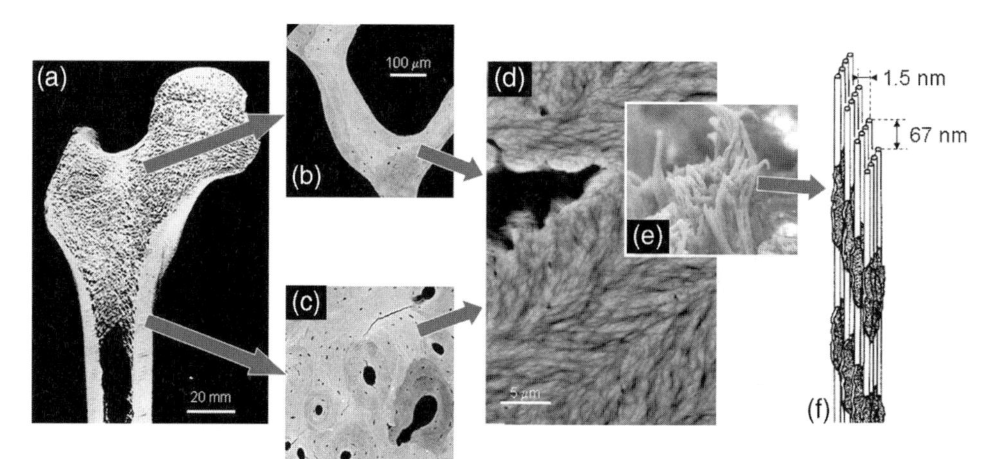

Figure 12.1 Hierarchical structure of a human femur. The femoral head (a) is filled with cancellous bone, consisting of individual trabeculae (b). The cortical bone shell contains osteons (c) where the central Haversian canal is surrounded by concentric lamellae of bone tissue. Lamellar bone (d) consists of thin layers of parallel collagen fibrils with rotating orientation similar to plywood. Collagen fibrils are constituted by parallel collagen molecules with a longitudinal stagger of 67 nm (e) and are reinforced with plate-like mineral particles located inside and on the surface of the fibrils.

and later grow in thickness [15, 16]. Among bone tissues from several different mammalian and nonmammalian species, the bone mineral crystals have thicknesses ranging from 1.5 to 4.5 nm [2, 7, 16–20]. While bone mineral is based mainly on hydroxyapatite ($Ca_5(PO_4)_3OH$), it also typically contains additional elements that replace either the calcium ions or the phosphate or hydroxyl groups; one of the most common such occurrences is replacement of the phosphate group by carbonate [1, 2].

12.3
Mechanical Behavior of Bone at the Nanoscale

The fracture resistance of bone results from the ability of its microstructure to dissipate deformation energy, without the propagation of large cracks leading to eventual material failure [21–23]. Different mechanisms have been reported for the dissipation of energy [24], including: the formation of nonconnected microcracks ahead of the crack tip [25, 26]; crack deflection and crack blunting at interlamellar interfaces and cement lines [27]; and crack bridging in the wake zone of the crack [28–30], which was attributed a dominant role [28].

One striking feature of the fracture properties in compact bone is the anisotropy of the fracture toughness, which differs by almost two orders of magnitudes between a crack that propagates parallel or perpendicular to the collagen fibrils [24]. This results in a zig-zag pattern of the crack path, when it needs to propagate perpendicular to the fibril direction (Figure 12.2a). This dependence of fracture properties on collagen orientation underlines the general importance of the organic matrix and

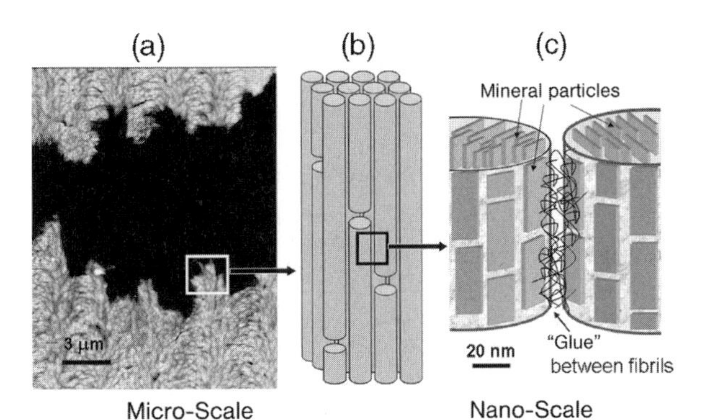

Figure 12.2 Some structural features of bone at the microscale and nanoscale that are responsible for dissipating energy during deformation and fracture. (a) Cracks propagating perpendicularly to the lamellar structure are forced into a zig-zag path, which increases the dissipated energy by about a factor of 30 [24]; (b) Each layer consists of parallel collagen fibrils arranged in a plywood-like structure where the fibril direction rotates along the direction perpendicular to the layer [47, 48]. About one-half of the deformation in a fibril bundle occurs in a glue layer between fibrils [40, 44, 45]. (c) The fibrils are stiffened by mineral particles inside and on the surface of fibrils. The 'glue' layer [43] may contain proteoglycans and phosphorylated proteins, perhaps coordinated by divalent ions, such as calcium [44].

its organization for bone toughness. The organic matrix varies with genetic background, age and disease, and this will clearly influence bone strength and toughness [2, 31–39].

The dominant structural motif at the nanoscale is the *mineralized collagen fibril*. Important contributions to the fracture resistance and defect tolerance of bone composites are believed to arise from these nanometer-scale structural motifs. In recent studies [40], it has been shown that both mineral nanoparticles and the mineralized fibrils deform at first elastically, but to different degrees, in a ratio of 12 : 5 : 2 between tissue, fibrils and mineral particles. These different degrees of deformation of different components arranged in parallel manner within the tissue can be explained by a shear deformation between the components [41]. This means that there is shear deformation within the collagen matrix inside the fibril to accommodate for the difference between the strain in mineral particles and fibrils. In addition, there must also be some shear deformation between adjacent collagen fibrils to accommodate the residual tissue strain. This shear deformation occurs presumably in a 'glue' layer between fibrils (Figure 12.2c), which may consist of proteoglycans and noncollagenous phosphorylated proteins [40, 42–45]. The existence of a glue layer was originally proposed as a consequence of investigations using scanning force microscopy [43]. Beyond the regime of elastic deformation, it is likely that the glue matrix is partially disrupted, and that neighboring fibrils move past each other, breaking and reforming the interfibrillar bonds. An alternative explanation could be the debonding between organic matrix and hydroxyapatite particles (Figure 12.2c) and a modification of the frictional stress between fibril structures [46].

The maximum strain seen in mineral nanoparticles (0.15–0.2%) can reach up to twice the fracture strain calculated for bulk apatite. The origin of this very high strength (~200 MPa) of the mineral particles may result directly from their extremely small size [49]. The strength of brittle materials is known to be controlled by the size of the defects, and of course it can be argued that a defect in a mineral particle cannot be larger than the particle itself. Under such conditions, the strength approaches the theoretical value determined by the chemical bonds between atoms rather than by the defects [49]. Although the nanoparticles in bone are still a way off this value ($\sim E/10$ or 10 GPa), it is believed that the trend towards higher strengths is related to their small size.

As a consequence, it must be concluded that the mechanical properties of bone material are determined by a number of structural features, including:

- the mineral concentration inside the organic matrix, the 'bone mineral density distribution (BMDD)
- the size of mineral particles
- the quality of the collagen, in terms of its amino-acid sequence, crosslinks and hydration
- the quality and composition of the extrafibrillar organic matrix between the collagen fibrils (consisting mostly of noncollagenous organic molecules)
- the orientation distribution of the mineralized collagen fibrils.

Assuming that these parameters are typically optimized in healthy bone material, it is likely that any variation from normal might affect the mechanical performance of the bone. Although these material characteristics cannot typically be determined in a noninvasive manner for patients, they are accessible when studying biopsies using different – and in some cases well-established – technology.

12.4
Bone Mineral Density Distribution in Osteoporosis and Treatments

The mineral concentration inside the organic bone matrix is a major determinant of bone stiffness and strength [2, 33, 50, 51]. However, the mineral content within both, the trabecular and the cortical bone motifs, is far from homogeneous (Figure 12.3). At least two processes that occur in bones over the whole lifetime of an adult individual are responsible for this situation [52]:

- Bone remodeling: The cortical and trabecular bone compartments are continuously remodeled. This means that, during a cycle of about 200 days, areas of bone are resorbed by specific bone cells (*osteoclasts*); this results in *resorption lacunae* which are re-filled with new bone matrix [53] produced by other bone cells (*osteoblasts*). Thus, the bone tissue of an individual adult is on average younger than that adult's chronological age, because the bone turnover time is about

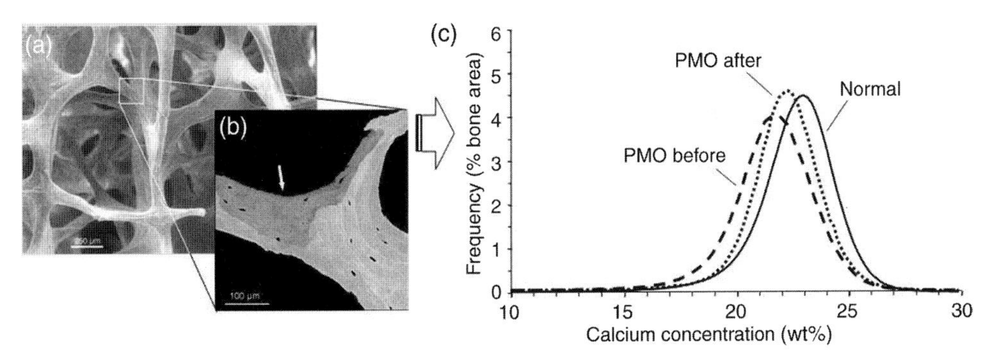

Figure 12.3 Bone is composed of packets of mineralized bone matrix with different mineral concentrations. The distribution of mineral is described by a histogram, called the bone mineralization density distribution (BMDD). (a) Scanning electron microscopy image (secondary electron emission) showing the 3-D structure of trabecular bone; (b) Backscattered electron image of a single trabecula in a bone section, revealing several bone packets with different mineral contents. The dark gray region indicates low mineralization, and the light gray high mineralization of the bone matrix. The arrow shows a newly forming bone packet with a lower mineral content than adjacent packets; (c) Examples of BMDD curves deduced from calibrated backscattered electron images of trabecular bone: NORMAL = healthy individual; PMO before = post-menopausal osteoporotic women before any treatment; PMO after = post-menopausal osteoporotic women after bisphosphonate treatment [57].

five years [54]. In addition, the more such remodeling sites act on the bone surface, the higher will be the bone turnover rate, and more bone packets will be present at a younger stage.

- Kinetics of matrix mineralization: The newly formed bone matrix is initially unmineralized (*osteoid*). However, after an initial maturation time of about 14 days the bone goes through a stage of rapid mineralization, where 70% of the full matrix mineral content is achieved in a few days (*primary mineralization*). Later on, the mineral content increases very slowly to reach full mineralization within years (*secondary mineralization*) [55, 56].

As a consequence of these processes, bone is composed of bone packets – also known as basic structural units (BSUs) – all of which have a different age and mineral content (Figure 12.3). The BSUs generate a characteristic mineralization pattern, sometimes referred to as the bone mineralization density distribution (BMDD) [57], which reflects the bone turnover status and the kinetics of mineralization in an individual [52].

The BMDD can best be measured and quantified using a backscattered electron method (quantitative backscattered electron imaging; qBEI), as described elsewhere [58] and recently reviewed [57]. In contrast to the noninvasive and widely used technique of dual X-ray absorptometry (DXA), which provides an estimate of the total amount of mineral in a scanned area of bone (BMD), the measurement of BMDD requires bone biopsies to be taken. However, the BMDD can also be determined using undecalcified resin-embedded bone blocks, as are prepared for histological examinations. The physical principle of the technique is based on a quantification of the intensity of electrons that are backscattered from a polished bone surface and yield a signal which is proportional to the local concentration of mineral (calcium). Thus, the resulting backscattered electron image visualizes regions of low and high mineral content in dark and light gray, respectively (Figure 12.3b). A suitable calibration of gray levels allows the deduction of frequency distributions of the Ca-concentrations that occur in the scanned bone area (BMDD) with a spatial resolution of 1–4 μm and a sensitivity of 0.17 wt% Ca (Figure 12.3c). The BMDD curve visualizes potential differences in mineralization status of bone between individuals with a high sensitivity (see Figure 12.3c).

With this technique at hand, it has become possible to study the mineral distribution in bone as well as its disease-related changes. The trabecular bone of normal (healthy) adults was found to exhibit minimal variations in BMDD between different skeletal sites, and due to other biological factors such as age, gender or ethnicity. Hence, the BMDD of adult trabecular bone may reflect an evolutionary optimum in bone matrix mineralization as a result of the bone cells' activity and mechanical loading, which most likely represents a compromise between optimum stiffness (which increases with mineral content) and toughness (which decreases with mineral content) of the bone material [2, 3]. It follows that deviations from the normal BMDD, as are observed for example in osteoporosis of post-menopausal women, are most likely of mechanical relevance.

12.4.1 Osteoporosis

Osteoporosis is a disease of enormous socioeconomic impact that is characterized by increased bone fragility [33, 59]. Such fragility is generally associated with an abnormal loss in bone volume, a deterioration in the quality of the bone micro-architecture, an increased bone turnover rate, and also a shift of BMDD towards a lower mineralization density (Figure 12.3c). Interestingly, basic treatment with Ca and vitamin D can have a beneficial effect on bone matrix mineralization and shift the BMDD curve back towards the normal peak position [60, 61]. Additional treatment with antiresorptive agents (e.g. bisphophonates such as alendronate, risedronate or zolendronate) results in a further increase in mineralization, as well as in a higher homogeneity of mineralization within three years of treatment [60, 62, 63]. A prolonged treatment with bisphosphonates over five and 10 years, for example, seemed to restore the BMDD to normal [64]. The treatment effect on BMDD can be explained by a reduction of the remodeling rate, together with a restoration of sufficiently high levels of Ca and vitamin D, thus allowing a more complete mineralization of the BSUs. A combined analysis of bone density (BMD), measured using DXA, and of BMDD as determined by qBEI, revealed that an 8% increase in BMD by bisphosphonate plus Ca/vitamin D treatment was due to a 5% contribution in the improvement of matrix mineralization and a slight (3%) increase in bone volume [61]. The beneficial effect on bone volume might indicate that the therapy also positively affects the negative net balance between bone formation and resorption, as characteristically is the case for osteoporosis. Both effects likely contribute to the sustained anti-fracture efficacy of about 30–50% provided by this anti-resorptive treatment.

Another therapeutic approach in the treatment of osteoporosis is to stimulate bone formation (anabolic treatment). Such an effect on bone can, in principle, be achieved by using sodium fluoride (NaF), which has been used at a daily dose level of 60 mg in several European countries, although it failed to exhibit any anti-fracture efficacy [65]. Interestingly, the BMDD showed a shift to a higher mineralization density with fluoride treatment, and the bone matrix was partly hypermineralized. Changes in the nanocomposite structure of bone (as described in Section 12.5) are most likely responsible for this [66]. Another anabolic agent, parathyroid hormone (PTH), when provided intermittently for a limited period of about 18 months, proved to be successful and has now been approved for the treatment of osteoporosis worldwide. The anabolic effect is clearly reflected in the changes of the BMDD [55], which shows a slight shift to a lower mineralization density and a remarkable broadening of the distribution peak, indicating an increased formation of new bone matrix. This therapy has proven especially useful when the bone loss is already severe. In order to preserve the bone mass gained in such anabolic treatment, possible combinations with anti-resorptive treatments are presently under investigation [67]. It is expected that this might also be beneficial to the mineralization status of the bone, as the prolonged time for secondary mineralization during the anti-resorptive treatment would also normalize the BMDD.

In summary, the BMD in a healthy bone matrix seems to approach an optimum which represents the best compromise between stiffness requiring a high mineral content and toughness, which decreases with mineral content. Biopsies may be used to assess the status of mineral concentration for individual patients at risk of bone fractures in a number of diseases.

12.5 Examples of Disorders Affecting the Structure of Bone Material

As discussed above, the mechanical performance of bone tissue depends on all levels of hierarchy, and several diseases are characterized by modifications at the nanostructural level. In this section, we will detail three examples: (i) *osteogenesis imperfecta,* which is based on mutations of the collagen gene; (ii) *pycnodysostosis,* which originates from a mutation of the cathepsin K gene; and (iii) *fluorosis,* which is caused by higher doses of fluoride. Whilst all three conditions are characterized by a modification at the nanoscale either of the organic matrix or of the mineral particles, adaptation processes during bone remodeling [3] may lead to a partial compensation of the original defect, sometimes at higher hierarchical levels. This means that the modification of bone structure may 'spread' over different hierarchical levels, making it more difficult to pin down the actual origin of the defect.

12.5.1 Osteogenesis Imperfecta

Osteogenesis imperfecta (OI) is a genetic disease that generally affects the collagen gene and leads to brittle bone with different degrees of severity [68–70]. The origin of the brittleness of the tissue is not fully understood, but must be linked to a mutation of the collagen molecule and the resultant changes in tissue quality. Generally, OI also leads to a reduced bone mass and cortical thickness [70] which additionally increases bone fragility. It has been shown that anti-resorptive treatment of affected children with the bisphosphonate pamidronate leads to an increase in cortical thickness and to a concomitant reduction of fracture incidence [70]. At the nanostructural level, an increased mineral content was found in the bone matrix [71, 72], which leads to increased stiffness and hardness of the bone tissue [73]. However, the significance of bone fragility for this increased mineralization is not yet fully clear, as it is not affected by bisphosphonate treatment [73].

More detailed information on the bone matrix nanostructure and the disease-related changes of its properties were obtained in a mouse model of OI [74, 76–82]. This model, which is known as osteogenesis imperfecta murine (oim), is characterized by an absence of the α_2 procollagen molecule, leading to the formation of collagen α_1 homotrimers. The mechanical properties of bone tissue were found to be altered, with a reduced failure load [83] and toughness [82] in oim compared to controls. The mineral content was, however, increased in oim [75], leading to a stiffer matrix [75, 78] (see Figure 12.4). In agreement with observations in humans, this

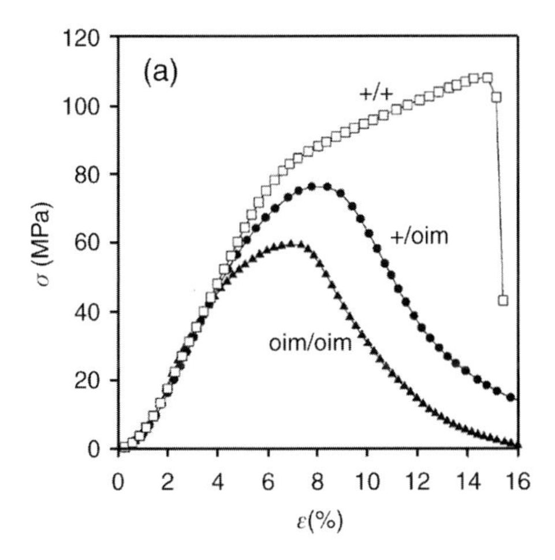

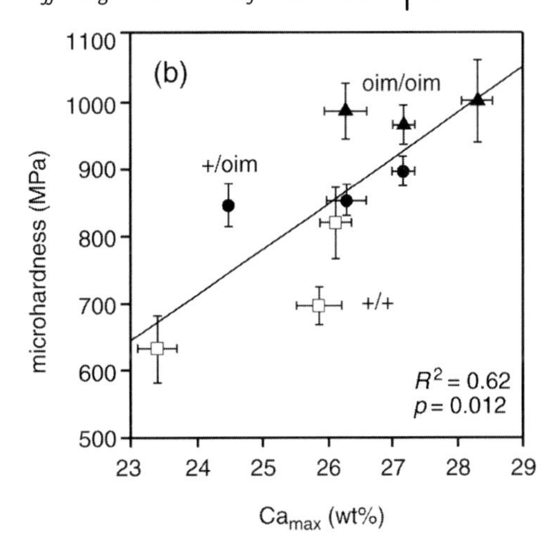

Figure 12.4 Collagen and mineral properties in the osteogenesis imperfecta murine (oim) model. Homozygote oim/oim (full triangles) and heterozygote +/oim (full circles) are compared to normal littermates +/+ (open squares). (a) Stress–strain curve of collagen from the mouse tail (from Ref. [74]). The strength of collagen (the maximum stress σ) is reduced by half from +/+ to oim; (b) Mineral content (Ca_{max}) and microhardness of bone tissue (from Ref. [75]). Both parameters increase from +/+ to oim/oim.

increased tissue mineralization was preserved in treatment with bisphosphonates [84]. The increased brittleness of the tissue is most likely due to a weakness of the collagen-matrix, associated with an increase in mineralization. Indeed, the collagen fibrils seem to break at only half the load in oim homozygotes [74] (see Figure 12.4). The reason for this inherent weakness of collagen might be a modified crosslink pattern in the fibrils [81], due to the fact that normal crosslinks between α_2 and α_1 chains cannot form due to an absence of α_2.

12.5.2
Pycnodysostosis

Pycnodysostosis is an extremely rare human genetic syndrome characterized by an increased bone mass (osteosclerosis), short stature and high bone fragility. Despite very small numbers of cases (about 100), pycnodysostosis was best known for its suggested affliction of the French painter, Henri de Toulouse-Lautrec.

The disorder is caused by a mutation in the gene encoding for cathepsin K, a key enzyme of osteoclastic degradation [85]. Indeed, patients affected by the disease [86], as well as mice mutants lacking cathepsin K activity [87–89], have differentiated osteoclasts that are able to demineralize the bone matrix but not to degrade the remaining organic matrix. As a consequence, the loss of cathepsin K activity leads to a fundamental defect of bone resorption and, subsequently, to an increase in bone mass. The bone resorption is not completely inhibited, however, and can occur to a limited extent via an alternative pathway. The degradation of any unmineralized

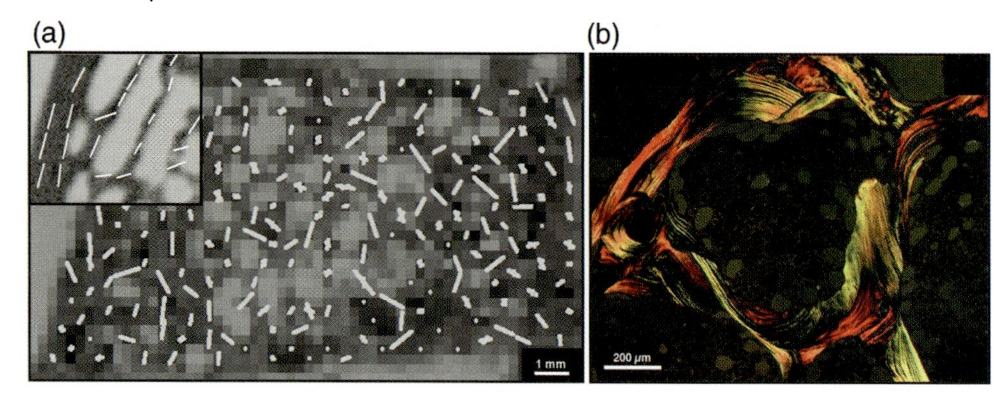

Figure 12.5 Disturbed lamellar organization in pycnodysostosis [92]. (a) The orientation of mineral particles in a biopsy is much less aligned than in normal bone (see inset (i)). The orientation and the length of the white bars indicate the direction and the local degree of alignment of the elongated plate-like mineral nanoparticles in bone, as measured with scanning small angle X-ray scattering; (b) Disturbed organization of the lamellar architecture of trabeculae, as revealed by polarized light microscopy.

collagen left by the dysfunctional osteoclasts may also be possible by the action of matrix-metalloproteinases synthesized by the bone lining-cells, which are members of the osteoblastic lineage [90]. It appears that these two pathways are not equivalent, however, as the lack of cathepsin K activity leads not only to a disturbed bone resorption [91] but also to a decreased bone formation activity [92, 93]. Bone tissue analyses of two affected patients also revealed defects at the nanostructural level, with mineral crystals being increased in size and reflecting a less-remodeled 'older' bone tissue. Moreover, the trabecular architecture appeared to be severely disturbed, with an unusually large variability in the orientation of the mineral particles and a highly disturbed lamellar organization, with the main collagen fibrils not oriented in the principal stress direction (see Figure 12.5). Thus, the absence of functional cathepsin K activity has a profound effect on bone quality at the nanoscale, and leads most likely to an observed increase in bone fragility.

12.5.3
Fluorosis

Fluoride has an anabolic effect on bone and is known to increase cancellous bone mass. During the 1980s and 1990s this led to fluoride being considered as a potential treatment of osteoporosis [94, 95], although clinical trials failed to confirm the anticipated anti-fracture efficacy [65, 96]. One reason for this is that fluoride clearly not only stimulates osteoblasts to form new bone but also has a direct effect on bone material quality. Studies involving small-angle X-ray scattering (SAXS) and back-scattered electron imaging [66, 97, 98] revealed that bone formed under the influence of fluoride has a quite different microscopic structure (see Figure 12.6). Moreover, the collagen–mineral nanocomposite was seen to be massively disturbed. Indeed, the

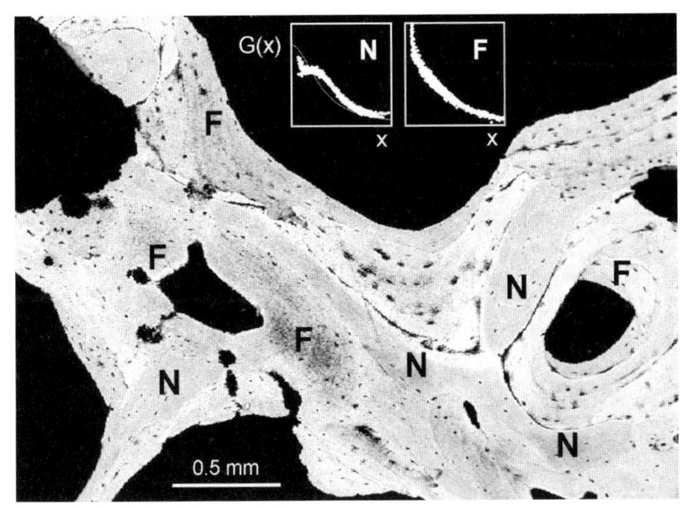

Figure 12.6 Bone biopsy after long-term fluoride treatment (3 years therapy; 50 mg NaF per day) [99]. 'N' indicates areas with normal bone and 'F' with fluorotic bone. The two insets at the top show the shape of the SAXS curves G(x) in normal and fluorotic bone [66, 98], indicating a severe modification of bone mineral nanoparticles during treatment.

strongly modified SAXS signal from bone areas newly formed under the influence of fluoride revealed the presence of mineral crystals much larger than in normal bone (Figure 12.6). This implied that the collagen and mineral in fluorotic bone did not form a well-organized nanocomposite, but that the large mineral crystals simply coexisted with the collagen fibrils. The result was a bone material of lower quality that would most likely be more brittle than usual. The images in Figure 12.6, which show a bone biopsy of a patient treated with sodium fluoride, also indicate that old bone with a normal structure coexists with newly formed fluorotic bone material. Due to a constant bone turnover, the old normal bone is gradually replaced by new bone with a fluorotic structure. This gradually compensates the positive mechanical effect of the bone mass increase and finally leads to a deterioration of bone stability against fracture [66, 97, 98].

12.6 Conclusions

Fractures – the clinical endpoint of disorders affecting the structure of bone material – are associated with increased morbidity, mortality and high socioeconomic costs [100, 101]. Today, due to an increased life expectancy for the general population, the incidence of fractures is also increasing, and the assessment of fracture risk and identification of those individuals who might benefit from the prevention and treatment of skeletal disorders represent major challenges in modern medicine.

New analyses of epidemiologic data provide strong evidence for the view that all (or better still, the overwhelming majority of) fractures – regardless of when they occur and of the level of trauma that precipitates them – may be based upon bone fragility [59], thus focusing all aspects of pathophysiology, diagnosis and treatment of skeletal diseases to the central question of mechanical competence and bone fragility. Following Robert Marcus' thoughts on "... the nature of osteoporosis" [102], 'bone fragility' might be defined most appropriately from the pathophysiological point of view as "... the consequence of a stochastic process, that is, multiple genetic, physical, hormonal and nutritional factors acting alone or in concert to diminish skeletal integrity."

Based on the fact that skeletal integrity is determined by the outstanding mechanical properties of bone at all hierarchical levels of its structure and organization [2], it becomes increasingly evident that a simple diagnostic parameter such as lumbar spine or hip BMD [103–105], although frequently used as a noninvasive diagnostic tool in clinical routine, does not have the diagnostic power to reflect the complex pathophysiological mechanisms that determine bone fragility. Thus, the availability of new diagnostic tools developed by materials scientists, coupled with a possible combinatorial approach using different methods to define the material qualities of bone from the micrometer to the nanometer scale, should introduce a renaissance of bone biopsies as diagnostic tools in clinical osteology. For example, the BMDD of trabecular human bone (as described above) was shown to be evolutionarily optimized within relative small variations (ca. 3%), independently of different skeletal regions for healthy adults aged between 25 and 95 years. Until now, no differences have been identified for BMDD-derived parameters with regards to gender or ethnicity. As shown in several examples, deviations from the normal BMDD seem to be associated with skeletal disorders, and in many examples indicate 'bone fragility' [57]. BMDD can be determined by using qBEI on a transiliac biopsy, as routinely occurs for histomorphometry, and combined with a variety of techniques based on spectroscopy, light scattering or biomechanical testing [2, 57].

When investigating the treatment of post-menopausal osteoporosis with the anti-resorptives alendronate [62] or risedronate [60] and the anabolic intermittent PTH [55], slight – but significant – deviations in BMDD indicated a lower mineralization for all placebo groups, and this was confirmed for idiopathic osteoporosis in pre-menopausal women. An example of learning from a materials science perspectives was that of fluorosis, and the fluoride treatment of post-menopausal osteoporosis [66, 97]. Yet, despite sodium fluoride being used widely to treat post-menopausal osteoporosis, no anti-fracture efficacy was reported. Rather, the bone quality revealed extensive and pathologic mineralization at both micro- and nanoscale, leading to a more brittle material with increased fragility.

Two classical genetic bone diseases – pycnodysostosis [92] and OI [68, 70, 71, 74–76, 78, 84] – point to a genetically related diminution of skeletal integrity. In OI, which often is fatal, the primary pathology was shown as brittle bones, inefficient repair mechanisms and a high bone turnover, whereas in pycnodysostosis the effects were caused by nonfunctioning osteoclasts due to mutations of the essential enzyme cathepsin K [85]. However, an inability to optimize structure by bone remodeling results in a sclerosing bone disease with high bone mass and fragility fractures due to a disorganized structure at several hierarchical levels.

In conclusion, a wealth of evidence has been accumulated during the past years supporting the concept that the study of bone micro- and nanostructures will not only improve our understanding of the mechanisms that underlie bone fragility but also help to identify the effects of treatments. Nanomedicine, and its application to bone research, will in time undoubtedly broaden our knowledge of pathopysiology and improve the diagnoses, prevention and treatment of bone diseases. The availability of new techniques to investigate bone biopsies will surely challenge clinical osteologists and bone pathologists in the near future.

References

1 Weiner, S. and Wagner, H.D. (1998) *Annual Review of Materials Science*, **28**, 271.

2 Fratzl, P., Gupta, H.S., Paschalis, E.P. and Roschger, P. (2004) *Journal of Materials Chemistry*, **14**, 2115.

3 Fratzl, P. and Weinkamer, R. (2007) *Progress in Materials Science*, **52**, 1263.

4 Canty, E.G. and Kadler, K.E. (2002) *Comparative Biochemistry and Physiology. Part A, Molecular & Integrative Physiology*, **133**, 979.

5 Kadler, K.E., Holmes, D.F., Trotter, J.A. and Chapman, J.A. (1996) *The Biochemical Journal*, **316**, 1.

6 Hodge, A.J. and Petruska, J.A. (1963) *Aspects of Protein Structure* (ed G.N. Ramachandran), Academic Press, New York, p. 289.

7 Rubin, M.A., Rubin, J. and Jasiuk, W. (2004) *Bone*, **35**, 11.

8 Fantner, G.E., Hassenkam, T., Kindt, J.H., Weaver, J.C., Birkedal, H., Pechenik, L., Cutroni, J.A., Cidade, G.A.G., Stucky, G.D., Morse, D.E. and Hansma, P.K. (2005) *Nature Materials*, **4**, 612.

9 Landis, W.J. (1996) *Connective Tissue Research*, **35**, 1.

10 Weiner, S. and Traub, W. (1992) *The FASEB Journal*, **6**, 879.

11 Hassenkam, T., Fantner, G.E., Cutroni, J.A., Weaver, J.C., Morse, D.E. and Hansma, P.K. (2004) *Bone*, **35**, 4.

12 Sodek, J., Ganss, B. and McKee, M.D. (2000) *Critical Reviews in Oral Biology and Medicine*, **11**, 279.

13 Landis, W.J., Hodgens, K.J., Song, M.J., Arena, J., Kiyonaga, S., Marko, M., Owen, C. and McEwen, B.F. (1996) *Journal of Structural Biology*, **117**, 24.

14 Traub, W., Arad, T. and Weiner, S. (1992) *Matrix (Stuttgart, Germany)*, **12**, 251.

15 Roschger, P., Grabner, B.M., Rinnerthaler, S., Tesch, W., Kneissel, M., Berzlanovich, A., Klaushofer, K. and Fratzl, P. (2001) *Journal of Structural Biology*, **136**, 126.

16 Fratzl, P., Fratzl-Zelman, N., Klaushofer, K., Vogl, G. and Koller, K. (1991) *Calcified Tissue International*, **48**, 407.

17 Glimcher, M.J. (1984) *Philosophical Transactions of the Royal Society of London. Series B, Biological Sciences*, **304**, 479.

18 Grynpas, M., Bonar, L.C. and Glimcher, M.J. (1985) *Journal of Materials Science*, **19**, 723.

19 Posner, A.S. (1985) *Clinical Orthopaedics*, **200**, 87.

20 Fratzl, P., Groschner, M., Vogl, G., Plenk, H., Eschberger, J., Fratzl-Zelman, N., Koller, K. and Klaushofer, K. (1992) *Journal of Bone and Mineral Research*, **7**, 329.

21 Currey, J.D. (1999) *The Journal of Experimental Biology*, **202**, 3285.

22 Currey, J.D. (2003) *Journal of Bone and Mineral Research*, **18**, 591.

23 Taylor, D., Hazenberg, J.G. and Lee, T.C. (2007) *Nature Materials*, **6**, 263.

24 Peterlik, H., Roschger, P., Klaushofer, K. and Fratzl, P. (2006) *Nature Materials*, **5**, 52.

25 Zioupos, P. and Currey, J.D. (1994) *Journal of Materials Science*, **29**, 978.

26 Vashishth, D., Tanner, K.E. and Bonfield, W. (2003) *Journal of Biomechanics*, **36**, 121.

27 Liu, D.M., Weiner, S. and Wagner, H.D. (1999) *Journal of Biomechanics*, **32**, 647.

28 Nalla, R.K., Kruzic, J.J. and Ritchie, R.O. (2004) *Bone*, **34**, 790.

29 Nalla, R.K., Kinney, J.H. and Ritchie, R.O. (2003) *Nature Materials*, **2**, 164.

30 Nalla, R.K., Kruzic, J.J., Kinney, J.H. and Ritchie, R.O. (2005) *Biomaterials*, **26**, 217.

31 Viguet-Carrin, S., Garnero, P. and Delmas, P.D. (2006) *Osteoporosis International*, **17**, 319.

32 Chavassieux, P., Seeman, E. and Delmas, P.D. (2007) *Endocrine Reviews*, **28**, 151.

33 Seeman, E. and Delmas, P.D. (2006) *The New England Journal of Medicine*, **354**, 2250.

34 Landis, W.J. (1995) *Bone*, **16**, 533.

35 Zioupos, P. and Currey, J.D. (1998) *Bone*, **22**, 57.

36 Zioupos, P., Currey, J.D. and Hamer, A.J. (1999) *Journal of Biomedical Materials Research*, **45**, 108.

37 Zioupos, P. (2001) *Journal of Biomaterials Applications*, **15**, 187.

38 Wang, X.D., Bank, R.A., Te Koppele, J.M. and Agrawal, C.M. (2001) *Journal of Orthopaedic Research*, **19**, 1021.

39 Wang, X., Shen, X., Li, X. and Agrawal, C.M. (2002) *Bone*, **31**, 1.

40 Gupta, H.S., Seto, J., Wagermaier, W., Zaslansky, P., Boesecke, P. and Fratzl, P. (2006) *Proceedings of the National Academy of Sciences of the United States of America*, **103**, 17741.

41 Jager, I. and Fratzl, P. (2000) *Biophysical Journal*, **79**, 1737.

42 Thompson, J.B., Kindt, J.H., Drake, B., Hansma, H.G., Morse, D.E. and Hansma, P.K. (2001) *Nature*, **414**, 773.

43 Fantner, G., Hassenkam, T., Kindt, J.H., Weaver, J.C., Birkedal, H., Pechenik, L., Cutroni, J.A., Cidade, G.A.G., Stucky, G.D., Morse, D.E. and Hansma, P.K. (2005) *Nature Materials*, **4**, 612.

44 Gupta, H.S., Fratzl, P., Kerschnitzki, M., Benecke, G., Wagermaier, W. and Kirchner, H.O.K. (2007) *Journal of the Royal Society Interface*, **4**, 277.

45 Gupta, H.S., Wagermaier, W., Zickler, G.A., Aroush, D.R.B., Funari, S.S., Roschger, P., Wagner, H.D. and Fratzl, P. (2005) *Nano Letters*, **5**, 2108.

46 Tai, K., Ulm, F.J. and Ortiz, C. (2006) *Nano Letters*, **6**, 2520.

47 Giraud-Guille, M.M. (1988) *Calcified Tissue International*, **42**, 167.

48 Weiner, S., Arad, T., Sabanay, I. and Traub, W. (1997) *Bone*, **20**, 509.

49 Gao, H.J., Ji, B.H., Jager, I.L., Arzt, E. and Fratzl, P. (2003) *Proceedings of the National Academy of Sciences of the United States of America*, **100**, 5597.

50 Currey, J.D. (2001) *Calcified Tissue International*, **68**, 205.

51 Currey, J.D. (2002) *Bones - Structure and Mechanics*, Princeton University Press, Princeton.

52 Ruffoni, D., Fratzl, P., Roschger, P., Klaushofer, K. and Weinkamer, R. (2007) *Bone*, **40**, 1308.

53 Eriksen, E.F., Axelrod, D.W. and Melsen, F. (1994) *Bone histomorphometry*, Raven Press: New York.

54 Eriksen, E.F., Melsen, F., Sod, E., Barton, I. and Chines, A. (2002) *Bone*, **31**, 620.

55 Misof, B.M., Roschger, P., Cosman, F., Kurland, E.S., Tesch, W., Messmer, P., Dempster, D.W., Nieves, J., Shane, E., Fratzl, P., Klaushofer, K., Bilezikian, J. and Lindsay, R. 2003 *The Journal of Clinical Endocrinology and Metabolism*, **88**, 1150.

56 Boivin, G. and Meunier, P.J. (2003) *Osteoporosis International*, **14**, S22.
57 Roschger, P., Paschalis, E.P., Fratzl, P. and Klaushofer, K. (2008) *Bone*, **42**, 456.
58 Roschger, P., Fratzl, P., Eschberger, J. and Klaushofer, K. (1998) *Bone*, **23**, 319.
59 Mackey, D.C., Lui, L.Y., Cawthon, P.M., Bauer, D.C., Nevitt, M.C., Cauley, J.A., Hillier, T.A., Lewis, C.E., Barrett-Connor, E. and Cummings, S.R. (2007) *The Journal of the American Medical Association*, **298**, 2381.
60 Zoehrer, R., Roschger, P., Paschalis, E.P., Hofstaetter, J.G., Durchschlag, E., Fratzl, P., Phipps, R. and Klaushofer, K. (2006) *Journal of Bone and Mineral Research*, **21**, 1106.
61 Fratzl, P., Roschger, P., Fratzl-Zelman, N., Paschalis, E.P., Phipps, R. and Klaushofer, K. (2007) *Calcified Tissue International*, **81**, 73.
62 Roschger, P., Rinnerthaler, S., Yates, J., Rodan, G.A., Fratzl, P. and Klaushofer, K. (2001) *Bone*, **29**, 185.
63 Haas, M., Leko-Mohr, Z., Roschger, P., Kletzmayr, J., Schwarz, C., Mitterbauer, C., Steininger, R., Grampp, S., Klaushofer, K., Delling, G. and Oberbauer, R. (2003) *Kidney International*, **63**, 1130.
64 Roschger, P., Mair, G., Fratzl-Zelman, N., Fratzl, P., Kimmel, D., Klaushofer, K., LaMotta, A. and Lombardi, A. (2007) *Journal of Bone and Mineral Research*, **22**, S129.
65 Riggs, B.L., Hodgson, S.F., Ofallon, W.M., Chao, E.Y.S., Wahner, H.W., Muhs, J.M., Cedel, S.L. and Melton, L.J. (1990) *The New England Journal of Medicine*, **322**, 802.
66 Fratzl, P., Roschger, P., Eschberger, J., Abendroth, B. and Klaushofer, K. (1994) *Journal of Bone and Mineral Research*, **9**, 1541.
67 Finkelstein, J.S., Hayes, A., Hunzelman, J.L., Wyland, J.J., Lee, H. and Neer, R.M. (2003) *The New England Journal of Medicine*, **349**, 1216.
68 Prockop, D.J. (1992) *The New England Journal of Medicine*, **326**, 540.
69 Sillence, D.O., Senn, A. and Danks, M.D. (1979) *Journal of Medical Genetics*, **16**, 101.
70 Rauch, F. and Glorieux, F.H. (2004) *Lancet*, **363**, 1377.
71 Jones, S.J., Glorieux, F.H., Travers, R. and Boyde, A. (1999) *Calcified Tissue International*, **64**, 8.
72 Roschger, P., Fratzl-Zelman, N., Misof, B.M., Glorieux, F.H., Klaushofer, K. and Rauch, F. (2008) *Calcified Tissue International*, **82**, 263.
73 Weber, M., Roschger, P., Fratzl-Zelman, N., Schoberl, T., Rauch, F., Glorieux, F.H., Fratzl, P. and Klaushofer, K. (2006) *Bone*, **39**, 616.
74 Misof, K., Landis, W.J., Klaushofer, K. and Fratzl, P. (1997) *The Journal of Clinical Investigation*, **100**, 40.
75 Grabner, B., Landis, W.J., Roschger, P., Rinnerthaler, S., Peterlik, H., Klaushofer, K. and Fratzl, P. (2001) *Bone*, **29**, 453.
76 Fratzl, P., Paris, O., Klaushofer, K. and Landis, W.J. (1996) *The Journal of Clinical Investigation*, **97**, 396.
77 Camacho, N.P., Hou, L., Toledano, T.R., Ilg, W.A., Brayton, C.F., Raggio, C.L., Root, L. and Boskey, A.L. (1999) *Journal of Bone and Mineral Research*, **14**, 264.
78 Mehta, S.S., Antich, P.P. and Landis, W.J. (1999) *Connective Tissue Research*, **40**, 189.
79 Grabner, B., Landis, W.J., Roschger, P., Rinnerthaler, S., Peterlik, H., Klaushofer, K. and Fratzl, P. (2001) *Bone*, **29**, 453.
80 Miles, C.A., Sims, T.J., Camacho, N.P. and Bailey, A.J. (2002) *Journal of Molecular Biology*, **321**, 797.
81 Sims, T.J., Miles, C.A., Bailey, A.J. and Camacho, N.P. (2003) *Connective Tissue Research*, **44**, 202.
82 Miller, E., Delos, D., Baldini, T., Wright, T.M. and Pleshko Camacho, N. (2007) *Calcified Tissue International*, **81**, 206.
83 Camacho, N.P., Hou, L., Toledano, T.R., Ilg, W.A., Brayton, C.F., Raggio, C.L., Root, L. and Boskey, A.L. (1999) *Journal of Bone and Mineral Research*, **14**, 264.

84 Misof, B.M., Roschger, P., Baldini, T., Raggio, C.L., Zraick, V., Root, L., Boskey, A.L., Klaushofer, K., Fratzl, P. and Camacho, N.P. (2005) *Bone*, **36**, 150.

85 Gelb, B.D., Shi, G.P., Chapman, H.A. and Desnick, R.J. (1996) *Science*, **273**, 1236.

86 Hou, W.S., Bromme, D., Zhao, Y.M., Mehler, E., Dushey, C., Weinstein, H., Miranda, C.S., Fraga, C., Greig, F., Carey, J., Rimoin, D.L., Desnick, R.J. and Gelb, B.D. 1999 *The Journal of Clinical Investigation*, **103**, 731.

87 Saftig, P., Hunziker, E., Wehmeyer, O., Jones, S., Boyde, A., Rommerskirch, W., Moritz, J.D., Schu, P. and von Figura, K. (1998) *Proceedings of the National Academy of Sciences of the United States of America*, **95**, 13453.

88 Gowen, M., Lazner, F., Dodds, R., Kapadia, R., Feild, J., Tavaria, M., Bertoncello, I., Drake, F., Zavarselk, S., Tellis, I., Hertzog, P., Debouck, C. and Kola, I. 1999 *Journal of Bone and Mineral Research*, **14**, 1654.

89 Li, C.Y., Jepsen, K.J., Majeska, R.J., Zhang, J., Ni, R.J., Gelb, B.D. and Schaffler, M.B. (2006) *Journal of Bone and Mineral Research*, **21**, 865.

90 Everts, V., Delaisse, J.M., Korper, W., Jansen, D.C., Tigchelaar-Gutter, W., Saftig, P. and Beertsen, W. (2002) *Journal of Bone and Mineral Research*, **17**, 77.

91 Chen, W., Yang, S.Y., Abe, Y., Li, M., Wang, Y.C., Shao, J.Z., Li, E. and Li, Y.P. (2007) *Human Molecular Genetics*, **16**, 410.

92 Fratzl-Zelman, N., Valenta, A., Roschger, P., Nader, A., Gelb, B.D., Fratzl, P. and Klaushofer, K. (2004) *The Journal of Clinical Endocrinology and Metabolism*, **89**, 1538.

93 Chavassieux, P., Asser Karsdal, M., Segovia-Silvestre, T., Neutzsky-Wulff, A.V., Chapurlat, R., Boivin, G. and Delmas, P.D. (2008) *Journal of Bone and Mineral Research*, **23**, 1076.

94 Kleerekoper, M. (1998) *Endocrinology and Metabolism Clinics of North America*, **27**, 441.

95 Kleerekoper, M. (1996) *Critical Reviews in Clinical Laboratory Sciences*, **33**, 139.

96 Rubin, M.R. and Bilezikian, J.P. (2003) *Endocrinology and Metabolism Clinics of North America*, **32**, 285.

97 Roschger, P., Fratzl, P., Klaushofer, K. and Rodan, G. (1997) *Bone*, **20**, 393.

98 Fratzl, P., Schreiber, S., Roschger, P., Lafage, M.H., Rodan, G. and Klaushofer, K. (1996) *Journal of Bone and Mineral Research*, **11**, 248.

99 Fratzl, P., Rinnerthaler, S., Roschger, P. and Klaushofer, K. (1998) *Osteologie*, **7**, 130.

100 Ray, N.F., Chan, J.K., Thamer, M. and Melton, L.J. 3rd (1997) *Journal of Bone and Mineral Research*, **12**, 24.

101 Melton, L.J. 3rd (1993) *Bone*, **14** (Suppl 1), S1.

102 Marcus, R. (1996) *The Journal of Clinical Endocrinology and Metabolism*, **81**, 1.

103 Cummings, S.R., Bates, D. and Black, D.M. (2002) *The Journal of the American Medical Association*, **288**, 1889.

104 Schuit, S.C., van der Klift, M., Weel, A.E., de Laet, C.E., Burger, H., Seeman, E., Hofman, A., Uitterlinden, A.G., van Leeuwen, J.P. and Pols, H.A. (2004) *Bone*, **34**, 195.

105 Miller, P.D. (2006) *Reviews in Endocrine and Metabolic Disorders*, **7**, 75.

13
Nanoengineered Systems for Tissue Engineering and Regeneration

Ali Khademhosseini, Bimal Rajalingam, Satoshi Jinno, and Robert Langer

13.1
Introduction

In recent years, tissue engineering has emerged as a potentially powerful approach for the treatment of a variety of diseases by merging the principles of life sciences and engineering to generate biological substitutes that restore, maintain and enhance human tissue function [1]. In a typical tissue engineering approach, cells are seeded within biodegradable scaffolds. Then as the scaffolds degrade, the cells deposit their own matrices and self-assemble into tissue-like structures. This reassembly and degradation process eventually results in the formation of three-dimensional (3-D) tissue structures.

Over the past few years, tissue engineering has generated much excitement for fabricating a renewable source of transplantable tissues. Advances in the field have resulted in the engineering of clinically usable skin substitutes. In addition, other engineered tissues such as cartilage and bone are at various stages of clinical trials. Research groups have also attempted to engineer nerve tissue, pancreas, bladder and other organs. However, despite success in clinical studies, the dream of 'off-the-shelf' organs has eluded scientists and clinicians alike. This can be attributed to our inability to direct the behavior of cells in a desired manner, as well as to generate 3-D tissues with sufficient complexity and structural integrity to perform the function of the native tissues. Other concerns include the transmission of infection through the implanted tissue-engineered substrate, the possibility of immune reaction against the implanted cells, the variability of engineered products, the introduction of genetically modified, unwanted and potentially harmful cells into the body as well as regulatory and ethical aspects.

In order to generate functional tissues it is important to mimic the biological microenvironment by developing approaches and tools that can control materials and cells at the nanoscale. This is because many structural elements such as the extracellular matrix (ECM), as well as biological processes such as receptor clustering, are at the nanoscale. Thus, the ability to engineer the cellular environment and tissue

Nanotechnology, Volume 5: Nanomedicine. Edited by Viola Vogel

ISBN: 978-3-527-31736-3

structure at the nanoscale is a potentially powerful approach for generating biomimetic tissues. These processes will be critical not only for generating tissue engineered constructs but also for engineering *in vitro* systems that can be used for various drug discovery and diagnostics applications.

Nanotechnology is an emerging field that is concerned with the design, synthesis, characterization and application of materials and devices that have a functional organization in at least one dimension on the nanometer scale, ranging from a few to about 100 nm [1]. Due to this ability to control features at small length scales, nanotechnology is becoming more commonly used in a number of biomedical endeavors ranging from drug delivery [2–5] to *in vivo* imaging [6].

In this chapter we will discuss the application of nanotechnology to tissue engineering as an enabling tool. Specifically, we will provide an overview of two different types of nanoengineered system that are used in tissue engineering. First, we will focus on various approaches that are used to generate nanoscale modifications to existing polymers and materials. These nanoengineered systems, such as nanopatterned substrates and electrospun scaffolds, provide structures that influence cell behavior and the subsequent tissue formation. Furthermore, we will discuss the use of other nanoscale structures such as controlled-release nanoparticles for tissue engineering. In the second part of the chapter we will discuss the use of nanotechnology for the synthesis of novel materials that behave differently as bulk compared to their nanoscale versions. Such materials include self-assembled materials, carbon nanotubes and quantum dots. Throughout the chapter we will discuss the use of nanomaterials for controlling the cellular microenvironment and for generating 3-D tissues. We will also detail the potential limitations and emerging topics of interest and challenges in this area of research. Clearly, the application of nanotechnology to tissue engineering and cell culture is an 'exploding' field, and hopefully in this chapter we will provide a glimpse into the various applications. Throughout the chapter, when applicable, the reader is directed to more extensive reviews to provide further detail regarding specific topics.

13.2 Nanomaterials Synthesized Using Top-Down Approaches

In 1959, Richard Feynman introduced the significant benefits associated with manipulating materials atom by atom. Since then, extensive research has been conducted in nanotechnology owing to the advances in technologies that enable the manipulation and characterization of nanoscale objects such as scanning tunneling microscopy (STM), atomic force microscopy (AFM) and other related technologies. Nanomaterials have generated interest in a variety of fields such as clinical medicine, defense, pharmaceuticals, aerospace, energy and biological research. Today, engineered nanomaterials are increasingly utilized for various tissue engineering applications due to their controllable and unique properties. Furthermore, nanostructures can aid in mimicking Nature, as many biological structures have features that are on the order of few to hundreds of nanometers. Nanoengineered materials can be created by tailoring the properties of existing materials, for example by controlling the 3-D structure or surface roughness. In general, nanomaterials can be produced by either 'top-down' or 'bottom-up' approaches. Top-down approaches involve the miniaturization of materials to nano length scales, and have been enabled due to increased technological capability for miniaturizing materials either by using novel approaches or by making improvements to existing techniques. In this section, we will describe two techniques used to prepare nanomaterials using top-down approaches, namely the use of electrospinning and nanopatterned substrates to engineer cell behavior.

13.2.1 Electrospinning Nanofibers

Electrospinning is a technique that is used to generate nanofibrous scaffolds (Figure 13.1a). These scaffolds are highly porous (i.e. a large surface area-to-volume ratio), and thus mimic many of the properties of natural tissues and also provide cells with a pseudo 3-D environment [7]. Electrospinning is also relatively inexpensive and capable of producing nanofibers from a variety of biodegradable synthetic polymers, such as poly(lactic-*co*-glycolic acid) (PLGA) [8], polycaprolactone (PCL) [9] and poly(L-lactic acid) (PLLA) [10], as well as natural polymers such as collagen [11]. The nanoscale features of electrospun scaffolds promote cell proliferation and also guide cell growth. For example, aligned nanofibers have been shown to induce the growth and proliferation of cardiac cells into contractile spindle structures [12]. The seeding of neural stem cells (NSCs) on aligned PLLA nano/micro fibrous scaffolds has also resulted in neurite outgrowth along with the fiber direction for the aligned scaffolds [10]. Furthermore, fiber diameter and orientation have been shown to influence the cell morphology, while cell proliferation is not sensitive to the above-mentioned parameters. Goldstein and colleagues, while testing a range of diameters (0.14–3.6 μm) and angular standard deviations (31–60°), found that an increasing fiber diameter and degree of fiber orientation resulted in increased projected cell area and aspect ratio [8].

Electrospinning has been used to fabricate scaffolds for various tissues such as bone [9], cartilage [13–15] and cardiac muscle [16]. An example of bone tissue fabrication is shown in Figure 13.1b. These scaffolds have also been used to direct the differentiation of stem cells; for example, the coating of electrospun nanofibers composed of polyamide have been used to promote the proliferation and self-renewal of mouse embryonic stem cells (ESCs). These ESCs maintained their ability to differentiate into various lineages, which showed that such fibers could be used not only for *in vivo* applications as tissue engineered scaffolds but also as tools for *in vitro* cell culture. Electrospun nanofibers have also been shown to influence cell shape, actin cytoskeleton and matrix deposition, both *in vitro* [17, 18] and *in vivo* [19].

Despite its versatility, one disadvantage with electrospinning process is that the range of the resulting fibers is usually limited to the upper range of natural ECM fibers. In addition, it is difficult to control the complex architecture and intricate

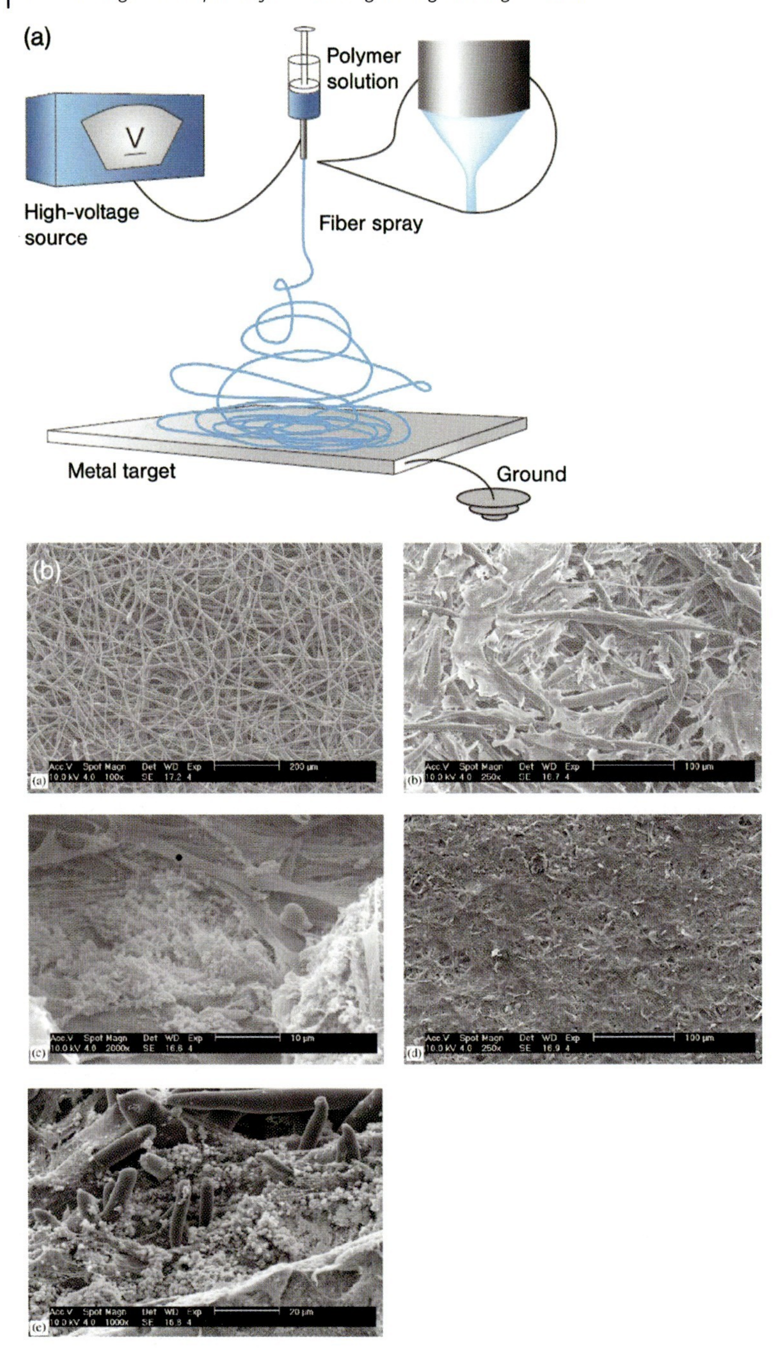

(a)
Polymer solution
High-voltage source
Fiber spray
Metal target
Ground
(b)

cell–cell, cell–ECM and cell-soluble factors in the resulting scaffolds. This has prompted the use of other techniques to fabricate nanofabricated scaffolds. For example, nanofabricated PLLA scaffolds that supported the differentiation and outgrowth of NSCs have been fabricated by a liquid–liquid phase separation method [20]. Yet, despite these limitations, electrospinning is a powerful technology for the fabrication of 3-D nanoscaffolds.

13.2.2 Scaffolds with Nanogrooved Surfaces

Both, micro- and nanotextured substrates can significantly influence cell behavior such as adhesion, gene expression [15] and migration [16]. This is because the interaction of cells with a biomaterial results in the localization of focal adhesions, actin stress fibers and microtubules [21]. Focal contacts are involved in signal transduction pathways which in turn can regulate a wide array of cell function [22] (Figure 13.2). As nanofiber scaffolds have a larger surface area, they have more potential binding sites to cell membrane receptors, thus affecting the cell behavior in unique ways. It has also been observed that the cells display more *filopodia* when they come in contact with the nanoscale surfaces, presumably to sense the external topography [23]. Although the mechanism of cellular response to nanotopography is not entirely understood, it is suggested that an interaction of the cellular processes and interfacial forces results in peculiar cellular behavior [7]. Both, micro- and nanotextured substrates can be engineered either on tissue culture substrates or within 3-D tissue scaffolds. Within tissue engineering scaffolds, nanotextures provide physical cues to seeded cells and regulate the interaction of host cells with the scaffold. For example, surfaces that have desired roughness have been shown to increase osteoblast adhesion for orthopedic replacement/augmentation applications [24].

Nanotextures can be generated using a variety of techniques, depending on the material as well as the dimension and shapes of the desired structures. For example, features less than 100 nm may be produced by a range of techniques including chemical etching in metals [25], the embedding of carbon nanofibers in composite

Figure 13.1 (a) Schematic of the electrospinning apparatus. The set-up for electrospinning consists of a spinneret with a metallic needle, a syringe pump, a high-voltage power supply and a grounded collector. By using the electrospinning apparatus, uniform fibers with nanometer-scale diameters can be fabricated. These scaffolds are highly porous (large surface area-to-volume ratio), thus mimicking many of the properties of natural tissues and providing cells with a pseudo 3-D environment. The nanoscale features of electrospun scaffolds promote cell proliferation and guide cell growth; (b) Panel (a) shows an electrospun poly(ε-caprolactone) scaffold, prior to cell seeding. Panels (b) and (c) are images of mesenchymal stem cells (MSCs) seeded on the scaffold after 7 days of culture, with low and high magnification, respectively. Panels (d) and (e) show the MSCs after four weeks of culture, again with low and high magnification, respectively. After 7 days, osteoblast-like cells developed, indicating bone-like formation. (Adapted from Ref. [9].)

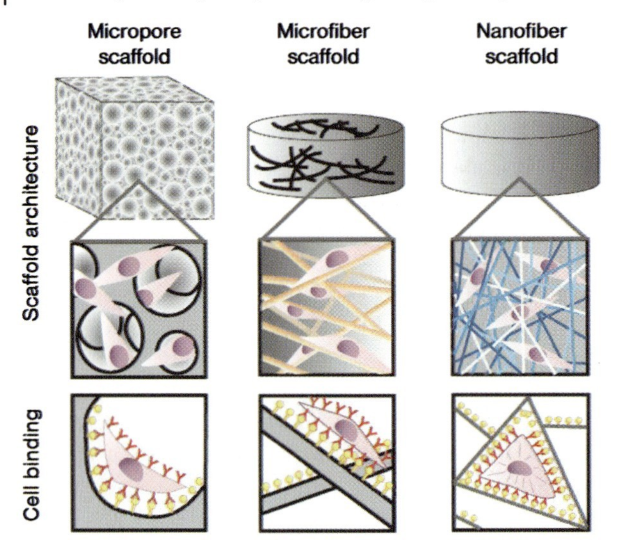

Figure 13.2 The influence of scaffold architecture on cell attachment and spreading. The attachment of cells on the micropore or microfiber scaffold is similar to that of cells cultured on flat surfaces. The larger surface area provided by the nanofiber scaffold results in many more binding sites to cell membrane receptors, thus influencing cell behavior and the subsequent tissue formation in unique ways. Focal contacts are involved in signal transduction pathways, which in turn can regulate a wide array of cell function. (Adapted from Ref. [7].)

materials [26], casting polymer replicas from ECM [27], or the embedding of constituent nanoparticles in materials ranging from metals to ceramics to composites [28–31]. Electron-beam lithography (EBL) technology has been used to fabricate well-defined nanostructures at sub-50 nm length scales [21, 32]. These tools can be used to form structures at the same length scales as the native ECM, and thus enable the systematic study of cell behavior (for a review, see Ref. [24]).

The shape of the nanostructures influences the cell behavior and phenotype. For example, nanogrooves [33–37] result in an alignment of cells parallel to the direction of the grooves [33, 38] as well as the alignment of actin, microtubules and other cytoskeletal elements [39–41]. Interestingly, both the pitch and depth of the grooves influence cell behavior. For example, typically the orientation increases with increased depth of the nanogrooves [42]. Another shape that has been shown to influence cell behavior is the natural roughness of tissues. For example, endothelial cells that were cultured on the ECM-textured replicas spread faster and had an appearance more like the cells in their native arteries than did cells grown on nontextured surfaces [27, 43]. Fibroblasts cultured on nanopatterned ε-PCL surfaces were also less spread compared to those on a planar substrate [44]. Furthermore, human mesenchymal stem cells (MSCs) and ESCs align on nanofabricated substrates and differentiate in a specific manner [45, 46]. Therefore, by controlling the nanotopography of tissue engineering scaffolds inductive signals can be delivered to enhance tissue formation and function.

13.3 Nanomaterials Synthesized using Bottom-Up Approaches

In this section we describe the synthesis and application of nanomaterials that are built by nanoscale assembly of molecules with properties that are often different from their individual components and their bulk material. These materials include self-assembled peptide hydrogels, quantum dots, carbon nanotubes and layer-by-layer deposited films.

13.3.1 Self-Assembled Peptide Scaffolds

A promising approach in tissue engineering is to use nanoengineered materials made from synthetic peptides or peptide amphiphiles (PA) that self-assemble. (An extensive review of this topic is provided in Chapter 14.) Self-assembled PA hydrogels can be generated by linking a carbon alkyl tail to functional peptides; these PA molecules then self-assemble based on hydrophobic interactions of the alkyl tail and thus form nanofibers that can form hydrogels either alone or by mixing with a cell suspension [47]. An example of a self-assembled peptide is shown in Figure 13.3. In addition, self-assembled peptide hydrogels made purely from peptides that self-assemble based on hydrophobic interactions have also been demonstrated [48–50]. In this approach, self-assembled beta-sheets are formed which can assemble into hydrogels. Both of these approaches have shown promising results in various tissue engineering, stem cell differentiation and cell culture applications [47, 51–54]. An example of this is the recent investigation of the proliferation and differentiation of MSCs in PA hydrogels. When rat MSCs were seeded into the PA nanofibers, with or without RGD, a larger number of cells attached to the PA nanofibers that contained RGD. Furthermore, upon examination of the osteogenic differentiation of MSCs, the alkaline phosphatase (ALP) activity and osteocalcin content increased for the PA nanofibers that contained RGD compared with those without RGD. In another study, the use of MSC-seeded hybrid scaffolds prepared from PAs and a collagen sponge reinforced with poly(glycolic acid) (PGA) fibers was examined to show increased osteogenic differentiation of MSCs and ectopic bone formation.

13.3.2 Layer-by-Layer Deposition of Nanomaterials

The ability to control the surface properties of biological interfaces is useful in various aspects of tissue engineering. One means of obtaining such controlled surfaces is by the layer-by-layer (LBL) deposition of the charged biopolymers. LBL deposition uses the electrostatic interaction between the surface and the polyelectrolyte solutions to generate films with nanoscale dimensions. LBL has been used extensively to control the cellular microenvironment *in vitro*. For example, we have generated patterned cellular cocultures using the LBL deposition of ionic biopolymers hyaluronic acid (HA), poly-L-lysine (PLL) [55] and collagen. In this approach, micropatterns of

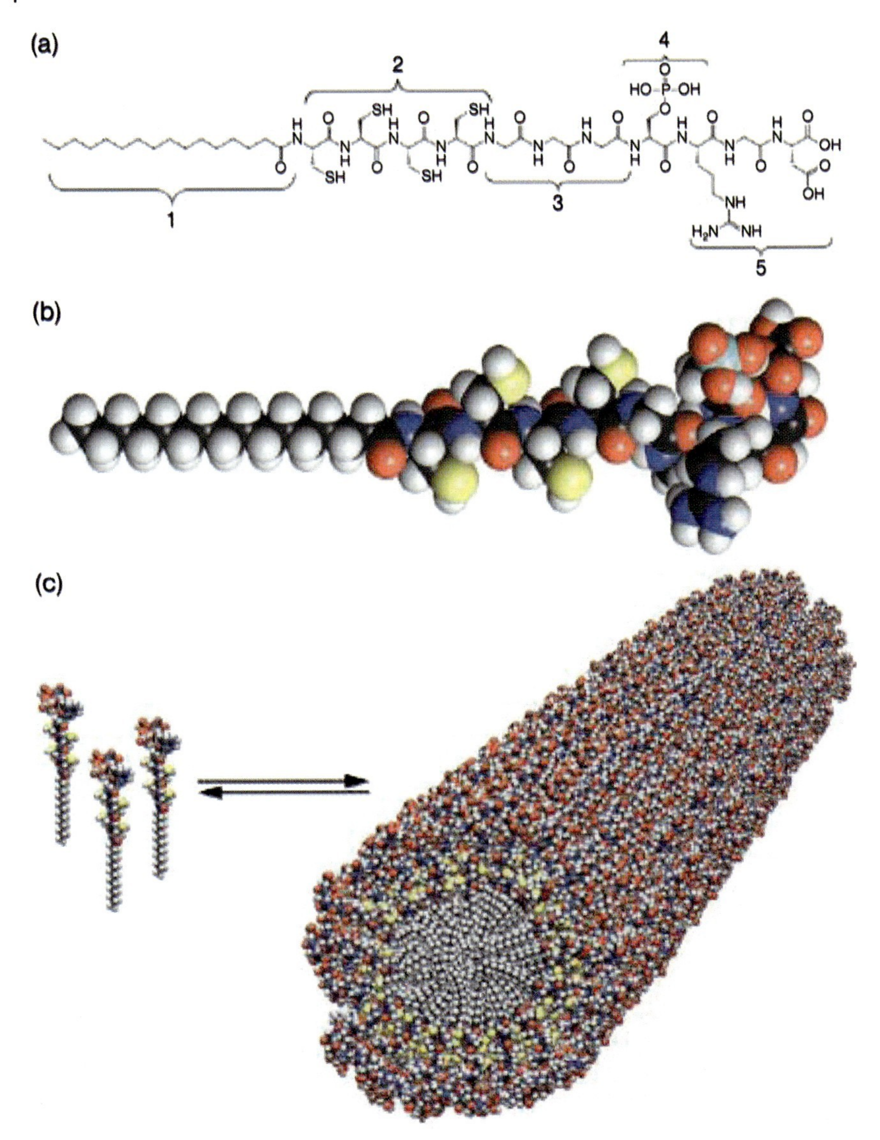

Figure 13.3 Self-assembled peptide scaffold. (a) The chemical structure of the peptide amphiphile. This is composed of a long alkyl tail (region 1), four consecutive cysteine residues (region 2), a flexible linker region of three glycine residues (region 3), a single phosphorylated serine residue (region 4) and cell adhesion ligand RGD (region 5); (b) The molecular model of (a); (c) These peptide amphiphiles are self-assembling into a cylindrical micelle. (Adapted from Ref. [47].)

nonbiofouling anionic HA were used to pattern cells on glass substrates. The subsequent adsorption of the cation PLL to HA pattern resulted in an adherent surface promoting the attachment of a second cell type. In order to minimize toxicity, instead of PLL other positively charged molecules such as ECM components (i.e. collagen) have also been used (Figure 13.4) [56]. In a related experiment, cultured human endothelial cells have been patterned using LBL on a polyurethane surface. Here, it was observed that the cells did not spread on the negatively charged surface due to an electrostatic repulsion, whereas inversing the surface charge by adding positively charged collagen increased the cell spreading and proliferation. Thus, cell

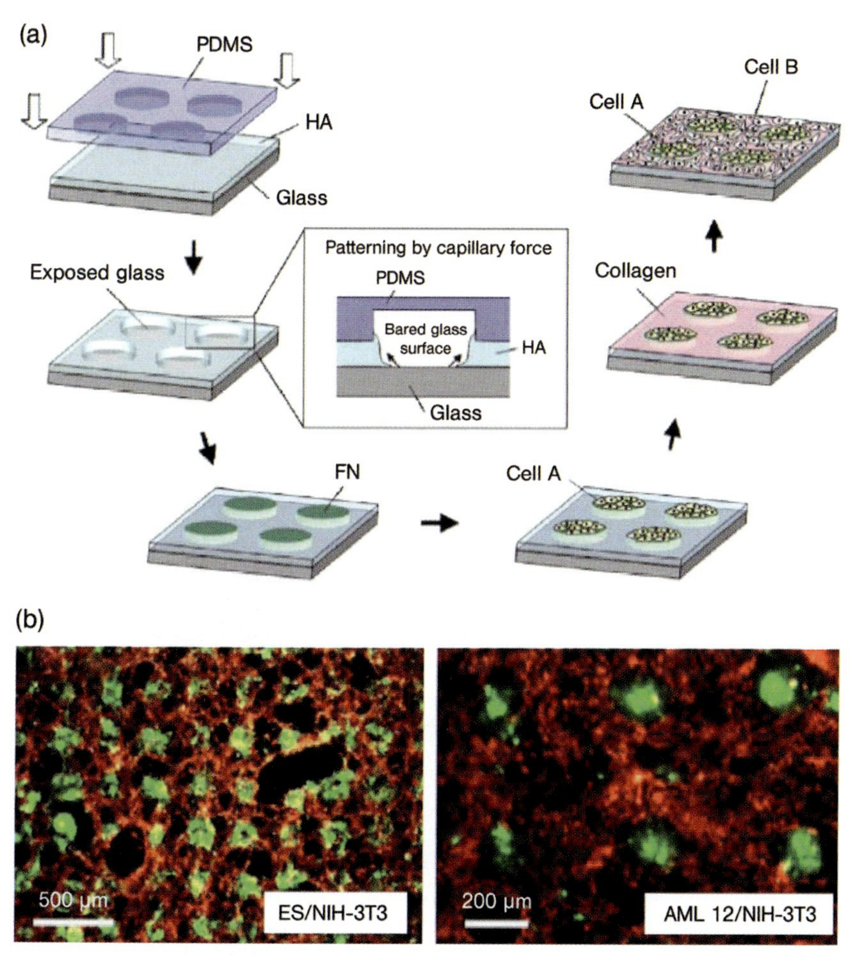

Figure 13.4 (a) The generation of cocultures using layer-by-layer deposition of the ECM materials, hyaluronic acid (HA) and collagen. A polydimethylsiloxane (PDMS) mold was placed on a glass slide coated with a thin layer of HA. Due to capillary forces, the HA in the exposed space of the PDMS mold receded, thus exposing the underlying glass surface. The exposed region of a glass substrate was coated with fibronectin (FN), where primary cells (cell A) could be selectively adhered. Subsequently, the PDMS mold was removed and collagen layered on the HA surface to make the surface adherent to secondary cells (cell B); (b) The fluorescent images of the patterned coculture after three days of culture, generated by a layer-by-layer coculture approach. The ESCs and hepatocytes (AML 12) were cocultured with the fibroblasts (NIH-3T3).

attachment on multilayer thin films may depend on the charge of the terminal polyion layer [57].

The number of layers in the LBL films may play a role in the cell attachment behavior. It was reported that increasing the number of layers of titanium oxide nanoparticle thin films increased the surface roughness, cell attachment and the rate of cell spreading. Although this may be due to the increased surface roughness, it also demonstrates the potential of this technology in controlling cell-surface properties [58]. Furthermore, the LBL assembly of PLL and dextran sulfate could be used to increase the rate of fibronectin deposition and the subsequent cell adhesion relative to the control substrates [59].

The LBL deposition of materials has been used in a variety of tissue engineering applications. For example, the LBL assembly of HgTe has been used to fabricate a hybrid device where the absorption of light by quantum nanoparticles stimulates neural cells by a sequence of photochemical and charge-transfer reactions [60]. These devices may be of potential use in tissue engineering applications in which it is desired to stimulate nerve cells using external cues. The LBL assemblies of nanoparticles have also been explored as a means of protecting arteries damaged during revascularization procedures. It has been reported that the deposition of self-assembled nanocoatings comprising alternating depositions of HA and chitosan onto aortic porcine arteries, led to a significant inhibition of the growth of thrombus on the damaged arterial surfaces. Clearly, this technique has the potential for clinical application to protect damaged arteries and to prevent subsequent restenosis [61]. Therefore, by properly choosing the LBL materials it is possible to modify the surface properties of materials for tissue engineering as well as for biosensing [62, 63] and drug delivery applications [64].

Mironov and colleagues have used the LBL technique for organ printing, with precise control over the spatial position of the deposited cell [65]. LBL deposition can also be used for the fabrication of immunosensors [66], islet cell encapsulation [67] and polyelectrolyte capsules for drug release [68].

13.3.3 Carbon Nanotubes

Carbon nanotubes are nanomaterials with unique mechanical and chemical properties. They have been used for cell tracking, for the delivery of desired molecules to cells, and as components of tissue engineering scaffolds [69]. Carbon nanotubes, depending on the number of carbon walls, can range from 1.5 to 30 nm in diameter and may be hundreds of nanometers in length.

Within tissue engineering scaffolds carbon nanotubes can be used to modify the mechanical and chemical properties. Furthermore, carbon nanotubes can be functionalized with biomolecules to signal the surrounding cells, or they may be electrically stimulated due to their high electrical conductivity to excite tissues such as muscle cells and nerve cells. One potentially powerful method of integrating carbon nanotubes into tissue engineering is by generating composite materials which comprise a biocompatible material such as collagen with embedded, single-walled carbon nanotubes. As an example, smooth muscle cells (SMCs) have been encapsulated within collagen–carbon nanotube composite matrices with high cell viability (>85%) for at least 7 days [70]. Single-walled carbon nanotubes can also be used for culturing excitable tissues such as neuronal and muscle cells [71]. It has also been suggested that the growth of the neuronal circuits in carbon nanotubes might result in a significant increase in the network activity and an increase in neural transmission, perhaps due to the high electric conductivity of carbon nanotubes [72]. Furthermore, the electrical stimulation of osteoblasts cultured in nanocomposites comprising PLA and carbon nanotubes increased their cell proliferation and the deposition of calcium after 21 days. These data show that the use of novel current-conducting nanocomposites would be valuable for enhancing the function of the osteoblasts, and also provide useful avenues in bone tissue engineering [73].

Carbon nanotubes have also been used to delivery pharmaceutical drugs [74–76], genetic material [77–79] and biomolecules such as proteins [80, 81] to various cell types. For example, carbon nanotubes have been used to deliver Amphotericin B to fungal-infected cells. Here, the Amphotericin B was found to be bonded covalently to carbon nanotubes and was uptaken by mammalian cells, without significant toxicity, while maintaining its antifungal activity [82]. Thus, carbon nanotubes may be used for the delivery of antibiotics to specific cells.

The influence of carbon nanotubes on cells varies, depending on the type and their surface properties. For example, it has been reported that rat osteosarcoma (ROS) 17/2.8 cells, when cultured on carbon nanotubes carrying a neutral electric charge, proliferated to a greater extent than did other control cells [83]. Chemically modifying the surface of carbon nanotubes can also be used to enhance their cytocompatibility. For example, carbon nanotubes have been coated with bioactive 4-hydroxynonenal in order to culture embryonic rat brain neurons that promote neuronal branching compared to unmodified carbon nanotubes [84]. Despite such promise, however, the cytotoxicity of carbon nanotubes remains unclear. It is well known that various properties such as surface modifications and size greatly influence the potential toxicity of these structures. For example, long carbon nanotubes have been shown to generate a greater degree of inflammation in rats than shorter carbon nanotubes ($\sim$200 nm), which suggests that the smaller particles may be engulfed more easily by macrophages [85]. Other studies have also shown that carbon nanotubes may not only inhibit cell growth [86] but also induce pulmonary injury in the mouse model [87], when sequential exposure to carbon nanotubes and bacteria enhanced pulmonary inflammation and infectivity. Thus, more extensive and systematic studies must be conducted to ensure that the use of these nanomaterials in tissue engineering does not result in long-term toxicity.

13.3.4 MRI Contrast Agents

Nanotechnology may also permit the high-resolution imaging of tissue-engineered constructs. Specifically, the use of imaging contrast agents in magnetic resonance imaging (MRI) can be used to track cells *in vivo* and visualize constructs [88–90].

Although MRI contrast agents take many forms, nanoparticle systems have emerged in recent years as one of the most promising as nanoparticles not only provide an enormous surface area but can also be functionalized with targeting ligands and magnetic labels. Moreover, their smaller size provides an easy permeability across the blood capillaries.

Iron oxide nanoparticles have shown great promise for use in MRI to track cells, because they can be uptaken without compromising cell viability and are relatively safe. A wide variety of iron oxide-based nanoparticles have been developed that differ in hydrodynamic particle size and surface-coating material (dextran, starch, albumin, silicones) [91]. In general, these particles are categorized based on their diameter into superparamagnetic iron oxides (SPIOs) (50–500 nm) and ultrasmall superparamagnetic iron oxides (USPIOs) (<50 nm), with the size dictating their physico-chemical and pharmacokinetic properties. It has also been shown that clearance of the iron oxide nanoparticles in the rat liver depends on the outer coating [92].

Iron oxide nanoparticles have been used for imaging various organs, including the gastrointestinal tract, liver, spleen and lymph nodes. Furthermore, smaller-sized particles can also be used for angiography and perfusion imaging in myocardial and neurological diseases. Iron oxide particles can be coated with various molecules to increase their circulation and targeting. For example, dextran-coated iron oxide nanoparticles have been used for labeling cells, while anionic magnetic nanoparticles can be used to target positively charged tissues by using electrostatic interactions [93] (Figure 13.5). In addition, iron oxide nanoparticles can be used to track cells *in vivo* after transplantation. For example, MSCs and other mammalian cells labeled with SPIO nanoparticles were used to track cells in both experimental and clinical settings [94, 95]. Furthermore, green fluorescent protein (GFP^+) ESCs that were

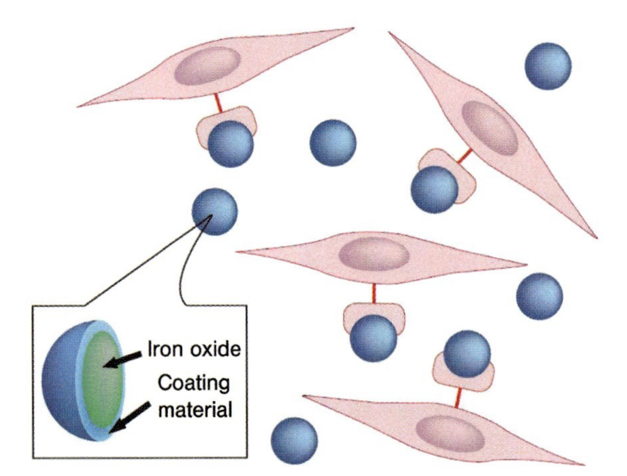

Figure 13.5 Schematic of the magnetic resonance molecular imaging. The tracking of magnetic nanoparticles to cancer cells is based on their static and dynamic magnetism, along with an ability to impart cell-specific functionality. The biocompatibility of coating materials (i.e., biscarboxyl-terminated poly(ethylene glycol)) allows for *in vivo* applications in animals and humans. Therefore, iron oxide nanoparticles show great promise for use in MRI to track cells *in vivo* and to visualize constructs, without compromising cell viability.

labeled with dextran-coated iron oxide nanoparticles and implanted into the brains of rats with brain stroke showed that the cells could be tracked for at least three weeks [96]. The *in vivo* tracking of iron oxide nanoparticle-labeled rat bone marrow MSCs and mouse ESCs and human $CD34^+$ hematopoietic progenitor cells in rats with a cortical or spinal cord lesion has also shown that cells may remain visible in the lesion for at least 50 days [97, 98]. Taken together, these and other results [99] indicate that magnetic nanoparticles are well-suited for the noninvasive analysis of cell migration, engraftment and morphological differentiation at high spatial and temporal resolution.

In order to target desired cells or to modify the rate of cellular uptake, nanoparticles may be engineered with specific molecules on their surfaces. For example, in order to increase their internalization, NSCs and $CD34^+$ bone marrow cells were labeled with superparamagnetic nanoparticles that were conjugated with short HIV-Tat peptides. This increased the internalization of the particles by the cells, without affecting their viability, differentiation or proliferation. The localization and retrieval of cell populations *in vivo* enabled a detailed analysis of specific stem cell and organ interactions that were critical for advancing the therapeutic use of stem cells [100]. In addition to iron oxide nanoparticles, other types of nanoparticle have also been used for tissue imaging, notably with applications in tissue engineering. As an example, fluoroscein isothiocyanate (FITC) -conjugated mesoporous silica nanoparticles (MSNs) have been used to label human bone marrow MSCs and 3T3-L1 cells. The FITC-MSNs were efficiently internalized into MSCs and 3T3-L1 cells, even with short-term incubation (2–4 h), without affecting cell viability [101]. Thus, it seems that nanoparticles can be used potentially not only to track cells but also to image tissues which may be useful for the noninvasive imaging of tissue-engineered constructs.

13.3.5 Quantum Dots

Nanoscale probes can also be used in tissue engineering applications for the study of various biological processes, as well as for real-time cell detection and tracking. Fluorescent dyes, which traditionally have been used to image cells and tissues, have several drawbacks including photobleaching and a lack of long-term stability. Quantum dots (QDs) are nanoparticles that have several advantages over conventional fluorophores for imaging, including tunable properties and a resistance to photobleaching [6]. QDs are semiconductor nanostructures that confine the motion of conduction band electrons, valence band holes or excitons in all three spatial directions. The band gap energy of the QD is the energy difference between the valence band and the conduction band. For nanoscale semiconductor particles such as QDs, the bandgap is dependent on the size of the nanocrystal, which results in a size-dependent variation in emission. A single light source can also be used for the simultaneous excitation of a spectrum of emission wavelength, which makes the method useful for multicolor, multiplexed biological detection and imaging applications.

QDs can be used for the ultrasensitive imaging of molecular targets in deep tissue and living animals (Figure 13.6). Here, they are used as specific markers for cellular structures [102, 103] and molecules [104], for monitoring physiological events in live cells [105–107], for measuring cell motility [108], and for monitoring RNA delivery and tracking cells [109] *in vivo*. As an example, QDs have been used for locating multiple distinct endogenous proteins within cells, thus determining the precise protein distribution in a high-throughput manner [110]. Peptide ligand-conjugated QDs have also been used for imaging G-protein-coupling receptors in both whole-cells and as single-molecules [111]. Cellular events such as the transport of lipids and proteins across membranes have also been tracked using QDs with molecular resolution in live cells [112]. Furthermore, QDs conjugated to immunoglobulin G (IgG) and streptavidin have been used to label the breast cancer marker Her2 on the surface of fixed and live cancer cells [113].

QDs have significant potential in analyzing the mechanisms of cell growth, apoptosis, cell–cell interactions, cell differentiations and inflammatory responses. For example, QDs have been used to study the signaling pathways of mast cells during an inflammatory response [114], as well as to quantify changes in organelle morphology during apoptosis [115]. In addition, the photostability and biocompatibility of QDs make them the preferred agents for the long-term tracking of live cells [116]. QDs are internalized into cells by endocytosis [117], by receptor-mediated uptake [118], by peptide-mediated transportation [119, 120] or microinjection [121]. An example of this was recently demonstrated in studies in which ligand-conjugated QDs were used to monitor antigen binding, entry and trafficking in dendritic cells [122]. QDs, when conjugated to a transporter protein, have also been used to label malignant and nonmalignant hematological cells and to track cell division, thus enabling lineage tracking [109].

Despite the remarkable potential for the application of QDs in clinical medicine, their toxicity and long-term adverse effects are still not clearly understood. The metabolism, excretion and toxicity of QDs may depend on multiple factors such as size, charge, concentration and outer coating bioactivity, as well as their oxidative, photolytic and mechanical stabilities and other unknown factors [123]. Importantly, these issues must be addressed before QDs can be used for *in vivo* applications in humans.

13.4 Future Directions

During a relatively short period of time, nanomaterials have spawned a number of new approaches to address important challenges in tissue engineering. These challenges range from understanding the mechanisms of stem cell differentiation to generating functional vasculature within tissue engineering constructs. Despite these advancements, further investigations are required to analyze the true potential and clinical viability of these technologies. Our current lack of knowledge regarding the long-term toxicity of many nanoengineered materials represents a

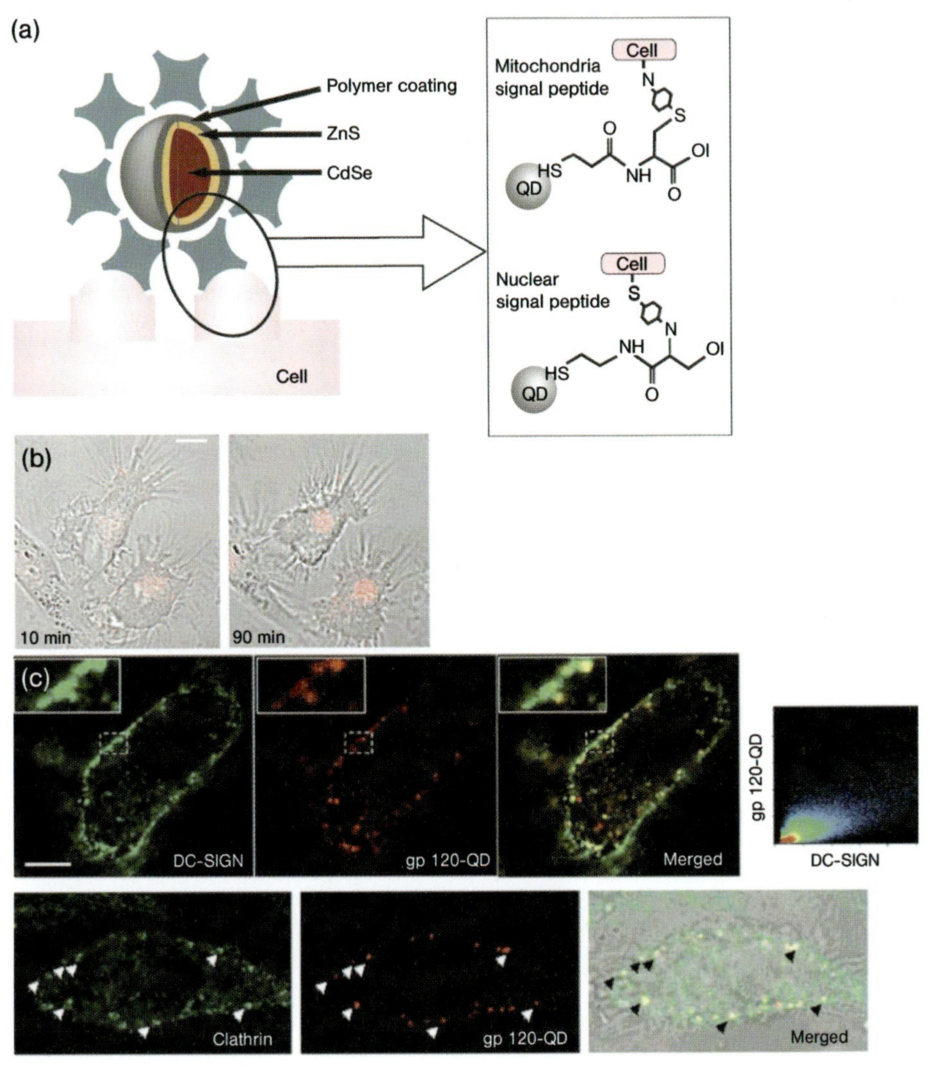

Figure 13.6 (a) Schematic of quantum dots (QDs). QDs are typically made from nanocrystals of a semiconductor material (CdSe), which has been coated with an additional semiconductor shell (ZnS) to improve the material's optical properties. This material is coated with a polymer shell that allows the materials to be conjugated to biological molecules and to retain their optical properties. These nanocrystals have been coupled to various biomolecules directly or indirectly. The inset at the right shows a schematic of peptide-conjugated QDs for organelle targeting and imaging. The amino acid-coated QDs are conjugated with target peptides by coupling. QDs can reveal the transduction of proteins and peptides into specific subcellular compartments as a powerful tool for studying intracellular analysis *in vitro* and even *in vivo*. (Adapted from Ref. [124].); (b) Ligand-conjugated QDs internalized by dendritic cells (DCs) via their specific binding protein (DC-SIGN). Ligand-coated QDs bind to DC-SIGN and are endocytosed into DCs; (c) DCs were incubated with the HIV-1 envelope glycoprotein gp120-QDs (red). After washing of unbound QDs, DCs were fixed and labeled with DC-SIGN marker (green). Data were obtained using confocal microscopy. The right-hand panel shows a 2-D histogram of DC-SIGN signal versus gp120-QDs signal. This result indicated that the small amount of dispersion leads to high colocalization. (Adapted from Ref. [122].)

critical barrier for their use in humans. Traditionally, tissue engineers have favored materials that have a long history of medical application (i.e. FDA approved), although many such materials have limitations to be overcome, perhaps through rational design enabled by nanotechnology. Thus, there remains a clear need to develop nanoengineered materials capable of addressing the various challenges of tissue engineering. In addition, systematic toxicity studies must be conducted in order to fully optimize and characterize not only the function but also the long-term behavior of nanomaterials *in vivo*. Yet, many of the traditional methods used to analyze previous generations of biomaterials do not apply to nanoscale materials, and a clear paradigm shift is required in these analytical and standardization procedures. This will range from how we study material–cell interactions *in vitro* and *in vivo*, to the standardization requirements of regulatory bodies such as the FDA. Clearly, these modifications will require extensive discussion amongst the scientists, the patients, the general public, the clinicians and the regulatory officers.

13.5 Conclusions

Today, nanotechnology offers a wide variety of tools in tissue engineering, biomedical imaging, biosensing, diagnostics and drug delivery. Nanotechnology-based applications are valuable in the research and development of viable substitutes that may restore, maintain or even improve the function of human tissues. Today, these materials have not only opened up novel applications but have also addressed a number of limitations associated with traditional approaches and materials. Nonetheless, much research is required to further demonstrate the long-term stability and clinical utility of nanoengineered materials.

Acknowledgments

The authors greatly appreciate the helpful discussions with Drs Hossein Hosseinkhani and Hossein Baharvand. They also acknowledge generous funding from the Draper laboratory, the CIMIT, NIH and Coulter Foundation.

References

1 Khademhosseini, A. and Langer, R. (2006) Nanobiotechnology for Tissue Engineering and Drug Delivery. *Chemical Engineering Progress*, **102**, 38–42.

2 Sengupta, S. and Sasisekharan, R. (2007) Exploiting nanotechnology to target cancer. *British Journal of Cancer*, **96**, 1315–1319.

3 Bisht, S., Feldmann, G., Soni, S., Ravi, R., Karikar, C., Maitra, A. and Maitra, A. (2007) Polymeric nanoparticle-encapsulated curcumin (nanocurcumin):

a novel strategy for human cancer therapy. *Journal of Nanobiotechnology*, **5**, 3.
4 Allen, T.M. and Cullis, P.R. (2004) Drug delivery systems: entering the mainstream. *Science*, **303**, 1818–1822.
5 Pfeifer, B.A., Burdick, J.A., Little, S.R. and Langer, R. (2005) Poly(ester-anhydride): poly(beta-amino ester) micro- and nanospheres: DNA encapsulation and cellular transfection. *International Journal of Pharmaceutics*, **304**, 210–219.
6 Alivisatos, A.P., Gu, W. and Larabell, C. (2005) Quantum dots as cellular probes. *Annual Review of Biomedical Engineering*, **7**, 55–76.
7 Stevens, M.M. and George, J.H. (2005) Exploring and engineering the cell surface interface. *Science*, **310**, 1135–1138.
8 Bashur, C.A., Dahlgren, L.A. and Goldstein, A.S. (2006) Effect of fiber diameter and orientation on fibroblast morphology and proliferation on electrospun poly(D,L-lactic-*co*-glycolic acid) meshes. *Biomaterials*, **27**, 5681–5688.
9 Yoshimoto, H., Shin, Y.M., Terai, H. and Vacanti, J.P. (2003) A biodegradable nanofiber scaffold by electrospinning and its potential for bone tissue engineering. *Biomaterials*, **24**, 2077–2082.
10 Yang, F., Murugan, R., Wang, S. and Ramakrishna, S. (2005) Electrospinning of nano/micro scale poly(L-lactic acid) aligned fibers and their potential in neural tissue engineering. *Biomaterials*, **26**, 2603–2610.
11 Matthews, J.A., Wnek, G.E., Simpson, D.G. and Bowlin, G.L. (2002) Electrospinning of collagen nanofibers. *Biomacromolecules*, **3**, 232–238.
12 Xu, C.Y., Inai, R., Kotaki, M. and Ramakrishna, S. (2004) Aligned biodegradable nanofibrous structure: a potential scaffold for blood vessel engineering. *Biomaterials*, **25**, 877–886.
13 Fertala, A., Han, W.B. and Ko, F.K. (2001) Mapping critical sites in collagen II for rational design of gene-engineered proteins for cell-supporting materials. *Journal of Biomedical Materials Research*, **57**, 48–58.
14 Li, W.J., Laurencin, C.T., Caterson, E.J., Tuan, R.S. and Ko, F.K. (2002) Electrospun nanofibrous structure: a novel scaffold for tissue engineering. *Journal of Biomedical Materials Research*, **60**, 613–621.
15 Li, W.J., Cooper, J.A. Jr, Mauck, R.L. and Tuan, R.S. (2006) Fabrication and characterization of six electrospun poly (alpha-hydroxy ester)-based fibrous scaffolds for tissue engineering applications. *Acta Biomaterialia*, **2**, 377–385.
16 Zong, X., Bien, H., Chung, C.-Y., Yin, L., Fang, D., Hsiao, B.S., Chu, B. and Entcheva, E. (2005) Electrospun fine-textured scaffolds for heart tissue constructs. *Biomaterials*, **26**, 5330–5338.
17 Nur, E.K.A., Ahmed, I., Kamal, J., Schindler, M. and Meiners, S. (2005) Three dimensional nanofibrillar surfaces induce activation of Rac. *Biochemical and Biophysical Research Communications*, **331**, 428–434.
18 Schindler, M., Ahmed, I., Kamal, J., Nur-E-Kamal, A., Grafe, T.H., Chung, H.Y. and Meiners, S. (2005) A synthetic nanofibrillar matrix promotes *in vivo*-like organization and morphogenesis for cells in culture. *Biomaterials*, **26**, 5624–5631.
19 Nur, E.K.A., Ahmed, I., Kamal, J., Schindler, M. and Meiners, S. (2006) Three-dimensional nanofibrillar surfaces promote self-renewal in mouse embryonic stem cells. *Stem Cells (Dayton, Ohio)*, **24**, 426–433.
20 Yang, F., Murugan, R., Ramakrishna, S., Wang, X., Ma, Y.-X. and Wang, S. (2004) Fabrication of nano-structured porous PLLA scaffold intended for nerve tissue engineering. *Biomaterials*, **25**, 1891–1900.
21 Dalby, M.J., Marshall, G.E., Johnstone, H.J., Affrossman, S. and Riehle, M.O. (2002) Interactions of human blood and tissue cell types with 95-nm-high

nanotopography. *IEEE Transactions on NanoBioscience*, **1**, 18–23.

22 Mrksich, M., Chen, C.S., Xia, Y., Dike, L.E., Ingber, D.E. and Whitesides, G.M. (1996) Controlling cell attachment on contoured surfaces with self-assembled, monolayers of alkanethiolates on gold. *Proceedings of the National Academy of Sciences of the United States of America*, **93**, 10775–10778.

23 Curtis, A.S., Gadegaard, N., Dalby, M.J., Riehle, M.O., Wilkinson, C.D. and Aitchison, G. (2004) Cells react to nanoscale order and symmetry in their surroundings. *IEEE Transactions on NanoBioscience*, **3**, 61–65.

24 Sato, M. and Webster, T.J. (2004) Nanobiotechnology: implications for the future of nanotechnology in orthopedic applications. *Expert Review of Medical Devices*, **1**, 105–114.

25 de Oliveira, P.T. and Nanci, A. (2004) Nanotexturing of titanium-based surfaces upregulates expression of bone sialoprotein and osteopontin by cultured osteogenic cells. *Biomaterials*, **25**, 403–413.

26 Price, R.L., Waid, M.C., Haberstroh, K.M. and Webster, T.J. (2003) Selective bone cell adhesion on formulations containing carbon nanofibers. *Biomaterials*, **24**, 1877–1887.

27 Goodman, S.L., Sims, P.A. and Albrecht, R.M. (1996) Three-dimensional extracellular matrix textured biomaterials. *Biomaterials*, **17**, 2087–2095.

28 Webster, T.J., Siegel, R.W. and Bizios, R. (1999) Osteoblast adhesion on nanophase ceramics. *Biomaterials*, **20**, 1221–1227.

29 Webster, T.J. (2001) *Nanostructured Materials* (ed. J.Y. Ying), Academy Press, New York, pp. 126–166.

30 Webster, T.J., Ergun, C., Doremus, R.H., Siegel, R.W. and Bizios, R. (2000) Enhanced functions of osteoblasts on nanophase ceramics. *Biomaterials*, **21**, 1803–1810.

31 Webster, T.J., Ergun, C., Doremus, R.H., Siegel, R.W. and Bizios, R. (2001) Enhanced osteoclast-like cell functions on nanophase ceramics. *Biomaterials*, **22**, 1327–1333.

32 Suh, K.Y., Khademhosseini, A., Eng, G. and Langer, R. (2004) Single nanocrystal arrays on patterned poly(ethylene glycol) copolymer microstructures using selective wetting and drying. *Langmuir*, **20**, 6080–6084.

33 Chou, L., Firth, J.D., Uitto, V.J. and Brunette, D.M. (1998) Effects of titanium substratum and grooved surface topography on metalloproteinase-2 expression in human fibroblasts. *Journal of Biomedical Materials Research*, **39**, 437–445.

34 Rajnicek, A., Britland, S. and McCaig, C. (1997) Contact guidance of CNS neurites on grooved quartz: influence of groove dimensions, neuronal age and cell type. *Journal of Cell Science*, **110** (Pt 23), 2905–2913.

35 Meyle, J., Wolburg, H. and von Recum, A.F. (1993) Surface micromorphology and cellular interactions. *Journal of Biomaterials Applications*, **7**, 362–374.

36 van Kooten, T.G. and von Recum, A.F. (1999) Cell adhesion to textured silicone surfaces: the influence of time of adhesion and texture on focal contact and fibronectin fibril formation. *Tissue Engineering*, **5**, 223–240.

37 Wojciak-Stothard, B., Curtis, A., Monaghan, W., MacDonald, K. and Wilkinson, C. (1996) Guidance and activation of murine macrophages by nanometric scale topography. *Experimental Cell Research*, **223**, 426–435.

38 Meyle, J., Gultig, K. and Nisch, W. (1995) Variation in contact guidance by human cells on a microstructured surface. *Journal of Biomedical Materials Research*, **29**, 81–88.

39 Wojciak-Stothard, B., Curtis, A.S., Monaghan, W., McGrath, M., Sommer, I. and Wilkinson, C.D. (1995) Role of the cytoskeleton in the reaction of fibroblasts to multiple grooved substrata. *Cell Motility and the Cytoskeleton*, **31**, 147–158.

40 Wojciak-Stothard, B., Madeja, Z., Korohoda, W., Curtis, A. and Wilkinson, C. (1995) Activation of macrophage-like cells by multiple grooved substrata. Topographical control of cell behaviour. *Cell Biology International*, **19**, 485–490.

41 Oakley, C. and Brunette, D.M. (1995) Response of single, pairs, and clusters of epithelial cells to substratum topography. *Biochemistry and Cell Biology*, **73**, 473–489.

42 Webb, A., Clark, P., Skepper, J., Compston, A. and Wood, A. (1995) Guidance of oligodendrocytes and their progenitors by substratum topography. *Journal of Cell Science*, **108** (Pt 8), 2747–2760.

43 Flemming, R.G., Murphy, C.J., Abrams, G.A., Goodman, S.L. and Nealey, P.F. (1999) Effects of synthetic micro- and nano-structured surfaces on cell behavior. *Biomaterials*, **20**, 573–588.

44 Gallagher, J.O., McGhee, K.F., Wilkinson, C.D. and Riehle, M.O. (2002) Interaction of animal cells with ordered nanotopography. *IEEE Transactions on NanoBioscience*, **1**, 24–28.

45 Yim, E.K. and Leong, K.W. (2005) Significance of synthetic nanostructures in dictating cellular response. *Nanomedicine*, **1**, 10–21.

46 Yim, E.K., Wen, J. and Leong, K.W. (2006) Enhanced extracellular matrix production and differentiation of human embryonic germ cell derivatives in biodegradable poly(epsilon-caprolactone-*co*-ethyl ethylene phosphate) scaffold. *Acta Biomaterialia*, **2**, 365–376.

47 Hartgerink, J.D., Beniash, E. and Stupp, S.I. (2001) Self-assembly and mineralization of peptide-amphiphile nanofibers. *Science*, **294**, 1684–1688.

48 Zhang, S. (2002) Emerging biological materials through molecular self-assembly. *Biotechnology Advances*, **20**, 321–339.

49 Zhang, S. (2003) Fabrication of novel biomaterials through molecular self-assembly. *Nature Biotechnology*, **21**, 1171–1178.

50 Zhang, S., Marini, D.M., Hwang, W. and Santoso, S. (2002) Design of nanostructured biological materials through self-assembly of peptides and proteins. *Current Opinion in Chemical Biology*, **6**, 865–871.

51 Hosseinkhani, H., Hosseinkhani, M., Tian, F., Kobayashi, H. and Tabata, Y. (2006) Ectopic bone formation in collagen sponge self-assembled peptide-amphiphile nanofibers hybrid scaffold in a perfusion culture bioreactor. *Biomaterials*, **27**, 5089–5098.

52 Hosseinkhani, H., Hosseinkhani, M., Tian, F., Kobayashi, H. and Tabata, Y. (2006) Osteogenic differentiation of mesenchymal stem cells in self-assembled peptide-amphiphile nanofibers. *Biomaterials*, **27**, 4079–4086.

53 Hartgerink, J.D., Beniash, E. and Stupp, S.I. (2002) Peptide-amphiphile nanofibers: a versatile scaffold for the preparation, of self-assembling materials. *Proceedings of the National Academy of Sciences of the United States of America*, **99**, 5133–5138.

54 Niece, K.L., Hartgerink, J.D., Donners, J.J. and Stupp, S.I. (2003) Self-assembly combining two bioactive peptide-amphiphile molecules into nanofibers by electrostatic attraction. *Journal of the American Chemical Society*, **125**, 7146–7147.

55 Khademhosseini, A., Suh, K.Y., Yang, J.M., Eng, G., Yeh, J., Levenberg, S. and Langer, R. (2004) Layer-by-layer deposition of hyaluronic acid and poly-l-lysine for patterned cell co-cultures. *Biomaterials*, **25**, 3583–3592.

56 Fukuda, J., Khademhosseini, A., Yeh, J., Eng, G., Cheng, J., Farokhzad, O.C. and Langer, R. (2006) Micropatterned cell co-cultures using layer-by-layer deposition of extracellular matrix components. *Biomaterials*, **27**, 1479–1486.

57 Zhu, Y. and Sun, Y. (2004) The influence of polyelectrolyte charges of polyurethane membrane surface on the growth of human endothelial cells. *Colloids and Surfaces. B, Biointerfaces*, **36**, 49–55.

58 Kommireddy, D.S., Sriram, S.M., Lvov, Y.M. and Mills, D.K. (2006) Stem cell attachment to layer-by-layer assembled TiO_2 nanoparticle thin films. *Biomaterials*, **27**, 4296–4303.

59 Wittmer, C.R., Phelps, J.A., Saltzman, W.M. and Van Tassel, P.R. (2007) Fibronectin terminated multilayer films: protein adsorption and cell attachment studies. *Biomaterials*, **28**, 851–860.

60 Pappas, T.C., Wickramanyake, W.M., Jan, E., Motamedi, M., Brodwick, M. and Kotov, N.A. (2007) Nanoscale engineering of a cellular interface with semiconductor nanoparticle films for photoelectric stimulation of neurons. *Nano Letters*, **7**, 513–519.

61 Thierry, B., Winnik, F.M., Merhi, Y. and Tabrizian, M. (2003) Nanocoatings onto arteries via layer-by-layer deposition: toward the *in vivo* repair of damaged blood vessels. *Journal of the American Chemical Society*, **125**, 7494–7495.

62 Zhao, L., Liu, H. and Hu, N. (2006) Assembly of layer-by-layer films of heme proteins and single-walled carbon nanotubes: electrochemistry and electrocatalysis. *Analytical and Bioanalytical Chemistry*, **384**, 414–422.

63 Wu, B.Y., Hou, S.H., Yin, F., Li, J., Zhao, Z.X., Huang, J.D. and Chen, Q. (2007) Amperometric glucose biosensor based on layer-by-layer assembly of multilayer films composed of chitosan, gold nanoparticles and glucose oxidase modified Pt electrode. *Biosensors and Bioelectronics*, **22**, 838–844.

64 Fan, Y.F., Wang, Y.N., Fan, Y.G. and Ma, J.B. (2006) Preparation of insulin nanoparticles and their encapsulation with biodegradable polyelectrolytes via the layer-by-layer adsorption. *International Journal of Pharmaceutics*, **324**, 158–167.

65 Mironov, V., Kasyanov, V., Drake, C. and Markwald, R.R. (2008) Organ printing: promises and challenges. *Regenerative Medicine*, **3**, 93–103.

66 Pastorino, L., Soumetz, F.C. and Ruggiero, C. (2007) Nanostructured thin films for the development of piezoelectric immunosensors. *29th Annual International Conference of the IEEE Engineering in Medicine and Biology Society (EMBS 2007), Volume 1*, pp. 2257–2260.

67 Teramura, Y., Kaneda, Y. and Iwata, H. (2007) Islet-encapsulation in ultra-thin layer-by-layer membranes of poly(vinyl alcohol) anchored to poly(ethylene glycol)-lipids in the cell membrane. *Biomaterials*, **28**, 4818–4825.

68 De Geest, B.G., Sanders, N.N., Sukhorukov, G.B., Demeester, J. and De Smedt, S.C. (2007) Release mechanisms for polyelectrolyte capsules. *Chemical Society Reviews*, **36**, 636–649.

69 Harrison, B.S. and Atala, A. (2007) Carbon nanotube applications for tissue engineering. *Biomaterials*, **28**, 344–353.

70 MacDonald, R.A., Laurenzi, B.F., Viswanathan, G., Ajayan, P.M. and Stegemann, J.P. (2005) Collagen-carbon nanotube composite materials as scaffolds in tissue, engineering. *Journal of Biomedical Materials Research Part A*, **74**, 489–496.

71 Liopo, A.V., Stewart, M.P., Hudson, J., Tour, J.M. and Pappas, T.C. (2006) Biocompatibility of native and functionalized single-walled carbon nanotubes for neuronal interface. *Journal of Nanoscience and Nanotechnology*, **6**, 1365–1374.

72 Lovat, V., Pantarotto, D., Lagostena, L., Cacciari, B., Grandolfo, M., Righi, M., Spalluto, G., Prato, M. and Ballerini, L. (2005) Carbon nanotube substrates boost neuronal electrical signaling. *Nano Letters*, **5**, 1107–1110.

73 Supronowicz, P.R., Ajayan, P.M., Ullmann, K.R., Arulanandam, B.P., Metzger, D.W. and Bizios, R. (2002) Novel current-conducting composite substrates for exposing osteoblasts to alternating current stimulation. *Journal of Biomedical Materials Research*, **59**, 499–506.

74 Bianco, A., Kostarelos, K. and Prato, M. (2005) Applications of carbon nanotubes in drug delivery. *Current Opinion in Chemical Biology*, **9**, 674–679.

75 Venkatesan, N., Yoshimitsu, J., Ito, Y., Shibata, N. and Takada, K. (2005) Liquid filled nanoparticles as a drug delivery tool for protein therapeutics. *Biomaterials*, **26**, 7154–7163.

76 LaVan, D.A., McGuire, T. and Langer, R. (2003) Small-scale systems for *in vivo* drug delivery. *Nature Biotechnology*, **21**, 1184–1191.

77 Singh, R., Pantarotto, D., McCarthy, D., Chaloin, O., Hoebeke, J., Partidos, C.D., Briand, J.P., Prato, M., Bianco, A. and Kostarelos, K. (2005) Binding and condensation of plasmid DNA onto functionalized carbon nanotubes: toward the construction of nanotube-based gene delivery vectors. *Journal of the American Chemical Society*, **127**, 4388–4396.

78 Dobson, J. (2006) Gene therapy progress and prospects: magnetic nanoparticle-based gene delivery. *Gene Therapy*, **13**, 283–287.

79 Gao, L., Nie, L., Wang, T., Qin, Y., Guo, Z., Yang, D. and Yan, X. (2006) Carbon nanotube delivery of the GFP gene into mammalian cells. *Chembiochem: A European Journal of Chemical Biology*, **7**, 239–242.

80 Shi Kam, N.W., Jessop, T.C., Wender, P.A. and Dai, H. (2004) Nanotube molecular transporters: internalization of carbon nanotube-protein conjugates into mammalian cells. *Journal of the American Chemical Society*, **126**, 6850–6851.

81 Kam, N.W. and Dai, H. (2005) Carbon nanotubes as intracellular protein transporters: generality and biological functionality. *Journal of the American Chemical Society*, **127**, 6021–6026.

82 Wu, W., Wieckowski, S., Pastorin, G., Benincasa, M., Klumpp, C., Briand, J.P., Gennaro, R., Prato, M. and Bianco, A. (2005) Targeted delivery of amphotericin B to cells by using functionalized carbon nanotubes. *Angewandte Chemie - International Edition in English*, **44**, 6358–6362.

83 Zanello, L.P., Zhao, B., Hu, H. and Haddon, R.C. (2006) Bone cell proliferation on carbon nanotubes. *Nano Letters*, **6**, 562–567.

84 Mattson, M.P., Haddon, R.C. and Rao, A.M. (2000) Molecular functionalization of carbon nanotubes and use as substrates for neuronal growth. *Journal of Molecular Neuroscience*, **14**, 175–182.

85 Sato, Y., Yokoyama, A., Shibata, K., Akimoto, Y., Ogino, S., Nodasaka, Y., Kohgo, T., Tamura, K., Akasaka, T., Uo, M., Motomiya, K., Jeyadevan, B., Ishiguro, M., Hatakeyama, R., Watari, F. and Tohji, K. (2005) Influence of length on cytotoxicity of multi-walled carbon nanotubes against human acute monocytic leukemia cell line THP-1 in vitro and subcutaneous tissue of rats *in vivo*. *Molecular BioSystems*, **1**, 176–182.

86 Raja, P.M., Connolley, J., Ganesan, G.P., Ci, L., Ajayan, P.M., Nalamasu, O. and Thompson, D.M. (2007) Impact of carbon nanotube exposure, dosage and aggregation on smooth muscle cells. *Toxicology Letters*, **169**, 51–63.

87 Chou, C.C., Hsiao, H.Y., Hong, Q.S., Chen, C.H., Peng, Y.W., Chen, H.W. and Yang, P.C. (2008) Single-walled carbon nanotubes can induce pulmonary injury in mouse model. *Nano Letters*, **8**, 437–445.

88 Frangioni, J.V. and Hajjar, R.J. (2004) *In vivo* tracking of stem cells for clinical trials in cardiovascular disease. *Circulation*, **110**, 3378–3383.

89 Shapiro, E.M., Sharer, K., Skrtic, S. and Koretsky, A.P. (2006) *In vivo* detection of single cells by MRI. *Magnetic Resonance in Medicine*, **55**, 242–249.

90 Shapiro, E.M., Gonzalez-Perez, O., Manuel Garcia-Verdugo, J., Alvarez-Buylla, A. and Koretsky, A.P. (2006) Magnetic resonance imaging of the migration of neuronal precursors generated in the adult rodent brain. *NeuroImage*, **32**, 1150–1157.

91 Weissleder, R., Elizondo, G., Wittenberg, J., Rabito, C.A., Bengele, H.H. and Josephson, L. (1990) Ultrasmall superparamagnetic iron oxide: characterization of a new class of contrast agents for MR imaging. *Radiology*, **175**, 489–493.

92 Briley-Saebo, K.C., Johansson, L.O., Hustvedt, S.O., Haldorsen, A.G., Bjornerud, A., Fayad, Z.A. and Ahlstrom, H.K. (2006) Clearance of iron oxide particles in rat liver: effect of hydrated particle size and coating material on liver metabolism. *Investigative Radiology*, **41**, 560–571.

93 Wilhelm, C., Billotey, C., Roger, J., Pons, J.N., Bacri, J.C. and Gazeau, F. (2003) Intracellular uptake of anionic superparamagnetic nanoparticles as a function of their surface coating. *Biomaterials*, **24**, 1001–1011.

94 Frank, J.A., Miller, B.R., Arbab, A.S., Zywicke, H.A., Jordan, E.K., Lewis, B.K., Bryant, L.H. Jr. and Bulte, J.W. (2003) Clinically applicable labeling of mammalian and stem cells by combining superparamagnetic iron oxides and transfection agents. *Radiology*, **228**, 480–487.

95 Frank, J.A., Anderson, S.A., Kalsih, H., Jordan, E.K., Lewis, B.K., Yocum, G.T. and Arbab, A.S. (2004) Methods for magnetically labeling stem and other cells for detection by *in vivo* magnetic resonance imaging. *Cytotherapy*, **6**, 621–625.

96 Hoehn, M., Kustermann, E., Blunk, J., Wiedermann, D., Trapp, T., Wecker, S., Focking, M., Arnold, H., Hescheler, J., Fleischmann, B.K., Schwindt, W. and Buhrle, C. (2002) Monitoring of implanted stem cell migration *in vivo*: a highly resolved, *in vivo* magnetic resonance imaging investigation of experimental stroke in rat. *Proceedings of the National Academy of Sciences of the United States of America*, **99**, 16267–16272.

97 Sykova, E. and Jendelova, P. (2005) Magnetic resonance tracking of implanted adult and embryonic stem cells in injured brain and spinal cord. *Annals of the New York Academy of Sciences*, **1049**, 146–160.

98 Stroh, A., Faber, C., Neuberger, T., Lorenz, P., Sieland, K., Jakob, P.M., Webb, A., Pilgrimm, H., Schober, R., Pohl, E.E. and Zimmer, C. (2005) *In vivo* detection limits of magnetically labeled embryonic stem cells in the rat brain using high-field (17.6 T) magnetic resonance imaging. *NeuroImage*, **24**, 635–645.

99 Zhu, M., Zhong, Y.M., Li, Y.H., Sun, A.M. and Jin, B. (2005) Congenital aortic arch anomalies: diagnosis using contrast enhanced magnetic resonance angiography. *Chinese Medical Journal*, **118**, 1751–1753.

100 Lewin, M., Carlesso, N., Tung, C.H., Tang, X.W., Cory, D., Scadden, D.T. and Weissleder, R. (2000) Tat peptide-derivatized magnetic nanoparticles allow *in vivo* tracking and recovery of progenitor cells. *Nature Biotechnology*, **18**, 410–414.

101 Huang, D.M., Hung, Y., Ko, B.S., Hsu, S.C., Chen, W.H., Chien, C.L., Tsai, C.P., Kuo, C.T., Kang, J.C., Yang, C.S., Mou, C.Y. and Chen, Y.C. (2005) Highly efficient cellular labeling of mesoporous nanoparticles in, human mesenchymal stem cells: implication for stem cell tracking. *The FASEB Journal*, **19**, 2014–2016.

102 Bouzigues, C., Levi, S., Triller, A. and Dahan, M. (2007) Single quantum dot tracking of membrane receptors. *Methods in Molecular Biology (Clifton, NJ)*, **374**, 81–92.

103 Courty, S., Luccardini, C., Bellaiche, Y., Cappello, G. and Dahan, M. (2006) Tracking individual kinesin motors in living cells using single quantum-dot imaging. *Nano Letters*, **6**, 1491–1495.

104 Courty, S., Bouzigues, C., Luccardini, C., Ehrensperger, M.V., Bonneau, S. and Dahan, M. (2006) Tracking individual proteins in living cells using single quantum dot imaging. *Methods in Enzymology*, **414**, 211–228.

105 Foster, K.A., Galeffi, F., Gerich, F.J., Turner, D.A. and Muller, M. (2006) Optical and pharmacological tools to investigate the role of mitochondria during oxidative stress and neurodegeneration. *Progress in Neurobiology*, **79**, 136–171.

106 Dahan, M., Levi, S., Luccardini, C., Rostaing, P., Riveau, B. and Triller, A. (2003) Diffusion dynamics of glycine receptors revealed by single-quantum dot tracking. *Science*, **302**, 442–445.

107 Schwartz, M.P., Derfus, A.M., Alvarez, S.D., Bhatia, S.N. and Sailor, M.J. (2006) The smart Petri dish: a nanostructured photonic crystal for real-time monitoring of living cells. *Langmuir*, **22**, 7084–7090.

108 Gu, W., Pellegrino, T., Parak, W.J., Boudreau, R., Le Gros, M.A., Gerion, D., Alivisatos, A.P. and Larabell, C.A. (2005) Quantum-dot-based cell motility assay. *Science's STKE: Signal Transduction Knowledge Environment*, **2005**, l5.

109 Garon, E.B., Marcu, L., Luong, Q., Tcherniantchouk, O., Crooks, G.M. and Koeffler, H.P. (2007) Quantum dot labeling and tracking of human leukemic, bone marrow and cord blood cells. *Leukemia Research*, **31**, 643–651.

110 Giepmans, B.N., Deerinck, T.J., Smarr, B.L., Jones, Y.Z. and Ellisman, M.H. (2005) Correlated light and electron microscopic imaging of multiple endogenous proteins using quantum dots. *Nature Methods*, **2**, 743–749.

111 Zhou, M., Nakatani, E., Gronenberg, L.S., Tokimoto, T., Wirth, M.J., Hruby, V.J., Roberts, A., Lynch, R.M. and Ghosh, I. (2007) Peptide-labeled quantum dots for imaging GPCRs in whole cells and as single molecules. *Bioconjugate Chemistry*, **18**, 323–332.

112 Bannai, H., Levi, S., Schweizer, C., Dahan, M. and Triller, A. (2006) Imaging the lateral diffusion of membrane molecules with quantum dots. *Nature Protocols*, **1**, 2628–2634.

113 Wu, X., Liu, H., Liu, J., Haley, K.N., Treadway, J.A., Larson, J.P., Ge, N., Peale, F. and Bruchez, M.P. (2003) Immunofluorescent labeling of cancer marker Her2 and other cellular targets with semiconductor quantum dots. *Nature Biotechnology*, **21**, 41–46.

114 Hernandez-Sanchez, B.A., Boyle, T.J., Lambert, T.N., Daniel-Taylor, S.D., Oliver, J.M., Wilson, B.S., Lidke, D.S. and Andrews, N.L. (2006) Synthesizing biofunctionalized nanoparticles to image cell signaling pathways. *IEEE Transactions on NanoBioscience*, **5**, 222–230.

115 Funnell, W.R. and Maysinger, D. (2006) Three-dimensional reconstruction of cell nuclei, internalized quantum dots and sites of lipid peroxidation. *Journal of Nanobiotechnology*, **4**, 10.

116 Jaiswal, J.K. and Simon, S.M. (2007) Optical monitoring of single cells using quantum dots. *Methods in Molecular Biology (Clifton, NJ)*, **374**, 93–104.

117 Jaiswal, J.K., Mattoussi, H., Mauro, J.M. and Simon, S.M. (2003) Long-term multiple color imaging of live cells using quantum dot bioconjugates. *Nature Biotechnology*, **21**, 47–51.

118 Chan, W.C. and Nie, S. (1998) Quantum dot bioconjugates for ultrasensitive nonisotopic detection. *Science*, **281**, 2016–2018.

119 Mattheakis, L.C., Dias, J.M., Choi, Y.J., Gong, J., Bruchez, M.P., Liu, J. and Wang, E. (2004) Optical coding of mammalian cells using semiconductor quantum dots. *Analytical Biochemistry*, **327**, 200–208.

120 Lagerholm, B.C. (2007) Peptide-mediated intracellular delivery of quantum dots. *Methods in Molecular Biology (Clifton, NJ)*, **374**, 105–112.

121 Dubertret, B., Skourides, P., Norris, D.J., Noireaux, V., Brivanlou, A.H. and Libchaber, A. (2002) *In vivo* imaging of quantum dots encapsulated in phospholipid micelles. *Science*, **298**, 1759–1762.

122 Cambi, A., Lidke, D.S., Arndt-Jovin, D.J., Figdor, C.G. and Jovin, T.M. (2007) Ligand-conjugated quantum dots monitor antigen uptake and processing by dendritic cells. *Nano Letters*, **7**, 970–977.

123 Hardman, R. (2006) A toxicologic review of quantum dots: toxicity depends on physicochemical and environmental factors. *Environmental Health Perspectives*, **114**, 165–172.

124 Hoshino, A., Fujioka, K., Oku, T., Nakamura, S., Suga, M., Yamaguchi, Y., Suzuki, K., Yasuhara, M. and Yamamoto, K. (2004) Quantum dots targeted to the assigned organelle in living cells. *Microbiology and Immunology*, **48**, 985–994.

14
Self-Assembling Peptide-Based Nanostructures for Regenerative Medicine

Ramille M. Capito, Alvaro Mata, and Samuel I. Stupp

14.1
Introduction

The goal of regenerative medicine is to develop therapies that can promote the growth of tissues and organs in need of repair as a result of trauma, disease or congenital defects. For most of the patient population this means regeneration of our bodies in adulthood, although there are also many critical pediatric needs in regenerative medicine. One specific target that would deeply impact the human condition is regeneration of the central nervous system (CNS). This would bring a higher quality of life to individuals paralyzed as a result of spinal cord injury, brought into serious dysfunction by stroke, afflicted with Parkinson's and Alzheimer's diseases, or those blind as a result of macular degeneration or retinitis pigmentosa. Another area that would benefit from regenerative medicine is heart disease, which continues to be one of the most dominant sources of premature death in humans. Here, the potential to regenerate myocardium would have a great impact on clinical outcomes. Many additional important targets exist. The regeneration of insulin-producing pancreatic β cells would bring a higher quality of life to individuals suffering from diabetes. Damage to cartilage – a critical tissue in correct joint function – is an enormous source of pain and compromised agility for many individuals, especially in societies that value a physically active life style for as long as possible. Other musculoskeletal tissues such as bone, intervertebral disc, tendon, meniscus and ligament all remain major therapeutic challenges in regenerative medicine. Another emerging target that could have an enormous impact is the regeneration of teeth, as this would prevent the need for dentures and other dental implants. All of these important targets in regenerative medicine would not only raise the quality of life for many individuals worldwide, but they would also have, for obvious reasons, a significant economic impact.

The development of effective regenerative medicine strategies generally includes the use of cells, soluble regulators (e.g. growth factors or genes) and

Nanotechnology, Volume 5: Nanomedicine. Edited by Viola Vogel

ISBN: 978-3-527-31736-3

scaffold technologies. In their natural environment, mammalian cells live surrounded by a form of solid or fluid matrix composed of structural protein fibers (i.e. collagen and elastin), adhesive proteins (i.e. fibronectin and laminin), soluble proteins (i.e. growth factors) and other biopolymers (i.e. polysaccharides), all of which have specific inter-related roles in the structure and normal function of the extracellular matrix (ECM). The creation of biomimetic artificial matrices represents a common theme in designing materials for regenerative medicine therapies, and stems from the idea that providing a more natural three-dimensional (3-D) environment can preserve cell viability and encourage cell differentiation and matrix synthesis. The nanoscale design of biomaterials, with particular attention to dimension, shape, internal structure and surface chemistry, may more effectively emulate the very sophisticated architecture and signaling machinery of the natural ECM for improved regeneration.

Strategies utilizing self-assembled supramolecular aggregates, macromolecules and even inorganic particles could be used to design a signaling machinery *de novo* that initiates regeneration events which do not occur naturally in mammalian biology. Self-assembly – a bioinspired phenomenon which involves the spontaneous association of disordered components into well-defined and functionally organized structures [1] – can play a major role in creating sophisticated and biomimetic biomaterials for regenerative therapies [2–7]. In molecular systems, self-assembly implies that molecules are programmed by design to organize spontaneously into supramolecular structures held together through noncovalent interactions, such as electrostatic or ionic interactions, hydrogen bonding, hydrophobic interactions and van der Waals interactions. Large collections of these relatively weak bonds compared to covalent bonds can result in very stable structures.

The first fundamental reason for a link between self-assembly and regenerative medicine is the potential to create multifunctional artificial forms of an ECM starting with liquids. Such liquids could contain dissolved molecules or pre-assembled nanostructures, and they could then be introduced by injection at a specific site or targeted through the circulation. Following self-assembly, a solid matrix could mechanically support cells and also signal them for survival, proliferation, differentiation or migration. Alternatively, the self-assembled solid matrix could be designed to recruit specific types of cells in order to promote a regenerative biological event, or serve as cell delivery vehicles by localizing them in 3-D environments within tissues and organs. These self-assembling molecules could also be used to modify the surfaces of solid implants in order to render them bioactive [1, 8, 9]. The 'bottom-up' approach that is possible using self-assembly can permit the creation of an architecture that multiplexes signals or tunes their concentration per unit area. This versatility makes self-assembling systems ideal for creating optimal materials for regenerative medicine therapies.

In this chapter, we focus on the use of self-assembling nanostructures – in particular, peptide-based molecules – which are currently being developed for regenerative medicine applications. Although many of these technologies are relatively new, much very promising biological data – both *in vitro* and *in vivo* – demonstrating the promise of these systems in regenerative medicine is already available. Two currently important research areas in this field include:

- The development of self-assembling injectable bioactive scaffolds that have the ability to mimic the natural 3-D ECM of cells.
- The nanoscale surface modification of surfaces or 3-D tissue engineering scaffolds using self-assembling molecules to create bioactive implants and devices.

14.2 Self-Assembling Synthetic Peptide Scaffolds

Peptides are among the most useful building blocks for creating self-assembled structures at the nanoscale; they possess the biocompatibility and chemical versatility that are found in proteins, yet they are more chemically and thermally stable [10]. They can also be easily synthesized on a large scale by using conventional chemical techniques, and designed to contain functional bioactive sequences. A variety of short peptide molecules have been shown to self-assemble into a wide range of supramolecular structures including nanofibers, nanotubes, nanospheres and nanotapes. Some self-assembling nanostructures have been used successfully to generate injectable scaffolds with an extremely high water content and architectural features that mimic the natural structure of the ECM. These self-assembling scaffolds show great potential as 3-D environments for cell culture and regenerative medicine applications, and also as vehicles for drug, gene or protein delivery.

14.2.1 β-Sheet Peptides

Aggeli and colleagues demonstrated that the biological peptide β-sheet motif can be used to design oligopeptides that self-assemble into semi-flexible β-sheet nanotapes [11]. Depending on the intrinsic chirality of the peptides and concentration, these nanostructures can further assemble into twisted ribbons (double tapes), fibrils (twisted stacks of ribbons) and fibers (fibrils entwined edge-to-edge) (Figure 14.1a) [12]. The assembly process is principally driven by hydrogen bonding along the peptide backbone and interactions between specific side chains [13]. At sufficiently high peptide concentrations, these structures can become entangled to form gels, the viscoelastic properties of which can be altered by controlling the pH, by applying a physical (shear) stress, or by altering the peptide concentration. As in other peptide self-assembling systems, the hierarchical assembly can be altered by the addition and position of charged amino acids within the peptide sequence that is highly controlled by changes in pH [12] (Figure 14.1b and c). It has also been shown that mixing aqueous solutions of cationic and anionic peptides that have complementary charged side chains and a propensity to form antiparallel β-sheets, results in the spontaneous self-assembly of fibrillar networks and hydrogels that are robust to variations in pH and peptide concentration [14].

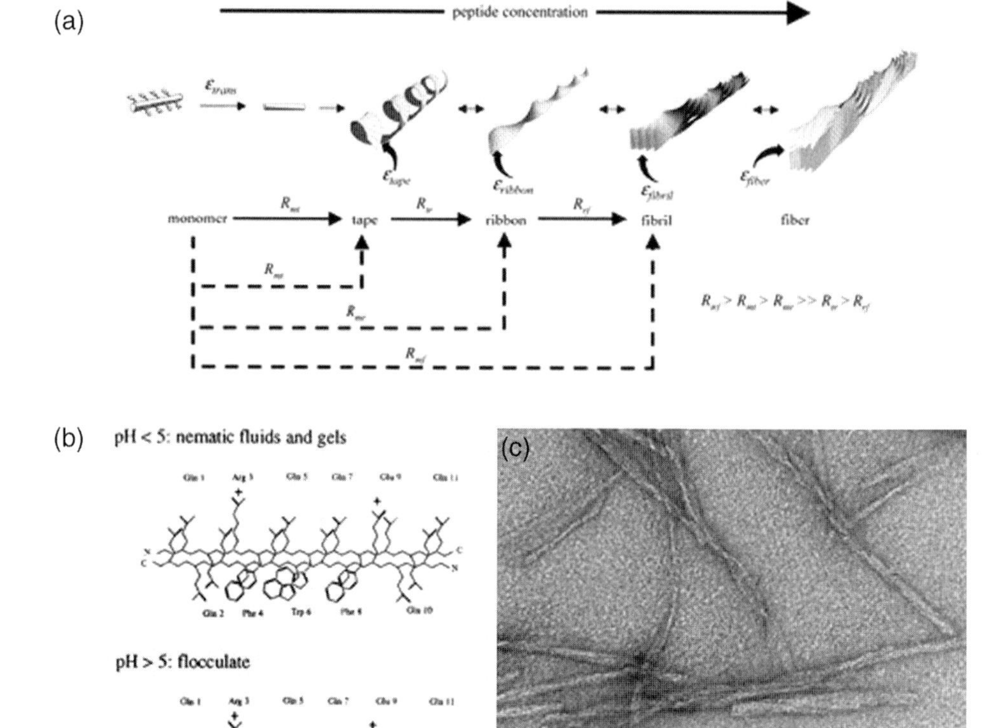

Figure 14.1 (a) The global equilibrium configurations obtained by the hierarchical self-assembly of β-sheet-forming peptides. The set of energy parameters ε_j correspond to the free energy differences per peptide molecule between successive structures. The 'critical' peptide concentrations at which each new configuration begins to appear is determined by the ε_j. The R_{ij} are the conversion rates both between and to the various configurations. The process depicted by solid arrows represents the dissolution route of lyophilized solid at constant pH, while the dashed arrows represent the direct and simultaneous conversion of monomer to tapes, ribbons, fibrils and fibers when the respective critical concentrations governing their self-assembly are instantaneously switched by pH change to values above the absolute peptide concentration in solution; (b) Electrostatic charge distribution on P11-2 dimer in an antiparallel β-sheet tape-like substructure: top, pH < 5; bottom, pH > 5; (c) Transmission electron microscopy image of a gel. (Reproduced with permission from Ref. [12]; © 2003, American Chemical Society.)

These β-sheet peptide nanostructures have been studied for the treatment of enamel decay [13]. *In vitro*, extracted human premolar teeth (containing caries-like lesions) were exposed to several cycles of demineralizing (acidic conditions) and remineralizing solutions (neutral pH conditions). Application of the self-assembling peptides to the defects significantly decreased demineralization during exposure to acid and increased remineralization at neutral pH, resulting in a net mineral gain of the lesions compared to untreated controls [13]. Furthermore, when the peptide gels were incubated for one week in mineralizing solutions, *de novo* nucleation of hydroxyapatite by the nanostructures was observed [13].

In another application, these peptides were evaluated as an alternative injectable joint lubricant to hyaluronic acid (HA) for the treatment of osteoarthritis [15]. A β-sheet peptide designed to have molecular, mesoscopic and rheological properties that most closely resembled HA, performed similarly to HA in healthy static and dynamic friction tests, but not as well in friction tests with damaged cartilage [15]. The optimization of these peptides may result in a new alternative viscosupplementation treatment for degenerative osteoarthritis.

14.2.2
β-Hairpin Peptides

Another peptide design that exploits β-sheet nanostructure formation into a hydrogel network is composed of strands of alternating hydrophilic and hydrophobic residues flanking an intermittent tetrapetide [16–21] (Figure 14.2a and b). These peptides are designed so that they are fully dissolved in aqueous solutions in random coil conformations. Under specific stimuli, the molecules can be triggered to fold into a β-hairpin conformation that undergoes rapid self-assembly into a β-sheet-rich, highly crosslinked hydrogel. The molecular folding event – where one face of the β-hairpin structure is lined with hydrophobic valines and the other with hydrophilic lysines – is governed by the arrangement of polar and nonpolar residues within the sequence. Subsequent self-assembly of the individual hairpins occurs by hydrogen bonding between distinct hairpins and hydrophobic association of the hydrophobic faces. One such peptide was designed to self-assemble under specific pH conditions [16]. Under basic conditions, the peptide intramolecularly folds into the hairpin structure and forms a hydrogel. Unfolding of the hairpins and dissociation of the hydrogel structure can be triggered when the pH is subsequently lowered below the pK_a of the lysine side chains, where unfolding is a result of the intrastrand charge repulsion between the lysine residues [16]. Rheological studies indicate that these β-hairpin hydrogels are both rigid and shear-thinning; however, the mechanical strength is quickly regained after cessation of shear [16] (Figure 14.2a).

These gels can also be triggered to self-assemble when the charged amino acid residues within the sequence are screened by ions [22]. If a positively charged side chain of lysine is replaced by a negatively charged side chain of glutamic acid, the overall peptide charge is decreased and the peptide can be more easily screened, resulting in a much faster self-assembly [21]. The kinetics of hydrogelation were found to be significant for the homogeneous distribution of encapsulated cells within these types of self-assembling gels [21] (Figure 14.2c). Thermally reversible, self-assembling peptides were also synthesized by replacing specific valine residues with threonines to render the peptides less hydrophobic [17]. At ambient temperature and slightly basic pH, the peptide is unfolded; however, upon heating the peptide

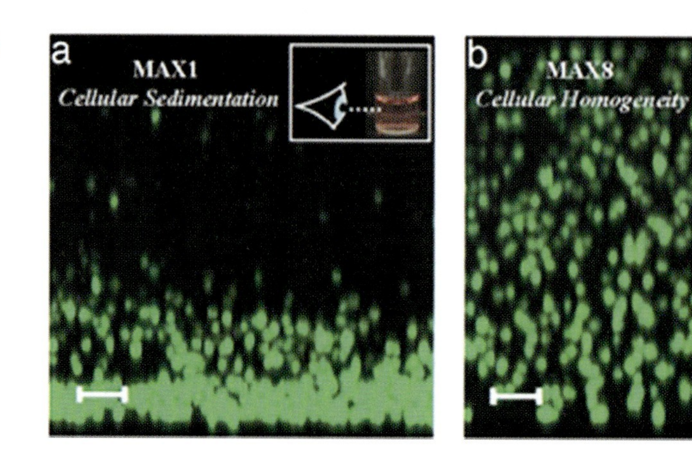

Figure 14.2 Self-assembly, shear-thinning and self-healing mechanism allowing rapid formation of β-hairpin hydrogels that can be subsequently syringe-delivered. (a) Addition of Dulbecco's modified Eagle medium (DMEM; pH 7.4, 37 °C) to a buffered solution of unfolded peptide induces the formation of a β-hairpin structure that undergoes lateral and facial self-assembly affording a rigid hydrogel with a fibrillar supramolecular structure. Subsequent application of shear stress disrupts the noncovalently stabilized network, leading to the conversion of hydrogel to a low-viscosity gel. Upon cessation of shear stress, the network structure recovers, converting the liquid back to a rigid hydrogel; (b) Peptide sequences of MAX8 and MAX1; (c) Encapsulation of mesenchymal C3H10t1/2 stem cells in 0.5 wt% MAX1 and MAX8 hydrogels. Shown are LSCM z-stack images (viewed along the y-axis) showing the incorporation of cells into a MAX1 gel leading to cell sedimentation (panel a) and into a MAX8 gel resulting in cellular homogeneity (panel b). Cells are prelabeled with cell tracker green to aid visualization (scale bars = 100 μm). (Reproduced with permission from Ref. [21]; © 2007, National Academy of Sciences.)

folds and assembles via hydrophobic collapse as the temperature dehydrates the nonpolar residues within the peptide [17]. Yet another rendition of this peptide self-assembles via light activation [20]. In this design, a photocaged peptide is incorporated within the peptide sequence, with β-hairpin folding and subsequent hydrogelation occurring only when the photocage is released upon irradiation of the sample [20].

In vitro studies have shown that these types of β-hairpin hydrogels can support the survival, adhesion and migration of NIH 3T3 fibroblasts [20–22], and can be used to encapsulate mesenchymal stem cells (MSCs) and hepatocytes [21]. Another study also showed that these gels have an inherent antibacterial activity, with selective toxicity to bacterial cells versus mammalian cells [23].

14.2.3 Block Copolypeptides

Deming and colleagues have developed diblock copolypeptide amphiphiles containing charged and hydrophobic segments that self-assemble into rigid hydrogels and can remain mechanically stable even at high temperatures (up to 90 °C) [24, 25] (Figure 14.3). These hydrogels were also found to recover rapidly after an applied stress, attributed to the relatively low molecular mass of the copolypetides, enabling rapid molecular organization. Their amphiphilic characteristics, architecture (diblock versus triblock) and block secondary structure (e.g. α-helix, β-strand or random coil) were found to play important roles in the gelation, rheological and morphological properties of the hydrogel [24–26]. One type of block copolypeptide consists of a hydrophobic block of poly-L-lysine and a shorter hydrophobic block of poly-L-leucine [26]. The helical secondary structure of the poly-L-leucine blocks was shown to be instrumental for gelation, while the hydrophilic polyelectrolyte segments helped to stabilize the twisted fibril assemblies by forming a corona around the hydrophobic core [26] (Figure 14.3).

In vitro studies using mouse fibroblasts revealed that, at concentrations below gelation, lysine-containing diblocks were cytotoxic to the cells, whereas glutamic

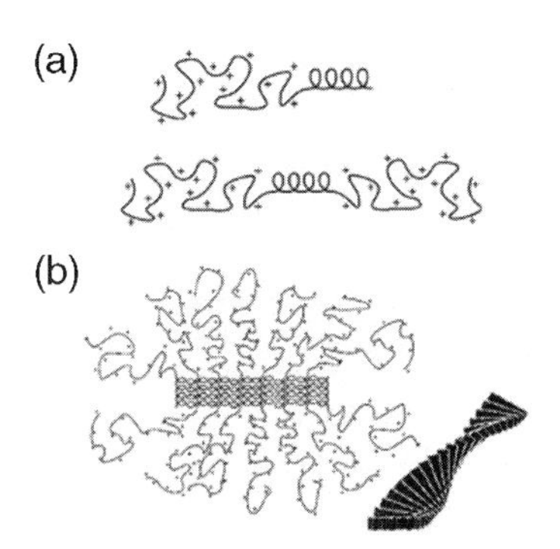

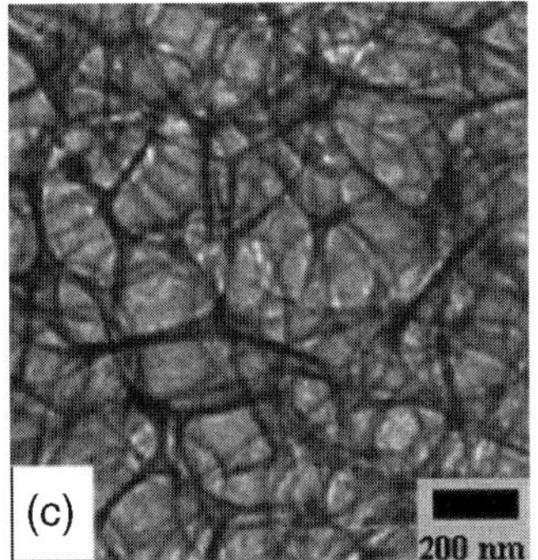

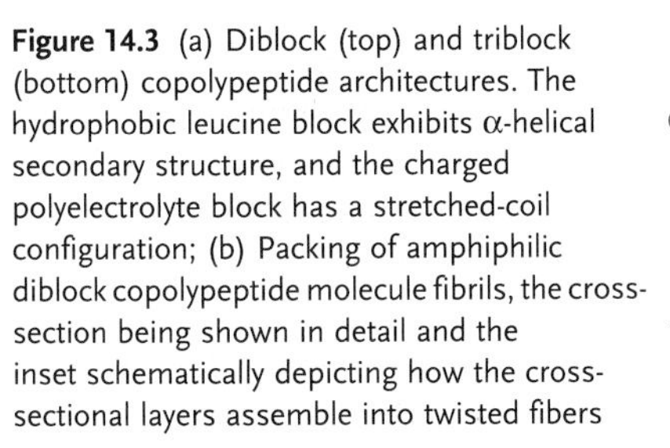

Figure 14.3 (a) Diblock (top) and triblock (bottom) copolypeptide architectures. The hydrophobic leucine block exhibits α-helical secondary structure, and the charged polyelectrolyte block has a stretched-coil configuration; (b) Packing of amphiphilic diblock copolypeptide molecule fibrils, the cross-section being shown in detail and the inset schematically depicting how the cross-sectional layers assemble into twisted fibers (for clarity, only the helices are drawn). (Reproduced with permission from Ref. [26]; © 2004, American Chemical Society.); (c) Cryogenic transmission electron microscopy image of 5.0 wt% $K_{180}(LV)_{20}$ showing the interconnected membrane, cellular nanostructure of gel matrix (dark) surrounded/filled by vitreous water (light). (Reproduced with permission from Ref. [25]; © 2002, American Chemical Society.)

acid-containing peptides were not cytotoxic [27]. In gel form, however, both lysine and glutamic acid-based diblocks were noncytotoxic, although the scaffolds did not support cell attachment or proliferation. This demonstrates how molecular design and charge can significantly affect the cytoxicity and biological activity of the resulting self-assembled material. Future research is directed towards covalently incorporating bioactive sites within these hydrogels in order to increase cellular attachment and enhance the biological response [27].

14.2.4 Ionic Self-Complementary Peptides

Another class of self-assembling peptide molecules developed by Zhang *et al.* was designed to include alternating positive and negative amino acid repeats within the peptide sequence [28, 29]. These oligopeptides associate to form stable fibrillar nanostructures in aqueous solution through β-sheet formation, due to their hydrophilic and hydrophobic surfaces and complementary ionic bonding between the oppositely charged residues. Upon addition of monovalent cations or physiological media, they form hydrogels composed of interwoven nanofibers (Figure 14.4a) [29, 30]. Studies have shown that oligopeptide length [31] and side-chain hydrophobicity [32] were important variables that affected the self-assembly and the resulting gel properties.

Several *in vitro* and *in vivo* studies have been conducted investigating the ability of these scaffolds to support cell attachment [29], survival, proliferation and differentiation for neural [30, 33–35], blood vessel [36–38], myocardial [39–42], liver [43], cartilage [44] and bone tissue regeneration [45, 46]. For the treatment of neural defects, primary mouse neuron cells encapsulated within the hydrogels were able to attach to the nanofiber matrix and showed extensive neurite outgrowth [30]. Furthermore, peptides implanted *in vivo* did not elicit a measurable immune response or tissue inflammation [30]. In a hamster model, the peptide scaffolds were shown to regenerate axons and reconnect target tissues in a severed optic tract that resulted in the restoration of visual function [33]. Likewise, the peptide scaffolds caused an effective promotion of cell migration, blood vessel growth and axonal elongation when implanted with neural progenitor cells and Schwann cells in the transected dorsal column of the rat spinal cord [47] (Figure 14.4b).

For the treatment of myocardial infarction, Davis *et al.* injected peptide scaffolds into rat myocardium and observed the recruitment of endothelial progenitor cells and vascular smooth muscle cells into the injection site that appeared to form functional vascular structures [40]. Biotinylated nanofibers were subsequently used to deliver insulin-like growth factor 1 (IGF-1) *in vivo* over prolonged periods (28 days) and were shown to significantly improve systolic function after myocardial infarction when delivered with transplanted cardiomyocytes [39]. Other *in vivo* studies, which delivered platelet-derived growth factor (PDGF)-BB with the self-assembling nanofibers in a rat myocardial infarct model, showed decreased cardiomyocyte death, reduced infarct size and a long-term improvement in cardiac performance after infarction, without systemic toxicity [41, 42].

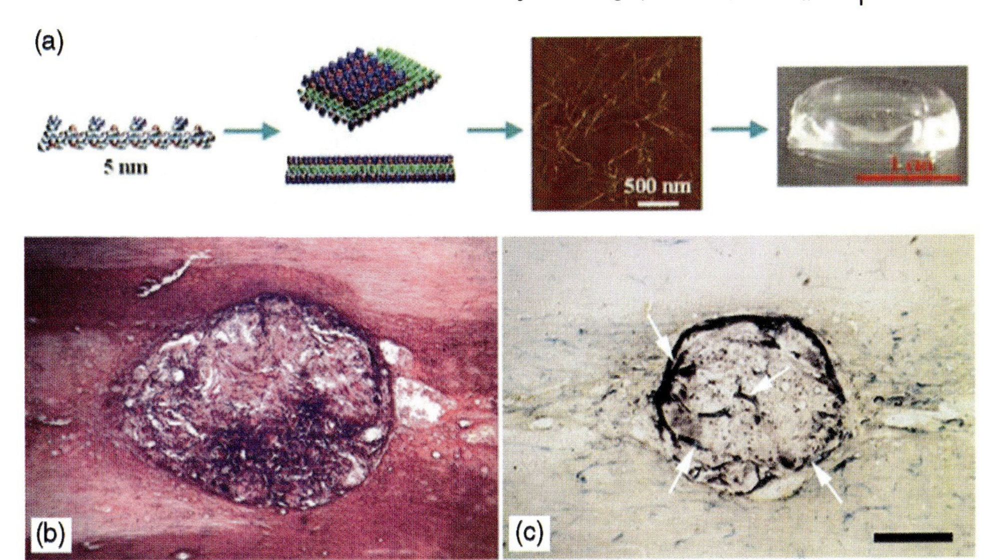

Figure 14.4 (a) Ionic self-complementary peptide consisting of 16 amino acids, ~5 nm in size, with an alternating polar and nonpolar pattern. These peptides form stable β-strand and β-sheet structures the side chains of which partition into two sides, one polar and the other nonpolar. These undergo self-assembly to form nanofibers with the nonpolar residues inside (green) and + (blue) and − (red) charged residues forming complementary ionic interactions, like a checkerboard. These nanofibers form interwoven matrices that further form a scaffold hydrogel with very high water content (>99.5%). (Reproduced with permission from Zhang, S. (2003) Fabrication of novel biomaterials through molecular self-assembly. *Nature Biotechnology*, **21**, 1171–8; © Wiley-VCH Verlag GmbH & Co. KGaA.); (b, c) Implantation of precultured peptide gels into the injured spinal cord of GFP-transgenic rats. (b) Hematoxylin and eosin staining showed a high level of integration between the implants and host, although in most cases a few small cysts were found near the implants; (c) Alkaline phosphatase histochemistry staining showed that blood vessels grew into the implants (arrows). Scale bar = 500 μm. (Reproduced with permission from Ref. [34]; © 2007, Elsevier Limited.)

14.2.5
Fmoc Peptides

A more recently developed class of self-assembling peptides that uses fluorenylmethyloxycarbonyl (Fmoc) -protected di- and tri-peptides have been shown to form highly tunable hydrogel structures (Figure 14.5a). The formation of these gels can be achieved either by pH adjustment [48] (Figure 14.5b) or by a reverse-hydrolysis enzyme action [49] (Figure 14.5c). Assembly occurs via hydrogen bonding in β-sheet arrangement and by π–π stacking of the fluorenyl rings that also stabilize the system [48] (Figure 14.5b). A number of sheets then twist together to form nanotubes (Figure 14.5d).

The results of *in vitro* studies indicated that these hydrogels can support chondrocyte survival and proliferation in both two and three dimensions [48]. It was also observed that cell morphology varied according to the nature of the molecular structure [48].

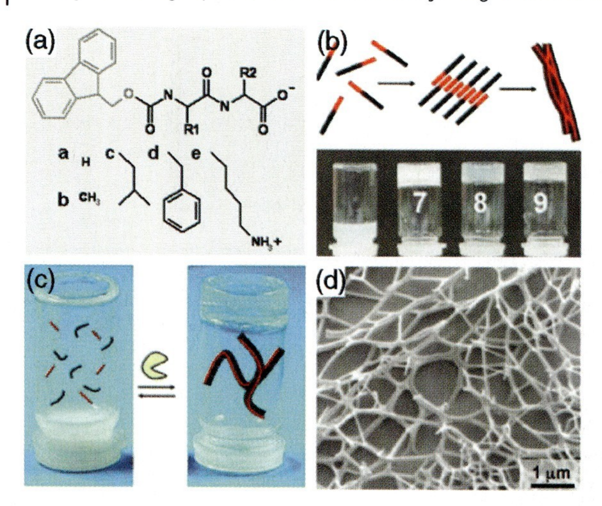

Figure 14.5 (a) Molecular structure of Fmoc–dipeptides. The R groups are the amino acids Gly (a), Ala (b), Leu (c), Phe (d) and Lys (e); (b) Proposed self-assembly mechanism (top): Fmoc groups stack through π–π interactions, and the resulting molecular stacks further assemble to form nanofibers. Self-supporting gels can be formed by manipulation of pH or by reverse-hydrolysis enzyme action (c); (d) Cryogenic scanning electron microscopy image of nanofibrous material obtained by self-assembly. (Panels (a)–(c) reproduced with permission from Ref. [49]; © 2006, American Chemical Society; panel (d) reproduced with permission from Ref. [48]; © 2007, The Biochemical Society.)

14.2.6
Peptide Amphiphiles

Peptide amphiphiles (PAs) are self-assembling molecules that also use hydrophobic and hydrophilic elements to drive self-assembly. There are different types of peptide amphiphiles that can assemble into a variety of nanostructures such as spherical micelles, fibrils, tubes or ribbons [50]. One unique PA design, which forms high-aspect ratio cylindrical nanofibers, has been exclusively studied during the past decade for regenerative medicine applications. These molecules are particularly distinguished from the other peptide systems described above, in that their amphiphilic nature derives from the incorporation of a hydrophilic head group and a hydrophobic alkyl tail, as opposed to molecules consisting of all amino acid residues with resultant hydrophilic and hydrophobic faces.

Stupp and colleagues have developed a family of amphiphilic molecules that can self-assemble from aqueous media into 3-D matrices composed of supramolecular nanofibers [4–6, 9, 51–59]. These molecules consist of a hydrophilic peptide segment which is bound covalently to a highly hydrophobic alkyl tail found in ordinary lipid molecules. The alkyl tail can be located at either the C or N terminus [51], and can also be constructed to contain branched structures [55]. In Stupp *et al.*'s specific design, the peptide region contains a β-sheet-forming peptide domain close to the hydrophobic segment and a bioactive peptide sequence (Figure 14.6a). Upon changes in

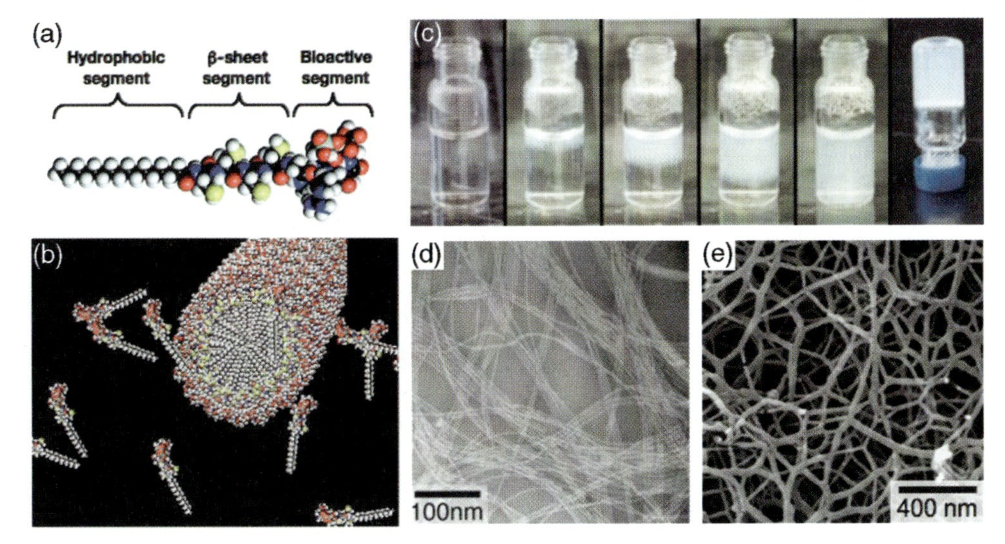

Figure 14.6 (a) General structure of PA molecules; (b) Illustrated self-assembly of PA molecules into nanofibers with hydrophobic cores; (c) Time sequence of pH-controlled PA self-assembly. From left to right: PA molecule dissolved in water at a concentration of 0.5% by weight at pH 8 is exposed to HCl vapor. As the acid diffuses into the solution a gel phase is formed, which self-supports upon inversion; (d) Transmission electron microscopy image of PA nanofibers. (Reproduced with permission from Ref. [5]; © 2002, National Academy of Sciences.); (e) Scanning electron microscopy image of PA nanofiber network. (Reproduced with permission from Ref. [7]; © 2005, Materials Research Society.)

pH or the addition of multivalent ions, the structure of these molecules drives their assembly into cylindrical nanofibers through hydrogen bonding into β-sheets and hydrophobic collapse of alkyl tails away from the aqueous environment to create nanofibers with a hydrophobic core (Figure 14.6b). This cylindrical nanostructure allows the presentation of high densities of bioactive epitopes at the surface of the nanofibers [6], whereas, if peptides were assembled into twisted sheets or tubes, this type of orientational biological signaling would not be possible [7]. These systems can also be used to craft nanofibers containing two or more PA molecules that can effectively coassemble, thus offering the possibility of multiplexing different biological signals within a single nanofiber [56].

The presence of a net charge in the peptide sequence ensures that the molecules or small β-sheet aggregates remain dissolved in water, inhibiting self-assembly through coulombic repulsion. Self-assembly and gelation is subsequently triggered when the charged amino acid residues are electrostatically screened or neutralized by pH adjustment, or by the addition of ions (Figure 14.6c–e). The growth of nanofibers can therefore be controlled by changing the pH or raising the concentration of screening electrolytes in the aqueous medium [7]. Growth and bundling of the nanofibers eventually lead to gelation of the PA solution. *In vivo*, ion concentrations present in physiological fluids can be sufficient to induce the formation of PA nanostructures [54]. Thus, a minimally invasive procedure could be designed with these systems through a simple injection of the PA solution that spontaneously self-assembles into a bioactive scaffold at the desired site. Over time, the small molecules composing the nanofibers should biodegrade into amino acids and lipids, thus minimizing the potential problems of toxicity or immune response [54].

The results of both *in vitro* and *in vivo* studies have shown that these PA systems can serve as an effective analogue of the ECM by successfully supporting cell survival and attachment [60, 61], mediating cell differentiation [6] and promoting regeneration *in vivo* [57]. In efforts to address neural tissue regeneration for the repair of a spinal cord injury or treatment of extensive dysfunction as a result of stroke or Parkinson's disease, Stupp and colleagues have designed PAs to display the pentapeptide epitope isoleucine-lysine-valine-alanine-valine (IKVAV). This particular peptide sequence is found in the protein laminin, and has been shown to promote neurite sprouting and to direct neurite growth [6]. When neural progenitor cells were encapsulated within this PA nanofiber network, the cells more rapidly differentiated into neurons compared to using the protein laminin or the soluble peptide (Figure 14.7a and b). The PA scaffold was also found to discourage the development of astrocytes, a type of cell in the CNS which is responsible for the formation of glial scars that prevent regeneration and recovery after spinal cord injury. In this same study, the

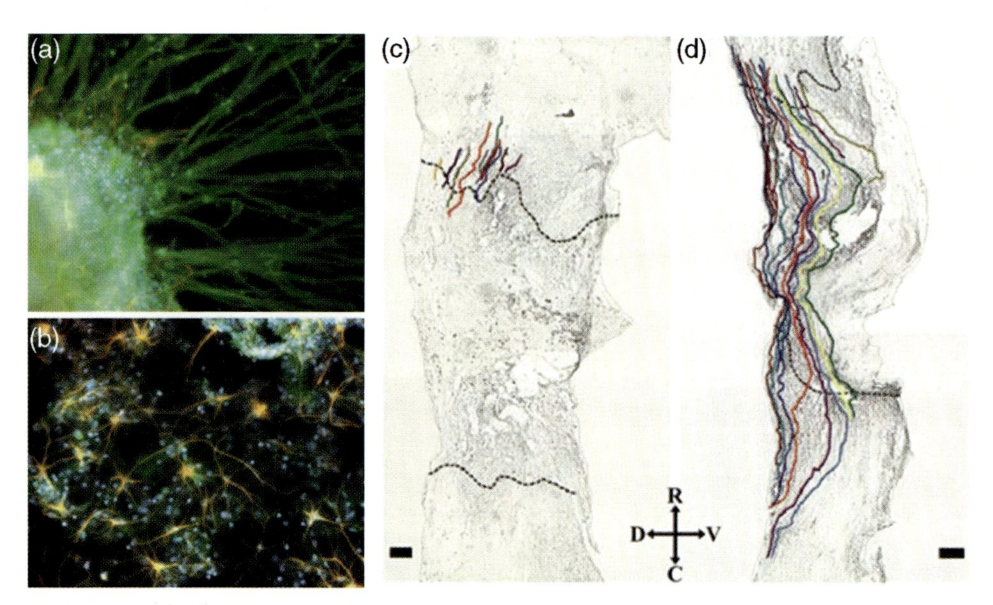

Figure 14.7 (a) Immunocytochemistry of a neuroprogenitor cell neurosphere encapsulated in an IKVAV-PA nanofiber network at 7 days, showing a large extent of neurite outgrowth; (b) Neural progenitor cells cultured on laminin-coated coverslips at 7 days. The prevalence of astrocytes is apparent. (Reproduced with permission from Ref. [6]; © 2004, American Association for the Advancement of Science.); (c, d) Representative Neurolucida tracings of labeled descending motor fibers within a distance of 500 μm rostral of the lesion in vehicle-injected (c) and IKVAV PA-injected (d) animals. The dotted lines demarcate the borders of the lesion. Scale bars = 100 μm. (Reproduced with permission from Ref. [62]; © 2008, Society for Neuroscience.)

density of epitopes displayed on the nanofibers proved to be a significant variable in the ability to induce rapid and selective differentiation of cells encapsulated within the PA gels. Furthermore, *in vivo* studies in which this self-assembling neural nanofiber scaffold was injected within a spinal cord injury in a rat model showed better functional improvement and axonal elongation through the injury site compared to controls [62] (Figure 14.7c and d).

Another PA molecule was designed to self-assemble upon the addition of heparin, a biopolymer that binds to angiogenic growth factors [57]. The resultant nanofibers displayed heparin chains on the periphery, which orient proteins on the surface for cell signaling. In an *in vivo* rabbit corneal model, the heparin-binding PA nanostructures, administered with only nanogram quantities of angiogenic growth factors, was sufficient to stimulate extensive neovascularization compared to controls (Figure 14.8). When using the same PA system, Kapadia *et al.*, was also able to create self-assembling nitric oxide (NO)-releasing nanofiber gels for the prevention of neointimal hyperplasia [63]. Using a rat carotid artery model, the group showed that the NO-releasing PA gels significantly reduced neointimal hyperplasia, inhibited inflammation and stimulated re-endothelialization compared to controls.

The value of this nanotechnology lies in its self-assembly code, which yields nanofibers that can be designed to have a great diversity of bioactive signals [64, 65]. For example, PAs have been successfully synthesized to contain binding groups for growth factors by phage display technology [51]. The inclusion of growth factor binding domains enables a greater retention of incorporated growth factors within the scaffold, or even the 'capture' of desired endogenous growth factors localized

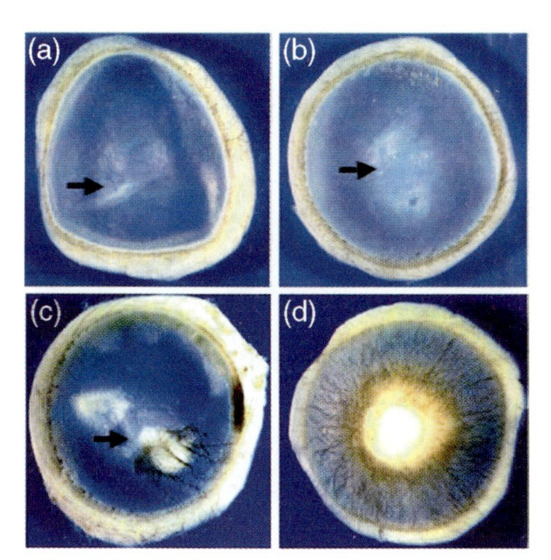

Figure 14.8 *In vivo* angiogenesis assay in a rat cornea 10 days after the placement of various materials at the site indicated by the black arrow. Growth factors alone (a) and heparin with growth factors (b) showed little to no neovascularization. Collagen, heparin and growth factors (c) showed some neovascularization; (d) Heparin-nucleated PA nanofiber networks with growth factors showed extensive neovascularization. (Reproduced with permission from Ref. [57]; © 2006, American Chemical Society.)

at the implant site, thus eliminating the need for exogenous growth factor supplementation altogether. Hartgerink *et al.* synthesized PAs with a combination of biofunctional groups including a cell-mediated enzyme-sensitive site, a calcium-binding site and a cell-adhesive ligand [66]. The incorporation of an enzyme-specific cleavage site allows cell-mediated proteolytic degradation of the scaffold for cell-controlled migration and matrix remodeling. *In vitro* studies demonstrated that these PA scaffolds do degrade in the presence of proteases, and that the morphology of cells encapsulated within the nanofiber scaffolds was dependent on the density of the cell-adhesive ligand, with more elongated cells observed in gels with a higher adhesive ligand density [66].

To date, hundreds of peptide amphiphile nanofibers designs are known, including those that nucleate hydroxyapatite with some of the crystallographic features found in bone [4], increase the survival of cultured islets for the treatment of diabetes, bind to various growth factors [51], contain integrin-binding sequences [61], incorporate contrast agents for fate mapping of PA nanostructures [52], and have pro-apoptotic sequences for cancer therapy, among many others. Research investigating the development of hybrid materials using these versatile PA systems is also emerging. For example, PA nanofibers were integrated within titanium foams as a means to promote bone ingrowth or bone adhesion for improved orthopedic implant fixation (Figure 14.9) [1]. Preliminary *in vivo* results implanting these PA–Ti hybrids within bone defects in a rat femur demonstrated *de novo* bone formation around and inside the implant, vascularization in the vicinity of the implant, and no cytotoxic response [1]. Another type of hybrid system developed by Hartgerink *et al.* includes hydrogels that contain a mixture of PA and phospholipid (Figure 14.10) [67]. The phospholipid inclusions within the PA nanostructure were found to modulate

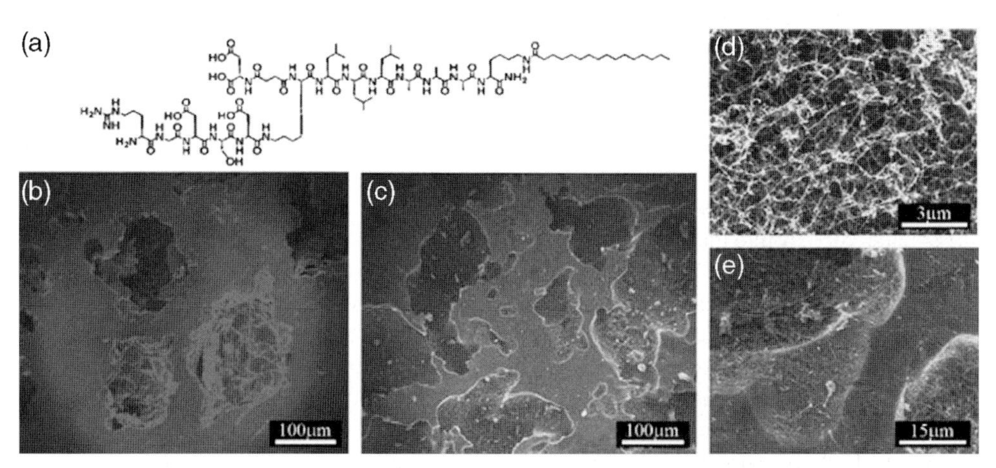

Figure 14.9 (a) Chemical structure of the peptide amphiphile (PA) used to infiltrate and fill the pores of the Ti–6Al–4V foam. Scanning electron microscopy (SEM) images of (b) the bare Ti–6Al–4V foam; (c) Ti–6Al–4V foam filled with PA gel; (d) higher magnification of the self-assembled PA nanofibers forming a 3-D matrix within the pores; (e) Higher magnification of the PA coating the Ti–6Al–4V foam surface and filling the pores. (Reproduced with permission from Ref. [1]; © 2008, Elsevier Limited.)

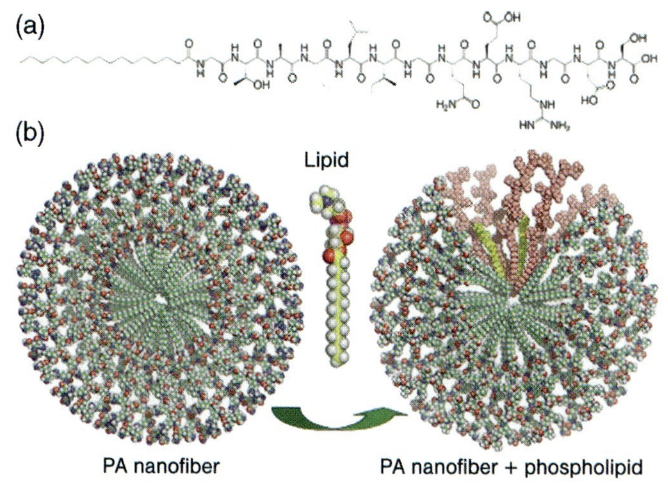

Figure 14.10 (a) Chemical structure of the PA and (b) cross-section of a PA fiber and a PA fiber containing 6.25 mol% of lipid (yellow). Highlighted in pink are the PA molecules situated adjacent to the lipid molecules. (Reproduced with permission from Ref. [67]; © 2006, American Chemical Society.)

the peptide secondary structure as well as the mechanical properties of the hydrogel, with little change in the nanostructure. This composite system enables the optimization of mechanical and chemical properties of the hydrogel by simple adjustment of the PA to phospholipid ratios [67].

The ability to access new mechanisms to control self-assembly across the scales, and not just at the nanostructure level, offers new possibilities for regenerative therapies as bioactive functions can be extended by design into microscopic – and even macroscopic – dimensions. One system involves the self-assembly of hierarchically ordered materials at an aqueous interface resulting from the interaction between small, charged self-assembling PA molecules and oppositely charged high-molar mass biopolymers [68]. A PA–polymer sac can be formed instantly by injecting the polymer directly into the PA solution (Figure 14.11a). The interfacial interaction between the two aqueous liquids allows the formation of relatively robust membranes with tailorable size and shape (Figure 14.11b), self-sealing and suturable sacs (Figure 14.11c–f), as well as continuous strings (Figure 14.11g). The membrane structure grows to macroscopic dimensions with a high degree of hierarchical order across the scales. Studies have demonstrated that the sac membrane is permeable to large proteins, and therefore can be successfully used to encapsulate cells (Figure 14.11h). *In vitro* studies of mouse pancreatic islets (Figure 14.11i) and human MSCs (Figure 14.11j) cultured within the sacs showed that these structures can support cell survival and can be effective 3-D environments for cell differentiation. The unique structural and physical characteristics of these novel systems offer significant potential in cell therapies, drug diagnostics and regenerative medicine applications.

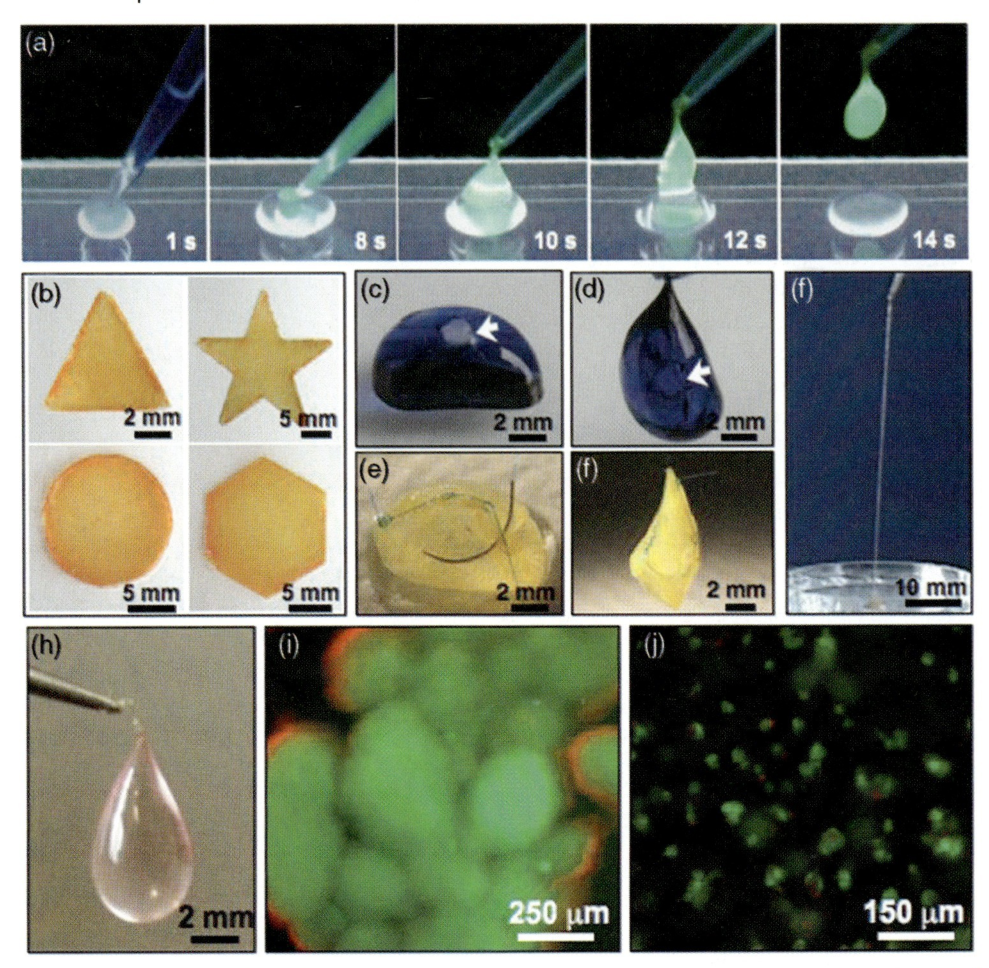

Figure 14.11 (a) Time-lapse photography of sac formation. A sample of a charged biopolymer solution is injected into an oppositely charged peptide amphiphile (PA) solution. The self-assembled sac is formed instantly; (b) PA-polymer membranes of different shapes created by interfacing the large- and small-molecule solutions in a very shallow template (~1 mm thick); (c) Hierarchically ordered sac formed with polydiacetylene PA containing a macroscopic defect within the membrane (arrow); (d) Sac in (c) after the defect is repaired and the sac resealed by triggering additional self-assembly with a drop of PA (arrow). Sacs are robust enough to withstand suturing (e) and can hold their weight without further tearing of the membrane (f); (g) Continuous string pulled from the interface between the PA and polymer solutions; (h) A sac encapsulating cells (sac is a pink color from cell media). Live/dead assay of (i) mouse pancreatic islets and (j) human mesenchymal stem cells (hMSCs) cultured within the sacs (green cells are live, red cells are dead). The islets remained viable up to a week and the hMSCs up to a month in sac culture. (Reproduced with permission from Ref. [68]; © 2008, American Association for the Advancement of Science.)

14.3
Self-Assembling Systems for Surface Modification

Implantable materials are the essence of today's regenerative medicine. The ability to control these materials at the nanoscale has moved them from simple inert materials to biocompatible and bioactive materials [69]. The surface of a biomaterial is particularly important in regenerative medicine as it is the first point of contact with the body. Whether it is presenting a biomimetic atmosphere, disguising a foreign body, or activating specific biological processes, the surface of an implant plays a crucial role and can determine its success or failure. One key advantage of self-assembly is the possibility to modify and tailor surfaces to elicit a specific biological response. In the following, we discuss the use of self-assembly to modify the properties of surfaces and 3-D structures.

14.3.1
Coatings on Surfaces

Within the scope of regenerative medicine, the molecular self-assembly of peptides represents a promising tool to modify the surfaces of medical implants or regenerative scaffolds. This technique facilitates the presentation of bioactive surface chemistries in a controlled, ordered fashion to mimic those of natural extracellular matrices. While the bioactivity of surfaces can be highly modulated by the presentation of specific ECM proteins, the effectiveness of this approach depends greatly on the appropriate conformation of the protein to expose the bioactive epitopes. Self-assembling materials offer the possibility to incorporate small peptide sequences as part of the self-assembling molecule. This approach avoids complications associated with intact proteins, such as undesirable protein folding and immune reactions, and also increases the specificity and efficiency of the bioactive epitope [70]. The use of small peptide sequences such as RGDS for cell adhesion [61] or IKVAV for neuronal differentiation [6], in combination with the capacity to self-assemble molecules in unique and specific conformations, makes this a powerful tool to modify and functionalize surfaces of materials used in regenerative medicine.

Self-assembled monolayers (SAMs) are single layers of molecules that react with and spontaneously order on solid surfaces. SAMs of alkanethiols on gold have been used extensively to study peptide and protein adsorption on surfaces [71, 72], as well as their effect on cell behavior [73, 74]. Recently, the modification of traditional SAM techniques has increased the level of surface chemistry manipulation and complexity that can be achieved. For example, the introduction of soft lithographic techniques such as microcontact printing has facilitated the use and significantly increased the potential of peptide-containing SAMs [70, 75]. This approach has been used to create self-assembled surface patterns to control and study a variety of cell behaviors such as cell adhesion, growth and apoptosis (Figure 14.12a) [76, 77]. Another approach that takes advantage of SAMs, and has been used in combination with soft lithographic techniques, consists of developing dynamically controlled and regulated surfaces, which offer a unique opportunity to recreate and study dynamic biological processes. These types of surface can be achieved by controlling SAMs through different switching mechanisms (i.e. electrical, electrochemical, photochemical, thermal, and mechanical transduction [78, 79]) that organize specific ligands and peptides. By using these techniques, SAMs of peptides such as EG3- and RGD-terminated peptides have been used to study dynamic mechanisms controlling the adhesion and migration of bovine capillary endothelial cells [80] and fibroblasts [81], respectively (Figure 14.12b). Another modification of traditional SAM patterning takes advantage of dip-pen nanolithography (DPN), which uses atomic force microscopy (AFM) tips dipped into alkanethiol inks to transfer molecules by capillary force on

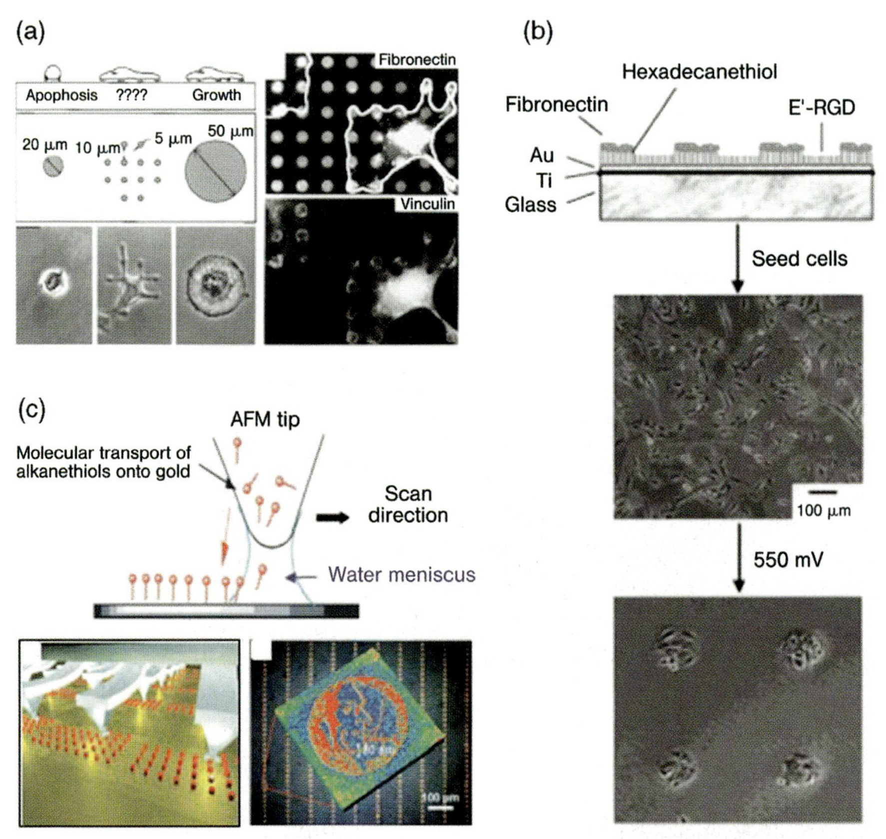

Figure 14.12 Approaches to create complex self-assembled monolayers (SAMs) including: (a) Micro-contact printing to create adhesive patterns of SAMs to study cell mechanisms such as growth and apoptosis. (Reproduced with permission from Ref. [76]; © 1997, American Association for the Advancement of Science.); (b) Patterns of SAMs that can be electroactively controlled and regulated to study cell adhesion and migration. (Reproduced with permission from Ref. [81]; © 2003, The American Chemical Society.); (c) Schematic of SAMs generated through dip-pen nanolithography (DPN). (Reproduced with permission from Ref. [82]; © 2007, Wiley-VCH Verlag GmbH & Co. KGaA.)

the gold surface (Figure 14.12c) [82, 83]. A major advantage of this technique is that it can create patterns of SAMs down to 15 nm in lateral dimension, significantly surpassing the resolution of soft lithographic techniques [82]. These types of studies not only provide reproducible tools to engineer biomimetic cell microenvironments, but also offer great promise for a deeper understanding of cell behaviors that could then be applied to the design of materials and implants in regenerative medicine [69].

While SAMs rely on individual molecules or peptides to create single-layer coatings, more complex self-assembled structures such as tubes or fibers are also being used as surface modifiers. One such example is a class of organic self-assembled fibers, referred to as helical rosette nanotubes (HRNs), that have been used to coat and functionalize bone prosthetic biomaterials (Figure 14.13). This approach was recently used to coat titanium surfaces, and caused a significant enhancement of osteoblast adhesion *in vitro* [84]. These molecules self-assemble from a single bicyclic block resulting from the complementary hydrogen-bonding arrays of both guanine (G) and cytosine (C). This C/G motif serves as the building block that self-assembles in water to form a six-membered supermacrocycle (rosette) maintained by 18 H-bonds. Subsequent assembly of these rosettes forms hollow nanotubes that are 1.1 nm in diameter and up to millimeters in length [85]. The outer surface of the G/C motif could further be modified to present specific physical and chemical properties.

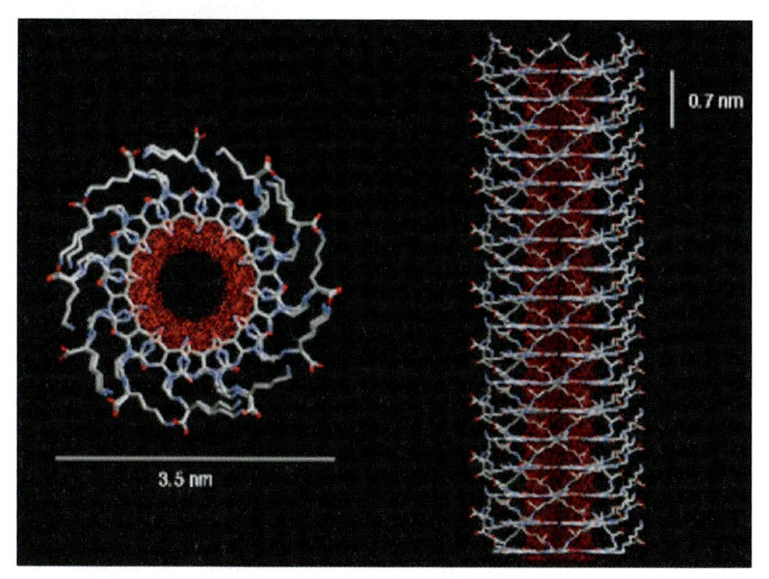

Figure 14.13 Illustration depicting the molecular structure of helical rosette nanotubes (HRNs) used for coating titanium surfaces with potential use in functionalizing the surface of bone prosthetic biomaterials. (Reproduced with permission from Ref. [84]; © 2005, Elsevier Limited.)

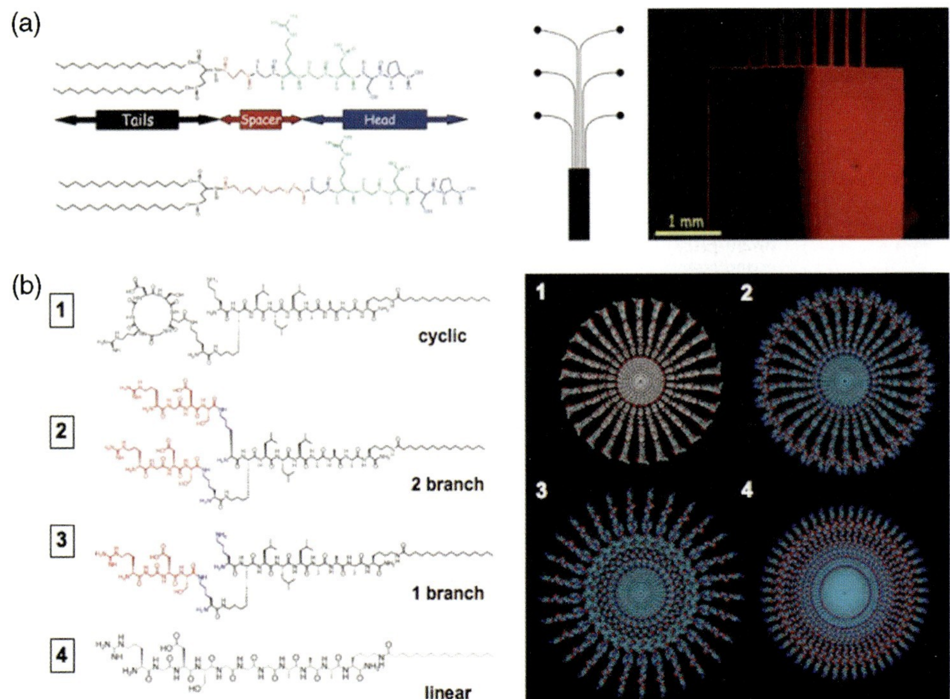

Figure 14.14 RGD-containing peptide amphiphile (PA) molecules used to control and modulate surface cell adhesion. (a) Surface-patterning techniques using microfluidic devices. (Reproduced with permission from Ref. [87]; © 2007, American Chemical Society.); (b) Optimum epitope presentation through specific molecular architectures. (Reproduced with permission from Ref. [61]; © 2007, Elsevier Limited.)

The self-assembly of PAs has also been used to functionalize two-dimensional (2-D) surfaces. As mentioned previously PAs are highly bioactive and biocompatible materials that have been used to develop 3-D scaffolds for tissue engineering and regenerative medicine. The diversity of design and robustness of these molecules has permitted a wide range of approaches for their use as surface-functionalizing coatings. For example, PAs containing cell-adhesive and triple-helical or α-helical structural motifs have been used to influence adhesion and signal transduction of human melanoma cells *in vitro* [86]. By utilizing microfluidic systems, Stroumpoulis *et al.* generated self-assembling patterns of RGD-containing PAs that directed mouse fibroblast adhesion (Figure 14.14a) [87]. In an attempt to optimize the presentation of PA coatings, Storrie *et al.* investigated the effect of PA molecular architecture and epitope concentration on nanofiber self-assembly, epitope presentation and fibroblast recognition for cell adhesion and spreading on surfaces (Figure 14.14b) [61].

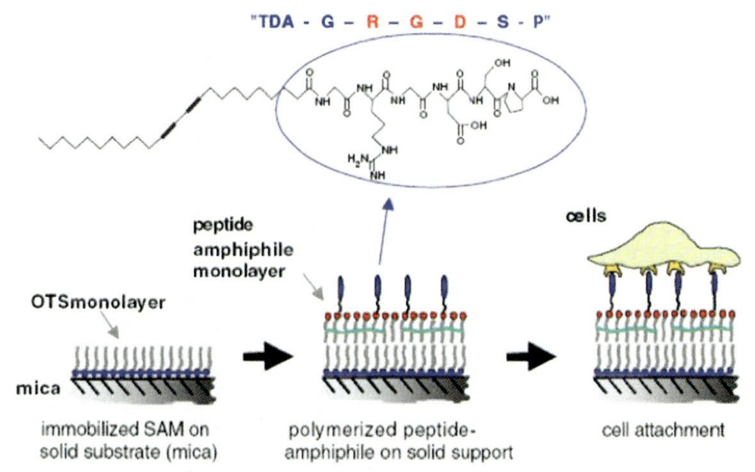

Figure 14.15 Peptide amphiphile (PA) molecules and self-assembling mechanism used to functionalize surfaces with improved molecular properties. A diacetylene-photosensitive segment in the hydrophobic tail promotes PA polymerization and subsequent monolayer stability. (Reproduced with permission from Ref. [89]; © 2006, Elsevier Limited.)

A number of strategies have been investigated to improve the chemical stability of PA coatings on implant surfaces. For example, Sargeant *et al.* has developed a method to covalently attach PA nanofibers to the surface of nickel–titanium (NiTi) shape memory alloys [88]. Here, the group used an RGDS containing PA and demonstrated its capacity to form robust PA coatings capable of promoting cell adhesion, proliferation and differentiation. This method significantly improves the potential of using PA materials for *in vivo* applications such as vascular stents, bone plates and artificial joints. Another approach used to improve PA stability was reported by Biesalski *et al.*, who developed a PA molecule comprising a diacetylene photosensitive segment, which promotes PA polymerization (Figure 14.15) [89]. This molecule was used to develop a stable polymerized monolayer of RGD-terminated PA molecules that significantly enhanced fibroblast adhesion.

The vast majority of investigations related to self-assembling peptides for surface modification has been dedicated to developing functional and bioactive surface chemistries to affect or elicit specific cell behaviors. However, in addition to surface chemistry, surface topography has also been shown to significantly affect cell and tissue behavior [90–92]. An ideal surface modification treatment for regenerative medicine would permit the fine-tuning of both surface chemistry and surface topography across different size scales. One approach to achieve this topographical/biochemical integration is to create SAMs of peptides on microfabricated surfaces comprising topographical features [93, 94]. The integration of microfabrication with molecularly designed self-assembling PAs also represents a unique opportunity to develop physical and biochemical environments with hierarchical

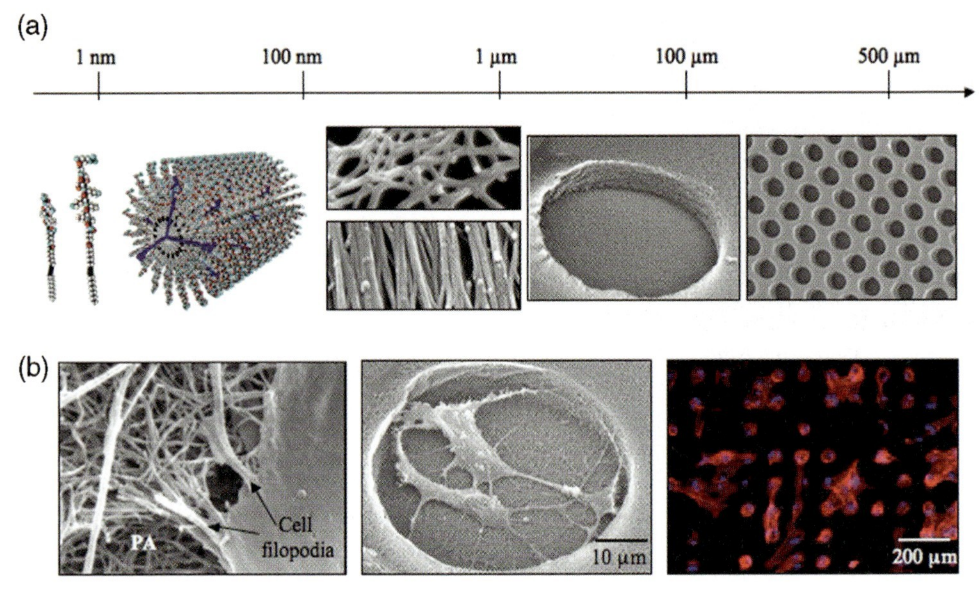

Figure 14.16 (a) Fabrication approach that combines bottom-up (self-assembling peptide-amphiphiles) with top-down (microfabrication) techniques to create biomimetic environments for stem cell manipulation. This method integrates precise topographical patterns and specific biochemical signals within a hierarchical structure that expands from the molecular to the macro scale; (b) Topographical patterns made from self-assembled PA nanofibers have demonstrated the capacity to guide the growth and differentiation of human mesenchymal stem cells (red = actin cytoskeleton, blue = cell nuclei). (with permission from *Soft Matter*, DOI:10.1039/b819002j).

organization (Figure 14.16a). With this approach, it may be possible to create biomimetic scaffolds made from PA molecules with the capacity to elicit specific cell behaviors using both topographical features and bioactive epitopes at different size scales (Figure 14.16b). Recent studies performed by Stupp *et al.* have demonstrated the capacity to promote osteoblastic differentiation of human MSCs by using topographical patterns made from self-assembled PA nanofibers. (*Soft Matter*, DOI:10.1039/b819002j).

14.3.2
Coatings on 3-D Scaffolds

Three-dimensional scaffolds prepared with synthetic materials such as poly(glygolic acid) (PGA) and poly (L-lactic acid) (PLLA) provide porous and biodegradable materials that have found extensive use in regenerative medicine applications [95]. The surface characteristics of these materials, however, do not have any specific or desired bioactivity. Therefore, self-assembling peptides may be used to functionalize surfaces to further improve bioactivity and tissue integration. For example, Harrington *et al.* used an RGDS-containing PA to self-assemble into well-defined

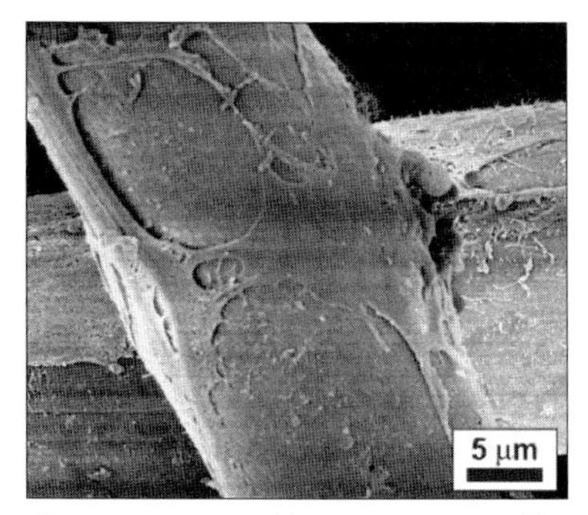

Figure 14.17 Osteoblasts growing on a fibrous poly(L-lactic acid) (PLLA) scaffold coated with molecules comprising cholesterol and lysine moieties. (Reproduced with permission from Ref. [8]; © 2004, Elsevier Limited.)

nanofibers on the surface of PGA porous scaffolds [95]. These RGDS-coated scaffolds significantly improved human bladder smooth muscle cell adhesion. Stendahl *et al.* self-assembled a triblock molecule comprising cholesterol and lysine moieties to coat and modify PLLA fiber scaffolds [8]. These amphiphilic molecules improved the adhesion and overall growth of osteoblastic cells (Figure 14.17). Another examples of a recent surface modification technique used within 3-D architectures includes studies conducted by Zhu *et al.*, who used poly(ethylenimine) (PEI) to activate the surface of poly(lactide) (PLA) scaffolds, which was subsequently modified with gelatin using electrostatic self-assembly [96, 97]. This treatment successfully promoted the growth of seeded osteoblasts.

14.4 Clinical Potential of Self-Assembling Systems

As discussed above, several preclinical studies have already shown great promise for the use of self-assembling biomaterials in regenerative medicine. Particularly, *in vivo* experiments using self-assembling peptide amphiphiles by Stupp and colleagues have shown that these bioactive matrices can be specifically designed to: (i) promote angiogenesis (rat cornea model); (ii) promote regeneration of axons in a spinal cord injury model (mouse and rat models), of cartilage (rabbit model), and of bone (rat model); (iii) promote recovery of cardiac function after infarct (mouse model); and (iv) show promise as treatments for Parkinson's disease (mouse models). It is expected that the self-assembly of supramolecular systems will, in time, lead to many effective regenerative medicine therapies providing an excellent platform to design for bioactivity, harmless degradation with appropriate half-lives after providing a function, and noninvasive methods for clinical delivery:

- Design for bioactivity: it is possible to engineer these peptide-based, self-assembling systems to include various combinations of amino acid sequences that are bioactive and can enhance the regeneration process – that is, deliver growth factors, contain cell adhesion sequences, mimic the bioactivity of growth factors, and so on.

- Harmless degradation: peptide-based, self-assembling systems are capable of being degraded by enzymes in the body into basic amino acids that can be metabolized naturally, with appropriate half-lives after providing a function. Their degradation characteristics can be manipulated through molecular design. Ideally, such bioactive materials would provide a specific function (i.e. deliver growth factors, attract and adhere desired cell types, etc.) to enhance the regenerative process, and simultaneously degrade as the tissue starts to regenerate. Over time, what is left would be the completely regenerated tissue. The challenge would be to determine the 'appropriate' degradation rate for optimal regeneration.
- Noninvasive methods for clinical delivery: the ability of these peptide-based molecules to self-assemble spontaneously allows for the administration of materials through noninvasive methods. For example, a solution of the self-assembling molecules could be injected into the defect site, after which gelation in vivo could be triggered by ions within the body.

14.5 Conclusions

Today, research into the development of self-assembling biomaterials for regenerative medicine continues to expand–the main aim being to achieve real improvements in the quality of life for mankind. Without strategies for regeneration, genomic data and personalized medicine will not have the significant impact that is being promised. It is important that therapies for regenerative medicine must be not only highly effective and predictable, but also as noninvasive as possible, with the capacity to reach deep into problem areas of the heart, brain, skeleton, skin and other vital organs. It is for this reason that self-assembly at the nanoscale appears as the most sensible technological strategy, to signal and recruit the organism's own cells, or to manage the delivery of cell therapies to the correct sites after effective *in vitro* manipulation. The ability to design at both the nanoscale and macroscale will open the door to vast possibilities for biomaterials and regenerative medicine, with materials that can be designed to multiplex the required signals, can be delivered in a practical and optimal manner, and can reach targets across barriers via the blood circulation. In addition, molecular self-assembly on the surfaces of implants may enhance the bioactivity and predictable biocompatibility of metals, ceramics, composites and synthetic polymers. Self-assembly is at the root of structure versus function in biology and, in the context of regenerative medicine technology, is the 'ultimate inspiration from Nature'.

References

1 Sargeant, T.D., Guler, M.O., Oppenheimer, S.M., Mata, A., Satcher, R.L., Dunand, D.C. and Stupp, S.I. (2008) *Biomaterials*, **29**, 161.

2 Hwang, J.J., Iyer, S.N., Li, L.S., Claussen, R., Harrington, D.A. and Stupp, S.I. (2002) *Proceedings of the National Academy of Sciences of the United States of America*, **99**, 9662.

3 Klok, H.A., Hwang, J.J., Iyer, S.N. and Stupp, S.I. (2002) *Macromolecules*, **35**, 746.

4 Hartgerink, J.D., Beniash, E. and Stupp, S.I. (2001) *Science*, **294**, 1684.

5 Hartgerink, J.D., Beniash, E. and Stupp, S.I. (2002) *Proceedings of the National Academy of Sciences of the United States of America*, **99**, 5133.

6 Silva, G.A., Czeisler, C., Niece, K.L., Beniash, E., Harrington, D.A., Kessler, J.A. and Stupp, S.I. (2004) *Science*, **303**, 1352.

7 Stupp, S.I. (2005) *MRS Bulletin*, **30**, 546.

8 Stendahl, J.C., Li, L., Claussen, R.C. and Stupp, S.I. (2004) *Biomaterials*, **25**, 5847.

9 Harrington, D.A., Cheng, E.Y., Guler, M.O., Lee, L.K., Donovan, J.L., Claussen, R.C. and Stupp, S.I. (2006) *Journal of Biomedical Materials Research Part A*, **78**, 157.

10 Gazit, E. (2007) *Chemical Society Reviews*, **36**, 1263.

11 Aggeli, A., Bell, M., Boden, N., Keen, J.N., Knowles, P.F., McLeish, T.C., Pitkeathly, M. and Radford, S.E. (1997) *Nature*, **386**, 259.

12 Aggeli, A., Bell, M., Carrick, L.M., Fishwick, C.W., Harding, R., Mawer, P.J., Radford, S.E., Strong, A.E. and Boden, N. (2003) *Journal of the American Chemical Society*, **125**, 9619.

13 Kirkham, J., Firth, A., Vernals, D., Boden, N., Robinson, C., Shore, R.C., Brookes, S.J. and Aggeli, A. (2007) *Journal of Dental Research*, **86**, 426.

14 Aggeli, A., Bell, M., Boden, N., Carrick, L.M. and Strong, A.E. (2003) *Angewandte Chemie - International Edition in English*, **42**, 5603.

15 Bell, C.J., Carrick, L.M., Katta, J., Jin, Z., Ingham, E., Aggeli, A., Boden, N., Waigh, T.A. and Fisher, J. (2006) *Journal of Biomedical Materials Research Part A*, **78**, 236.

16 Schneider, J.P., Pochan, D.J., Ozbas, B., Rajagopal, K., Pakstis, L. and Kretsinger, J. (2002) *Journal of the American Chemical Society*, **124**, 15030.

17 Pochan, D.J., Schneider, J.P., Kretsinger, J., Ozbas, B., Rajagopal, K. and Haines, L. (2003) *Journal of the American Chemical Society*, **125**, 11802.

18 Ozbas, B., Rajagopal, K., Schneider, J.P. and Pochan, D.J. (2004) *Physical Review Letters*, **93**, 268106.

19 Ozbas, B., Rajagopal, K., Haines-Butterick, L., Schneider, J.P. and Pochan, D.J. (2007) *The Journal of Physical Chemistry B*, **111**, 13901.

20 Haines, L.A., Rajagopal, K., Ozbas, B., Salick, D.A., Pochan, D.J. and Schneider, J.P. (2005) *Journal of the American Chemical Society*, **127**, 17025.

21 Haines-Butterick, L., Rajagopal, K., Branco, M., Salick, D., Rughani, R., Pilarz, M., Lamm, M.S., Pochan, D.J. and Schneider, J.P. (2007) *Proceedings of the National Academy of Sciences of the United States of America*, **104**, 7791.

22 Kretsinger, J.K., Haines, L.A., Ozbas, B., Pochan, D.J. and Schneider, J.P. (2005) *Biomaterials*, **26**, 5177.

23 Salick, D.A., Kretsinger, J.K., Pochan, D.J. and Schneider, J.P. (2007) *Journal of the American Chemical Society*, **129**, 14793.

24 Nowak, A.P., Breedveld, V., Pakstis, L., Ozbas, B., Pine, D.J., Pochan, D. and Deming, T.J. (2002) *Nature*, **417**, 424.

25 Pochan, D.J., Pakstis, L., Ozbas, B., Nowak, A.P. and Deming, T.J. (2002) *Macromolecules*, **35**, 5358.

26 Breedveld, V., Nowak, A.P., Sato, J., Deming, T.J. and Pine, D.J. (2004) *Macromolecules*, **37**, 3943.

27 Pakstis, L.M., Ozbas, B., Hales, K.D., Nowak, A.P., Deming, T.J. and Pochan, D. (2004) *Biomacromolecules*, **5**, 312.

28 Zhang, S., Holmes, T., Lockshin, C. and Rich, A. (1993) *Proceedings of the National Academy of Sciences of the United States of America*, **90**, 3334.

29 Zhang, S., Holmes, T.C., DiPersio, C.M., Hynes, R.O., Su, X. and Rich, A. (1995) *Biomaterials*, **16**, 1385.

30 Holmes, T.C., S., de Lacalle Su, X., Liu, G., Rich, A. and Zhang, S. (2000) *Proceedings of the National Academy of Sciences of the United States of America*, **97**, 6728.

31 Caplan, M.R., Schwartzfarb, E.M., Zhang, S., Kamm, R.D. and Lauffenburger, D.A. (2002) *Journal of Biomaterials Science, Polymer Edition*, **13**, 225.

32 Caplan, M.R., Schwartzfarb, E.M., Zhang, S., Kamm, R.D. and Lauffenburger, D.A. (2002) *Biomaterials*, **23**, 219.

33 Ellis-Behnke, R.G., Liang, Y.X., You, S.W., Tay, D.K., Zhang, S., So, K.F. and Schneider, G.E. (2006) *Proceedings of the National Academy of Sciences of the United States of America*, **103**, 5054.

34 Guo, J., Su, H., Zeng, Y., Liang, Y.X., Wong, W.M., Ellis-Behnke, R.G., So, K.F. and Wu, W. (2007) *Nanomedicine*, **3**, 311.

35 Semino, C.E., Kasahara, J., Hayashi, Y. and Zhang, S. (2004) *Tissue Engineering*, **10**, 643.

36 Genove, E., Shen, C., Zhang, S. and Semino, C.E. (2005) *Biomaterials*, **26**, 3341.

37 Narmoneva, D.A., Oni, O., Sieminski, A.L., Zhang, S., Gertler, J.P., Kamm, R.D. and Lee, R.T. (2005) *Biomaterials*, **26**, 4837.

38 Sieminski, A.L., Was, A.S., Kim, G., Gong, H. and Kamm, R.D. (2007) *Cell Biochemistry and Biophysics*, **49**, 73.

39 Davis, M.E., Hsieh, P.C., Takahashi, T., Song, Q., Zhang, S., Kamm, R.D., Grodzinsky, A.J., Anversa, P. and Lee, R.T. (2006) *Proceedings of the National Academy of Sciences of the United States of America*, **103**, 8155.

40 Davis, M.E., Motion, J.P., Narmoneva, D.A., Takahashi, T., Hakuno, D., Kamm, R.D., Zhang, S. and Lee, R.T. (2005) *Circulation*, **111**, 442.

41 Hsieh, P.C., Davis, M.E., Gannon, J., MacGillivray, C. and Lee, R.T. (2006) *The Journal of Clinical Investigation*, **116**, 237.

42 Hsieh, P.C., MacGillivray, C., Gannon, J., Cruz, F.U. and Lee, R.T. (2006) *Circulation*, **114**, 637.

43 Semino, C.E., Merok, J.R., Crane, G.G., Panagiotakos, G. and Zhang, S. (2003) *Differentiation; Research in Biological Diversity*, **71**, 262.

44 Kisiday, J., Jin, M., Kurz, B., Hung, H., Semino, C., Zhang, S. and Grodzinsky, A.J. (2002) *Proceedings of the National Academy of Sciences of the United States of America*, **99**, 9996.

45 Garreta, E., Gasset, D., Semino, C. and Borros, S. (2007) *Biomolecular Engineering*, **24**, 75.

46 Garreta, E., Genove, E., Borros, S. and Semino, C.E. (2006) *Tissue Engineering*, **12**, 2215.

47 Guo, J., Su, H., Zeng, Y., Liang, Y.X., Wong, W.M., Ellis-Behnke, R.G., So, K.F. and Wu, W. (2007) *Nanomedicine*, **3**, 311.

48 Jayawarna, V., Smith, A., Gough, J.E. and Ulijn, R.V. (2007) *Biochemical Society Transactions*, **35**, 535.

49 Toledano, S., Williams, R.J., Jayawarna, V. and Ulijn, R.V. (2006) *Journal of the American Chemical Society*, **128**, 1070.

50 Lowik, D.W. and van Hest, J.C. (2004) *Chemical Society Reviews*, **33**, 234.

51 Behanna, H.A., Donners, J.J., Gordon, A.C. and Stupp, S.I. (2005) *Journal of the American Chemical Society*, **127**, 1193.

52 Bull, S.R., Guler, M.O., Bras, R.E., Meade, T.J. and Stupp, S.I. (2005) *Nano Letters*, **5**, 1.

53 Claussen, R.C., Rabatic, B.M. and Stupp, S.I. (2003) *Journal of the American Chemical Society*, **125**, 12680.

54 Guler, M.O., Hsu, L., Soukasene, S., Harrington, D.A., Hulvat, J.F. and Stupp, S.I. (2006) *Biomacromolecules*, **7**, 1855.

55 Guler, M.O., Soukasene, S., Hulvat, J.F. and Stupp, S.I. (2005) *Nano Letters*, **5**, 249.

56 Niece, K.L., Hartgerink, J.D., Donners, J.J. and Stupp, S.I. (2003) *Journal of the American Chemical Society*, **125**, 7146.

57 Rajangam, K., Behanna, H.A., Hui, M.J., Han, X., Hulvat, J.F., Lomasney, J.W. and Stupp, S.I. (2006) *Nano Letters*, **6**, 2086.

58 Sone, E.D. and Stupp, S.I. (2004) *Journal of the American Chemical Society*, **126**, 12756.

59 Stendahl, J.C., Rao, M.S., Guler, M.O. and Stupp, S.I. (2006) *Advanced Functional Materials*, **16**, 499.

60 Beniash, E., Hartgerink, J.D., Storrie, H., Stendahl, J.C. and Stupp, S.I. (2005) *Acta Biomaterialia*, **1**, 387.

61 Storrie, H., Guler, M.O., Abu-Amara, S.N., Volberg, T., Rao, M., Geiger, B. and Stupp, S.I. (2007) *Biomaterials*, **28**, 4608.

62 Tysseling-Mattiace, V.M., Sahni, V., Niece, K.L., Birch, D., Czeisler, C., Fehlings, M.G., Stupp, S.I. and Kessler, J.A. (2008) *The Journal of Neuroscience*, **28**, 3814.

63 Kapadia, M.R., Chow, L.W., Tsihlis, N.D., Ahanchi, S.S., Eng, J.W., Murar, J., Martinez, J., Popowich, D.A., Jiang, Q., Hrabie, J.A., Saavedra, J.E., Keefer, L.K., Hulvat, J.F., Stupp, S.I. and Kibbe, M.R. 2008 *Journal of Vascular Surgery*, **47**, 173.

64 Jiang, H., Guler, M.O. and Stupp, S.I. (2007) *Soft Matter*, **3**, 454.

65 Palmer, L.C., Velichko, Y.S., Olvera De La Cruz, M. and Stupp, S.I. (2007) *Philosophical Transactions of the Royal Society of London. Series A: Mathematical and Physical Sciences*, **365**, 1417.

66 Jun, H., Yuwono, V., Paramonov, S.E. and Hartgerink, J.D. (2005) *Advanced Materials*, **17**, 2612.

67 Paramonov, S.E., Jun, H.W. and Hartgerink, J.D. (2006) *Biomacromolecules*, **7**, 24.

68 Capito, R.M., Azevedo, H.S., Velichko, Y.S., Mata, A. and Stupp, S.I. (2008) *Science*, **319**, 1812.

69 Stupp, S.I., Donners, J.J.J.M., Li, L.S. and Mata, A. (2005) *MRS Bulletin*, **30**, 864.

70 Geim, A.K., Dubonos, S.V., Grigorieva, I.V., Novoselov, K.S., Zhukov, A.A. and Shapoval, S.Y. (2003) *Nature Materials*, **2**, 461.

71 Ruiz, S.A. and Chen, C.S. (2007) *Soft Matter*, **3**, 168.

72 Sniadecki, N.J., Tan, J., Anguelouch, A., Ruiz, S.A., Reich, D.H. and Chen, C.S. (2004) *Molecular Biology of the Cell*, **15**, 54.

73 Roberts, C., Chen, C.S., Mrksich, M., Martichonok, V., Ingber, D.E. and Whitesides, G.M. (1998) *Journal of the American Chemical Society*, **120**, 6548.

74 Houseman, B.T. and Mrksich, M. (2001) *Biomaterials*, **22**, 943.

75 Mrksich, M. and Whitesides, G.M. (1995) *Trends in Biotechnology*, **13**, 228.

76 Chen, C.S., Mrksich, M., Huang, S., Whitesides, G.M. and Ingber, D.E. (1997) *Science*, **276**, 1425.

77 Chen, C.S., Mrksich, M., Huang, S., Whitesides, G.M. and Ingber, D.E. (1998) *Biotechnology Progress*, **14**, 356.

78 Mrksich, M. (2005) *MRS Bulletin*, **30**, 180.

79 Lahann, J. and Langer, R. (2005) *MRS Bulletin*, **30**, 185.

80 Jiang, X.Y., Ferrigno, R., Mrksich, M. and Whitesides, G.M. (2003) *Journal of the American Chemical Society*, **125**, 2366.

81 Yeo, W.S., Yousaf, M.N. and Mrksich, M. (2003) *Journal of the American Chemical Society*, **125**, 14994.

82 Huck, W.T.S. (2007) *Angewandte Chemie - International Edition*, **46**, 2754.

83 Piner, R.D., Zhu, J., Xu, F., Hong, S.H. and Mirkin, C.A. (1999) *Science*, **283**, 661.

84 Chun, A.L., Moralez, J.G., Webster, T.J. and Fenniri, H. (2005) *Biomaterials*, **26**, 7304.

85 Chun, A.L., Moralez, J.G., Fenniri, H. and Webster, T.J. (2004) *Nanotechnology*, **15**, S234.

86 Fields, G.B., Lauer, J.L., Dori, Y., Forns, P., Yu, Y.C. and Tirrell, M. (1998) *Biopolymers*, **47**, 143.

87 Stroumpoulis, D., Zhang, H.N., Rubalcava, L., Gliem, J. and Tirrell, M. (2007) *Langmuir*, **23**, 3849.

88 Sargeant, T.D., Rao, M.S., Koh, C.Y. and Stupp, S.I. (2008) *Biomaterials*, **29**, 1085.

89 Biesalski, M.A., Knaebel, A., Tu, R. and Tirrell, M. (2006) *Biomaterials*, **27**, 1259.

90 Tirrell, M., Kokkoli, E. and Biesalski, M. (2002) *Surface Science*, **500**, 61.

91 Charest, J.L., Eliason, M.T., Garcia, A.J. and King, W.P. (2006) *Biomaterials*, **27**, 2487.

92 Kunzler, T.P., Huwiler, C., Drobek, T., Voros, J. and Spencer, N.D. (2007) *Biomaterials*, **28**, 5000.

93 Mrksich, M., Chen, C.S., Xia, Y.N., Dike, L.E., Ingber, D.E. and Whitesides, G.M. (1996) *Proceedings of the National Academy of Sciences of the United States of America*, **93**, 10775.

94 Lussi, J.W., Michel, R., Reviakine, I., Falconnet, D., Goessl, A., Csucs, G., Hubbell, J.A. and Textor, M. (2004) *Progress in Surface Science*, **76**, 55.

95 Harrington, D.A., Cheng, E.Y., Guler, M.O., Lee, L.K., Donovan, J.L., Claussen, R.C. and Stupp, S.I. (2006) *Journal of Biomedical Materials Research Part A*, **78**, 157.

96 Zhu, H., Ji, J. and Shen, J. (2004) *Biomacromolecules*, **5**, 1933.

97 Zhu, H., Ji, J., Barbosa, M.A. and Shen, J. (2004) *Journal of Biomedical Materials Research Applied Biomaterials*, **71**, 159.

Index

Nanotechnology, Volume 5: Nanomedicine. Edited by Viola Vogel

ISBN: 978-3-527-31736-3

t